Advances in Intelligent Systems and Computing

Volume 247

Series Editor

Janusz Kacprzyk, Warsaw, Poland

For further volumes:
http://www.springer.com/series/11156

Suresh Chandra Satapathy · Siba K. Udgata
Bhabendra Narayan Biswal
Editors

Proceedings of the International Conference on Frontiers of Intelligent Computing: Theory and Applications (FICTA) 2013

Springer

Editors

Suresh Chandra Satapathy
Department of Computer Science
    and Engineering,
Anil Neerukonda Institute of Technology
    and Sciences,
Vishakapatnam, Andhra Pradesh,
India

Siba K. Udgata
University of Hyderabad,
Hyderabad, Andhra Pradesh,
India

Bhabendra Narayan Biswal
Bhubaneswar Engineering College,
Bhubaneswar, Odisa,
India

ISSN 2194-5357   ISSN 2194-5365   (electronic)
ISBN 978-3-319-02930-6   ISBN 978-3-319-02931-3   (eBook)
DOI 10.1007/978-3-319-02931-3
Springer Cham Heidelberg New York Dordrecht London

Library of Congress Control Number: 2013951279

Springer is part of Springer Science+Business Media (www.springer.com)

Preface

This AISC volume contains the papers presented at the Second International Conference on Frontiers in Intelligent Computing: Theory and Applications (FICTA-2013) held during 14–16 November 2013 organized by Bhubaneswar Engineering College (BEC), Bhubaneswar, Odisa, India.

FICTA-2013 is the second edition of FICTA 2012 which is aimed to bring researchers from academia and industry to report, deliberate and review the latest progresses in the cutting-edge research pertaining to intelligent computing and its applications to various engineering fields. FICTA 2012 was conducted by BEC and was a big success in achieving its target.

FICTA-2013 had received a good number of submissions from the different areas relating to intelligent computing and its applications and after a rigorous peer-review process with the help of our program committee members and external reviewers finally we accepted 63 papers.

The conference featured many distinguished keynote address by eminent speakers like Dr. K.P.N. Murthy, University of Hyderabad, India, Dr. M. Prasad Babu, Andhra University, Vishakhapatnam, Dr. Naeem Hannoon, Multimedia University, Malaysia, Dr. Swagatam Das, ISI Kolkata.

We take this opportunity to thank authors of all submitted papers for their hard work, adherence to the deadlines and patience with the review process. The quality of a referred volume depends mainly on the expertise and dedication of the reviewers. We are indebted to the program committee members and external reviewers who not only produced excellent reviews but also did these in short time frames. Special thanks to Ms Pritee Parwekar, Jaypee Institute of Information Technology, Nodia for timely support in doing extensive review work.

We would also like to thank Bhubaneswar Engineering College (BEC), Bhubaneswar having coming forward to support us to organize the second edition of this conference in the series. Our heartfelt thanks are due to Er. Pravat Ranjan Mallick, Chairman, KGI, Bhubaneswar for the unstinted support to make the conference a grand success. Er. Alok Ranjan Mallick, Vice-Chairman, KGI, Bhubaneswar and Chairman of BEC deserve our heartfelt thanks for continuing

to support us from FICTA 2012 onwards. It is due to him FICTA 2013 could
be organized in BEC campus for second time. Needless to say Mr. Alok has
been the source of inspiration from the very conceptualization of this conference
series. We are amazed to note the enthusiasm of all faculty, staff and students
of this young college BEC to organize the conference in such a professional way.
Involvements of faculty co-ordinators and student volunteer are praise worthy in
every respect. We are hopeful that this intention of conducting such high level
of conferences will go a long way to make BEC to carve the niche to become a
prestigious engineering college in the country. We are confident that in future
too we would like to organize many more international level conferences in this
beautiful campus. We would also like to thank our sponsors for providing all the
support and financial assistance.

We thank Prof. P.K. Dash, SOA University, Bhubaneswar and Prof. Ganapati
Panda, Dy Director, IIT Bhubaneswar for providing valuable guidelines and
inspirations to overcome various difficulties in the process of organizing this
conference as Honorary General Chairs of this Conference. We extend our heartfelt thanks to Prof. P.N. Suganthan, NTU Singapore for guiding us being the
General chair of the conference. Dr. B.K. Panigrahi, IIT Delhi and Dr. Swagatam
Das, ISI Kolkota deserves special thanks for being with us from the beginning to
the end of this conference, without their support this conference could never have
been successful. We would also like to thank the participants of this conference,
who have considered the conference above all hardships. Finally, we would like to
thank all the volunteers who spent tireless efforts in meeting the deadlines and
arranging every detail to make sure that the conference can run smoothly. All
the efforts are worth and would please us all, if the readers of this proceedings
and participants of this conference found the papers and conference inspiring
and enjoyable.

Our sincere thanks to all press print & electronic media for their excellent
coverage of this conference.

Volume Editors

November 2013

Dr. Suresh Chandra Satapathy
Dr. Siba K. Udgata
Dr. Bhabendra Narayan Biswal

Organization

Organizing Committee

Chief Patron

Er. Pravat Ranjan Mallick,
Chairman, KGI, Bhubaneswar

Patron

Er. Alok Ranjan Mallick,
Vice-Chairman, KGI, Bhubaneswar
Chairman, BEC, Bhubaneswar

Organizing Secretary

Prof B.N. Biswal,
Director (A &A), BEC, Bhubaneswar

Honorary Chairs

Dr. P.K. Dask, SMIEEE,FNAE, Director(Research),
SOA University, Bhubaneswar, India
Dr. Ganapati Panda, SMIEEE, FNAE,
Deputy Director, IIT, Bhubaneswar, India

General Chairs

Dr. P.N. Suganthan,
NTU, Singapore
Dr. Swagatam Das.
ISI, Kolkota

Steering Committee Chair

Dr B.K. Panigrahi, IIT, Delhi, India

Program Chairs

Dr. Suresh Chandra satapathy,
ANITS, Vishakapatnam, India
Dr. S.K. Udgata,
University of Hyderabad, India
Dr B.N. Biswal,
Director (A &A), BEC, Bhubaneswar, India

International Advisory Committee/Technical Committee

P.K. Patra, India
Sateesh Pradhan, India
J.V.R Murthy, India
T.R. Dash, Kambodia
Maurice Clerc, France
Roderich Gross, England
Sangram Samal, India
K.K. Mohapatra, India
L. Perkin, USA
Sumanth Yenduri, USA
Carlos A. Coello Coello, Mexico
Dipankar Dasgupta, USA
Peng Shi, UK
Saman Halgamuge, Australia
Jeng-Shyang Pan, Taiwan
X.Z. Gao, Finland
Juan Luis Fernández Martínez,
    California
Oscar Castillo, Mexcico
Leandro Dos Santos Coelho, Brazil
Heitor Silvério Lopes, Brazil
Rafael Stubs Parpinelli, Brazil
S.S. Pattanaik, India
Gerardo Beni, USA
Namrata Khemka, USA
G.K. Venayagamoorthy, USA
K. Parsopoulos, Greece
Zong Woo Geem, USA
Lingfeng Wang, China
Athanasios V. Vasilakos, Athens

S.G. Ponnambalam, Malaysia
Pei-Chann Chang, Taiwan
Ying Tan, China
Chilukuri K. Mohan, USA
M.A. Abido, Saudi Arabia
Saeid Nahavandi, Australia
Almoataz Youssef Abdelaziz, Egypt
Hai Bin Duan, China
Delin Luo, China
M.K. Tiwari, India
A.Damodaram, India
Oscar Castillo, Mexcico
John MacIntyre, England
Rafael Stubs Parpinelli, Brazil
Jeng-Shyang Pan, Taiwan
P.K. Singh, India
Sachidananda Dehuri, India
P.S. Avadhani, India
G. Pradhan, India
Anupam Shukla, India
Dilip Pratihari, India
Amit Kumar, India
Srinivas Sethi, India
Lalitha Bhaskari, India
V. Suma, India
Pritee Parwekar, India
Pradipta Kumar Das, India
Deviprasad Das, India
J.R. Nayak, India

Organizing committee

Dr. Sangram Samal, HOD, (Aeronautical Engineering)
Dr. P.P. Das, HOD,(Civil Engineering)
Prof. R.K. Behuria, HOD, (Mechanical Engineering)
Prof. P.M. Dash, HOD, (Electrical and Electronics Engineering)
Prof. M.K. Swain, HOD, (Computer Science Engineering)
Prof. A.K. Sutar, HOD, (Electronics and Telecommunication Engineering)
Prof. Debasish Panda, HOD, (Dept. of Science and Humanities)
Miss. Ankita Kanungoo, Dept. of Electrical and Electronics Engineering
Miss. Sonam Mohapatra, Dept. of Mechanical Engineering
Miss. Rasmita Behera, Dept. of Electrical and Electronics Engineering
Miss. Priyanka Rath, Dept. of Mechanical Engineering
Miss. Sourya Snigdha Mohapatra, Dept. of Civil Engineering
Mrs. Susmita Nanda. Dept. of Electronics and Telecommunication Engineering

Contents

Fuzzy Systems

Application of Fuzzy c-Means Clustering for Polymer Data
Mining for Making SAW Electronic Nose 1
Prabha Verma, R.D.S. Yadava

Quantitative Identification of Volatile Organics by SAW
Sensor Transients – Comparative Performance Analysis
of Fuzzy Inference and Partial-Least-Square-Regression
Methods 9
Prashant Singh, R.D.S. Yadava

Effective Routing Protocol in Cognitive Radio Ad Hoc
Network Using Fuzzy-Based Reliable Communication
Neighbor Node Selection 17
Srinivas Sethi, Sangita Pal

A Self-tuning Fuzzy PI Controller for Pure Integrating
Processes 25
Dharmana Simhachalam, Rajani K. Mudi

Implementing Fuzziness in the Pattern Recognition Process
for Improving the Classification of the Patterns Being
Recognised 33
Sapna Singh, Daya Shankar Singh

Fuzzy Self-tuning of Conventional PID Controller
for High-Order Processes 41
Ritu Rani De (Maity), Rajani K. Mudi

Design of Fuzzy Logic Power System Stabilizer
in Multi-machine Power System 49
M. Megala, C. Cristober Asir Rajan

A Fuzzy C Mean Clustering Algorithm for Automated
Segmentation of Brain MRI   59
Geenu Paul, Tinu Varghese, K.V. Purushothaman, N. Albert Singh

Fuzzy Type – Ahead Keyword Search in RDF Data   67
Selvani Deepthi Kavila, Ravi Ravva, Rajesh Bandaru

Human Emotion Classification Using Fuzzy and PCA
Approach   75
*Soumya Ranjan Mishra, B. Ravikiran, K. Sai Madhu Sudhan,
N. Anudeep, G. Jagdish*

Machine Learning and Ann

Application of Artificial Neural Networks and Rough Set
Theory for the Analysis of Various Medical Problems
and Nephritis Disease Diagnosis   83
Devashri Raich, P.S. Kulkarni

Optimisation Using Levenberg-Marquardt Algorithm
of Neural Networks for Iris   91
Asim Sayed, M. Sardeshmukh, Suresh Limkar

Neural Networks – A Case Study in Gene Identification   99
V. Bhaskara Murthy, G. Pardha Saradhi Varma

Breast Cancer Diagnosis: An Intelligent Detection System
Using Wavelet Neural Network 111
V. Dheeba, N. Albert Singh, J. Amar Pratap Singh

Efficient UMTS Location Update and Management
Based on Wavelet Neural Networks 119
J. Amar Pratap Singh, J. Dheeba, N. Albert Singh

Discrimination between Alzheimer's Disease, Mild Cognitive
Impairment and Normal Aging Using ANN Based MR Brain
Image Segmentation 129
Tinu Varghese, R. Sheela Kumari, P.S. Mathuranath, N. Albert Singh

Modified Mean Square Error Algorithm with Reduced Cost
of Training and Simulation Time for Character Recognition
in Backpropagation Neural Network 137
Sapna Singh, Daya Shankar Singh, Shobhit Kumar

A Novel Approach for Odia Part of Speech Tagging Using
Artificial Neural Network 147
Bishwa Ranjan Das, Srikanta Patnaik

Fractional Fourier Transform Based Features for Musical
Instrument Recognition Using Machine Learning Techniques ... 155
D.G. Bhalke, C.B. Rama Rao, D.S. Bormane

Longitudinal Evaluation of Structural Changes in
Frontotemporal Dementia Using Artificial Neural Networks 165
*R. Sheela Kumari, Tinu Varghese, C. Kesavadas, N. Albert Singh,
P.S. Mathuranath*

Content Based Image Retrieval Using Machine Learning
Approach 173
Palepu Pavani, T. Sashi Prabha

Hand Gesture Detection and Recognition Using Affine-Shift,
Bag-of-Features and Extreme Learning Machine Techniques 181
M. Kranthi Kiran, T. ShyamVamsi

Comparative Study of Machine Learning Algorithm
for Intrusion Detection System 189
K. Sravani, P. Srinivasu

Evolutionary Computation Techniques and Its Applications

Grammatical Swarm Based-Adaptable Velocity Update
Equations in Particle Swarm Optimizer 197
Tapas Si, Arunava De, Anup Kumar Bhattacharjee

Performance of Teaching Learning Based Optimization
Algorithm with Various Teaching Factor Values for Solving
Optimization Problems 207
*M. Ramakrishna Murty, J.V.R. Murthy, P.V.G.D. Prasad Reddy,
Anima Naik, Suresh Chandra Satapathy*

Efficient Clustering of Dataset Based on Differential
Evolution 217
Anima Naik, Suresh Chandra Satapathy

Numerical Optimization of Novel Functions Using vTLBO
Algorithm 229
*S. Mohankrishna, Anima Naik, Suresh Chandra Satapathy,
K. Raja Sekhara Rao, B.N. Biswal*

Sensitivity Analysis of Load-Frequency Control of Power
System Using Gravitational Search Algorithm 249
Rabindra Kumar Sahu, Umesh Kumar Rout, Sidhartha Panda

Privacy Preserving Distributed Data Mining with
Evolutionary Computing.. 259
Lambodar Jena, Narendra Ku. Kamila, Sushruta Mishra

Boundary Searching Genetic Algorithm: A Multi-objective
Approach for Constrained Problems 269
Shubham J. Metkar, Anand J. Kulkarni

Security Analysis of Digital Stegno Images Using Genetic
Algorithm 277
G. Praneeta, B. Pradeep

Applications of Intelligent Techniques to Computer Networks, Security and Distributed Systems

Probability and Priority Based Routing Approach
for Opportunistic Networks..................................... 285
Kiran Avhad, Suresh Limkar, Anagha Kulkarni

D&PMV: New Approach for Detection and Prevention
of Misbehave/Malicious Vehicles from VANET 293
Megha Kadam, Suresh Limkar

On the Use of MFCC Feature Vector Clustering for Efficient
Text Dependent Speaker Recognition 305
Ankit Samal, Deebyadeep Parida, Mihir Ranjan Satapathy,
Mihir Narayan Mohanty

A Survey on Power Aware Routing Protocols for Mobile
Ad-Hoc Network 313
Samrat Sarkar, Koushik Majumder

A New Trust Based Secure Routing Scheme in MANET........ 321
Mousumi Sardar, Koushik Majumder

IWDRA: An Intelligent Water Drop Based QoS-Aware
Routing Algorithm for MANETs 329
Debajit Sensarma, Koushik Majumder

Design and Performance Analysis of D-STATCOM for
Non-linear Load Composite Compensation...................... 337
Gokulananda Sahu, Kamalakanta Mahapatra, Subrat Kumar Sahu

AODV Based Black-Hole Attack Mitigation in MANET 345
Subhashis Banerjee, Mousumi Sardar, Koushik Majumder

Path Planning Strategy for Mobile Robot Navigation Using
MANFIS Controller 353
Prases Kumar Mohanty, Dayal R. Parhi

Speech Emotion Recognition Using Regularized Discriminant
Analysis 363
*Swarna Kuchibhotla, B.S. Yalamanchili, H.D. Vankayalapati,
K.R. Anne*

An Extensive Selection of Features as Combinations
for Automatic Text Categorization 371
*Aamir Sohail, Chaitanya Kotha, Rishanth Kanakadri Chavali,
Krishna Meghana, Suneetha Manne, Sameen Fatima*

Performance Investigation of DMV (Detecting Malicious
Vehicle) and D&PMV (Detection and Prevention of
Misbehave/Malicious Vehicles): Future Road Map 379
Megha Kadam, Suresh Limkar

Performance Analysis of OFDM Based DAB Systems Using
BCH-FEC Coding Technique 389
Arun Agarwal, Kabita Agarwal

A Study of Incomplete Data – A Review 401
S.S. Gantayat, Ashok Misra, B.S. Panda

Application of Sensor in Shoe................................ 409
Pritee Parwekar, Akansha Gupta, Shrey Arora

Robot Assisted Emergency Intrusion Detection and Avoidance
with a Wireless Sensor Network 417
Pritee Parwekar, Rishabh Singhal

Interaction between Internet Based TCP Variants
and Routing Protocols in MANET 423
Sukant Kishoro Bisoy, Prasant Kumar Patnaik

Symbolic Data Analysis for the Development of Object
Oriented Data Model for Sensor Data Repository............... 435
Doreswamy, Srinivas Narasegouda

Fault Detection in Sensor Network Using DBSCAN
and Statistical Models 443
Doreswamy, Srinivas Narasegouda

Fast SSIM Index for Color Images Employing
Reduced-Reference Evaluation 451
Vikrant Bhateja, Aastha Srivastava, Aseem Kalsi

A Novel Weighted Diffusion Filtering Approach for Speckle
Suppression in Ultrasound Images 459
Vikrant Bhateja, Gopal Singh, Atul Srivastava

Detection of Progression of Lesions in MRI Using Change
Detection 467
Ankita Mitra, Arunava De, Anup Kumar Bhattacharjee

Optimal Covering Problem for a MANET through Geometric
Theory 475
Jagadish Gurrala, Jayaprakash Singampalli

Data Minning

Hierarchical Divisive Clustering with Multi View-Point Based
Similarity Measure 483
S. Jayaprada, Amarapini Aswani, G. Gayathri

Discovery of Knowledge from Query Groups 493
Sunita A. Yadwad, M. Pavani

D-Pattern Evolving and Inner Pattern Evolving for High
Performance Text Mining 501
B. Vignani, Suresh Chandra Satapathy

A Fast Auto Exposure Algorithm for Industrial Applications
Based on False-Position Method 509
B. Ravi Kiran, G.V.N.A. Harsha Vardhan

Web Technologies and Multi Media Applications

Legacy to Web 2.0 517
Hemant Kumbhar, Raj Kulkarni, Suresh Limkar

A Bio-Chaotic Block Cryptosystem for Privacy Protection
of Multimedia Data 527
Musheer Ahmad, Sonia Gupta, A.K. Mohapatra

Some Popular Usability Evaluation Techniques for Websites 535
Sharmistha Roy, Prasant Kumar Pattnaik

A Secure Lightweight and Scalable Mobile Payment
Framework 545
Shaik Shakeel Ahamad, Siba K. Udgata, Madhusoodhnan Nair

Service Oriented Architecture Based SDI Model for Education
Sector in India 555
Rabindra K. Barik, Arun B. Samaddar

Author Index 563

Application of Fuzzy *c*-Means Clustering for Polymer Data Mining for Making SAW Electronic Nose

Prabha Verma and R.D.S. Yadava

Sensor & Signal Processing Laboratory, Department of Physics, Faculty of Science,
Banaras Hindu University, Varanasi - 221005, India
{pverma.bhu,ardius}@gmail.com, ardius@bhu.ac.in

Abstract. Polymers provide chemical interfaces to electronic nose sensors for detection of volatile organic compounds. An electronic nose combines a properly selected set of sensors (array) with pattern recognition methods for generating information rich odor response patterns and extracting chemical identities. Each sensor in the array is functionalized by a different polymer for diversely selective sorption of target chemical analytes. Selection of an optimal set of polymers from a long list of potential polymers is crucial for cost-effective high performance development of electronic noses. In this paper we present an application of fuzzy *c*-means clustering on partition coefficient data available for target vapors and most prospective polymers for selection of minimum number of polymers that can create maximum discrimination between target chemical compounds. The selection method has been validated by simulating a polymer coated surface acoustic wave (SAW) sensor array for monitoring fish freshness.

Keywords: Fuzzy *c*-means clustering, polymer selection, SAW electronic nose, fish quality monitoring, pattern recognition.

1   Introduction

Electronic nose is an electronic instrument that sniffs odor causing volatile organic compounds (VOCs) in target samples and creates chemical fingerprints for identification. Its operational paradigm mimics mammalian smell sensing organ [1], [2]. An electronic nose instrument measures odor responses by using an array of chemical microelectronic sensors and combines multivariate data processing capability of pattern recognition methods [3], [4]. Microelectronic sensors such as surface acoustic wave (SAW) oscillators or MEMS cantilevers need polymeric interfaces for chemical selectivity. A proper selection of polymers is crucial for encoding information about the target chemicals in sensor array responses. The objective in polymer selection is that the selected set provides adequate chemical interaction possibilities with target chemical molecules without excessive overlap. Mutually orthogonal interaction diversity between vapor molecules and polymer set is vital for successful operation of electronic nose system [5], [6], [7], [8], [9].

In the present work we demonstrate application of fuzzy *c*-means (FCM) clustering technique for determining an optimal subset of polymers for SAW sensor array from a

S.C. Satapathy, S.K. Udgata, and B.N. Biswal (eds.), *FICTA 2013*,
Advances in Intelligent Systems and Computing 247,
DOI: 10.1007/978-3-319-02931-3_1, © Springer International Publishing Switzerland 2014

larger set of prospective polymers. The analysis is based on calculated partition coefficient data by using LSER relation and experimentally determined solvation parameters of polymers and target vapors. The analysis needs a prototype detection target. For this, we targeted establishing freshness or spoilage of fish. Fish is a highly consumed food product. Monitoring fish freshness or spoilage is of significant concern for ascertaining its fitness for consumption, nutritional content and taste [10]. As the fish meat loses its freshness it starts decomposing and emitting biogenic amines and other hazardous compounds dangerous to the health [11], [12], [13]. The validation of FCM method for polymer selection is done by simulating SAW sensor data based on selected polymers and analyzing response patterns by principal component analysis.

2   Volatile Organics for Fish Signature

Fish quality is assessed in terms of its freshness by various factors: appearance, color, texture and odor. These factors change with time due to biological processes in fish flesh. At initial stage of degradation there will be no significant change in appearance, color or texture. However, biological changes that occur after it is caught result in volatile emissions signaling biodegradation. Therefore, monitoring fish odor can be used to assess stage of degradation and fitness for consumption. The volatile organic compounds (VOCs) that emit from fresh fish are acetone, methyl-ethyl-ketone (MEK), toluene, ethyl-benzene (EB), m-xylene, p-xylene, o-xylene, styrene etc [11]. The biodegradation results in emission of several biogenic amines. Most significant of these are: ammonia, trimethylamine (TMA), dimethylamine (DMA) and methylamine (MA) [11]. Of these, TMA is the most dominant constituent [12], [14], [15], [16]. Table 1 lists these signature compounds and their LSER parameters (explained below) collected from [17], [18]. The table also shows concentrations of these vapors used for simulating sensor array responses in Section 6. The concentration levels of volatile organics from fresh fish are in accordance with reported experimental studies [11].

Table 1. Organic volatiles from fresh and spoiled fish and their LSER solvation parameters

S. No.	Volatile Organics	LSER Solvation Parameters					Concentration	Remarks
		R_2	π_2^*	α_2^H	β_2^H	$\log L^{16}$		
1	TMA	0.14	0.2	0	0.67	1.62	10-1000 ppm	Spoiled Fish
2	Ammonia	0.139	0.35	0.14	0.62	0.68	1-100 ppm	Spoiled Fish
3	MA	0.25	0.35	0.16	0.58	1.3	1-100 ppm	Spoiled Fish
4	DMA	0.189	0.3	0.08	0.66	1.6	1-100 ppm	Spoiled Fish
5	Acetone	0.179	0.7	0.04	0.49	1.696	91-910 ppb	Fresh Fish
6	MEK	0.166	0.7	0	0.51	2.287	44-440 ppb	Fresh Fish
7	Toluene	0.601	0.52	0	0.14	3.325	156-1560 ppb	Fresh Fish
8	EB	0.613	0.51	0	0.15	3.778	1.1-110 ppb	Fresh Fish
9	m-Xylene	0.623	0.52	0	0.16	3.839	1.1-111 ppb	Fresh Fish
10	p -Xylene	0.613	0.52	0	0.16	3.839	1.1-112 ppb	Fresh Fish
11	o-Xylene	0.663	0.56	0	0.16	3.939	0.6-60 ppb	Fresh Fish
12	Styrene	0.849	0.65	0	0.16	3.856	1.2-120 ppb	Fresh Fish

3   Prospective Polymers for Sensor Coating

Table 2 shows a set of 12 commercially available polymers whose LSER parameters are available and which have known chemical affinities for the target vapors listed in Table 1. These data were collected from published reports [17], [20], [21], [22].

Table 2. Potential polymers and their LSER solvation parameters

S. No.	Polymers	Solvation Parameter						Mass Density ρ_p (g.cm^{-3})
		c	r	s	a	b	l	
1	PIB	-0.77	-0.08	0.37	0.18	0	1.02	0.918
2	PECH	-0.75	0.1	1.63	1.45	0.71	0.83	1.36
3	OV202	-0.39	-0.48	1.3	0.44	0.71	0.81	1.252
4	OV25	-0.846	0.177	1.287	0.556	0.44	0.885	1.15
5	PEI	-1.58	0.495	1.516	7.018	0	0.77	1.05
6	PDMS	0.18	-0.05	0.21	0.99	0.1	0.84	0.965
7	PMCPS	-0.12	0.02	1.65	2.71	0.38	0.72	1
8	PMPS	-0.2	-0.03	0.97	1.11	0.1	0.86	1.12
9	ZDOL	-0.49	-0.75	0.61	1.44	3.67	0.71	1.8
10	PVPR	-0.57	0.67	0.83	2.25	1.03	0.72	1.01
11	PVA	-0.698	0.193	1.595	1.886	0	0.495	1.283
12	PMA	-0.617	0.257	1.472	1.656	0	0.475	1.263

4   LSER and K-Matrix

LSER equation for the partition coefficient $K = C_p / C_v$ for a given vapor and polymer combination is written as [17], [18]

$$\log K = c + rR_2 + s\pi_2^* + a\sum \alpha_2^H + b\sum \beta_2^H + lLog^{16} \tag{1}$$

where solubility parameters r, s, a, b and l are associated with the polymer and R_2, π_2^*, $\sum \alpha_2^H$, $\sum \beta_2^H$ and $\log L^{16}$ are the complimentary solubility parameters associated with the vapor. The logarithm on the left hand side is with respect to base 10. By using the LSER equation (1) and the solvation parameter values given in Tables 1 and 2, partition coefficients (K) for different vapor-polymer pairs were calculated as shown in Table 3. The table shows vapors in rows and polymers in columns. With 12 target vapors and 12 potential polymers this table defines a 12×12 multivariate data matrix which is commonly referred to as K-matrix. The K-matrix can be viewed as data for vapors as objects where each vapor is represented by a set of value realizations by 12 polymers as variables. Also, the transpose of K-matrix K^T can be interpreted as data for the polymers as objects represented by 12 vapors as variables.

Table 3. K-matrix for all vapor-polymer combinations

Vapors	Polymers											
	PIB	PECH	OV 202	0V 25	PEI	PDMS	PMCPS	PMPS	ZDOL	PVPR	PVA	PMA
TMA	8.81	25.72	38.97	14.61	1.09	43.93	43.03	28.18	1368.04	35.18	2.82	3.04
Ammonia	1.17	11.02	11.24	3.81	3.35	10.43	36.718	8.66	379.36	18.06	3.07	3.08
MA	4.95	37.01	30.21	13.91	15.78	35.43	112.85	30.59	657.66	60.45	7.14	6.99
DMA	9.38	46.83	51.18	21.11	5.78	52.81	99.63	41.32	1677.45	65.68	5.75	5.83
Acetone	16.29	166.91	149.05	66.89	14.34	67.81	358.34	106.07	729.14	88.46	23.33	21.44
MEK	64.34	465.49	451.74	215.11	21.10	195.34	756.41	310.34	2032.22	196.86	38.25	34.86
Toluene	583.20	1039.56	616.84	858.70	116.43	1164.53	1574.78	1445.01	178.24	634.56	78.21	76.26
EB	1671.94	2425.72	1397.78	2130.10	254.48	2785.93	3242.80	3470.48	392.85	1376.16	127.02	121.85
m-xylene	1942.41	2883.17	1622.37	2520.15	296.99	3153.63	3761.32	4010.98	470.80	1613.95	141.89	135.54
p-xylene	1945.99	2876.54	1640.40	2509.90	293.62	3157.26	3759.59	4013.75	479.01	1589.24	141.26	134.74
o-xylene	2523.13	4093.27	2108.38	3535.79	426.75	3883.38	5177.74	5331.51	547.32	2187.21	187.48	177.31
Styrene	2166.20	5110.70	1925.0	4205.2	623.41	3381.30	6406.2	5460.0	393.29	3016.10	257.83	245.24

5   Fuzzy *c*-Means Clustering for Polymer Selection

Fuzzy c-means clustering (FCM) is a clustering technique that reveals data structure according to some measure of association between objects. The objects are seen to be divided into '*c*' fuzzy clusters where each object has been assigned specific membership grades for measuring its association with different clusters. In a given data, the FCM algorithm seeks iteratively groupings of objects into c clusters such that intracluster associations between objects are maximized and intercluster associations are minimized [23], [24]. For doing this, the cluster centers and membership grades are successively iterated that minimizes a nonlinear objective function defined as

$$O_m = \sum_{i=1}^{n} \sum_{k=1}^{c} M_{ik}^m \|X_i - C_k\|^2 \quad \text{where} \quad X_i = \{x_{ij}\} \quad \text{with} \quad i = 1, 2, ..., n \quad \text{and} \quad j = 1, 2, ..., d$$

denote d-dimensional n pattern vectors associated with objects in $n \times d$ data matrix, C_k $(k = 1, 2, ..., c)$ denote centers of c fuzzy clusters, and M_{ik} $(i = 1, 2, ..., n; k = 1, 2, ..., c)$ denote grades of membership of n objects into c clusters. In this equation, $\|.\|^2$ represents the distance of datum X_i from the cluster center C_k, and the exponent m is any real number greater than 1. Ideally a polymer belonging to only one cluster will have membership grade equal to 1 for association with this centre, and zero for the others. We monitored membership grades of all polymers by inputting the number of clusters c successively increasing from 2 to 11 in the *fcm* function. It is found that five polymers (PIB, OV 25, PMCPS, ZDOL and PVA) always appear to have highest membership grade value irrespective of the value of c for $c \geq 5$. For $c > 5$, certain other polymers also could be declared as cluster centers on the basis of maximum value of membership grade but the repeatable declaration of these five polymers for any value of $c > 5$ suggested that this subset presents maximally diverse interaction possibilities with target vapors. Therefore, we selected this subset of polymers for the desired sensor array. Table 4 shows the membership grades for $c = 5$.

Table 4. Fuzzy *c*-means (*fcm*) computed membership grades with respect to the clusters

S. No.	Polymer	Membership Grade for Cluster Center 1	Membership Grade for Cluster Center 2	Membership Grade for Cluster Center 3	Membership Grade for Cluster Center 4	Membership Grade for Cluster Center 5
1	PIB	0.02825	**0.940784**	0.007084	0.010225	0.013657
2	PECH	0.784312	0.048669	0.142692	0.012046	0.012281
3	OV 202	0.022733	0.930195	0.00689	0.016952	0.02323
4	OV 25	**0.906908**	0.049135	0.028475	0.007433	0.008049
5	PEI	0.002724	0.009332	0.001314	0.011735	0.974895
6	PDMS	0.781305	0.092235	0.090118	0.017802	0.01854
7	PMCPS	0.034484	0.008987	**0.949612**	0.003508	0.003408
8	PMPS	0.040485	0.009586	0.942963	0.003496	0.003469
9	ZDOL	1.18E-06	2.91E-06	6.27E-07	**0.999989**	5.84E-06
10	PVPR	0.053821	0.885239	0.013585	0.020344	0.02701
11	PVA	0.000523	0.001604	0.000264	0.002549	**0.99506**
12	PMA	0.000651	0.001986	0.000328	0.003175	0.99386

6   SAW Sensor Array Simulation

Calculations of artificial SAW sensor array response is based on SAW velocity perturbation model under polymer loading developed by Martin, Frye and Senturia (MFS) [25]. The model allows calculation of changes in SAW velocity due to polymer film on sensor surface in terms of the polymer mass density, viscoelastic coefficients and film thickness. For the present calculations we used a simplified analytical expression for the MFS model written as [26]

$$\frac{\Delta v}{v_0} = \frac{\Delta f_p}{f_0} = -\omega h \rho_p \left[(c_1 + c_2 + c_3) - \frac{c_1 + 4c_3}{\rho_p v_0^2} G' + \frac{1}{2}(\omega h)^2 \rho_p (c_1 + c_2) \frac{G'}{|G|^2} \right] \qquad (2)$$

where f_0 and f_p respectively are uncoated and polymer coated SAW Oscillator frequencies, $\Delta f_p = f_p - f_0$, h polymer film thickness, ρ_p polymer mass density, and $\omega = 2\pi f$. Assuming SAW sensors on STX-quartz substrate the following parameters were used: $f_0 = 100$ MHZ, $h = 100$ nm, $v_0 = 3.158 \times 10^5$ cms^{-1} for unperturbed SAW velocity, and $c_1 = 0.013 \times 10^{-7}$, $c_2 = 1.142 \times 10^{-7}$ and $c_3 = 0.615 \times 10^{-7}$ (cm^{2}secg^{-1}) for substrate elastic constants, $G' \approx 1 \times 10^{10}$ and $G'' \cong 0.3G'$ for high frequency limiting values of polymer shear mode [27]. The sensor response arises due to changes in polymer mass density produced by vapor sorption, $\Delta \rho = C_p = KC_v$. This produces a change in SAW sensor frequency. Let this new frequency be f_v which is obtained by inserting $\rho_p = \rho_p + \Delta \rho$ in eq. (2). The steady state sensor response is then defined as $\Delta f = f_v - f_p$.

The 5-element SAW sensor array based on the selected set of polymers PIB, OV-25, PMCPS, ZDOL and PVA were used for generating the synthetic sensor responses.

An additive Gaussian noise with mean distributed randomly over ±30 Hz and variance 50 Hz is incorporated in SAW sensor array response. Total 20 samples of each VOC class were generated for concentration range specified in Table 1. The sensor array responses were preprocessed by normalization with respect to vapor concentration and scaled logarithmically in view of some earlier studies [8], [28] which had shown that this procedure improves vapor class differentiability. The preprocessed data is then analyzed by the principal component analysis (PCA).

7   Results

Fig. 1(a) shows principal component (PC) score plot of the vapor response patterns generated by 12-element SAW sensor array defined by all polymers in the original list, that is, without making any selection. In comparison, Fig 1(b) shows the PC score plots based on the 5-element SAW sensor array responses for the same vapors which were defined by using the selected set of five polymers. Even though the 12-element sensor array seems to broadly discriminate spoiled and fresh fish up to certain extent, there is significant overlap of acetone and toluene samples from the spoiled class with ammonia and DMA from the fresh class. The 5-element sensor array clearly removes this overlap in Fig. 1(b), hence improves discrimination ability of the nose. The main spoilage marker TMA also gets very well separated in Fig. 1(b). In Fig. 2 (a) and (b) the PC1-PC2 and PC1-PC3 plots based on this 5 element SAW sensor array shows this result more clearly. The role of polymer selection in sensor array making is quite apparent. It helps not only in improving the sensor array performance but also reduces the development effort, time and cost.

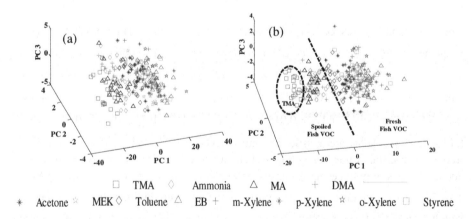

Fig. 1. Principal component score plots based on (a) 12-element sensor array without polymer selection, and (b) 5-element sensor array after FCM based polymer selection

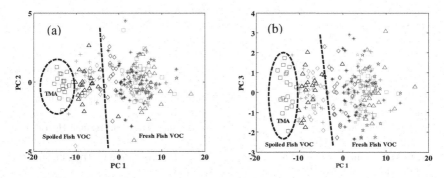

Fig. 2. Principal component score plots by using the 5-element sensor array

8   Conclusion

Fuzzy *c* means clustering is proven to be effective data mining technique in this polymer selection context. Five potential polymers: PIB, OV 25, PMCPS, ZDOL and PVA were selected from the list of 12 commercially available polymers. The SAW sensor array simulated by these five polymers successfully differentiates between fresh and spoiled fish marker VOC.

Acknowledgement. Author Prabha Verma is thankful to CSIR, Government of India for providing the Senior Research Fellowship.

References

1. Firestein, S.: How the Olfactory System Makes Sense of Scents. Nature 413, 211–218 (2001)
2. Persaud, K., Dodd, G.: Analysis of Discrimination Mechanisms in the Mammalian Olfactory System using a Model Nose. Nature 299, 352–355 (1982)
3. Gardner, J.W., Bartlett, P.N.: Electronic Noses: Principles and Applications. Oxford University Press, New York (1999)
4. Hines, E.L., Biolet, P., Gardner, J.W., Gongora, M.A.: Pattern Analysis for Electronic Nose. In: Pearce, T.C., Schiffman, S.S., Nagle, H.T., Gardner, J.W. (eds.) Handbook of Machine Olfaction: Electronic Nose Technology. Wiley-VCH, Weinheim (2003)
5. Carey, W.P., Beebe, K.R., Kowalski, B.R.: Selection of Adsorbates for Chemical Sensor Arrays by Pattern Recognition. Anal. Chem. 58, 149–153 (1986)
6. Grate, J.W.: Acoustic Wave Microsensor Arrays for Vapor Sensing. Chem. Rev. 100, 2627–2648 (2000)
7. Park, J., Groves, W.A., Zellers, E.T.: Vapor Recognition with Small Arrays of Polymer-Coated Microsensors. Anal. Chem. 71, 3877–3886 (1999)
8. Yadava, R.D.S., Chaudhary, R.: Solvation Transduction and Independent Component Analysis for Pattern Recognition in SAW Electronic Nose. Sens. Actuators B 113, 1–21 (2006)
9. Yadava, R.D.S.: Modeling, Simulation, and Information Processing for Development of a Polymeric Electronic Nose System. In: Korotcenkov, G. (ed.) Chemical Sensors – Simulation and Modeling, vol. 3, pp. 411–502. Momentum Press, New York (2012)

10. Mahboob, S., Hussain, B., Alkahem, H.F., Al-Akel, A.S., Iqbal, Z.: Volatile Aroma Comounds and Organoleptic Comparisons of Meat from Wild and Culured Cirrhina Mrigala and cyprinus carpio. Adv. in Nat. and App. Sci. 3, 113–126 (2009)

11. Phan, N.T., Kim, K.H., Jeon, E.C., Kim, U.H., Sohn, J.R., Pandey, S.K.: Analysis of Volatile Organic Compounds Released during Food Decaying Processes. Env. Monit. Assess. 184, 1683–1692 (2012)

12. Sadok, S., Uglow, R.F., Haswell, S.J.: Determination of Trimethylamine in Fish by Flow Injection Analysis. Anal. Chim. Acta 321, 69–74 (1996)

13. Shalaby, A.R.: Significance of Biogenic Amines to Food Safety and Human Health. Food Res. Inter. 29, 675–690 (1996)

14. Dyer, W.J.: Amines in Fish Muscle: I. Colorometric Determination of Trimethylamine as the Picrate Salt. J. Fish. Res. Bd. Can. 6, 351–358 (1945)

15. Tarr, H.L.A.: Trimethylamine Formation in Relation to the Viable Bacterial Population of Spoiling Fish Muscle. Nature 142, 1078–1078 (1938)

16. Bota, G.M., Harrington, P.B.: Direct Detection of Trimethylamine in Meat Food Products using Ion Mobility Spectrometry. Talanta 68, 629–635 (2006)

17. Abraham, M.H.: Scales of Solute Hydrogen Bonding: Their Construction and Application to Physicochemical and Biochemical Process. Chem. Soc. Rev. 22, 73–83 (1993)

18. Grate, J.W., Abraham, M.H., Wise, B.M.: Design and Information Content of Arrays of Sorption-based Vapor Sensors using Solubility Interactions and Linear Solvation Energy Relationships. In: Ryan, M.A., Shevade, A.V., Taylor, C.J., Homer, M.L., Blanco, M., Stetter, J.R. (eds.) Computational Methods for Sensor Material Selection, pp. 193–218. Springer Science, Business Media, New York (2009)

19. Ho, C.K., Lindgren, E.R., Rawlinson, K.S., McGrath, L.K., Wright, J.L.: Development of a Surface Acoustic Wave Sensor for In-Situ Monitoring of Volatile Organic Compounds. Sensors 3, 236–247 (2003)

20. Matatagui, D., Marti, J., Fernandez, M.J., Fontecha, J.L., Guturrez, J., Gracia, I., Cane, C., Horrilio, M.C.: Chemical Warfare Agents Simulants Detection with an Optimized SAW Sensor Array. Sens. and Actu. B 154, 199–205 (2011)

21. Santiuste, J.M., Dominguez, J.A.G.: Study of Retention Interactions of Solute And Stationary Phase in the Light of the Solvation Model Theory. Ana. Chi. Acta 405, 335–346 (2000)

22. Sanchez, I.C., Rodgers, P.A.: Solubility of Gases in Polymers. Pure & App. Chem. 62, 2107–2114 (1990)

23. Bezdek, J.C.: Pattern Recognition with Fuzzy Objective Function Algorithm. Plenum Press, New York (1981)

24. Chiu, S.: Fuzzy Model Identification Based on Cluster Estimation. J. of Inte. Fuzzy Syst. 2, 267–278 (1994)

25. Martin, S.J., Frye, G.C., Senturia, S.D.: Dynamics and Response of Polymer-Coated Surface Acoustic Wave Devices: Effect of Viscoelastic Properties and Film Resonance. Ana. Chem. 66, 2201–2219 (1994)

26. Yadava, R.D.S., Kshetrimayum, R., Khaneja, M.: Multifrequency Characterization of Viscoelastic Polymers and Vapor Sensing based on SAW Oscillators. Ultrasonics 49, 638–645 (2009)

27. Ferry, J.D.: Viscoelastic properties of polymer. John Wiley and Sons, Newyork (1980)

28. Jha, S.K., Yadava, R.D.S.: Preprocessing of SAW Sensor Array Data and Pattern Recognition. IEEE Sen. J. 9, 1202–1208 (2009)

Quantitative Identification of Volatile Organics by SAW Sensor Transients – Comparative Performance Analysis of Fuzzy Inference and Partial-Least-Square-Regression Methods

Prashant Singh[1] and R.D.S. Yadava[2]

[1] Department of Physics, College of Engineering, Teerthanker Mahaveer University,
Moradabad - 244001, India
p8singh@gmail.com

[2] Sensors and Signal Processing Laboratory, Department of Physics, Faculty of Science,
Banaras Hindu University, Varanasi-221005, India
ardius@bhu.ac.in, ardius@gmail.com
http://www.ardius.org

Abstract. We present a comparative performance analysis between three methods (fuzzy c-means and fuzzy subtractive clustering based fuzzy inference systems and partial-least-square regression) for simultaneous determination of vapor identity and concentration in gas sensing applications. Taking polyisobutylene coated surface acoustic wave sensor transients for measurements we analyzed simulated data for seven volatile organic compounds by applying these methods as a function of polymer film thickness. The sensor transients were represented by discrete wavelet approximation coefficients. It is found that PLS regression performs most optimally for both discrimination between vapor identities and simultaneous estimation of their concentration.

Keywords: SAW sensor transients, quantitative recognition, discrete wavelet decomposition, fuzzy inference system, partial-least-square regression.

1 Introduction

Measurements of single sensor transients provide an alternate method for vapor detection and identification to the conventional electronic nose systems based on an array of sensors functionalized for broad range of chemical selectivities [1-7]. Detection of volatile organic compounds (VOCs) is important for several applications like bomb detection, environment monitoring, chemical hazard detection, breath analysis for disease biomarkers, food quality monitoring etc. [8-11]. The conventional electronic nose is analogous functioning of human nose where olfactory receptor neurons generate response patterns at vapor inhalation which is processed by brain for identification. Transient sensors do not have a biological parallel, but sensor transients carry richer information compared to electronic nose array sensors as they capture complete signal generation kinetics (from first contact to steady state) in contrast to array sensors record only equilibrium responses. Besides, a single transient sensor may discriminate several vapors by utilizing vapor specific variations in sorption kinetics.

This feature may help in reducing system hardware complexity. Both approaches however are faced with similar data processing challenges for extraction of chemical signatures from multivariate time series and equilibrium response patterns [1].

Recently, we carried out a series of simulation studies based on polymer coated SAW sensor transients for vapor recognition [12-16] as well as for simultaneous concentration estimation [17]. Earlier, we had applied fuzzy c-means clustering based Sugeno-type fuzzy inference system (FCM-FIS) for quantitative identification of seven volatile organic compounds by polyisobutylene (PIB) coated SAW sensor transients [17]. It was found that the accuracy of the results depends on the polymer film thickness and its viscoelastic properties. There is an optimum region for film thickness ($0.18 < \varphi_3 / \pi < 0.36$) where both the classification rate and the concentration estimates were most accurate. The deterioration occurred at both (higher and lower) thickness ends. This was analyzed to be related to the large range of partition and diffusion coefficients associated with the vapors. By eliminating the fastest diffusing vapor data from the analysis the results improved substantially. This raised a question whether FCM-FIS can be relied for analyzing data with wide variations in vapor sorption and diffusion parameters. With motivation to determine most optimal method for quantitative recognition by SAW sensor transients we present in this paper a comparison of three methods. These are: fuzzy c-means clustering and fuzzy subtractive clustering based Sugeno-fuzzy inference systems (FCM-FIS and FSC-FIS), and partial-least-square-regression (PLSR).

2   Transient Data and Preprocesssing

SAW sensor parameters, target volatile organics and transient data generation used in the present analysis are the same as described in [17]. In brief, a polyisobutylene (PIB) coated SAW oscillator sensor operating at 100 MHz was used for calculating transient responses to step-function like vapor exposure. Detection of 7 volatile organic compounds: chloroform (200, 260), chlorobenzene (4680, 230), o-dichlorobenzene (22500, 5.49), n-heptane (1200, 48), toluene (1000, 35), n-hexane (180, 160), n-octane (955, 38) with concentrations varied over [10-300] ppm at step of 10 ppm were targeted for study. The numbers in parenthesis are partition coefficient K and diffusion coefficient ($D \times 10^{-11}$ cm^{2}s^{-1}) of these vapors in PIB. The vapor sorption and desorption times are 5 and 10 sec respectively with each time series containing 31 time points at 0.5 sec interval. The film thickness was taken to be $h = 524$ nm which is equivalent to $\varphi_3 / \pi = 0.24$. The data generation included an additive noise source with uniform frequency fluctuations within 2 ppm. The transient responses are shown in Fig. 1.

Data preprocessing was done by wavelet transform method. Discrete wavelet transform with *Daubechies-4* basis was used for decomposition up to 2nd level. The coefficients of approximation were used as features and the coefficients of detail were discarded as noise. The original transient data in each response contained 35 points. After 2nd level decomposition each transient is represented by 11 approximation coefficients. The wavelet decomposition is known to be an effective method for reducing data size and noise for time varying signals [19]. We have used this method for representation in almost all our previous studies on transient analysis [12-17].

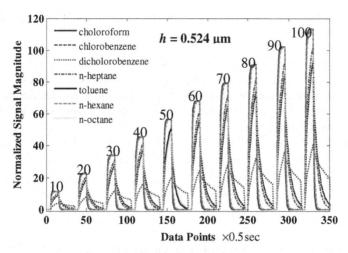

Fig. 1. Transient responses of a polyisobutylene coated SAW sensor for exposures to 7 vapors at 10 ppm steps concentrations in the range 10-100 ppm

In addition to comparing performance of the three methods (FCM-FIS, FSC-FIS and PLSR) for simultaneous qualification and quantification of analyte vapors our goal here is also to find out optimum condition of film thickness which yields best results. The film thickness here is represented in normalized form by φ_3 / π which is defined as $\varphi_3 = \mathrm{Re}(\beta_3 h)$ where β_3 represents the wave-vector of shear waves radiated towards the film surface from the film-substrate interface and h denotes the film geometrical thickness. For this purpose, the analyses were carried out for film thicknesses over $0.05 \le \varphi_3 / \pi \le 0.5$ (below film resonance). The film resonance occurs at $\varphi_3 / \pi = 0.5$ [20,21]. The film thicknesses were varied over $h=[0.022, 1.092]$ µm, equivalent to $\varphi_3/\pi=[\,0.05, 0.5]$ at step of $\Delta(\varphi_3/\pi)=0.01$. This range of film thickness variation extends from the acoustically thin region where sensor responses are linear to near resonance condition where responses become highly nonlinear [12, 17].

3   Data Processing

3.1   Fuzzy Clustering and Fuzzy Inference

Fuzzy clustering is a clustering method that groups the members of a data set into clusters according to some fuzzy measure of association between them. The data set constitutes an agglomerate of objects (members) where each object is described by a set of values of attributes (or features) it takes in multidimensional feature space. In the present context, each transient record is an object, and the DWT approximation coefficients define the feature space. The data set is collection of all transient records represented by the associated DWT coefficients. Thus, a vapor sample is a data point in the hyperspace defined by DWT coefficients. The fuzzy c-means clustering (FCM) and the fuzzy subtractive clustering (FSC) are the two important fuzzy clustering methods. The clustering algorithm results in cluster centres and membership values for different data points to belong to different clusters.

Fuzzy c-Means Clustering. The fuzzy c-means clustering is an iterative method where it selects an initial fuzzy pseudo-partition and repeats the process, then computes the centroid of each cluster using fuzzy partition and it updates the fuzzy partition until it gets no change in centroid [21]. This method is based on minimizing the objective function defined as

$$
u_{ik}^{(t+1)} = \left[\sum_{j=1}^{c} \left(\frac{\left\| x_k - v_i^{(t)} \right\|^2}{\left\| x_k - v_j^{(t)} \right\|^2} \right)^{\frac{2}{m-1}} \right]^{-1}
\tag{1}
$$

where u_{ik} is the membership value of x_k in cluster i ($i = 1,...c$) and $v_i^{(t)}$ ($i=1,...c$) is the cluster center [21-22].

Fuzzy Subtractive Clustering. Fuzzy clustering is a clustering technique that combines the data in feature space by certain fuzzy logic. The fuzzy clusters are defined as accumulation of different classes in a specific region in data space. In subtractive clustering method, each data point is specified to a potential (p_i) and calculating its vector distance $\left\| x_i - x_j \right\|$ from all other data points, assuming that it belongs to a cluster of radius r_a, and then assigning the potential p_i as

$$
p_i = \sum_j e^{-\alpha \left\| x_i - x_j \right\|^2} \quad \text{with} \quad \alpha = \frac{4}{r_a^2}
\tag{2}
$$

where the positive constant α represents cluster radii in probability metric and summation extends over all data points. The data point with highest potential is selected as the first cluster center. Next, the second cluster center is determined in similar manner after excluding the data points associated to the first cluster. The process is continued until all residual data points are within r_a of a cluster center [23-24].

Fuzzy Inference System. FIS system makes use of the cluster centres and membership grades for building fuzzy *if* (premise) - *then* (consequence) rules for inference making where both *premise* and *consequence* are characterized by fuzzy or linguistic elements [24]. The FIS utilizes the fuzzy logic, fuzzy *if-then* rules and membership function for decision making. The output of the FIS system is obtained as the weighted average of all the output rules

$$
\text{Output} = \sum_{i=1}^{N} w_i \, z_i \left/ \sum_{i=1}^{N} w_i \right.
\tag{3}
$$

where N is the number of rules [23-24].

3.2   Partial-Least-Square Regression

Regression analysis is to develop a mathematical model for quantitative relationship between sensor response as independent variable and vapor concentration as dependent variable so that by observing the sensor output the concentration of vapor could be predicted. In effect, regression analysis generates calibration curves from the measured data [25]. A regression analysis method aims to achieve best fit model as per some criteria for 'goodness of fit'. The regression relation is defined as a matrix equation $Y = BX$ between dependent variable matrix Y (concentration in rows and vapor classes in columns) and response matrix X (a time series associated with each element in Y) where the matrix of regression coefficients B are determined by some least square solution [26-28].

3.3   MatLab Implementation

The total data is divided into the training and test sets with equal number of samples for each. The data processing was done in MatLab environment by using functions 'genfis2' for FSC, 'genfis3' for FCM and 'plsregress' for PLSR based on the wavelet transformed data matrices. The output data (Y) structure is defined in the form that the first column represents concentration and the remaining columns represent class information in [0, 1] format (1 for true, 0 for false). The input data (X) contains DWT approximation coefficients as predictors with rows corresponding to samples and columns to variables (wavelet approximation coefficients). The class and concentration data in Y matrix is used as response variables.

The input arguments for genfis2 and genfis3 were set for Sugeno type FIS with the input from FSC and FCM clustering. The input data matrix in the training phase consisted of 105 samples × 11 DWT coefficients corresponding to 15 samples from each of 7 vapor classes. The output matrix consists of 105 samples × 8 (concentration + class) parameters. The 1st column of the output matrix contained estimated vapor concentrations, and the rest 7 columns contained vapor class assignments. The output in training phase is an FIS model. The test data is evaluated by using function 'evalfis' which takes the FIS model created by genfis2 or genfis3 as its input. The output from evalfis is 105×8 matrix with the first column containing concentration and the next 7 columns containing class membership grades. The class membership grades were then converted to crisp class label assignments according to the highest grade value.

The plsregress computes a partial-least-squares regression of Y on X in terms of latent variables (or PLS components) XS such that its covariance with response scores YS is maximum. The structure 'stats' in the plsregress output contains the vapor information according to specified data matrix except the first row which define the intercepts and equivalent to B_{PLS} [29].

The relative system performances in regard to concentration estimation were evaluated on the basis of root average relative-mean-squared-error (hereafter referred to as RMSE) defined as

$$RMSE = \frac{1}{C} \sum_{i=1}^{C} \left[\frac{1}{M_i} \sum_{k=1}^{M_i} \left(\frac{c_{ik}(\text{estimated}) - c_{ik}(\text{actual})}{c_{ik}(\text{actual})} \right)^2 \right]^{1/2} \tag{4}$$

where M_i denotes the number of samples of i-th class, C denotes the number classes, and c_{ik} (estimated) and c_{ik} (actual) are the estimated and the actual concentrations respectively. Eq. (4) defines average over all test samples of all vapor classes.

4   Results

Figure 2 shows the results for 105 test samples of 7 vapor classes with actual concentrations over [10, 300] ppm obtained by all three methods (FCM-FIS, FSC-FIS and PLSR). The results are plotted as a function of polymer film thickness expressed in terms of φ_3 / π. Figure 2(a) shows the RMSE errors in concentration estimations Fig. 2 (b) shows the true classification rate in percentage.

Figure 2 contains the earlier report result by using FCM-FIS [17] in addition to the new results based on FSC-FIS and PLSR. It can be seen that FSC-FIS performance is lowest in all regions of the film thickness in regard to both the concentration estimation and the classification accuracy. The results by FCM-FIS and PLSR are comparable in most regions of the film thickness except for $\varphi_3 / \pi < 0.18$ where FCM-FIS

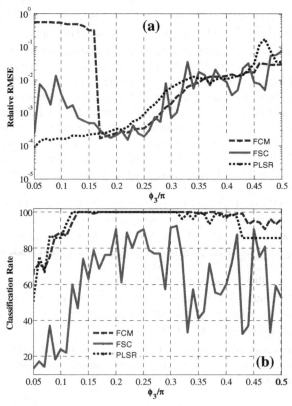

Fig. 2. Results for FCM-FIS, FSC-FIS and PLSR based concentration estimation and true classification rate: (a) RMSE error in concentration estimation, and (b) true classification rate in percentage for polymer film thickness variation over $0.05 < \varphi_3 / \pi < 0.5$.

performance deteriorates. It was found in [17] that this was because of vapors having large variation in partition coefficients and diffusion coefficients. In contrast, we notice here that the PLSR performance consistently well in this region too. This leads us to conclude that PLSR is more robust and accurate under wide variations in analyte characteristics.

As regards optimum film thickness we can see that for fuzzy methods yield best results in the film thickness region $0.18 < \varphi_3 / \pi < 0.36$ as found before in [17], whereas PLSR shows consistently high performance for film thicknesses only a little below resonance condition. Near film resonance, the performances of fuzzy methods also deteriorate. Figure 2 contains the earlier report result by using FCM-FIS [17] in addition to the new results based on FSC-FIS and PLSR. It can be seen that FSC-FIS performance is lowest in all regions of the film thickness in regard to both the concentration estimation and the classification accuracy. The results by FCM-FIS and PLSR are comparable in most regions of the film thickness except for $\varphi_3 / \pi < 0.18$ where FCM-FIS performance deteriorates.

5   Conclusion

In conclusion, we find that for simultaneous determination of vapor identities and estimates of their concentration by using SAW sensor transients the PLSR should be the method of choice as it yields consistently high results and is robust against variations in film and analytes parameters. FCM-FIS yields comparable results for film thicknesses beyond $\varphi_3 / \pi > 0.18$, whereas FSC-FIS performs poorly.

References

1. Yadava, R.D.S.: Modeling, Simulation and Information Processing for Development of a Polymeric Electronic Nose System. In: Korotcenkov, G. (ed.) Chemical Sensors – Simulation and Modelling, ch. 10, pp. 411–502. Momentum Press, New York (2012)
2. Vilanova, X., Llobet, E., Alcubilla, R., Sueiras, J.E., Correig, X.: Analysis of the Conductance Transient in Thick-Film Tin Oxide Gas Sensors. Sens. Actuat. B 31, 175–180 (1996)
3. Llobet, E., Brezmes, J., Vilanova, X., Sueiras, J.: Qualitative and Quantitative Analysis of Volatile Organic Compounds Using Transient and Steady-State Responses of Thick-Film Tin Oxide Gas Sensor Array. Sens. Actuat. B 41, 13–21 (1997)
4. Hines, E.L., Llobet, E., Gardner, J.W.: Electronic noses: A review of signal processing techniques. IEE Proc. Circuits, Devices System 156, 297–309 (1999)
5. Osuna, R.G., Nagle, H.T., Schiffman, S.S.: Transient Response Analysis of an Electronic Nose Using Multi-Exponential Models. Sens. Actuat. B 61, 170–182 (1999)
6. Hoyt, S., McKennoch, S., Wilson, D.M.: Transient Response Chemical Discrimination Module. Proc. of IEEE Sensors 1, 376–381 (2002)
7. Phaisangittisagul, E., Nagle, H.T.: Sensor Selection for Machine Olfaction Based on Transient Feature Extraction. IEEE Trans. Instrum. Meas. 57, 369–378 (2008)
8. Pearce, T.C., Schiffman, S.S., Nagle, H.T., Gardner, J.W.: Handbook of Machine Olfaction. Wiley-VCH, Weinheim (2003)
9. Francesco, F.D., Fuoco, R., Trivella, M.G., Ceccarini, A.: Breath Analysis: Trends in Techniques and Clinical Applications. Microchem. J. 79, 405–410 (2005)
10. Tothil, I.E.: Rapid and On-Line Instrumentation for Food Quality Assurance. CRC Press (2003)

11. Rogers, E.K.: Handbook of Biosensors and Electronic Noses: Medicine, Food and the Environment. CRC Press (1997)
12. Singh, P., Yadava, R.D.S.: Effect of Film Thickness and Viscoelasticity on Separability of Vapour Classes by Wavelet and Principal Component Analyses of Polymer-Coated Surface Acoustic Wave Sensor Transients. Meas. Sci. Technol. 22, 025202 (15pp) (2011)
13. Singh, P., Yadava, R.D.S.: Enhancing Chemical Identification Efficiency by SAW Sensor Transients Through a Data Enrichment and Information Fusion Strategy—A Simulation Study. Meas. Sci. Technol. 24, 055109 (13pp) (2013)
14. Singh, P., Yadava, R.D.S.: A Fusion Approach to Feature Extraction by Wavelet Decomposition and Principal Component Analysis in Transient Signal Processing of SAW Odor Sensor Array. Sens. Transducers J. 126, 64–73 (2011)
15. Singh, P., Yadava, R.D.S.: Feature Extraction by Wavelet Decomposition of Surface Acoustic Wave Sensor Array Transients. Def. Sci. J. 60, 377–386 (2010)
16. Singh, P., Yadava, R.D.S.: Discrete Wavelet Transform and Principal Component Analysis Based Vapor Discrimination by Optimizing Sense-and-Purge Cycle Duration of SAW Chemical Sensor Transients. In: IEEE Conf. on Computational Intelligence and Signal Proc-essing (CISP 2012), pp. 71–75 (2012)
17. Singh, P., Yadava, R.D.S.: Wavelet Based Fuzzy Inference System for Simultaneous Identification and Quantitation of Volatile Organic Compounds Using SAW Sensor Transients. In: Panigrahi, B.K., Suganthan, P.N., Das, S., Satapathy, S.C. (eds.) SEMCCO 2011, Part II. LNCS, vol. 7077, pp. 319–327. Springer, Heidelberg (2011)
18. Burrus, C.S., Gopinath, R.A., Guo, H.: Introduction to Wavelets and Wavelet Transforms: A Primer. Prentice-Hall, New Jersey (1998)
19. Martin, S.J., Frye, G.C., Senturia, S.D.: Dynamics and Response of Polymer-Coated Surface Acoustic Wave Devices: Effect of Viscoelastic Properties and Film Resonance. Analyt. Chem. 66, 2201–2219 (1994)
20. Yadava, R.D.S., Kshetrimayuma, R., Khaneja, M.: Multifrequency Characterization of Viscoelastic Polymers and Vapor Sensing Based on SAW Oscillators. Ultrasonics 49, 638–645 (2009)
21. Bezdek, J.: Pattern Recognition with Fuzzy Objective Function Algorithms. Plenum Press, New York (1981)
22. Bezdek, J.C., Ehrlich, R., Full, W.: FCM: The Fuzzy c-Means Clustering Algorithm. Comp. & Geosciences 10, 191–203 (1984)
23. Chiu, S.: Fuzzy Model Identification Based on Cluster Estimation. J. of Intelligent & Fuzzy Syst. 2, 267–278 (1994)
24. Chiu, S.: Extracting Fuzzy Rules from Data for Function Approximation and Pattern Classification. In: Dubois, D., Prade, H., Yager, R. (eds.) Fuzzy Information Engineering: A Guided Tour of Applications, ch. 9. John Wiley & Sons, New York (1997)
25. Adams, M.J.: Chemometrics in Analytical Spectroscopy. The Royal Society of Chemistry, Cambridge (2004)
26. Geladi, P., Kowalski, R.: Partial Least Squares Regression (PLS): A Tutorial. Analytica Chimica Acta 185, 1–17 (1986)
27. Wold, S., Sjöström, M., Eriksson, L.: PLS-Regression: A Basic Tool of Chemometrics. Chemometrics and Intelligent Laboratory Systems 58, 109–130 (2001)
28. Abdi, H.: Partial Least Squares Regression. In: Lewis, B.M., Bryman, A., Futing, T. (eds.) Encyclopaedia of Social Science Research Methods Thousand Oaks (CA), Sage (2003)
29. Tan, Y., Shi, L., Tong, W., Hwang, G.T.G., Wang, C.: Multi-Class Tumor Classification by Discriminant Partial Least Squares. Comput. Bio. and Chem. 28, 235–244 (2004)

Effective Routing Protocol in Cognitive Radio Ad Hoc Network Using Fuzzy-Based Reliable Communication Neighbor Node Selection

Srinivas Sethi and Sangita Pal

Department of CSEA,
IGIT, Saranag, Orissa
srinivas_sethi@igitsarang.ac.in,
sangitapalmtech.cet@gmail.com

Abstract. Cognitive Radio Ad Hoc NETwork (CRAHN) is a burning technology in the wireless communication area and has the advanced features like self-healing, self-configuring and robustness with low deployment cost. In this environment routing has an important role to establish the path and transmit the data from source node to destination node. So the cutting edge research on the efficient and effective route construction in the networks is paying more interest. The efficient and effective route can be made by selecting the best neighbor node for transmitting the packets. In this paper, it has been established the efficient and effictive route, using fuzzy logic based on node's energy and signal strength of anteena of a node with better spectrum management which are important constraints for routing the packets.

Keywords: CRAHN, Routing Protocol, Fuzzy, Channel Utilization.

1   Introduction

Cognitive radio technology has opened new doors to emerging applications. This technology has been widely used in several application scenarios including military and mission-critical networks consumer-based applications smart grid net-works, public safety networks, post-disaster situations and wireless medical networks. In emergency situations, with the help of multi-interface or Software Defined Radio (SDR), this technology can serve as a facilitator of communications for other devices which may operate in different band and/or have incompatible wireless interfaces. Similarly, this technology can also be used to provide opportunistic access to large parts of the underutilized spectrum in cellular networks.

Cognitive Radio (CR) technology can help Delay Tolerant Networks (DTNs) to provide reliable, delay-sensitive opportunities for communication. For instance, DTNs and CR technology could be used in urban scenario, where high density of wireless devices causes delay in communication due to contention over the link. Cognitive radios thus help in finding free channels for opportunistic use and ensure timely delivery of messages.

S.C. Satapathy, S.K. Udgata, and B.N. Biswal (eds.), *FICTA 2013*,   17
Advances in Intelligent Systems and Computing 247,
DOI: 10.1007/978-3-319-02931-3_3, © Springer International Publishing Switzerland 2014

CR technology achieves Dynamic Spectrum Access of licensed bands of the spectrum by taking the advantage of spectrum utilization with an access to the unlicensed users [1]. Originally CR referred to software radios extended with self-awareness about their characteristics and requirements, in order to determine appropriate radio protocol to be used. According to the spectrum bands the secondary users are accessing, spectrum sharing and allocation schemes can be divided into two types they are: open spectrum sharing and licensed spectrum sharing [2][3].

The present scenario envisages that the secondary users (SUs) communicate over certain bandwidth originally allocated to a primary network, have drawn great research interests. Specifically, the primary users (PUs) are the licensed users in the primary network, who have the absolute right to access their spectrum bands, and yet would be willing to share the spectrum with the unlicensed SUs. The SUs in CRAHN are allowed to access the licensed bands as long as they do not disturb or generate harmful interference to the PUs.

The available spectrum varies considerably and depends heavily on how the primary users are expected to tolerate interference on their spectrum. Novel problem settings arise for the secondary users also from the fact that the instantaneous information on the varying resources may be incomplete, or obtaining the information may cause significant delays. So spectrum or channel management and its neighbor node selection important role in CRAHN, which is our objective to select the efficient neighbor node for better route formation.

The rest of the paper describe as follow. Related works discussed in section 2 and Neighbor Node Selection for a Channel described in section 3. Section 4 give the description of proposed routing protocol in CRAHN. Simulation environment and result discussion have been placed in section 5. Section 6 concludes the paper.

2   Related Works

To use free spectrum in TV bands proposed a centralized and distributed Cognitive Radio Protocol for dynamic spectrum allocation in [4], which enable the nodes to share the free portion of spectrum in optimal way. The authors in it are also introduced the concept of using a Time-Spectrum block which represents the time reserved for use of portion of spectrum by the cognitive user. This minimizes spectrum allocation problem.

In [5], the authors describe a carrier sense multiple access (CSMA) based Medium Access Control (MAC) protocols for the Cognitive Radio Network simultaneously transmit packet for a CR during the transmission of MS of Primary System (PS). The PS and CRN are all carrier sensing based systems. Before transmission, each MS of the PS will sense the carrier for a period of time. If the channel is free then the MS of the PS can transmit packets. When transmissions of a CR or other MSs are taking place, the MS of the PS transmit no packet because the channel is busy. MSs of the PS should have a higher channel access priority than CRs. The CR can transmit with its feasible power and rate if interference to MS of the PS can be maintained at an acceptable level.

In [6], the authors discuss a negotiation mechanism which able to exchange game information among game players (nodes).Then the nodes updates their information in the game timing. A MAC protocol is a multiple access protocol which able to communicate multiple nodes within the same transmission medium. Dynamic Open Spectrum Sharing (DOSS) protocol is a MAC protocol for wireless ad hoc networks which utilizes more spectrums by allowing using dynamic control channels and arbitrary data channels [7]. The control channels are dynamic in both center frequency and bandwidth. It also allots to interference like jamming, primary user's activities from outside of networks and traffic loads from inside of the networks. The data channels are used for flexible utilization of the radio spectrum which is achieved by performing coordination over the control channels.

A MAC Protocol called Hardware-Constrained cognitive MAC (HC-MAC) protocol utilizes multiple channels to improve the cognitive radio network throughput and overall spectrum utilization [8]. The cognitive radio used by secondary users takes two types of hardware constraints that are sensing constraints and transmission constraints. This protocol identifies the problem of spectrum sensing and spectrum access decision tradeoff and formulates an optimal stopping problem. The primary users have certain specifications of their maximum tolerable interference from the secondary users

3   A Fuzzy-Based Neighbor Node Selection for a Channel

In a CRAHN channel management is important factor for routing protocol and more than one node can transmit multiple possible different data over a shared channel to one or more destination node. In this environment if a node communicates with less reliable in terms of node energy and its antenna signal strength through a channel, the route generated by destination or intermediate node (either PUs or SUs) may be break before transmission of data. Further the communication may not be proper as the signal strength of antenna of a node. So, we try to find out reliable communication neighbor node for a channel with the help of fuzzy logic using two parameters, they are; energy of the node and its signal strength of antenna.

Fuzzy logic is used to approximate functions and can be used to model any continuous function or system and has been used to solve several routing protocols and handover problems efficiently in wireless networks [9] [10]. In this paper, communication neighbor node selection process is evaluated using a fuzzy logic approach with the inputs of energy level of the node and its signal strength of antenna. These two parameters make the node's ability to selection the communication neighbor node. The fuzzifier performs the fuzzification process that converts two input, energy level of the node and its signal strength of antenna of the node which are used to generate reliable node as a output which are needed in the inference system.

3.2.1   Fuzzification of Inputs and Output

It obtains two inputs are energy level of the node and its signal strength of antenna of the node and determines the degree to which they belong to each of the appropriate fuzzy sets via membership functions and the reliable value of a node evaluated through the inputs. After creating fuzzy sets from all inputs, output fuzzy sets for reliable node are evaluated by rule evaluation where, the rule evaluation consists of "if-then"-statements that give evidence of how to proceed with a node with certain fuzzy-sets. The fuzzy sets as in Fig.-1, Fig.-2, Fig.-3, are used to consider each as a value between {0,1}. These evaluations are passed to a fuzzy inference system that applies a set of fuzzy rules that determines the node is reliable or not.

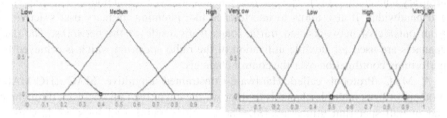

Fig. 1. Energy Level of Node   **Fig. 2.** Signal Strength of antenna of Node

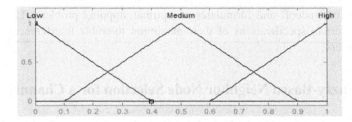

Fig. 3. Reliability of Node

3.2.2   Inference System with Rule Base and Membership Function

Fuzzy Inference System is the system that simulates human decision-making based on the fuzzy control rules and the related input linguistic parameters and the low-high inference method is used to associate the outputs of the inferential rules [11][12]. Using the rule-based structure of fuzzy logic, a series of IF-THEN rules are defined for the output with the help of the input conditions. There are twelve (3x4) possible logical-product output response conclusions, as shown in Fig.-4.

```
1. If (Signal_Strength_of_antenna is Very_Low) and (Energy_Level_of_Node is Low) then (Reliability_of_Node is Low) (1)
2. If (Signal_Strength_of_antenna is Very_Low) and (Energy_Level_of_Node is Medium) then (Reliability_of_Node is Low) (1)
3. If (Signal_Strength_of_antenna is Very_Low) and (Energy_Level_of_Node is High) then (Reliability_of_Node is Low) (1)
4. If (Signal_Strength_of_antenna is Low) and (Energy_Level_of_Node is Low) then (Reliability_of_Node is Low) (1)
5. If (Signal_Strength_of_antenna is Low) and (Energy_Level_of_Node is Medium) then (Reliability_of_Node is Medium) (1)
6. If (Signal_Strength_of_antenna is Low) and (Energy_Level_of_Node is High) then (Reliability_of_Node is Medium) (1)
7. If (Signal_Strength_of_antenna is High) and (Energy_Level_of_Node is Low) then (Reliability_of_Node is Medium) (1)
8. If (Signal_Strength_of_antenna is High) and (Energy_Level_of_Node is Medium) then (Reliability_of_Node is High) (1)
9. If (Signal_Strength_of_antenna is High) and (Energy_Level_of_Node is High) then (Reliability_of_Node is High) (1)
10. If (Signal_Strength_of_antenna is Very_High) and (Energy_Level_of_Node is Low) then (Reliability_of_Node is Low) (1)
11. If (Signal_Strength_of_antenna is Very_High) and (Energy_Level_of_Node is Medium) then (Reliability_of_Node is High) (1)
12. If (Signal_Strength_of_antenna is Very_High) and (Energy_Level_of_Node is High) then (Reliability_of_Node is High) (1)
```

Fig. 4. Fuzzy control rules for Reliability of Node

3.2.3   Defuzzification

The Defuzzification is the process of conversion of fuzzy output set into a single number in which the input for the defuzzification process is a fuzzy set and the method used for the defuzzification is smallest of minimum (SOM). Defuzzifier adopts the aggregated linguistic values from the inferred fuzzy control action and generates a non-fuzzy control output, which represents the reliable node adapted to node conditions [11][12].

The reliable communication neighbor node can be detected by using fuzzy logic with the help of energy and signal strength of a node may be PUs or SUs in CRAHN. This may help to make a better route formation from source and destination node then transfer the data between them, which is main objective of routing protocol.

4   Description of Proposed Protocol

In this section, we discuss the adaptation of the fuzzy based reliable node with better channel management for Effective Routing Protocol in CRAHN. The Cognitive Radio Ad hoc On-demand Distance Vector (CRAODV) is a routing protocol is designed based on ad hoc on-demand distance vector (AODV) [13] with secondary users. The proposed protocol Cognitive Radio Ad hoc On-demand Distance Vector for Reliable Node (CRAODV-RN) is designed for effective and efficient route discovery and reliability of neighbor node. It has been implemented the essential functionality of control packet like route request (RREQ), route reply (RREP) and HELLO messages. Apart from these it also need to implement the proper utilization of spectrum in CRAHN. In Multiple access channel, multiple sender nodes transmit multiple possible different data over a shared physical medium to one or more destination nodes. This requires a channel access scheme, including MAC protocol combined with a multiplexing scheme and has applications in the uplink of the network systems.

The HELLO messages are broadcasted periodically from each node to inform its presence in the network to all its neighbors. Similarly each node in the network system expects to receive messages periodically from each of its outgoing nodes. The node is assumed to be no longer reachable in the network, if a node has received no message from some outgoing nodes for a definite period of time and it removes all affected route entries. During the HELLO message the node is sending the information of energy available with the node and its signal strength to its neighbor nodes and it is maintain in the routing table. Through this information each node can be determined the better life period of its neighbor nodes and its reliability either a PUs or SUs, so that the route will be consistent in CRAHN. This implies better channel utilization and particularly the SUs can sense the licensed channel without interference with PUs.

The RREQ is broadcasted by flooding and propagated from intermediate nodes to other nodes in the network to find the route information during the route discovery process. The initiating node transmits a RREQ message to its neighboring nodes, which transmit the message to their neighbor nodes, and this process is continuing till to destination node. A RREQ message is initiated with predefined time to live (TTL) with a specified hop number by the source node. Intermediate nodes receive the RREQ message. If none of them has the route information to the destination node, the intermediate nodes retransmit the RREQ with an next hop number and this process is continue with the same hop number from the generated by source node. The route node would stop retransmission the RREQ if the destination node is found. Then it sends a RREP message to the initiating node with the complete route information consisting of the stored route in itself, which establish the route from source to destination node and transfer the data through the path. Nodes along with the route can be updated their routing table entries with the latest destination sequence number.

The reallocation of channel may be started if the running channel accessed by PUs. In this way the SUs sense and access the channel to transfer the data in CRAHN.

5   Simulation and Result Discussion

The performances of cognitive radio ad hoc network routing protocol is evaluated by simulations through NS-2 [14], based on the CRCN integrated simulator [15]. The number of secondary users (SUs) in the network is varying from 10 to 100 in the simulation. It has been adopted the IEEE 802.11b protocol to demonstrate the performance evaluation of proposed routing protocol and carried out the simulation as per table-1.

Table 1. Simulation Parameters for CRAODV-RN and CRAODV

S.No	Parameters	Values
1	Area size	**500m x500 m.**
2	Transmission range	**250 m.**
3	Simulation time	**100 s.**
4	Nodes speed	**5 m/s**
5	Pause times	**10 s.**
6	Data rate	**5 Kbps**
7	Mobility model	**Random any point.**
8	Interference	**1**
9	Numbers of SU Node	**10,20,30,40,50,60,70,80,90,100**
10	Numbers of PU Node	**10**
11	No. of Simulation	**5**

The performance evaluations parameters can be obtained through the NS2 Trace Analyzer are as follows, which are the main parameters of interest for QOS (Quality of services).

Normalized Routing Load (NRL): This is the total amount of routing packets per total delivery packets and each transmission over one hop in the network, the packets transmitted count as one transmission.

End-to-End Delay: It refers to the average time period taken for a packet to be transmitted across a network from source node to destination node.

Packet Deliver Ratio (PDR): This is used to calculate the efficiency, effectiveness and reliability of a routing protocol. This can be measured the percentage of the ratio between the number of received packets at destination and the number of packets sent by sources.

Throughput: It is the rate at which a network receives data. It is a channel capacity of network connections in terms of good and rated in terms bits per second (bit/s).

In this paper, reliability of the node in the cognitive radio ad hoc network is calculated by using fuzzy logic to make the protocol more effective. At the same time it also improves the PDR which denotes the effectiveness, efficiency and reliability of routing protocol.

As per fig.-5 our propose routing model (CRAODV-RN) have less NRL as compare to CRAODV. This show overhead is less as compared to conventional routing protocol and this implies the collision is less in our proposed model in CRAHN. This is batter significant of routing protocols to transfer the data from sender to receiver node via intermediate node between them including SUs.

As per fig.-6 our propose routing model consume less time to transmitted data through intermediate node as compare to CRAODV, which is conventional one from sender to receiver node including SUs. This is because of formation of better route using fuzzy based communication neighbor node is used in CRAHN. End-to-end delays minimize the collision in the network and because of this it takes less time to transmit the data in our proposed model.

With some exception our proposed model CRAODV-RN shows batter results in term of PDR as compared to CRAODV which is displayed in fig.-7. Since end-to-end delay minimize with less NRL the efficiency and effectiveness of our proposed routing model is batter as compare to conventional routing protocol. Similarly the system throughput per second by using CRAODV-RN is also better in all cases, as displayed in the fig-8. So we observed through different figure from 5 to 8, that the proposed routing protocol CRAODV-RN is better than conventional routing protocol i.e., CRAODV.

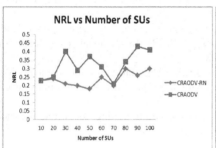

Fig. 5. NRL vs Number of SUs

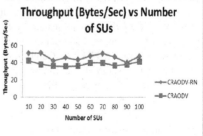

Fig. 6. Delay vs Number of SUs

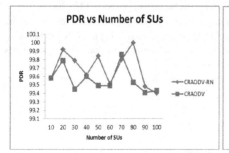

Fig. 7. PDR vs Number of SUs

Fig. 8. Throughput vs Number of SUs

6   Conclusions

In this paper an effective and efficient routing protocol CRAODV-RN is proposed to improve the performance of the cognitive radio ad hoc network. The proposed routing protocol CRAODV-RN in the cognitive radio ad hoc network selects the efficient neighbor node for the route construction. Due to this efficient neighbor node selection the route formation by using proposed model CRAODV-RN is better and more numbers of packet are transmitted. It increases the PDR and throughput with decrease the end-to-end delay and normalized routing load as compared to CRAODV, an existing routing protocol.

References

1. Federal Communications Commission: Facilitating opportunities forflexible,efficient,and reliable spectrum use employing cognitive radiotechnologies, FCC Report, ET Docket 03-322 (2003)
2. Ghasemi, A., Sousa, S.: Spectrum Sensing in Cognitive RadioNetworks: Requirements, Challenges and Design Trade-offs. Cognitive Radio Communicationsand Networks. IEEE Communications Magazine (2008)
3. Rout, A., Sethi, S.: Reusability of Spectrum in Congnitive Radio Network using Threshold. In: IEEE International Conference on Communication, Device and Intelligent System (CODIS), Kolkota (2012)
4. Yuan, Y., Bahl, P., Chandra, R., Moscibroda, T., Narlanka, S., Wu, Y.: Allocating Dynamic Time-Spectrum Blocks in Cognitive Radio Networks. In: Proceedings of ACM MobiHoc, Montreal, Canada (September 2007)
5. Lien, S.-Y., Tseng, C.-C., Chen, K.-C.: Carrier sensing based multiple access protocols for cognitive radio networks. In: Proceedings of IEEE International Conference on Communications (ICC), pp. 3208–3214 (2008)
6. Zhou, C., Chigan, C.: A game theoretic DSA-driven MAC framework for cognitive radio networks. In: Proceedings of IEEE International Conference on Communications (ICC), pp. 4165–4169 (2008)
7. Ma, L., Han, X., Shen, C.-C.: Dynamic open spectrum sharing for wireless adhoc networks. In: Proceedings of IEEE DySPAN, pp. 203–213 (2013)
8. Jia, J., Zhang, Q., Shen, X.: HC-MAC: A hardware-constrained cognitive MAC for efficient spectrum management. IEEE J. Selected Areas Commun. 26, 106–117 (2008)
9. Wong, Y.F., Wong, W.C.: A fuzzy-decision-based routing protocol for mobile ad hoc networks. In: 10th IEEE International Conference on Network, pp. 317–322 (2002)
10. Raju, G.V.S., Hernandez, G., Zou, Q.: Quality of service routing in ad hoc networks. In: IEEE Wireless Communications and Networking Conference, vol. 1, pp. 263–265 (2000)
11. Pedrycz, W., Gomide, F.: An introduction to fuzzy sets: analysis and design (complex adaptive systems). MIT Press, Cambridge (1998)
12. Buckley, J.J., Eslami, E., Esfandiar, E.: An introduction to fuzzy logic and fuzzy sets (advances in soft computing). Physica Verlag (2002)
13. Perkins, C.E., Belding-Royer, E.M., Das, S.R.: Ad Hoc On-Demand Distance Vector (AODV) Routing. Internet Draft, draft-ietf-manet-aodv-13.txt (February 2003)
14. Ns-2 Manual, internet draft (2009),
   http://www.isi.edu/nsnam/ns/nsdocumentation.html
15. CRCN integration, http://stuweb.ee.mtu.edu/~ljialian/

A Self-tuning Fuzzy PI Controller for Pure Integrating Processes

Dharmana Simhachalam and Rajani K. Mudi

Dept. of Inst. & Electronics Engg, Jadavpur University, Kolkata, India
chalamju10@gmail.com, rkmudi@yahoo.com

Abstract. We propose a self-tuning fuzzy PI controller (STFPIC) for pure integrating process with dead time (IPD) whose output scaling factor (SF) is continuously adjusted in each sampling instant by a non-fuzzy gain modifying factor γ, which is defined on the normalized error and change of error of the controlled variable and the number of their uniform fuzzy partitions. Performance of the proposed controller is studied for a pure integrating process with dead-time under both set-point change and load disturbance. Performance comparisons with other FLCs are provided in terms of various performance indices along with transient responses. Simulated results justify the effectiveness of the proposed STFPIC.

Keywords: Integrating process, Fuzzy logic controller, Self-tuning fuzzy control, Scaling factor.

1   Introduction

In the process industries, the integrating processes are frequently encountered and many chemical processes [1] can be modeled as a pure integrating plus dead time processes. In many control applications overshoots and/or undershoots are highly undesirable. For this special class of processes effective controller needs to be designed. The Fuzzy Logic Controllers (FLCs) have been successfully used, and proved to be superior to the conventional non-fuzzy controllers for a complex processes [2, 3] and are found to be less sensitive to parametric variations than conventional controllers [4]. Among the various forms, PI-type FLCs is the most common and practical [5, 6] due to their offset eliminating property. For integrating, non-linear and higher order systems, like conventional PI-controllers, the performance of a conventional PI-type FLC is not acceptable due to large overshoot and excessive oscillation [7-9].

The number of fuzzy *if-then* experts' rules defined on error (e) and change of error (Δe) of the controlled variable is used in FLC control policy. The membership functions (MFs) of the input and output linguistic variables are usually defined on a common normalized domain. The selection of proper input and output scaling factors (SFs) are very important for FLC design, which in many cases are done through trial and error or based on experimental data [10, 11]. For practical systems including integrating plus dead time processes, the performance of a conventional FLC with a limited number linguistic rules and simple MFs, may not fulfill the requirement.

S.C. Satapathy, S.K. Udgata, and B.N. Biswal (eds.), *FICTA 2013*,
Advances in Intelligent Systems and Computing 247,
DOI: 10.1007/978-3-319-02931-3_4, © Springer International Publishing Switzerland 2014

To overcome such limitations many research works have been reported where either the input-output SFs or the definitions of MFs and sometimes the control rules are tuned to improve the close-loop performance [5-6, 10-15].

The controller in [6] is tuned by dynamically adjusting its output SF in each sampling instant by a gain updating factor (α), which is further augmented by a fixed multiplicative factor (K) chosen empirically. The value of α is determined by 49 fuzzy rules defined on process error e and change of error Δe, and derived from the knowledge of process control engineering. Without 49 expert's defined fuzzy rules, here, we propose a self-tuning scheme for fuzzy PI controller (STFPIC) with non-fuzzy dynamic output scaling factor. The output SF is dynamically modified by a single heuristic rule defined on the normalized error, e_N, the normalized change of error, Δe_N, and the number of their fuzzy partitions or membership functions. As the output SF incorporated the dynamics of the process under control, the proposed controller is expected to improve the performance. It is worthwhile to mention that while developing improved auto-tuning PI/PID controllers [16-18], the authors have embedded such knowledge and information. The performance of the proposed STFPIC is tested by simulation experiments on pure integrating plus dead time process (IPD) with different values of dead time. Also the proposed method is tested with 49 as well as 25 fuzzy rules for controller. Results show a significantly improved performance of the proposed STFPIC compared to its conventional fuzzy (FPIC), RSTFPIC [6], and non-fuzzy ZNPI/ ZNPID [20, 21] controllers in transient response and load regulation.

2   Design of the Proposed Controller – STFPIC

Figure 1 shows the simplified block diagram of a non-fuzzy STFPIC. The output SF of the controller is dynamically modified by a non-fuzzy heuristic relation as shown. The various design aspects of the STFPIC are briefly discussed below.

MFs for inputs, (*i.e.*, e_N and Δe_N) and output, (*i.e.*, Δu_N) of the controller (shown in Fig. 2) are defined on the common normalized domain [-1, 1]. Except at the two extreme ends, symmetric triangular MFs are used. The relationships between the SFs (G_e , $G_{\Delta e}$ and G_u), and the input and output variables of the STFPIC are as follows:

$$e_N = G_e \times e \tag{1}$$

$$\Delta e_N = G_e \times \Delta e \tag{2}$$

$$\Delta u = (\gamma G_u) \times \Delta u_N \tag{3}$$

$$where, \quad \gamma = K(1 + \frac{1}{m} \times \Delta e_n \times Sign(e_N)) \tag{4}$$

In Eqn. 4, γ is the on-line gain updating factor for the output SF (G_u), K is a constant, which will make required variation in γ, and m is the number of uniform input (e_N and Δe_N) fuzzy partitions. In this study, we have used $m = 7$ and m =5 as shown in Fig. 2a and Fig.2b. Unlike fuzzy PI controllers (FPIC), which uses only G_u to generate the incremental output (*i.e.*, $\Delta u = G_u \times \Delta u_N$), the incremental output for STFPIC is obtained by using the effective SF, i.e., the product of the gain updating factor γ and G_u as shown in Fig. 1. In RSTFPIC [6], the 49 fuzzy Self tuning rules are used to compute the on-line gain updating factor and another 49 fuzzy rules for controller, i.e., total 98 rules are used. But in our proposed STFPIC, we used only 49 rules for controller and the γ is computed on-line using a *single* model independent non-fuzzy relation defined by Eqn. (4). Hence we reduced the 50% of the rules when compared to RSTFPIC [9]. Moreover, the 49 fuzzy rules for the controller reduced to only 25 rules for STFPIC and hence further 50% reduction of the rules. Therefore, there is an almost 75% rule reduction in the proposed STFPIC with 25 rules compared to the well known RSTFPIC [6], which is designed with total 98 rules (49 control rules + 49 gain update rules).

A PI-type FLC in its velocity form can be described by

$$u(k) = u(k\text{-}1) + \Delta u(k) . \tag{5}$$

In Eqn. (5) k is the sampling instance and Δu is the incremental change in controller output, which is determined by the rules of the form, R_{PI} : If e is E and Δe is ΔE then Δu is ΔU. The rule-base for computing Δu for 49 rules is shown in Fig. 3a, which is derived following the principle of sliding mode control [2, 18], and that of 25 rules is shown in Fig. 3b. The dynamic gain updating factor (γ) is calculated using relation (4), which is formulated keeping in mind the following logic with a view to improving the close-loop performance of the controller during both set point response and load rejection:

When the controlled or process variable is far from the set point and moving towards it, control action should be made weak to prevent possible large overshoot and/or undershoot; on the other hand, when the controlled or process variable is far from the set point and moving further away from the set-point, control action should be made aggressive to restrict such deviations thereby making a quick recovery of the process to its desired value.

Following this above strategy, we attempt to dynamically modify the output SF of the proposed STFPIC through the real-time gain updating parameter γ.

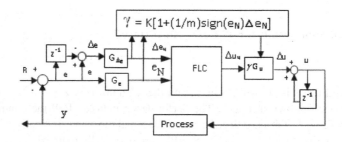

Fig. 1. Block Diagram of the proposed STFPIC

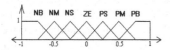

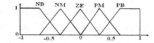

Fig. 2a. MFs of e_N, Δe_N and Δu_N for $m=7$ **Fig. 2b.** MFs of e_N, Δe_N and Δu_N for $m=5$

Δe/e	NB	NM	NS	ZE	PS	PM	PB
NB	NB	NB	NB	NM	NS	NS	ZE
NM	NB	NM	NM	NM	NS	ZE	PS
NS	NB	NM	NS	NS	ZE	PS	PM
ZE	NB	NM	NS	ZE	PS	PM	PB
PS	NM	NS	ZE	PS	PS	PM	PB
PM	NS	ZE	PS	PM	PM	PM	PB
PB	ZE	PS	PS	PM	PB	PB	PB

Δe\e	NB	NM	ZE	PM	PB
NB	NB	NB	NB	NM	ZE
NM	NB	NB	NM	ZE	PM
ZE	NB	NM	ZE	PM	PB
PM	NM	ZE	PM	PB	PB
PB	ZE	PM	PB	PB	PB

Fig. 3a. Fuzzy Rules for Computation of Δu for $m= 7$

Fig. 3b. Fuzzy Rules for Computation of Δu for $m= 5$

3   Results

The performance of the proposed STFPIC is compared with conventional fuzzy PI controller (FPIC) and RSTFPIC [6] and non-fuzzy PI/PID controllers, *i.e.*, Ziegler-Nichols (ZN) [20, 21] under both the set point change and load disturbance for pure integrating process with dead time (IPD) as given in (6), where L is the dead time.

$$G_P(s) = \frac{e^{-Ls}}{s} \tag{6}$$

To establish the robustness of the proposed scheme we tested the process for different values of dead time (L). Control performance is evaluated in terms of peak overshoot (%OS), settling time (t_s), rise time (t_r), integral absolute error (IAE), and integral time absolute error (ITAE). We have used Mamdani type inferencing and Centroid method of defuzzification [2]. We have simulated the proposed method on Matlab simulink software for both 49 fuzzy rules, i.e., m = 7, and 25 fuzzy rules, i.e., m = 5, for the controller and we have used K = 4.0 to calculate γ in (4).

3.1   Simulation Results for 49 Fuzzy Rules (m = 7)

Performance comparison of the conventional FLC (FPIC), RSTFPIC [6] and proposed STFPIC with 49 fuzzy rules under both set point change and load disturbance for different dead times is shown in Fig. 4, and that of non-fuzzy ZNPI and ZNPID [20, 21] controllers is shown in Fig. 5. The performance indices of all the controllers are recorded in Table 1. Observe that the Figs. 4 & 5 and the Table 1, clearly indicate the performance indices for the proposed STFPIC are drastically reduced when compared to other controllers. This is true for all the dead times, i.e., for L= 0.1, 0.2, 0.3, and 0.4. Hence, from the Figs. 4 & 5 and Table 1, we can conclude that the performance of the proposed STFPIC is improved when compared with the other controllers. The most important observation is that the proposed method significantly improves the load regulation, which is the main control objective for most of the continuous process plants.

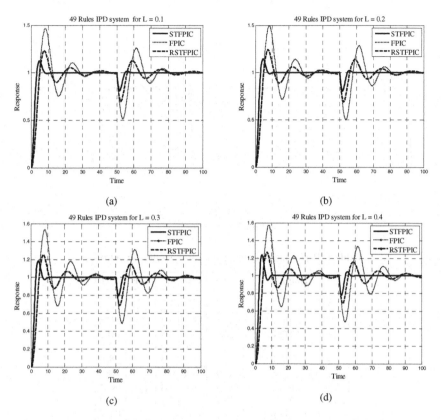

Fig. 4. Controller responses of the RSTFPIC, FPIC and STFPIC with 49 rules for; (a) L = 0.1, (b) L = 0.2, (c) L = 0.3, (d) L = 0.4

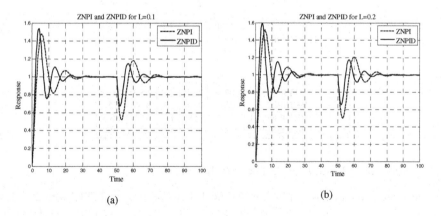

Fig. 5. Controller responses of the Non-Fuzzy ZNPI and ZNPID Controllers with; (a) L = 0.1, (b) L = 0.2, (c) L = 0.3, (d) L = 0.4

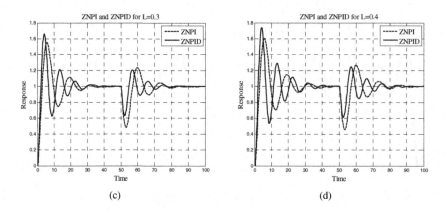

Fig. 5. (*continued*)

Table 1. Performance comparison for the various controllers

L	controller	%OS	t_s(s)	t_r	IAE	ITAE
0.1	ZNPIC	48.0	21.7	2.2	8.43	235.5
	ZNPIDC	54.1	19.1	1.6	6.05	130.7
	FPIC	46.4	26.7	3.2	11.60	330.3
	RSTFPIC	23.3	23.4	3.4	7.38	184.2
	STFPIC(49Rules)	12.2	6.5	2.2	2.68	32.4
	STFPIC(25Rules)	7.53	5.5	2.3	2.36	18.7
0.2	ZNPIC	51.7	21.8	2.1	9.1	258.3
	ZNPIDC	59.9	19.6	1.5	6.9	154.3
	FPIC	49.7	32.5	3.1	12.65	371.4
	RSTFPIC	24.3	23.5	3.3	7.68	196.2
	STFPIC(49Rules)	14.5	6.4	2.1	2.74	32.8
	STFPIC(25Rules)	8.64	5.35	2.2	2.38	19.0
0.3	ZNPIC	55.6	26.6	2.1	9.98	287.1
	ZNPIDC	66.1	23.7	1.4	8.2	190.2
	FPIC	53.4	33.3	3.0	13.9	422.5
	RSTFPIC	25.4	23.2	3.1	8.0	209.8
	STFPIC(49Rules)	18.4	6.3	1.9	2.87	33.9
	STFPIC(25Rules)	10.0	5.15	2.1	2.41	19.7
0.4	ZNPIC	60.6	27.4	2.0	11.13	326.1
	ZNPIDC	74.3	28.1	1.4	10.2	252.2
	FPIC	57.4	33.8	2.9	15.3	483.8
	RSTFPIC	26.6	28.6	3.1	8.39	225.5
	STFPIC(49Rules)	23.7	7.9	1.7	3.07	36.7
	STFPIC(25Rules)	12.3	5.0	1.9	2.51	22.2

3.2   Simulation Results for 25 Fuzzy Rules (m = 5)

Performance comparisons of FPIC, RSTFPIC [6] and proposed STFPIC with 25 rules under both set point change and load disturbance for different dead times is shown in Fig. 6, and that of non-fuzzy ZNPI and ZNPID [20, 21] controllers is shown in Fig. 5. The performance indices of all the controllers are depicted in Table 1. Observe that Figs. 5 & 6 and Table 1, clearly indicate significantly improved performance of the

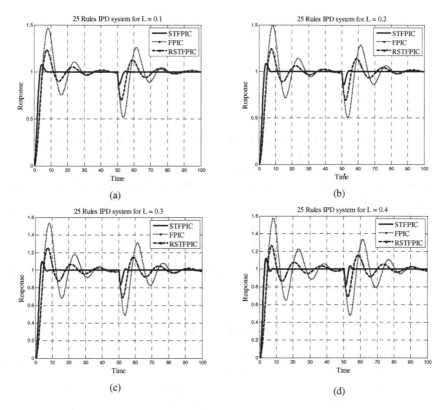

Fig. 6. Controller responses of the RSTFPIC, FPIC and STFPIC with 25 rules for; (a) $L = 0.1$, (b) $L = 0.2$, (c) $L = 0.3$, (d) $L = 0.4$

proposed STFPIC when compared to other controllers. Also when compared with proposed STFPIC with 49 fuzzy rules, the 25 rule based STFPIC is found to perform better. This is true for all the dead times, i.e., for $L= 0.1, 0.2, 0.3$, and 0.4. Hence, from the Figs. 4 & 6 and Table 1, we can conclude that the performance of the proposed STFPIC with 25 rules is more superior when compared to STFPIC with 49 rules in addition to the other controllers. Note that the proposed STFPIC with 25 rules significantly improved the load regulation.

4   Conclusion

We proposed a self-tuning fuzzy PI controller whose output scaling factor gets dynamically adjustment by a non-fuzzy gain updating parameter, which has been derived using information of the normalized error and change of error of the controlled variable and the number of partitions of the input linguistic variables of the proposed STFPIC. The performance of the proposed controller has been tested with 49 fuzzy rules as well as 25 fuzzy rules under both set point change and load disturbance for pure integrating process with different values of dead time. Detailed performance comparisons with other fuzzy logic controllers justified the effectiveness of the proposed controller. The results also proved that the performance of proposed controller with 25 fuzzy rules is more improved than controller with 49 fuzzy rules in both set point response and load regulation.

References

1. Chien, I.L., Fruehauf, P.S.: Consider IMC tuning to improve performance. Chem. Eng. Prog. 86(10), 33–41 (1990)
2. Sugeno, M.: Industrial Applications of Fuzzy Control. Elsevier Science, Amsterdam (1985)
3. Diarankov, D., Hellendoorn, H., Reinfrank, M.: An Introduction to Fuzzy Control. Springer, NY (1993)
4. Harris, C.J., Moore, C.G., Brown, M.: Intelligent Control - Aspects of Fuzzy Logic and Neural Nets. World Scientific, Singapore (1993)
5. Pal, A.K., Mudi, R.K.: Self-Tuning Fuzzy PI controller and its application to HVAC system. IJCC (US) 6 (2008)
6. Mudi, R.K., Pal, N.R.: A Robust Self-Tuning Scheme for PI and PD Type Fuzzy Controllers. IEEE Trans. on Fuzzy Systems 7, 2–16 (1999)
7. Ying, H., Siler, W., Buckley, J.J.: Fuzzy Control Theory: A Nonlinear Case. Automatica 26, 513–520 (1990)
8. Boverie, S.: Fuzzy Logic Control for High-order Systems. In: Proc. 2nd IEEE Int. Conf. on Fuzzy Systems, pp. 117–121 (1993)
9. Lee, J.: On Methods for Improving Performance of PI-Type Fuzzy Logic Controllers. IEEE Trans. on Fuzzy Syst. 1, 298–301 (1993)
10. Nomura, H., Hayashi, I., Wakami, N.: A Self-Tuning Method of Fuzzy Control by Decent Method. In: Proc. IFSA 1991, pp. 155–158 (1991)
11. Chung, H.-Y., Chen, B.-C., Lin, J.-J.: A PI-type Fuzzy Controller with Self-tuning Scaling Factors. Fuzzy Sets and Syst. 93, 23–28 (1998)
12. Palm, R.: Scaling of Fuzzy Controller Using the Cross-Correlation. IEEE Trans. on Fuzzy Syst. 3, 116–123 (1995)
13. Mudi, R.K., Pal, N.R.: A Self-Tuning Fuzzy PI Controller. Fuzzy Sets and Systems 115, 327–338 (2000)
14. Li, H.X., Gatland, H.B.: Conventional Fuzzy Control and Its Enhancement. IEEE Trans. on Syst., Man, Cybern. 26, 791–797 (1996)
15. Mudi, R.K., Pal, N.R.: A Self-Tuning Fuzzy PD Controller. IETE Journal of Research (Special Issue on Fuzzy Systems) 44, 177–189 (1998)
16. Mudi, R.K., Dey, C., Lee, T.T.: An improved auto-tuning scheme for PI controllers. ISA Transactions 47, 45–52 (2008)
17. Dey, C., Mudi, R.K.: An improved auto-tuning scheme for PID controllers. ISA Transactions 48, 396–409 (2009)
18. Mudi, R.K., Dey, C.: Performance improvement of PI controllers through dynamic set-point weighting. ISA Transactions 50, 220–230 (2011)
19. Palm, R.: Sliding Mode Fuzzy Control. In: Proc. 1st IEEE Int. Conf. on Fuzzy Systems, pp. 519–526 (1992)
20. Ziegler, J.G., Nichols, N.: Optimum settings for automatic controllers. ASME Trans. 64, 759–768 (1942)
21. Astrom, K.J., Hagglund, T.: Automatic tuning of PID controllers. Instrument Society of America (1988)

Implementing Fuzziness in the Pattern Recognition Process for Improving the Classification of the Patterns Being Recognised

Sapna Singh and Daya Shankar Singh

Computer Science & Engineering Department,
Madan Mohan Malaviya Engineering College
Gorakhpur, U.P., India
sapna2009singh@gmail.com, dss_mec@yahoo.co.in

Abstract. Correctly classifying and recognizing objects are essentially a knowledge based process. As the unpredictability of the objects to be identified increases, the process becomes increasingly difficult, even if the objects come from a small set. This variability has been taken into account by devising a fuzzy logic based approach using threshold value feature. In this paper, two methods of encoding knowledge in a system are covered-*neural network* and *fuzzy logic*-as they are currently applied to offline hand written character recognition, which is subject to high degrees of unpredictability. This paper proposes a recognition system that classifies a class of recognised patterns i.e. *"partially recognised"* applying fuzziness in the obtained patterns after training with backpropagation neural network and checks for the validation of the concept being proposed.

Index Terms: Backpropagation network, Artificial Neural Network, fuzzy logic, Knowledge Base.

1   Introduction

Intense research in the field of offline handwritten character recognition has led to the development of efficient algorithm from the last two decades. Variations due to writing style, orientation, stray marks in the vicinity of the character challenge the tough paradigm of conventional computing and highlighting the need for a computing method that can process the information more than the way the human brain can do. A natural approach for solving the handwritten character recognition problem is solved by the usage of neural network.

The trademark ability to handle ambiguity, fuzzy logic has also made some application to character recognition, generally as an enhancement to the initial feature extraction portion of an algorithm. Since unpredictability in a fuzzy system is handled through the application of linguistically defined sets, though some research has shown that it is simply possible to define an output rule set to perform the recognition task based on a given inputs membership to the various defined fuzzy sets.

The rest of the paper organized as follows: Section2 describes the related work, Section 3 describes the proposed protocol, Section 4 validation of the proposed protocol, and Section 5 conclusion of the work.

S.C. Satapathy, S.K. Udgata, and B.N. Biswal (eds.), *FICTA 2013*,
Advances in Intelligent Systems and Computing 247,
DOI: 10.1007/978-3-319-02931-3_5, © Springer International Publishing Switzerland 2014

Various approaches used for the design of OCR systems are discussed below:

1.1 Fuzzy Logic

Fuzzy logic is a multi-valued logic that allows intermediate values to be defined between conventional evaluations like yes/no, one/zero, black/white etc [16]. Fuzzy logic is used when answers do not have a distinct one or zero value and there is uncertainly involved.

1.2 Neural Networks

This strategy simulates the way the human neural system works. It samples the pixels in each image and matches them to a known index of character pixel patterns.

A control mechanism like fuzzy logic is used to make these neural networks intelligent. Fuzzy logic is basically the extension of crisp logic that includes the intermediate values between absolutely one and absolutely zero. It has the efficiency to solve the system uncertainties.

2   Proposed Work

The Backpropagation neural network technology is being implemented on set of input patterns. The input data is being trained using *gradient descent* training algorithm with a defined target value set. The output obtained after training the data is compared with the target value set for those particular data set, then the comparison of both is done. If the obtained output is nearest to the defined target value set then that particular pixel value is marked as recognised else unrecognised. A particular threshold value is set for the calculation of recognised training pattern. Here in this paper a fuzzy logic concept is being applied on the various threshold values in which a special case of "*partially recognised*" is being added. Implementation using the membership function for each predefined dataset is done. The data's are then categorised into *Fully Recognised, partially Recognised* and *Unrecognised.*

3   Proposed Protocol

The proposed protocol is a fuzzy logic based protocol for the selection of the recognised pattern obtained network output obtained after training the dataset using Backpropagation neural network. In this process of obtained output selection, two input functions such as *one* and *zero*, which are being termed by binary 1 and 0 respectively; and are transformed into fuzzy sets. A fuzzy set consists of degrees of membership. The One (O) and Zero (Z) Fuzzy sets are defined as,

$$O= \{(a, \mu_O (a))\}, a \in R \quad \text{and} \quad Z= \{(b, \mu_Z (b))\}, b \in U$$

Where,

R is a universe of discourse for one valued recognised patterns and U is a universe of discourse for zero valued recognized patterns, **a** and **b** are particular element of R and U respectively, $\mu_O (a)$, $\mu_Z (b)$ are membership functions, finds the degree of membership of the element in a given set.

$$\mu_O(a) = \left\{ \begin{array}{ll} 0, & \text{if } a \leq TH_1 \\ (a-TH_1)/\,TH_2 - TH_1\,, & \text{if } TH_1 < a < TH_2 \\ 1, & \text{if } a \geq TH_2 \end{array} \right\}$$

$$\mu_Z(b) = \left\{ \begin{array}{ll} 1, & \text{if } b \leq TH_2 \\ (TH_1-b)/\,TH_1 - TH_2, & \text{if } TH_2 < b < TH_1 \\ 0, & \text{if } b \geq TH_1 \end{array} \right\}$$

Where,

TH_1 = Threshold to unrecognized cases
TH_2 = Threshold which identifies the level of recognition of pattern
The value **1** is for true recognised value and **0** for false recognised value.
Membership functions for recognized cases and unrecognised cases are defined from Figure 1, as given below:

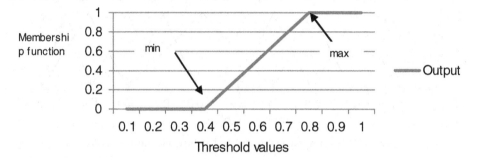

Fig. 1. Graph with minimum and maximum threshold for any input variable

A fuzzy relation is a relation between elements of O and elements of Z, described by a membership function,

$$\mu_{R \times U}\ (a, b), a \in R \text{ and } b \in U$$

Then we are applying AND fuzzy operator on fuzzy relation,

$$\mu_O(a) \wedge \mu_Z(b) = \min(\mu_O(a), \mu_Z(b)) = \left\{ \begin{array}{l} \mu_O(a), \text{ if and only if } \mu_O(a) \geq \mu_Z(b) \\ \mu_Z(b), \text{ if and only if } \mu_O(a) \leq \mu_Z(b) \end{array} \right\}$$

3.1 Rule Evaluation

The process of recognised pattern selection consists of One and Zero values. The Table 1 uses three membership functions to show the varying degrees of input variables. The varying probable output functions are shown in Table 2. The fuzzy rule set is defined in Table 3.

Table 1. Input Function

Input	Membership		
One	Unrecognised (UR)	Partially recognised (PR)	Fully Recognised (FR)
Zero	Weakly Recognised (WR)	Adequately Recognised (AR)	Strongly Recognised (SR)

Table 2. Output Function

Output	Membership
Probability	Lowest, Very Low, Low, Medium Low, Medium, Medium High, High, Very High, Highest

The varying probabilities of output function shown in figure 2.

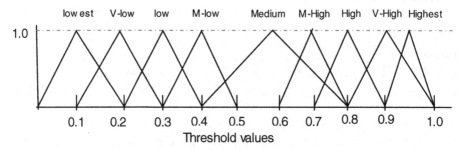

Fig. 2. Graph showing all possible probabilities of output variable

3.2   The Proposed Rule Set for Proposed Protocol

For choosing an optimal pattern following rule set has been considered. The fuzzy rule set is defined and shown in the following table 3.

Table 3. Rule Table

Rules	ONE	ZERO	Probability
1	UR	WR	Lowest
2	UR	AR	Very low
3	UR	SR	low
4	PR	WR	Medium low
5	PR	AR	Medium
6	PR	SR	Medium High
7	FR	WR	high
8	FR	AR	Very high
9	FR	SR	Highest

The various rule set can be described as below:

Rule 1: **IF** One is Unrecognised **AND** Zero is Weakly recognised **THEN** Probability is Lowest

Rule 2: **IF** One is Unrecognised **AND** Zero is Adequately recognised **THEN** Probability is Very low

Rule 3: **IF** One is Unrecognised **AND** Zero is Strongly recognised **THEN** Probability is Low

Rule 4: **IF** One is Partially recognised **AND** Zero is Weakly recognised **THEN** Probability is Medium Low

Rule 5: **IF** One is Partially recognised **AND** Zero is Adequately recognised **THEN** Probability is Medium

Rule 6: **IF** One is Partially recognised **AND** Zero is Strongly recognised **THEN** Probability is Medium High.

Rule 7: **IF** One is Fully Recognised **AND** Zero is Weakly recognised **THEN** Probability is High.

Rule 8: **IF** One is Fully Recognised **AND** Zero is Adequately recognised **THEN** Probability is Very High

Rule 9: **IF** One is Fully Recognised **AND** Zero is Strongly recognised **THEN** Probability is Highest.

4   Validation

Let us consider a network of 10 pixel value.

Case 1: The network is trained for a dataset and the network output is obtained. The target value is compared with the obtained output.

Target$_i$ = [0 1 0 1 0 1 0 1 0 1]

Output$_i$ = [0.8165, 0.62065, 0.203, 0.69945, 0.55131, 0.42276, 0.68634, 0.90259, 0.96061, 0.78761]

where $i = 1, 2.......n$

A membership function based on obtained output of each one pixel value of a character,

$$\mu_o(a) = \begin{cases} 0, & \text{if } O(a) \leq 0.4 \\ (O(a)-0.4)/0.4, & \text{if } 0.4 < O(a) < 0.8 \\ 1, & \text{if } O(a) \geq 0.8 \end{cases}$$

A membership function based on obtained output of each zero pixel value of a character,

$$\mu_z(b) = \begin{cases} 1, & \text{if } Z(b) \leq 0.4 \\ (Z(b)-0.4)/0.4, & \text{if } 0.4 < Z(b) < 0.8 \\ 0, & \text{if } Z(b) \geq 0.8 \end{cases}$$

Now calculate the degree of membership of One recognition and Zero recognition values using the above defined membership functions from figure 3 and figure 4. The degree of membership of One and Zero recognition is shown in Table 4 and Table 5 respectively.

Table 4. Degree of membership of One Recognition

One(a)	Rounded One(a)	Degree of Membership
0.62065	0.6	0.5
0.69945	0.7	0.75
0.4227	0.4	0
0.90259	0.9	1
0.78761	0.8	1

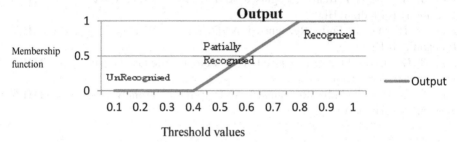

Fig. 3. Graph showing membership functions for input variable "One"

According to degree of membership of the One that can be represented by fuzzy variables from figure 3;

{0.4| Un-Recognised, 0.6|Partially Recognised, 0.7| Partially Recognised, 0.8|Recognised, 0.9|Recognised}

Table 5. Degree of membership of Zero Recognition Value

Zero (b)	Rounded Zero (b)	Degree of Membership
0.8165	0.8	0
0.203	0.2	1
0.55131	0.6	0.5
0.68634	0.7	0.75
0.96061	0.9	0

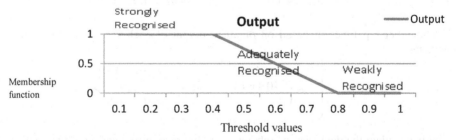

Fig. 4. Graph showing membership functions for input variable "Zero Recognition value"

According to degree of membership of the Zero recognition value that can be represented by fuzzy variables from figure 4,

{0.2|Strongly Recognised, 0.6| Adequately Recognised, 0.7| Adequately Recognised, 0.8|Weakly Recognised, 0.9|Weakly Recognised}

Table 6 shows a fuzzy relation in a relation between elements of O and elements of Z.

Table 6. Fuzzy relation on membership value of One and Zero

O / Z	0.6	0.7	0.4	0.9	0.8
0.8	0^0.5	0^0.75	0^0	0^1	0^1
0.2	1^0.5	1^0.75	1^0	1^1	1^1
0.6	0.5^0.5	0.5^0.75	0.5^0	0.5^1	0.5^1
0.7	0.75^0.5	0.75^0.75	0.75^0	0.75^1	0.75^1
0.9	0^0.5	0^0.75	0^0	0^1	0 ^1

Table 7. Result after AND fuzzy operation

O / Z	0.6	0.7	0.4	0.9	0.8
0.8	0	0	0	0	0
0.2	0.5	0.75	0	1	1
0.6	0.5	0.5	0.5	0.5	0.5
0.7	0.75	0.75	0	0.75	0.75
0.9	0	0	0	0	0

The result of operation process on membership values of One and Zero pixel value shown in Table 7. The possible combinations of one and zero with higher membership value shown in Table 8.The output with degree of membership is shown in table 9.

Table 8. Output Table

O / Z	0.9	0.8
0.2	1	1

Now applying Rule 3 from the above rule-set One of value 1 and Zero with value 1 is the best combination for optimal pattern set.

Now verifying above result from rule set,

Table 9. Output with degree of membership

Zero (b)	Degree of membership	One (a)	Degree of membership
0.2	Highest	0.9	Recognised
		0.8	Recognised

5   Conclusion

The current work is an attempt to simplify the existing manual data-recognition for the hand written documents for closely observing the recognised number of patterns. The proposed query processing methodology will help in further classification of the recognised patterns with effective use of fuzzy based systems. Still there is a lot of scope for improvement in this area. The future work will be descriptive classification on patterns for more improved system.

References

1. Pandey, M.K., Dasila, N.S.: Information retrieval using artificial intelligence and fuzzy logic for hand written documents through optical character recognition (ocr). Journal of Information and Operations Management 3(1), 976–7762 (2012) ISSN: 0976–7754 & E-ISSN: 0976–7762

2. Chaudhary, B.B., Pal, U.: OCR Error Detection and Correction of an Inflectional Indian Language Script. In: Chaudhary, B.B., Pal, U. (eds.) IEEE Proceeding of 13th International Conference on Pattern Recognition 1996, August 25-29, vol. 3, pp. 245–249 (1996)

3. Mani, N., Srinivasan, B.: Application of Artificial Network Model for Optical Character Recognition. In: 1997 IEEE International Conference on System, Man and Cybernetics, Computational Cybernetics and Simulation, October 12-15, pp. 2517–2520 (1997)

4. Bansal, V., Sinha, R.M.K.: Partitioning and Searching Dictionary for Correction of Optically Read Devnagari Character Strings. In: Proceedings of the Fifth International Conference on Document Analysis and Recognition, ICDAR 1999, September 20-22, pp. 653–656 (1999)

5. Plamondon, R., Srihari, S.: IEEE Transactions on Pattern Analysis and Machine Intelligence 22(1), 63–84 (2000)

6. Rakshit, S., Das, S.: International Journal of Computer Applications 6(11), 0975–8887 (2010)

7. Garain, U., Chaudhary, B.: IEEE Transaction on System, Man and Cybernetics- Part C: Applications and Reviews, 32 (2002)

8. Mori, S., et al.: IEEE 80(7), 1029–1058 (1992)

9. Singh, R., Yadav, C.S., Verma, P., Yadav, V.: Optical Character Recognition (OCR) for Printed Devnagari Script Using Artificial Neural Network. International Journal of Computer Science & Communication 1(1), 91–95 (2010)

10. Kim, P.: Improving Handwritten Numeral Recognition Using Fuzzy Logic. IEEE Speech and Image Technologies for Computing and Telecommunications Proc. (1997)

11. Fang, X., Alouani, A.T.: Unconstrained Handwritten Numeral recognition using fuzzy Rule-Based Classifier. In: Proc. of the IEEE (2002)

12. Chan, S.-C., Nah, F.-H.: Fuzzy Neural Logic Network and Its Learning Algorithms. In: Proceedings of the 24th Annual Hawaii International Conference on System Sciences: Neural Networks and Related Emerging Technologies, Kailua-Kona, Hawaii, vol. 1, pp. 476–485 (1991)

13. Bulsari, A., Saxena, H.: Fuzzy Logic Inferencing Using a Specially Designed Neural Network Architecture. In: Proceedings of the International Symposium on Artificial Intelligence Applications and Neural Networks, Zurich, Switzerland, pp. 57–60 (1991)

Fuzzy Self-tuning of Conventional PID Controller for High-Order Processes

Ritu Rani De (Maity)[1] and Rajani K. Mudi[2]

[1] Dept. of Electronics & Instrumentation Engineering
Dr. B.C. Roy Engineering College, Durgapur, India
`ritu_maity_8@yahoo.co.in`
[2] Dept. of Instrumentation & Electronics Engineering
Jadavpur University, Salt-lake, Kolkata – 700 098, India
`rkmudi@yahoo.com`

Abstract. We propose an improved auto-tuning Ziegler-Nichols PID (ZNPID) controller. The proposed controller is a Fuzzy self-tuning ZNPID controller (FST-ZNPID). Conventional Ziegler-Nichols tuned PI and PID controllers provide poor performance for non-linear and high-order systems. In FST-ZNPID enhanced performance is obtained by continuous modification of the proportional constant depending on the process trend. In the proposed FST-ZNPIDC an online auto-tuning factor 'α' is incorporated, which is derived by firing 25 fuzzy rules defined on the process error and change of error. Performance of the proposed controller (FST-ZNPID) is tested and compared with other reported works for high-order linear and non-linear dead time systems under both set point change and load disturbance. The developed controller is found to be robust with considerable changes in the process dead time.

Keywords: PID controller, Ziegler-Nichols tuning, Auto-tuning, Self-tuning fuzzy controller.

1 Introduction

PI and PID controllers are commonly used in process industries due to their simple design and tuning methods [1, 2]. However, PID controllers provide poor performance for high-order and nonlinear processes. In order to overcome the shortcomings of conventional controllers a lot of study and research are being carried out [3-8]. Many PID controllers have been designed on the basis of various adaptation mechanisms. The advancement of Fuzzy and neuro-Fuzzy concept has facilitated and improved those adaptation techniques. Using different auto-tuning mechanisms, either by simple parametric methods [5-8] or fuzzy inferencing schemes [9-12], the transient response of the close-loop system can be improved.

2 The Proposed Fuzzy Self-tuning PID Controller

Here, to start with, we have considered a simple ZNPID controller. Fig.1 shows the Block diagram of the proposed FST-ZNPID. The proposed controller is in velocity

S.C. Satapathy, S.K. Udgata, and B.N. Biswal (eds.), *FICTA 2013*,
Advances in Intelligent Systems and Computing 247,
DOI: 10.1007/978-3-319-02931-3_6, © Springer International Publishing Switzerland 2014

form and the proportional gain k_p of the conventional controller (ZNPID) is being online tuned by a gain updating factor 'α'. The value of α is determined using 25 fuzzy rules defined on two linguistic input variables, *i.e.*, the process error (e) and change of error (Δe). The next section includes the general form of the conventional Ziegler-Nichols (ZN) tuned Proportional-Integral-Derivative controller termed as ZNPID, the concept of tuning strategy and details of the proposed controller FST-ZNPID.

2.1   Design of FST-ZNPIDC

In this study, we consider the proportional-integral-derivative (PID) controllers. This resembles an adaptive conventional controller that fine tunes the k_p of an already working conventional controller (ZNPID). The discrete form of a PID Controller is:

$$u(k) = k_p[e(k) + \frac{\Delta t}{T_i}\sum_{i=0}^{k} e(i) + \frac{T_d}{\Delta t}\Delta e(k)] \qquad (1)$$

$$u(k-1) = k_p[e(k-1) + \frac{\Delta t}{T_i}\sum_{i}^{k-1} e(i) + \frac{T_d}{\Delta t}\Delta e(k-1)]$$

$$\Delta u(k) = u(k) - u(k-1) = k_p[\Delta e(k) + \frac{\Delta t}{T_i}\sum_{i}^{k-1} e(i) + \frac{T_d}{\Delta t}\{\Delta e(k) - \Delta e(k-1)\}]$$

$$\Delta u(k) = k_p[\Delta e(k) + \frac{\Delta t}{T_i} e(k) + \frac{T_d}{\Delta t}\Delta^2 e(k)]$$

$$\Delta u(k) = [k_p\Delta e(k) + k_i e(k) + k_d\Delta^2 e(k)] \qquad (2)$$

where, $k_i = k_p(\Delta t/T_i)$ and $k_d = k_p(T_d/\Delta t)$.

In (1), $e(k) = [r - y(k)]$ is the error, $\Delta e(k) = [e(k) - e(k-1)]$ is the change of error, $\Delta^2 e(k) = [\Delta e(k) - \Delta e(k-1)]$ is the change in change of error, r is the set value, $y(k)$ is the process output at the kth instant, and Δt is the sampling interval. Here, k_p, T_i, and T_d are the proportional gain, integral time, and derivative time, respectively. In (2), k_i is the integral gain and k_d is the derivative gain.

In case of the ZNPID, k_p, T_i, and T_d are obtained according to the ZN ultimate cycle tuning rules [13], *i.e*, $k_p = 0.6k_u$, $T_i = 0.5t_u$, and, $T_d = 0.12t_u$, where k_u and t_u are the ultimate gain and ultimate period respectively. The actual controller output at the kth sampling instance will be: $u(k) = u(k-1) + \Delta u(k)$.

In the proposed FST-ZNPID the online fine tuning using fuzzy logic is done on the nominal value (ZN-tuned) of k_p of an already working conventional controller, *i.e.*, ZNPID. Fig.1 shows the proposed close-loop control system with an existing ZNPID integrated with a fuzzy self-tuning scheme for real-time adjustments of k_p.

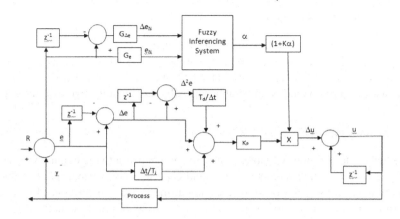

Fig. 1. Block diagram of the FST-ZNPID

2.2 Auto-tuning

In FST-ZNPID the ZN-tuned proportional gain k_p is proposed to modify by the gain updating factor α with the following empirical relation:

$$k_p^t = k_p(1+K.\alpha). \tag{3}$$

Therefore, the effective value of the proportional gain of FST-ZNPID becomes k_p^t. In (3), K is a multiplication factor. Observe that the value of k_p^t does not remain constant while the controller is in operation, rather, k_p^t is continuously modified at each sampling time by 'α', depending on the trend of the controlled process. We use 50% reduced rules for online gain modification compared to that which has been used in STFPIC [9] for achieving the desired variation in the updating factor. The following subsection includes the detail description of the fuzzy self-tuning strategy for k_p^t.

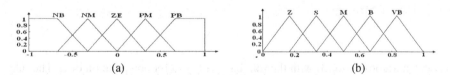

Fig. 2. Membership functions of (a) e and Δe, and (b) gain updating factor α

$\Delta e / e$	NB	NM	ZE	PM	PB
NB	VB	B	M	S	ZE
NM	B	M	B	M	S
ZE	S	M	ZE	M	S
PM	S	M	B	M	B
PB	ZE	S	M	B	VB

Fig. 3. Fuzzy rules for computation of α

2.3   Membership Function

The membership functions (MFs) for controller inputs, *i.e.*, error and change of error (e, and Δe) are defined on the common interval [-1, 1] whereas the MF's for α is defined within [0, 1]. Here as shown in Fig 2, we select symmetrical triangle with equal base and 50% overlap with neighboring MFs, except at the two extreme ends.

2.4   Scaling Factor

The actual values of input variables e and Δe are mapped into the common interval [-1,1] by the input scaling factors (SFs) G_e and $G_{\Delta e}$. Selection of appropriate values of SF's are done on the basis of knowledge about the process under control or by trial and error. Online computation of α is done on the basis of rule-base defined in terms of e and Δe. Following are the relationship of SFs and input variables:

$e_N = G_e.e$; $\Delta e_N = G_{\Delta e}.\Delta e$;

Now the incremental change in the proposed controller (FST-ZNPID) output is :

$$\Delta u(k) = k_p(1+K.\alpha)\{\Delta e(k)+(\Delta t/T_i).e(k)+(T_d/\Delta t.\Delta^2 e(k)\} \tag{4}$$

Observe that in STFPIC [9] the output scaling factor G_u is self-tuned by α. Similarly in FST-ZNPID, the k_p is continuously modified through the simple relation defined by (3).

2.5   Rule Base

In order to improve the overall performance of the ZNPID the gain updating factor is calculated online using the rules in Fig 3.The rules are in the form:

R_α : IF e is E and Δe is ΔE THEN α is $\boldsymbol{\alpha}$

2.6   Tuning Strategy

The main objective of the auto-tuning scheme is to bring the system under control in steady state with minimum oscillation within a shortest possible time. The value of k_p' is updated according to (3) with the rule-base of Fig. 3 for computation of α. The rule base is designed keeping in mind the following strategy shown in Fig 4.

e	+	-	+	-
Δe	-	-	+	+
α	small	large	large	small
k_p'	reduced	increased	increased	reduced

Fig. 4. Tuning strategy for computation of k_p'

3   Simulation Results

In order to prove the performance and the robustness, the proposed controller has been tested on the following second-order processes:

Linear system:
$$G(s) = \frac{e^{-Ls}}{(s+1)^2}.$$
(5)

Non-linear system:
$$\ddot{y} + \dot{y} + 0.2\,\dot{y}^2 = u(t-L)$$
(6)

Integrating system:
$$G_P(S) = e^{-Ls} / [s(s+1)]$$
(7)

Simulation study has been made with set point change and load disturbance. Detailed comparative performance study of the proposed FST-ZNPID is provided with ZNPID, AZNPID [5], and STFPIC [9] in Tables 1-3 and Figs. 5-7. The performance parameters include the percentage overshoot (%OS), settling time (t_s), rise time (t_r), integral absolute error (IAE) and integral-time absolute error (ITAE). For computation of FST-ZNPID the selected value of K is 2 and the same scaling factors are taken for the input variables e and Δe. We have considered different values of dead time (L) for all the processes in (5)-(7). Results in Tables 1-3 show that though the controller is tuned at a particular dead time for a given process, it shows similar performance for changed dead time. This fact proves the robustness of the proposed controller against model uncertainty. Both Tables 1-3 and Figs. 5-7 exhibit that FST-ZNPID significantly reduces the %OS compared to ZNPID. Note that the proposed controller produces higher value of %OS compared to AZNPID and STFPIC but lower values of settling, rise time, IAE, and ITAE. Tables 1-3 show that FST-ZNPID considerably reduces t_s, t_r, IAE, and ITAE in comparison to ZNPID, AZNPID [5] and STFPIC [9]. Performance analysis reveals that FST-ZNPID is capable of providing an outfitting performance with less sensitivity towards disturbances compared to both conventional (AZNPID [5]) and fuzzy (STFPIC [9]) controllers reported in the leading literature.

Table 1. Performance comparison of the second-order linear process (5)

L	Controller	%OS	t_s	t_r	IAE	ITAE
0.1	ZNPID	37.42	3.7	0.5	1.63	17.2
	AZNPID	0.00	9.9	3.9	3.24	60.8
	STFPIC	2.49	7.1	3.2	5.30	90.7
	FST-ZNPID	18.6	2.1	0.5	1.07	09.0
0.2	ZNPID	54.44	4.8	0.5	1.90	19.0
	AZNPID	0.00	11.1	5.0	3.57	62.6
	STFPIC	3.26	7.1	3.0	5.50	96.7
	FST-ZNPID	44.12	4.9	0.4	1.46	10.7

Table 2. Performance comparison of the second-order non-linear process (6)

L	Controller	%OS	t_s	t_r	IAE	ITAE
0.1	ZNPID	50.02	9.5	1.0	2.61	17.6
	AZNPID	7.37	7.8	1.4	2.33	36.7
	STFPIC	19.1	15.1	3.2	8.93	149.5
	FST-ZNPID	20.91	7.0	1.0	1.65	9.2
0.2	ZNPID	58.6	11.6	0.9	3.49	20.93
	AZNPID	9.97	7.8	1.3	2.47	37.4
	STFPIC	21.22	15.0	3.1	9.14	156.5
	FST-ZNPID	30.73	7.2	0.9	1.89	10.8
0.3	ZNPID	65.33	14.0	0.9	4.04	26.7
	AZNPID	12.76	11.2	1.2	2.60	38.5
	STFPIC	23.41	20.0	3.1	14.7	163.6
	FST-ZNPID	39.14	10.5	0.8	2.25	14.7

Table 3. Performance comparison of the second-order marginally stable process (7)

L	Controller	%OS	t_s	t_r	IAE	ITAE
0.1	ZNPID	70.81	14.9	0.9	4.25	50.7
	AZNPID	18.53	4.8	1.2	2.89	62.9
	STFPIC	15.46	21.6	2.6	5.57	110.8
	FST-ZNPID	31.9	7.3	0.9	2.02	19.4
0.2	ZNPID	84.3	14.4	0.9	5.62	72.7
	AZNPID	22.2	19.8	1.0	2.99	49.7
	STFPIC	22.1	29.5	2.8	7.45	168.4
	FST-ZNPID	46.52	10.7	0.8	2.54	25.7
0.3	ZNPID	95.44	24.8	0.7	8.59	137.3
	AZNPID	26.68	34.6	1.0	3.11	64.0
	STFPIC	73.00	06.5	0.8	5.45	274.0
	FST-ZNPID	60.28	19.6	0.8	3.98	50.9

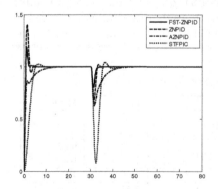

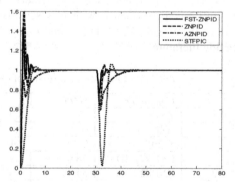

Fig. 5a. Response of (5) for FST-ZNPID(–), ZNPID(– –), AZNPID(– .), STFPIC(- - -) and L= 0.1

Fig. 5b. Responses of (5) for FST-ZNPID(–), ZNPID(– –), AZNPID(– .), STFPIC(- - -) and L= 0.2

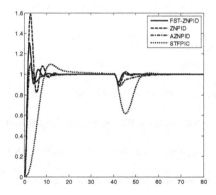

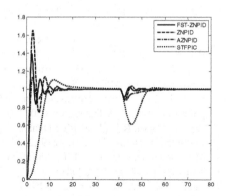

Fig. 6a. Responses of (6) for FST-ZNPID(–), ZNPID(– –), AZNPID(– .), STFPIC(- - -) and $L= 0.2$

Fig. 6b. Responses of (6) for FST-ZNPID(–), ZNPID(– –), AZNPID(– .), STFPIC(- - -) and $L= 0.3$

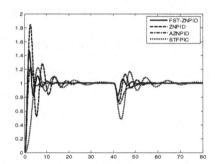

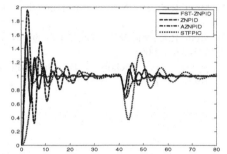

Fig. 7a. Responses of (7) for FST-ZNPID(–), ZNPID(– –), AZNPID(– .), STFPIC(- - -) and $L= 0.2$.

Fig. 7b. Responses of (7) for FST-ZNPID(–), ZNPID(– –), AZNPID(– .), STFPIC(- - -) and $L= 0.3$.

4   Conclusion

A fuzzy gain adaptive controller (FST-ZNPID) has been proposed. The gain modifying factor α continuously adjusted the proportional constant of the ZNPID. The proposed online self-tuning scheme was simple and model–independent. Simulation results for different types of second-order processes established the robustness and high performance features of the proposed FST-ZNPID.

References

[1] Shinsky, F.G.: Process control systems — application, design, and tuning. McGraw-Hill, New York (1998)
[2] Astrom, K.J., Hang, C.C., Person, P., Ho, W.K.: Towards intelligent PID control. Automatica 28(1), 1–9 (1992)

[3] Seborg, D.E., Edgar, T.F.: Adaptive control strategies for process control: A survey. AICHE J. 32(6), 881–913 (1986)

[4] Kristiansson, B., Lennartson, B.: Robust and optimal tuning of PI and PID controllers. IEE Proc. Control Theory Appl. 149(1), 17–25 (2002)

[5] Dey, C., Mudi, R.K.: An improved auto-tuning scheme for PID controllers. ISA Trans. 48(4), 396–409 (2009)

[6] Mudi, R.K., Dey, C.: Performance improvement of PI controllers through dynamic set-point weighting. ISA Transactions 50, 220–230 (2011)

[7] Dey, C., Mudi, R.K., Lee, T.T.: Dynamic set-point weighted PID controller. Control and Intelligent Systems 37(4), 212–219 (2009)

[8] Mudi, R.K., Dey, C., Lee, T.T.: An improved auto-tuning scheme for PI controllers. ISA Transactions 47, 45–52 (2008)

[9] Mudi, R.K., Pal, N.R.: A robust self-tuning scheme for PI and PD type fuzzy controllers. IEEE Trans Fuzzy Syst. 7(1), 2–16 (1999)

[10] Mudi, R.K., De Maity, R.R.: A Noble Fuzzy Self-Tuning Scheme for Conventional PI Controller. In: Satapathy, S.C., Udgata, S.K., Biswal, B.N. (eds.) Proceedings of Int. Conf. on Front. of Intell. Comput. AISC, vol. 199, pp. 83–91. Springer, Heidelberg (2013)

[11] Mudi, R.K., Pal, N.R.: A self-tuning fuzzy PD controller. IETE J. Res. 44(4-5), 177–189 (1998)

[12] Mudi, R.K., Pal, N.R.: A Self-Tuning Fuzzy PI Controller. Fuzzy Sets and Systems 115, 327–338 (2000)

[13] Ziegler, J.G., Nichols, N.B.: Optimum settings for automatic controllers. ASME Trans. 64, 759–768 (1942)

Design of Fuzzy Logic Power System Stabilizer in Multi-machine Power System

M. Megala and C. Cristober Asir Rajan

Department of EEE, Pondicherry Engg College, Puducherry, India
megalamuthusamy@yahoo.com, megu01984@gmail.com

Abstract. The Objective of the Paper is to design and implement Fuzzy logic power system stabilizer in multi-machine power system for reducing Integral Square Time Square Error (ISTSE). In this proposed fuzzy expert system, the generator speed deviation and acceleration are chosen as input signals to fuzzy logic power system stabilizer and the desired output is Integral Square Time Square Error. Finally the comparison of closed loop response of PID (Proportional Integral Derivative) with fuzzy logic controller is highlighted. The simulation results shows the superiority and robustness of fuzzy logic power system stabilizer as compared to conventionally tuned controllers for various Output disturbances

Keywords: Fuzzy Logic Power System Stabilizer (FLPSS), Integral Square Time Square Error (ISTSE), PID (Proportional Integral Derivative).

1 Introduction

The power systems are complex non-linear systems, which are often subjected to low frequency oscillations. The application of power system stabilizers for improving dynamic stability of power systems and damping out the low frequency oscillations due to disturbances has received much attention [1-3]. Power system is a highly nonlinear system and it is difficult to obtain exact mathematical model of the system. In recent years, adaptive self tuning, variable structure, artificial neural network based PSS, fuzzy logic based PSS have been proposed to provide optimum damping to the system oscillations under wide variations in operating conditions and system parameters [4-5].

Recently, fuzzy logic power system stabilizers have been proposed for effective damping of power system oscillations due to their robustness. Fuzzy logic controllers (FLC) are suitable for systems that are structurally difficult to model due to naturally existing non-linearities and other model complexities. Exact mathematical model is not required in designing a fuzzy logic controller. In contrast to conventional power system stabilizer, which is designed in frequency domain, a fuzzy logic power system stabilizer is designed in the time domain [6-10].

S.C. Satapathy, S.K. Udgata, and B.N. Biswal (eds.), *FICTA 2013*,
Advances in Intelligent Systems and Computing 247,
DOI: 10.1007/978-3-319-02931-3_7, © Springer International Publishing Switzerland 2014

Organization of the paper is as follows. The system model of is presented in section 2. Design methodology of PID and FLPSS are given in section 3, simulation results are discussed in section 4 finally, and section 5 concludes the paper.

2   System Model

The proposed algorithm has been tested on a 16 generator, 68-bus system that is a reduced order model of the New England (Old) and New York interconnected system. As per the phenomena of coherency, the five groups are classified as given in Table I. The reference generators are 15, 14,16, 5 and 13. The group 4 contains generators, which are simple generators of the New-England system, and the group 5 represents the New York system. The nonlinear model of multi-machine system is described by following bus system.

Table 1. Classification of Different Groups in 16 Machine, 68 Bus System

Group	Generator Numbers
1	15
2	14
3	16
4	1-9
5	11-13

Analysis of practical power system involves the simultaneous solution of equations consisting of synchronous machines and the associated excitation system and prime movers, interconnecting transmission network, static and dynamic load (motor) loads, and other devices such as HVDC converters, static var compensators. The dynamics of the machine rotor circuits, excitation systems, prime mover' and other devices are represented by differential equations.

The result is that the complete system model consists of large number of ordinary differential and algebraic equations.

Model 1 is assumed for synchronous machines by neglecting damper windings. In addition, the following assumptions are made for simplicity.

- The loads are represented by constant impedances.
- Transients saliency ignored by considering $Xq=xd_1$
- Mechanical power is assumed to be constant.
- Efd is single time constant AVR.

Non-linear model of Multi-machine power system:

i^{th} machine model :

$$\rho\omega_i = (\ T_{mi} - T_{ei}\ ) / 2H$$

$$\delta_i = \omega_o (\ \omega_i - 1)$$

$$\rho E'_{q\,i} = [E_{fd\,i} - (E'_{qi} + (X_{d\,i} - X'_{di})\ I_{di}) / T'_{doi}$$

$$\rho E_{fd\,i} = [K_{Ai} (V_{ref\,i} - V_{t\,i}) + v_{s\,i}) - E_{fd\,i}] / T_{ai}$$

$$T_e = E'd_i\ I_{di} + E'q_i\ I_{qi} - (X'_{qi} - X'_{di})\ I_{di} I_{qi}$$

$$E = E'_{qi} - (\ X_d - X'_d)\ I_{di}$$

$$\delta_{ij} = \delta_i - \delta_j$$

3   Design Methodology

The popularity of PID Controllers can be attributed partially to their robust performance over a wide range of operating conditions and partially to their functional simplicity which allows the process engineer to operate them in a simple and straight forward manner. The mathematical model of the system will be changed due to the change in operating conditions and environment [6]. The PID Controller exhibits poor performance because most of the process are non linear in nature.

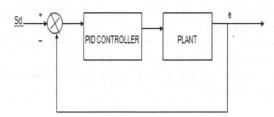

Fig. 1. Block diagram of Proportional Integral Derivative Controller

3.1   Fuzzy Logic Controller Design

In the conventional control, the amount of control is determined in relation to a number of data inputs using a set of equations to express the entire control process. Expressing human experience in the form of a mathematical formula is a very difficult task, if not an impossible one. Fuzzy logic provides a simple tool to interpret this experience into reality.

3.2   Process in Fuzzy Logic Control

Fuzzy logic controllers are rule-based controllers. The structure of the FLC resembles that of a knowledge based of the FLC resembles that of a knowledge based controller except that the FLC utilizes the principles of fuzzy set theory in its data representation and its logic. The basic configuration of the FLC can be simply represented in four parts, as shown in Fig.2.

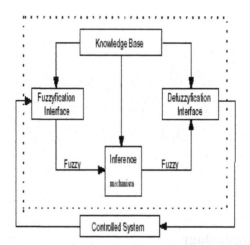

Fig. 2. Schematic Diagram of the FLC building blocks

Fuzzification module – the functions of which are first, to read, measure, and scale the control variable (speed, acceleration) and, second, to transform the measured numerical values to the corresponding linguistic (fuzzy variables with appropriate membership values)

Knowledge base - this includes the definitions of the fuzzy membership functions defined for each control variables and the necessary rules that specify the control goals using linguistic variables

Inference mechanism – it should be capable of simulating human decision making and influencing the control actions based on fuzzy logic

Defuzzufication module – which converts the inferred decision from the linguistic variables back the numerical values.

Power systems usually operate under highly uncertain and stress conditions. Moreover, load variations because the generator dynamic characteristics also vary so different operating conditions are obtained. Therefore, power system controllers should be designed to maintain the robust stability of the system. On the other hand, a CPSS is designed for a linear model representing the generator at a certain operating point and it often does not provide satisfactory results over a wide range of operating conditions. To overcome these drawbacks, fuzzy logic controller (FLC) is an effective tool, which has non-linear structure. In fuzzy controller design, there is no need to perfect model of the system, which is a significant advantage [7]. The design process of fuzzy logic controller may be split into five steps:

- The selection of control variables
- The membership function definition or "the fuzzification"

- The rule creation or "the knowledge base"
- The fuzzy interface engine and
- The de-fuzzification strategy or "the defuzzifier".

3.3   Steps Involved in FLC

The basic structure of the fuzzy controller is shown in Fig. 3. Also, in next Fig.4 is shown how to use fuzzy controller in a PSS structure. In the proposed method, two variables $\Delta\omega$ and ω_Δ are used as input signals in PSS. The coefficients Kin1 and Kin2 input stage, keep the input signals within allowable limit. These coefficients are called scaling factors Each of FLPSS input and output fuzzy variables membership functions have been chosen identical because of the normalization achieved on the physical variables. The normalization is important because it allows the controller to associate equitable weight to each of the rules and therefore, to calculate correctly the stabilizing signals [8].

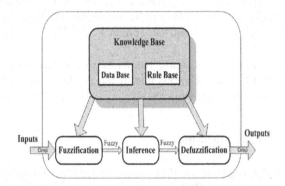

Fig. 3. The basic structure of the fuzzy controller

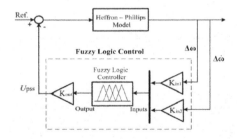

Fig. 4. Schematic Structure of FLPSS

3.4   Membership Functions

Each of the input and output fuzzy variables, y_i is assigned seven linguistic fuzzy subsets varying from Negative Big (NB) to Positive Big (PB). Each subset is associated with a triangular membership function to form a set of seven normalized and symmetrical triangular membership functions for fuzzy variables. The y_{max} and

y_{min} represent maximum and minimum variation of the input and output signals. These values are selected based on simulation data. The range of each fuzzy variable is normalized between -4 to 4 by introducing a scaling factor to represent the actual signal [9-10]. The membership function diagram is shown in Fig. 5.

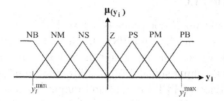

Fig. 5. Fuzzy Variable Seven Memberships Functions

The output signal was obtained by using the following principles

1) If the speed deviation is important, but tends to decrease, then the control must be moderated. In other words, when the machine decelerates, even though the speed is important, the system is capable, by itself, to return to steady state.

2) If the speed deviation is weak, but tends to increase, the control must be significant. In this case, it means that, if the machine accelerates, the control must permit to reverse the situation.

4   Results and Discussion

The Fig. 6 show the response of PID controller for the setpoint of 15 and Fig. 7 and Fig. 8 show the response of PID controller for various disturbances. Fig. 9 show the response of FLPSS and Fig. 10 shows the response of FLPSS for various Disturbances.

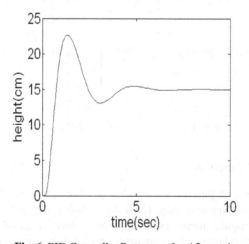

Fig. 6. PID Controller Response for 15 setpoint

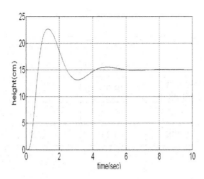

Fig. 7. PID Controller Response for 10% increase in set point

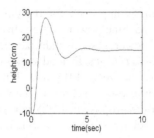

Fig. 8. Response of PID Controller for 10% decrease in set point

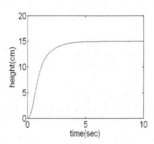

Fig. 9. FLPSS response for 15 set point

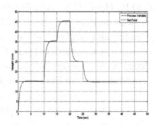

Fig. 10. FLPSS response for 10% Increase in step input (sp)

Table 2.

Controller	Integral square Time Square error	
	10% Set point	20% Set point
Without controllers	29.57	21.3
PID	25.12	18.46
FLPSS	10.23	12.93

5   Conclusion

This paper presents a systematic approach for the design of fuzzy logic power system stabilizers in a multi-machine power system. Investigations reveal that the performance of fuzzy logic power system stabilizer in a multi-machine power system is quite robust under variations in load conditions as compared to PID controller. The coordination between PID and Fuzzy logic power system stabilizer (FLPSS) has been presented for 16 machine 68 bus system. The simulation study reveals that the proposed FLPSS can provide good damping and reduces integral square time square error for various disturbances compared PID controller.

References

1. Larsen, E.V., Swann, D.A.: Applying power system stabilizers Part-I: general concepts. IEEE Transactions on Power Apparatus and Systems PAS-100(6), 3017–3024 (1981)
2. Larsen, E.V., Swann, D.A.: Applying power system stabilizers Part-II: performance objectives and tuning concepts. IEEE Transactions on Power Apparatus and Systems PAS-100(6), 3025–3033 (1981)
3. Larsen, E.V., Swann, D.A.: Applying power system stabilizers Part- III: Practical considerations. IEEE Transactions on Power Apparatus and Systems PAS-100(6), 3034–3046 (1981)
4. Zhang, Y., Chen, G.P., Malik, O.P., Hope, G.S.: An artificial neural based adaptive power system stabilizer. IEEE Transactions on Energy Conversion 8(1), 71–77 (1993)
5. Hoang, P., Tomsovic, K.: Design and analysis of an adaptive fuzzy power system stabilizer. IEEE Transactions on Energy Conversion 11(2), 97–103 (1996)
6. Cheng, S., Malik, O.P., Hope, G.S.: Design of self-tuning PID stabilizer for a multimachine power system. Transmission and Distribution IEE Proceedings - Generation, Part C 133(4), 176–185 (1986)
7. Abido, M.A., Abdel-Magid, Y.L.: Optimal design of power system stabilizers using evolutionaryprogramming. IEEE Transactions on Energy Conversion 17(4) (December 2002)

8. Kundur, P.: Power System Stability and Control. McGraw Hill Inc. (1993)
9. Dubey, M., Gupta, P.: Design of Genetic-Algorithm based robust power system stabilizer. International Journal of Computational Intelligence 2(1), 48–52 (2005)
10. Bhattacharya, K., Kothari, M.L., Nanda, J., Aldeen, M., Kalam, A.: Tuning of power system stabilizers in multi-machine system using ISE technique. Electric Power Systems Research 19(7), 449–458 (1997)

A Fuzzy C Mean Clustering Algorithm for Automated Segmentation of Brain MRI

Geenu Paul[1], Tinu Varghese[2], K.V. Purushothaman[3], and N. Albert Singh[2]

[1] Department of ECE, St Thomas Institute for Science and Technology
Thiruvananthapuram, Kerala, India
geenub@yahoo.com
[2] Noorul Islam University, Kumara coil, Thuckalay, Tamilnadu
tinuannevarghese@gmail.com, albertsingh@rediffmail.com
[3] Department of ECE, Heera College of Engineering and Technology
Thiruvananthapuram, Kerala, India
kvpuru@hotmail.com

Abstract. An automated scheme for MRI segmentation using fuzzy C means algorithm is proposed. Here a fuzzy C means algorithm is implemented in order to classify the brain voxel. The brain voxel are classified into three main tissue types: gray matter (GM), white matter (WM) and cerebro-spinal fluid (CSF). On studying the segmented image, the reduction in the GM in the brain image indicates the presence of degenerative disease. Segmentation procedure was done with real time data, ie in human .Automated segmented volumes were analyzed by the physician, by manually segmenting it.

Keywords: Magnetic Resonance Imaging, Partial Volume Effect, Grey Matter, White Matter, Cerebro-Spinal Fluid Automated Scheme, Segmentation, FCM algorithm, Magnetic resonance (MR) imaging, Fuzzy C Means(FCM).

1 Introduction

Magnetic resonance (MR) imaging is considered as an effective means for diagnosing some brain diseases. Volumetric analysis of different parts of the brain is useful in determining the progress of various diseases, such as Alzheimer's disease, epilepsy, multiple sclerosis, etc. Classifying the brain voxels into one of the main tissue type: grey matter (GM), white matter (WM), and cerebro spinal fluid (CSF) and the background can be done by automated MRI segmentation system. On dealing with MRI segmentation we have to be careful in handling the image deteriorating elements such as (a) acquisition noise,(b) partial volume effect (PVE) (c) bias field.

Several methods of supervised and unsupervised techniques of image segmentation and classification have been used for several applications such as future analysis, clustering and classifier designs in the fields such as anatomy, geology, medical imaging, target recognition and image segmentation. The FCM method classifies the images by grouping similar data that are present in the future points into clusters.FCM by varying the degree of membership allows the pixels to belong to the multiple

S.C. Satapathy, S.K. Udgata, and B.N. Biswal (eds.), *FICTA 2013*,
Advances in Intelligent Systems and Computing 247,
DOI: 10.1007/978-3-319-02931-3_8, © Springer International Publishing Switzerland 2014

classes. This method makes them more flexible to be used in the processing of magnetic resonances image (MRI)

On taking into account the literature that is devoted to the brain segmentation to MRI data we can see that it is so huge. A Maximum Likelihood (ML) or Maximum A Posteriori (MAP) approach and the Expectation- Maximization (EM) algorithm is used for the optimization process, here the statistical model parameters are usually estimated. Alternatively to statistical parametric methods, unsupervised non-parametric schemes have been recently proposed for adult brain MRI segmentation[9] One such approach is the mean-shift algorithm[6], whose key points include the fact that no initial clusters are required and that the number of distinct tissue clusters is estimated from the data. a regularization term taking into account local interactions between voxels is commonly added, In order to reduce the effect of noise [4]   or Hidden Markov Chains[2]

2   Magnetic Resonance Imaging

Here we are providing a brief description on the principle of Magnetic Resonance Imaging (MRI).It is a imaging technique used to produce high quality images of the inside of the human body

Brain MR Images provides rich information about the human soft tissue anatomy. It has several advantages over other imaging techniques enabling it to provide three-dimensional (3-D) data with high contrast between soft tissues. However, the amount of data is far too much for manual analysis/interpretation, and this has been one of the biggest obstacles in the effective use of MRI. For this reason, automatic or semi-automatic techniques of computer-aided image analysis are necessary..The contrast in images depends mainly on the way the image is acquired. High lightening different components in the object can be done by altering radio frequency (RF), by carefully choosing the relaxation timings and gradient pulses. MR images are not always high-contrast usually. The segmentation of medical images is based on information from a single image. The information within a single image is not sufficient to obtain a reliable segmentation due to various noise contributions. Therefore, in general, segmentation methods that can be implemented on multiple images that will be connected in space or time that will improve the final segmentation of each image. MRI, require the interpretation of tens to hundreds of slices.

3   Proposed Methodology

In the following section we describe the Proposed Methodology (fig 1). In this paper we describe the segmentation of real time data so that degeneration that has occurred to the brain can be studied .The segmentation technique, has accurately divided the entire brain images into its sub regions as GM(Grey Matter),WM(White Matter) and CSF (Cerebro Spinal Fluid).The purpose for our study is to examine the certain specific brain regions and to examine the structural characteristics at baseline and structural changes in them should be detected. So, that any detection of degenerative diseases can be detected earlier.

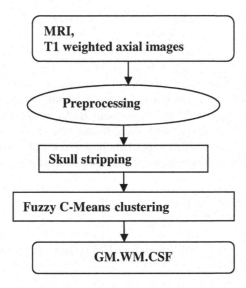

Fig. 1. Block diagram of the proposed system

3.1   MRI

T_1 weighted axial view of the DICOM MR Images were used as test images.

3.2   Preprocessing

These test images were under gone through the process called preprocessing .The obtained MR images undergo changes due to the interference of the noise. We undergo various filtering transactions for enhancing the MR images .This gives information of the edges and at the same time avoids blurring.

3.3   Skull Stripping

There are cortical tissues and non cortical tissues, for volumetric analysis of the brain we have to segment each separately .The unwanted tissues during segmentation are non cordial we have to remove it. The process of removing these non cortical tissues is called skull stripping.

The skull, scalp, fat, skin, eyeball, etc are the parts that are not required to be segmented, ie they come under the category of non brain tissues. The skull removed tissues of the MR image is used for further classification of brain tissues into White matter, Grey matter and cerebrospinal fluid. This can be differentiated using the primary colours such as Red, Green and Blue.

3.4   Fuzzy C-Means Clustering

Here we have used Fuzzy C-Means clustering (FCM). Fuzzy C-Means clustering (FCM), also known as ISODATA, is a data clustering algorithm in which each data point belongs to a cluster to a degree specified by a membership grade. FCM and its derivatives have been used very successfully in many applications, such as pattern recognition, classification, data mining, and image segmentation. FCM partitions a collection of n vector xi, I = 1,...,n into c fuzzy groups, and finds a cluster center in each group such that a cost function of dissimilarity measure is minimized. FCM employs fuzzy partitioning such that a given data point can belong to several groups with the degree of belongingness specified by membership grades between 0 and 1. To accommodate the introduction of fuzzy partitioning, the membership matrix U is allowed to have elements with values between 0 and 1. Imposing normalization stipulates that the summation of degrees of belongingness for a data set always be equal to unity

$$\sum_{i=1}^{c} u_{ij} = 1, \forall_j = 1,...,n \tag{1}$$

The cost function (or objective function) for FCM is then a generalization of Equation

$$J(U, c_1,..., c_c) = \sum_{i=1}^{c} J_i = \sum_{i=1}^{c} \sum_{j}^{n} u_{ij}^m d_{ij}^2 \tag{2}$$

Where ui j is between 0 and 1; ci is the cluster center of fuzzy group i; dij = ‖ci-xj‖ is the Euclidean distance between ith cluster center and jth data point; and m e [1,¥) is a weighting exponent. The necessary conditions for Equation (3) to reach a minimum can be found by forming a new objective function J as follows

$$J(U, c_1,..., c_c.\lambda_1,...,\lambda_n) = J(U, c_1,..., c_c) + \sum_{j=1}^{n} \lambda_j\left(\sum_{i=1}^{c} u_{ij} - 1\right)$$
$$= \sum_{i=1}^{c} \sum_{j}^{n} u_{ij}^m d_{ij}^2 + \sum_{j=1}^{n} \lambda_j\left(\sum_{i=1}^{c} u_{ij} - 1\right) \tag{3}$$

Where l j, j = 1 to n, are the Lagrange multipliers for the n constraints in Equation (1). By differentiating J(U, ci, ..., cc, l1, ...,ln) to all its inputs arguments, the necessary conditions for Equation (3) to reach its minimum are

$$c_i = \frac{\sum_{j=1}^{n} u_{ij}^m x_j}{\sum_{j=1}^{n} u_{ij}^m} \tag{4}$$

and

$$u_{ij} = \frac{1}{\sum_{k=1}^{c} \left(\dfrac{d_{ij}}{d_{kj}}\right)^{2/(m-1)}} \tag{5}$$

The fuzzy C-means algorithm is simply an iterated procedure through the preceding two conditions. In a batch-mode operation, FCM determines the cluster centers ci and the membership matrix U using the following steps:

Step 1: Initialize the membership matrix U with   random values between 0 and 1 such that the constraints in Equation (4.1) are satisfied.

Step 2: Calculates c fuzzy cluster center c_i, i = 1,...c., using   Equation (4).

Step 3: Compute the cost function (or objection function) by Equation (2). Stop if either it is below a certain tolerance value or its improvement over previous iteration is below a certain threshold.

Step 4: Compute a new U by Equation above Go to step 2. The cluster centers can also be first initialized and then the iterative procedure carried out. No guarantee ensures that FCM converges to an optimum solution. The performance depends on the initial cluster centers, thereby another fast algorithm to determine the initial cluster centers can be used. Or, FCM is run several times, each starting with a different set of initial cluster centers. The implementation of FCM algorithm using MATLAB.

4 Results on Real Time Database Images

In this section we present the performance of an unsupervised fuzzy segmentation method, which has been applied on real brain MR datasets. The FCM clustering segments all the brain tissues into WM, GM and CSF. In this analysis we have used slice no:44 as the template. On observing the segmented output we can observe the degeneration of the area of WM, GM and CSF.The figure shown represents the segmentation results using FCM of a real axial T1 weight MR Image. Since images from MRI database are commonly used for brain tissue segmentation assessment, The results presented in this paper are preliminary and further clinical evaluation is required. Segmented images are then differentiated with colours of red blue and green. The segmentation outputs of images after the Skull stripping are shown.

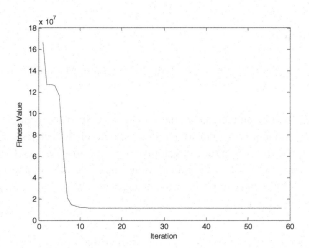

Fig. 2. Graphical relation between the Fitness value and the iteration

The graphical relation between the Fitness value and iteration is shown above .From the graph we can find that as the no of iteration is increased the amount of fitness is made more accurate. Iteration increasing means the number of times the learning is increased since FCM is an unsupervised learning technique the amount of iteration has an effect on the perfectness of the output.

The result of the FCM clustering is shown below. The original image that has to be segmented is skull stripped so that the unwanted areas are removed and the area that is n to be segmented is only obtained (Fig.2)

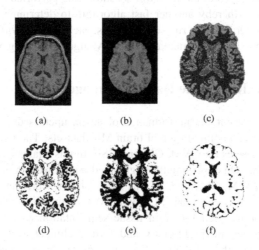

(a)   (b)   (c)

(d)   (e)   (f)

Fig. 3. Segmentation of MR brain images: (a)Original image: (b) skull stripped (c)Segmentation map (d) GM (e) WM (f) CSF.

5   Conclusion

On studying the segmented image, the reduction in the GM in the brain image indicates the presence of degenerative disease. To understand the disease progress in the early stage both quantitative and qualitative analysis is important. As there is a reduction in GM there will be an increase in the volume of CSF. The overall conclusion we can reach is that on identifying and studying the structural characteristics it is possible to function as predictors of future development of the disease. As Future work or suggestion we can say that the theoretical value that give the volume of WM, GM, and CSF makes it more easy to find out the changes that has occurred even those who are not expert in this field.

Acknowledgement. The authors like to express their thanks to Dr Kesavadas of Sree Chitra Tirunal Institute for Science and Technology, Trivandrum.

References

1. Van Leemput, K., Maes, F., Vandeurmeulen, D., Suetens, P.: Automated model-based tissue classification of MR images of the brain. IEEE Trans. Med. Imag. 18(10), 897–908 (1999)
2. Zhang, Y., Brady, M., Smith, S.: Segmentation of brain MR images through a hidden markov random field model and the expectation maximization algorithm. IEEE Trans. Med. Imag. 20(1), 45–57 (2001)
3. Greenspan, H., Ruf, A., Goldberger, J.: Constrained Gaussian mixture model framework for automatic segmentation of MR brain images. IEEE Trans. Med. Imag. 25(9), 1233–1245 (2006)
4. Comaniciu, D., Ramesh, V., Meer, P.: The variable bandwidth mean-shift and data-driven scale selection. In: IEEE Int. Conf. Comput. Vis. (ICCV), vol. 1, pp. 438–445 (2001)
5. Jimenez-Alaniz, J.R., Medina-Banuelos, V., Yanez-Suarez, O.: Data-driven brain MRI segmentation supported on edge confidence and a priori tissue information. IEEE Trans. Med. Imag. 25(1), 74–83 (2006)
6. Jimenez-Alaniz, J., Pohl-Alfaro, M., Medina-Banuelos, V., Yanez-Suarez, O.: Segmenting brain MRI using adaptive mean shift. In: IEEE EMBS, pp. 3114–3117 (2006)
7. Comaniciu, D.: An algorithm for data-driven bandwidth selection. IEEE Trans. Pattern Anal. Mach. Intell. 25(2), 281–288 (2003)
8. Shanthi, K.J., Sasi Kumar, M., Kesavadas, C.: Neural Network Model for Automatic Segmentation of Brain MRI. IEEE (2008)
9. Desikan, R.S., Howard, J.C.: Automated MRI measures identify individuals with MCI and AD. Brain 132, 2048–2057 (2009)
10. Greenspan, H., Ruf, A., Goldberger, J.: Constrained Gaussian mixture model framework for automatic segmentation of MR brain images. IEEE Trans. Med. Imag. 5(9), 1233–1245 (2006)
11. Jiménez-Alaniz, J.R., Medina-Bañuelos, V., Yáñez-Suárez, O.: Datadriven brain i segmentation supported on edge confidence and a priori tissue information. IEEE Trans. Med. Imag. 25, 74–83 (2006)
12. Spall, J.C.: Multivariate stochastic approximation using a simultaneous perturbation gradient approximation. IEEE Transactions on Automatic Control 37, 332–341 (1992)
13. Comaniciu, D.: An algorithm for data-driven bandwidth selection. IEEE Trans. Pattern Anal. Mach. Intell. 25(2), 281–288 (2003)
14. Silverman, B.W.: Density Estimation for Statistics and Data Analysis, 1st edn. Chapman & Hall/CRC, London (1998)
15. Meer, P., Georgescu, B.: Edge detection with embedded confidence. IEEE Trans. Pattern Anal. Mach. Intell. 23(12), 1351–1365 (2001)

Fuzzy Type – Ahead Keyword Search in RDF Data

Selvani Deepthi Kavila[1], Ravi Ravva[1], and Rajesh Bandaru[2]

[1] Department of Computer Science and Engineering
Anil Neerukonda Institute of Technology and Sciences, Visakhapatnam, India
selvanideepthi14@gmail.com, ravi4345@yahoo.co.in
[2] Department of Computer Science and Engineering,
Gitam University, Visakhapatnam, India
b.rajesh68@yahoo.com

Abstract. Keyword search is the process of identifying relevant data the matches with keyword. If it produces more than one document as a search results, there will be ranking process. This paper explain a novel keyword search method on RDF data model, and use fuzzy technique extract results according to the query we passed. Proposed model can improves the storage and querying characteristics of the underlying RDF store. In addition to the above, the problem of fuzzy search on RDF model while typing key words. The results have supported the objective of the work.

Keywords: Text Document, keyword search, Type a head fuzzy search.

1 Introduction

Searching is the process of identifying position of key in existing set of data. For Text databases, searching is done by using keywords. The keyword search is process searching is done by using keywords. The keyword search is process searching information in document repository by submitting keywords to system. Without having prior database knowledge and structure of the data. Due the reason that user has limited knowledge on existing document structure, the system should provide list of alternatives to user while typing part of keywords called type a head search. First disadvantage of "type a head search" system considered multiple words as single string. Second is the system should produce information as they type even its presence of minor errors (called fuzzy type ahead search). To solve the above problems we require to build data as hierarchal representation called trees. In this paper, we focus on fuzzy technique for keyword search on RDF data model. The task can be decomposed into following subtasks. (1) Keyword association (2) Build tree (3) Ranking (4) Summarization.

Keyword Association: Checks the keywords which are present in the database to extract the document which are relevant to the expected document in the repository.

Build Tree: which is the step to get the minimal root to find the required document [2].

S.C. Satapathy, S.K. Udgata, and B.N. Biswal (eds.), *FICTA 2013*,
Advances in Intelligent Systems and Computing 247,
DOI: 10.1007/978-3-319-02931-3_9, © Springer International Publishing Switzerland 2014

Ranking: If the database produces more than one document related to keyword query then the ranking function will apply to the search result and get the required document for the user.

Summarization: Summarization provides the possibility of finding the main points of text so that the user will spend less time on reading the whole document [5].

Fuzzy Search: Fuzzy is a technique which is used to retrieve the knowledge from the relevant information which is extracted from the large database [11].

RDF Data: RDF provides a way to set the loose relations and it has very flexible hierarchy concept. RDF uses the seamless technology to integrate the information on the web. RDF is a standard model for data interchange on the Web. The RDF data model is similar to classic conceptual modeling approaches such as entity–relationship or class diagrams, as it is based upon the idea of making statements about resources (in particular web resources) in the form of subject-predicate-object expressions. These expressions are known as triples in RDF terminology.

The rest of the paper is organized as follows. Section 2 introduces the related work of type a head fuzzy keyword search. Section 3 provides an overview and Problem Formulation. Issues on the design of type a head keyword search provided in Section 4, Section5 presents evaluation results. Finally, we conclude paper in Section 6.

2   Related Work

A. Balmin etal describes ObjectRank [1] system applied supervised ranking to query search in databases modeled as directional graphs. Conceptually, authority originates at the nodes (objects) containing the keywords and flows to objects according to their semantic connections. In the work of, E. Chu etal describes a common criticism [4] of database systems that they were hard to query. To solve this problem, form-based interfaces and keyword search have been proposed; they have investigated combining the two with the hopes of creating an approach that provides the best of both. Specifically, they have proposed to take as input a target database and then generate and index a set of query forms offline. At key word query time and a user with a question to answered issued standard keyword search queries but instead of returned records, the system returns forms relevant to the query. They have explored techniques to tackle these challenges, and present experimental results suggesting that the approach of combining keyword search and form-based interfaces is promising. In the work of, S. Agrawal etal describes the Internet search engines [7] have popularized the keyword-based search paradigm. , they have discussed DBXplorer, a system that enables query-based search in relational databases. DBXplorer has been implemented using a relational database. They have outlined the challenges and discussed the implementation of their system including results of extensive experimental evaluation. In the work of, S. Amer-Yahia etal based [8] on "Ranked XML Querying" that took place in Schloss Dagstuhl in Germany in March 2008. In the work of, S. Cohen etal describes a framework [10] for described relationships among nodes in documents is presented. A specific interconnection semantics in this framework could be defined explicitly or derived automatically. In the work of, Y. Chen etal Empowering users [12] to access various databases using simple keywords can relieve that the users from the steep training curve of mastering a structured query language and understood complex and possibly fast evolving schemas.

3   Problem Formulation

3.1   Architecture

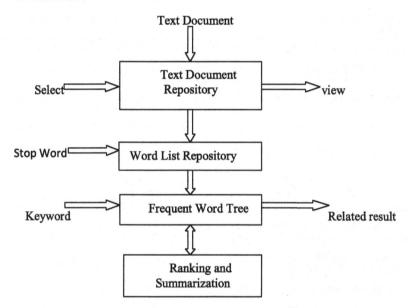

Fig. 1. Frame work for Keyword Search Engine

3.2   User Query

The user query Q is a set of keywords (k1, K2 ..., ki). The system queries Q are conjunctive queries. Where A conjunctive query is an expression of the form $(x1,...,xk).\exists x \ k+1,...xm$. andA1 $\wedge$ ... $\wedge$Ar, where $x1,...,xk$ are called distinguished variables $xk+1,...,xm$ are undistinguished variables and A1,...,Ar are query atoms.

3.3   Fuzzy Type Ahead Search

Fuzzy is a technique which is used to retrieve the knowledge from the relevant information which is extracted from the large database [11].

In the text Document set of words (Key word) can form as query Q = {w_1, w_2, w_3...., w_m} where m is number of keywords. The search process should produce results TR even of the minor errors has occurred.

4   Implementation Issues in Design of Proposed Model

This section describes issues related to the implementation of model.

4.1   Search for Minimal Matching Subgraph

The following is the algorithm to search text repository and return more appropriate results which takes Repository name and key word as input and produce summarized results.

Algorithm:

---

add cursor for each keyword element to Qi ∈ LQ
for Ki in K repeat
 for kin Ki repeat Qi
 add (newCursor (k, k,∅,0, k.w));
 end
end
while Qi not in LQ repeat
  c ← minCost (LQ);
  n ← c.n;
  if c.d higher than dmax then
    n.addCursor(c);
    for all neighbors except parent element of c
     Neighbor ← neighbors (n) \ (c.p).n;
      if there are no Neighbors then
       foreach n ∈ Neighbors do
       n already visited by c
       if n ∈ parents(c) then
         respective queueQi should be added
         .add (cursor (n, c.k,c.n,c.d+1,c.w +n.w));
       end
      end
     endQi
    .pop(c);
    R←Top-k(n, KlowC, LQ, k, R);
  end
end
return R;

4.2   Build Word Tree for Fuzzy Search

The first step is to read entire text document calculate the all the different word counts then create tree based the count of words and possible substitution (called fuzzy) while typing.   The outcome possibilities are more found after searching then the

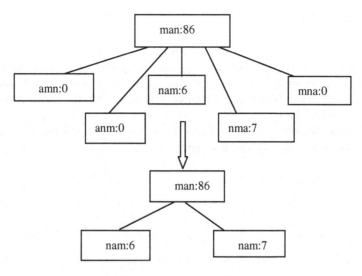

Fig. 2. Frequent word Tree

relevant links can be associated with the input query and the relevant search results will get. After this step relevant keywords will be extracts and then used to get the output documents. The following diagram represents tree for searching "man".

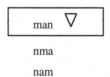

nma

nam

Fig. 3. Type ahead fuzzy Recommendation

4.3   Ranking and Summarization

A key word Query $Q = \{Q_1, Q2,\ldots\ldots, Qk\}$ is submitted to system to get search results. In the fuzzy search the path can be determined if exact key word not found it will check approximate results by checking prefixes similar to given key word. Suppose $Q_{i\,is}$ complete key word, K_i best similar prefix Q_i. Similarity can be calculated as follows.

$$\text{Sim}(K_i, Q_i) = r + \frac{1}{1+ed(Ki,ai)} + (1 - r) * \frac{|ai|}{|Qi|}$$

Where ed is edit distance.

r is constant is between 0 and 1 and a_i difference word.

5   Results

From graphs drawn above gives space and time required for search process. Fig. 4 represent graph drawn between Memory space required (MB) and Number of documents in repository. Fig. 5 represents searching time milliseconds and Number of documents graph drawn between in repository. By observing, the above curves give clarity about the scalability.

By using our algorithm on these datasets, the algorithm retrieves 90% accurate documents which are related to the user query.

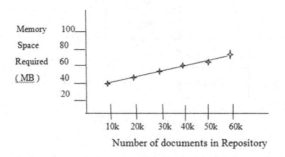

Fig. 4.

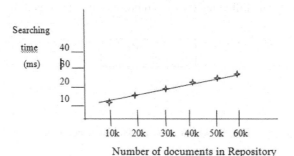

Fig. 5.

6   Conclusion

A new approach for keyword search on tree-structured data has presented, focusing on the RDF data model. The approach is also applicable to graph like data models. Query search can be applied on an frequent word tree that represents a summary of the original data and counts. Where query processing can leverage optimization capabilities of the text documents. Proposed model could improve the storage and querying characteristics of the underlying RDF store. In addition to the above, The problem of fuzzy search on RDF model while typing key words. The proposed work describes effective word list and better algorithms. The results have supported the objective of the work.

References

1. Balmin, A., Hristidis, V., Papakonstantinou, Y.: Objectrank: Authority-Based Keyword Search in Databases. In: Proc. Int'l Conf. Very Large Data Bases (VLDB), pp. 564–575 (2004)
2. Dalvi, B.B., Kshirsagar, M., Sudarshan, S.: Keyword Search on External Memory Data Graphs. In: Proc. Int'l Conf. Very Large Data Bases (VLDB), pp. 1189–1204 (2008)
3. Ding, B., Yu, J.X., Wang, S., Qin, L., Zhang, X., Lin, X.: Finding Top-k Min-Cost Connected Trees in Databases. In: Proc. Int'l Conf. Data Eng. (ICDE), pp. 836–845 (2007)
4. Chu, E., Baid, A., Chai, X., Doan, A., Naughton, J.F.: Combining Keyword Search and Forms for Ad Hoc Querying of Databases. In: Proc. ACM SIGMOD Int'l Conf. Management of Data, pp. 349–360 (2009)
5. Fagin, R., Lotem, A., Naor, M.: Optimal Aggregation Algorithms for Middleware. In: Proc. ACM SIGMOD-SIGACTSIGART Symp. Principles of Database Systems (PODS) (2001)
6. Al-Hashemi, R.: Text Summarization Extraction System (TSES) Using Extracted Keywords. International Arab Journal of e-Technology 1(4) (June 2010)
7. Agrawal, S., Chaudhuri, S., Das, G.: Dbxplorer: A System for Keyword-Based Search over Relational Databases. In: Proc. Int'l Conf. Data Eng. (ICDE), pp. 5–16 (2002)
8. Amer-Yahia, S., Hiemstra, D., Roelleke, T., Srivastava, D., Weikum, G.: Db&ir Integration: Report on the Dagstuhl Seminar 'Ranked Xml Querying'. SIGMOD Record 37(3), 46–49 (2008)
9. Cohen, S., Mamou, J., Kanza, Y., Sagiv, Y.: Xsearch: A Semantic Search Engine for Xml. In: Proc. Int'l Conf. Very Large Data Bases (VLDB), pp. 45–56 (2003)
10. Cohen, S., Kanza, Y., Kimelfeld, B., Sagiv, Y.: Interconnection Semantics for Keyword Search in Xml. In: Proc. Int'l Conf. Information and Knowledge Management (CIKM), pp. 389–396 (2005)
11. Ji, S., Li, G., Li, C., Feng, J.: Efficient Interactive Fuzzy Keyword Search. In: Proc. Int'l Conf. World Wide Web (WWW), pp. 371–380 (2009)
12. Chen, Y., Wang, W., Liu, Z., Lin, X.: Keyword Search on Structured and Semi-Structured Data. In: Proc. ACM SIGMOD Int'l Conf. Management of Data, pp. 1005–1010 (2009)

Human Emotion Classification Using Fuzzy and PCA Approach

Soumya Ranjan Mishra, B. Ravikiran, K. Sai Madhu Sudhan,
N. Anudeep, and G. Jagdish

Department of Computer Science & Engineering, ANITS,
Visakhapatnam, India

Abstract. The emotion recognition system has been a significant field in human-computer interaction. It is a considerably challenging field to generate an intelligent computer that is able to identify and understand human emotions for various vital purposes, e.g. security, society, entertainment. Many research studies have been carried out in order to produce an accurate and effective emotion recognition system. Emotion recognition methods can be classified into different categories along a number of dimensions: speech emotion recognition vs. facial emotion recognition; machine learning method vs. statistic method. Facial expression method can also be classified based on input data to a sequence video or static image. This report focuses on different types of human facial expressions, like different types of sad, happiness, and surprise moments. This is carried out by trying to extract unique facial expression feature among emotions using Fuzzy and the Principal Component Analysis (PCA) approach.

Keywords: facial expression recognition, face recognition, emotion recognition.

1 Introduction

Over the past decades, human-computer interaction together with computer vision has been an important field in computer study. It is concerned with the relationship of direct communication between the computer and human beings. Much research has been conducted on improving and developing the interaction between humans and the computer. Interested researchers worked on this area for different reasons. One of the significant factors that contributed to increasing and developing the interaction between the computer and humans is studying the computers' ability to distinguish emotions for humans. With emotion recognition systems, the computer will be able to assess the human expressions depending on their affective state in the same way that human's senses do. The intelligent computers will be able to understand, interpret and respond to human intentions, emotions and moods.

The emotion recognition applications have demonstrated their capabilities in different areas of life; for instance, in security and surveillance, they can predict the offender or criminal's behavior by analyzing the images of their faces that are

S.C. Satapathy, S.K. Udgata, and B.N. Biswal (eds.), *FICTA 2013*,
Advances in Intelligent Systems and Computing 247,
DOI: 10.1007/978-3-319-02931-3_10, © Springer International Publishing Switzerland 2014

captured by the control-camcorder. Furthermore, the emotion recognition system has been used in communication to make the answer machine more interactive with people. The emotion recognition system has had a considerable impact on the game and entertainment field besides its use to increase the efficiency of robots for specific tasks such as me-caring services, military tasks, medical robots, and manufacturing servicing. Generally, the intelligent computer with the emotion recognition system has been used to improve our daily lives. Scientists restrict the emotions of people in seven different feelings: Anger, Disgust, Fear, Happiness, Neutral, sad, and surprise. Generally, scientists have analyzed the human emotions and realized that Human emotion recognition can be achieved by analyzing the facial expressions.

Review of Existing System
It is an established fact that human to computer interaction will prove more naturally if the machines are able to distinguish and react to human non-verbal communication such as emotions. While numerous techniques have been studied to detect human emotions based on facial expressions or speech, there have been few research studies on combining two or more methods to enhance the exactness and potency of the emotion recognition system. Furthermore, the detection of accurate human emotions is of vital importance for efficient human-computer interaction. Numerous researches have explored this phenomenon which comprises 2D features, yet they are receptive to head pose, clutter, and variations in the lighting conditions. Researchers have applied various techniques to enhancing the interaction between humans and computers through the use of emotion recognition. The Principal Component Analysis (pca) and the Haar-like features been used for the classification of emotions. Overall, there are several techniques that have been used for the recognition of human emotions. These studies have provided support through such applications. The existing Emotion detection systems are not much accurate due to the following reasons 1.variation in lighting conditions. 2. Expression of emotion varies from time to time. We built a System that can give a better performance over the existing systems in various luminous conditions and also deals with the problem of varying time to time expressions.

2   Process Flow of Proposed System

Our system uses Viola Jones Face Detection algorithm for detecting the faces and build a system that would give the most efficient results when compared to the existing systems for recognizing the faces.

When a person's image is captured using camera, a face is detected and is stored in a temporary folder. The Eigen vectors of captured face are obtained and Eigen values are computed from them.

The principle components from above Eigen values are analyzed and saved into the database for further use. Now the image is processed for detecting the emotion. During this process the stored Eigen values of the face detected are compared against the training database values which are the average eigenfaces of different expression from training dataset. As we are taking average eigenfaces for a same expression so that it can identify different type of smile, sad, anger and surprise expression.

2.1   Face Detection and Tracking

Face detection is the first stage which is desired to be automated. In most of the research, face is already cropped and the analysis starts with feature extraction and tracking. In the rest, automated face detectors are used. These can be classified mainly into two classes: vision-based detection and detection using infrared (IR) cameras

There are also free face detection software available to researchers for usage and improvement. Most popular of these is the face detector of Open Source Computer Vision Library (OpenCV), which depends on Haar-like wavelet-based object detection proposed by Viola and Jones [6].

2.2   Fuzzy and PCA Approach

We have extended the concept of PCA for better results and named it as Fuzzy PCA. Instead of comparing with the obtained values with a single set; we defined a new set of values for each of the emotions by calculating the average of different expressions of different persons and obtained the mean value for each of the expressions. Then we calculated the deviations of the obtained values that change the data into a new coordinate system such that the variance is put in order from the greatest to the least. Each variance is the eigenvalue of the covariance matrix and has an eigenvector (or characteristic vector) associated with it. These vectors in this case are known as the principal components that will be used to analyze the data because they are simply a linear combination of the data in the original matrix [4].

Fig. 1. Eigenfaces of different expression

2.3   PCA Analysis for Average Eigenfaces

A 2-D facial image can be represented as 1-D vector by concatenating each row (or column) into a long thin vector. Let's suppose we have M vectors of size N (= rows of image × columns of image) representing a set of sampled images. pj's represent the pixel values[6].

$$xi = [p1 ..... pN]T ; i = 1,.....,M$$

The images are mean centered by subtracting the mean image from each image vector. Let m represent the mean image. m=$1/M\sum_{i=1}^{M} x_i$ And let wi be defined as mean centered image $wi = xi - m$, Our aim is to find a set of ei's which have the largest possible projection onto each of the wi's.

We wish to find a set of M orthonormal vectors ei for which the quantity $\lambda_{i=1/M} \sum_{n=1}^{M}(e_i^T wn)^2$ is maximized with the orthonormality constraint $e_i^T e_k = \delta_{lk}$[6].

It has been shown that the e_i's and λ_i's are given by the eigenvectors and eigenvalues of the covariance matrix $C = WWT$.

where W is a matrix composed of the column vectors wi placed side by side. The size of C is $N \times N$ which could be enormous. For example, images of size 64×64 create the covariance matrix of size 4096×4096. It is not practical to solve for the eigenvectors of C directly. A common theorem in linear algebra states that the vectors ei and scalars λ_i can be obtained by solving for the eigenvectors and eigenvalues of the $M \times M$ matrix $W^T W$. Let di and μ_i be the eigenvectors and eigenvalues of $W^T W$, respectively.

$$W^T W di = \mu i \, di$$

By multiplying left to both sides by W

$$WW^T (Wdi) = \mu i \, (Wdi)$$

which means that the first $M - 1$ eigenvectors ei and eigenvalues λ_i of WW^T are given by Wdi and μi , respectively. Wdi needs to be normalized in order to be equal to ei. Since we only sum up a finite number of image vectors, M, the rank of the covariance matrix cannot exceed $M - 1$ (The -1 come from the subtraction of the mean vector m). The eigenvectors corresponding to nonzero eigenvalues of the covariance matrix produce an orthonormal basis for the subspace within which most image data can be represented with a small amount of error. The eigenvectors are sorted from high to low according to their corresponding eigenvalues[6][7]. The eigenvector associated with the largest eigenvalue is one that reflects the greatest variance in the image. That is, the smallest eigenvalue is associated with the eigenvector that finds the least variance. They decrease in exponential fashion, meaning that the roughly 90% of the total variance is contained in the first 5% to 10% of the dimensions. A facial image can be projected onto M' ($<<M$) dimensions by computing $\Omega = [v1v2 ..... vM']^T$.

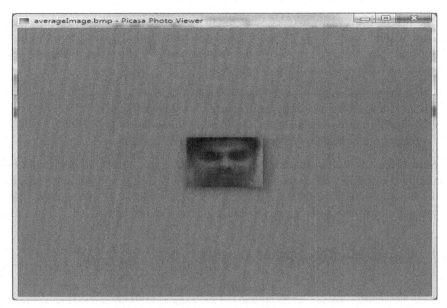

Fig. 2. Average Eigenfaces

2.4   Face Recognition

Once the eigenfaces have been computed, several types of decision can be made depending on the application.

1. Identification where the labels of individuals must be obtained.
2. Recognition of a person, where it must be decided if the individual has already been seen.
3. Categorization where the face must be assigned to a certain class.

PCA computes the basis of a space which is represented by its training vectors. These basis vectors, actually eigenvectors, computed by PCA are in the direction of the largest variance of the training vectors. Each eigenface can be viewed a feature. When particular face is projected onto the face space, its vector into the face space describe the importance of each of those features in the face. The face is expressed in the face space by its eigenface coefficients (or weights), Can handle a large input vector, facial image, only by taking its small weight vector in the face space. This means that can reconstruct the original face with some error, since the dimensionality of the image space is much larger than that of face space. In this report, consider face identification only. Each face in the training set is transformed into the face space and its components are stored in memory. The face space has to be populated with these known faces. An input face is given to the system, and then it is projected onto the face space. The system computes its distance from all the stored faces. However, two issues should be carefully considered:

1. What if the image presented to the system is not a face?
2. What if the face presented to the system has not already learned, i.e., not stored as a known face?

The first defect is easily avoided since the first eigenface is a good face filter which can test whether each image is highly correlated with itself. The images with a low correlation can be rejected. Or these two issues are altogether addressed by categorizing following four different regions:

1. Near face space and near stored face => known faces.
2. Near face space but not near a known face => unknown faces.
3. Distant from face space and near a face class => non-faces.
4. Distant from face space and not near a known class => non-faces.

Since a face is well represented by the face space, its reconstruction should be similar to the original; hence the reconstruction error will be small. Non-face images will have a large reconstruction error which is larger than some threshold θ^r. The distance ϵ_k determines whether the input face is near a known face.

3   Expression Recognition Accuracy

Table 1.

Detected Emotion	Actual Emotion			
	Happy	Sad	Anger	Surprise
Happy	97%	0	0	0
Sad	0	63%	12.5%	0
Anger	0	20%	75%	0
Surprise	0	0	0	100%
Confused or Neutral	0	15%	12.5%	0

4   Face Recognition Speed

The speed and accuracy of detection stage of the application was analyzed so that the algorithm providing the best performance could be used in the final implementation. The detection time for our algorithm was compared across 200 sample frames.

5   Conclusion

We were able to overcome many of the challenges of detecting facial expressions in real time by correctly identifying faces through a laptop camera input. Our future work includes development of a model that characterizes the further improvements of proposed system. Our system gives a better approximation in various luminous conditions. Here we accessed the face tracking device and processed the frames obtained from it. As a part of future enhancement we would try to work on the Auto exposure algorithms of the face tracking device so that we can have control over device exposures itself. So that whatever may be the lightening conditions we try to capture those exposures that would give more information about the facial features. The result of doing so is that we can have the better input and hence we can get further more efficient results. In addition, we would like to seek more efficient ways to perform the comparisons of input image with the training image so as to reduce the time taken for comparisons. Thus we can make our project more robust and reliable.

References

1. Belhumeur, P., Hespanha, J., Kriegman, D.: Eigenfaces vs. fisherface: Recognition using class specific linear projection. IEEE Transactions on Pattern Analysis and Machine Intelligence 19, 711–720 (1997)
2. Turk, M., Pentland, A.: Eigenfaces for recognition. Journal of Cognitive Neuroscience 3, 71–86 (1991)
3. Turk, M.A., Pentland, A.P.: Face Recognition Using Eigenfaces. In: Proceedings of the IEEE Conference on Computer Vision and Pattern Recognition (CVPR 1991), Maui, Hawaii, USA, June 3-6, pp. 586–591 (1991)
4. Turk, M.A., Pentland, A.P.: Face Recognition Using Eigenfaces. In: IEEE Conf. on Computer Vision and Pattern Recognition, pp. 586–591 (1991)
5. Kim, K.: Face Recognition using Principle Component Analysis Department of Computer Science University of Maryland, College Park MD 20742, USA
6. Fagertun, Face Recognition. Master Thesis, Technical University of Denmark (DTU) (2005)
7. Pentland, A., Moghaddam, B., Starner, T.: View-Based and Modular Eigenspaces for Face Recognition. In: IEEE Conf. on Computer Vision and Pattern Recognition. MIT Media Laboratory Tech. Report No. 245 (1994)
8. Biswas, B., Mukherjee, A.K., Konar, A.: Matching of digital images using fuzzy logic. AMSE Publication 35(2), 7–11 (1995)
9. Guo, Y., Gao, H.: Emotion Recognition System in Images Based On Fuzzy Neural Network and HMM. In: Proc. 5th IEEE Int. Conf. on Cognitive Informatics (ICCI 2006). IEEE (2006)
10. Etemad, K., Chellappa, R.: Discriminant Analysis for Recognition of Human Face Images. Journal of the Optical Society of America A 14(8), 1724–1733 (1997)
11. Kalita, J., Das, K.: Recognition of Facial Expression Using Eigenvector Based Distributed Features and Euclidean Distance Based Decision Making Technique. International Journal of Advanced Computer Science and Applications 4(2) (2013)
12. Yin, H., Fu, P., Meng, S.: Sampled FLDA for face recognition with single training image per person 69(16-18), 2443–2445 (2006)
13. Yang, J., Zhang, D., Frangi, A., Yang, J.-Y.: Two-dimensional PCA a new approach to appearance-based face representation and recognition. IEEE Transactions on Pattern Analysis and Machine Intelligence 26(1), 131–137 (2004)

Application of Artificial Neural Networks and Rough Set Theory for the Analysis of Various Medical Problems and Nephritis Disease Diagnosis

Devashri Raich and P.S. Kulkarni

Dept of CSE, RCERT, Chandrapur R.T.M. Nagpur University, India
{devashriraich,kulkarnips1811}@gmail.com

Abstract. Soft computing techniques are widely used for the research in various fields nowadays. Artificial Neural Networks and various other soft computing techniques can be used for handling large data for diagnosis of particular disease. The increasing demand of Artificial Neural Network application for predicting the disease shows better performance in the field of medical decision making. The rough set theory proposed by Pawlak is one of the widely used research area nowadays. Rough set theory can be used for handling impression and uncertainty in data; therefore it can be used for medical diagnosis systems .This paper represents the use of artificial neural networks in predicting disease i.e. diagnosis, and use of Rough Set Theory for finding the indicators in diagnosis of Nephritis. The proposed technique involves training a Multi Layer Perceptron with a BP learning algorithm to recognize a pattern for the diagnosis and prediction of Nephritis and calculating significance factor using Rough Set Theory to find indicators. In this paper, a brief introduction about soft computing techniques used nowadays for diagnosis of disease is given. The other part introduces Nephritis and the proposed method for diagnosis of Nephritis.

Keywords: Artificial Neural Networks, Rough Set Theory, back propagation, perceptron, medical diagnosis, nephritis, Soft computing techniques.

1 Introduction

Health of a human being is the most important aspect of being a good human being itself. Medical diagnosis mainly aims at health of a human being. The perfect diagnosis can help in fast recovery of a patient suffering from a disease. However clinical decision support system also plays a vital role in diagnosis. Medical professionals use these CDSS for categorizing the patients whether they need a special case or just can be cured with few oral medicines also. Medical Diagnosis can be stated as the process of determining or identifying a possible disease or a disorder.

This paper is organized as section 1 gives a brief introduction about medical diagnosis and various soft computing techniques used nowadays. Section 2 describes what nephritis is and how it can be diagnosed. Section 3 mainly aims at brief description about artificial neural networks and use of ANN for the diagnosis of nephritis.

S.C. Satapathy, S.K. Udgata, and B.N. Biswal (eds.), *FICTA 2013*,
Advances in Intelligent Systems and Computing 247,
DOI: 10.1007/978-3-319-02931-3_11, © Springer International Publishing Switzerland 2014

2   Literature Survey

2.1   Soft Computing Techniques for Medical Diagnosis

Medical diagnosis can be stated as the process of determining or identifying a possible disease or a disorder. A clinician uses several sources of data and classifies this data in order to find the disorder.[17]. Artificial intelligence has the ability to learn and take its decision.[5]. AI maps human intelligence into computer aided technology. Various soft computing techniques are used nowadays for diagnosis of particular disease. Some of them are: Neural networks, Fuzzy logic, Rough set theory, Genetic algorithms. ANN is one of the recent areas of research in field of biomedical sciences. [7]. Various diseases like Parkinson's disease, hepatitis B can be diagnosed using ANN. [18, 19]. Fuzzy logic creates rules by inferring the knowledge from imprecise, uncertain or unreliable information. It is also observed by the soft computing that hybrid systems combining different soft computing techniques into one system can often improve the quality of the constructed system. Rough set methods combined with neural networks, genetic algorithms, and statistical inference tools can give better solutions. Rough set based data reduction can be very useful in preprocessing of data input to neural networks. Moreover hybridization of rough set methods with classical methods like principal component analysis, Bayesian methods, 2DFFT or wavelets leads to classifiers of better quality.

2.2   Nephritis

Nephritis is the inflammation of kidney. Glomerulonephritis is the most prevalent form of acute nephritis. In general, acute refers to glomerular injury occurring over days or weeks, sub acute or rapidly progressive over weeks or a few months and chronic over many months or year. Acute Glomerulonephritis usually develops a few weeks after a strep infection of the throat or skin. Symptoms include fatigue, high blood pressure and swelling. Diagnosis of nephritis is based on:

1. The patient's symptoms and medical history.
2. Physical examination.
3. Laboratory tests.
4. Imaging studies such as ultra sound or X-rays to determine blockage and inflammation.

Urinanalysis can reveal the presence of:

1. Albumin and other proteins
2. Red and white blood cells
3. Pus, blood or bacteria in the urine.

Treatment
Treatment of Glomerulonephritis normally includes drugs such as cortisone or cytotoxic drugs (those that are destructive to certain cells or antigens). Diuretics may be prescribed to increase urination. If high blood pressure is present, drugs may be

prescribed to decrease the hypertension. Iron and vitamin supplements may be recommended if the patient becomes anemic. Acute pyelonephritis may require hospitalization or severe illness. Antibiotics will be prescribed, with the length of treatment based on the severity of the infection. In the case of chronic pyelonephritis, a six month course of antibiotics may be necessary to rid the infection. Surgery is sometimes necessary. Treatment of hereditary nephritis depends of the variety of the disease and severity at the time of treatment.

3   Proposed Methodology

3.1   ANN for Diagnosis

There have been developed different types of software systems based on rough set methods. Artificial Intelligence techniques consist of developing a computer based decision support system does somewhat that it were done by a human being. Several Neural Network Models are developed which helps doctors in diagnosing the patients more correctly and accurately. Neural networks provide a very general way of approaching problems. When the output of the network is categorical, it is performing prediction and when the output has discrete values, and then it is doing classification. Neural Network based Decision Support in medicine, has at least the role of enhancing the consistency of care.

3.2   Basic Terminologies of ANN

Artificial Neural Networks are relatively crude electronic models based on the neural structure of the brain. The brain basically learns from experience. These biologically inspired methods of computing are thought to be the next major advancement in the computing industry. Human brains store information as patterns. Some of these patterns are very complicated and allow us the ability to recognize individual faces from many different angles. This process of storing information as patterns, utilizing those patterns, and then solving problems encompasses a new field in computing. ANN involves the creation of massively parallel networks and the training of those networks to solve specific problems.

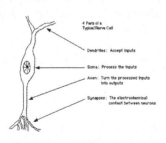

Fig. 1. A simple neuron

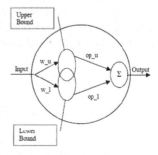

Fig. 2. Rough Neuron

The basic unit of neural networks, the artificial neurons, simulates the four basic functions of natural neurons. Various inputs to the network are represented by the mathematical symbol, x (n). Each of these inputs is multiplied by a connection weight. These weights are represented by w (n). In the simplest case, these products are simply summed, fed through a transfer function to generate a result, and then output.

3.3   Rough Set for Diagnosis

There have been developed different types of software systems based on rough set methods. There are numerous areas of successful application of rough set software systems.

Basic Terminologies of Rough Set
Rough Sets is one of the mathematical tools to process fuzzy and uncertain information. Rough sets process uncertain or high dimensional data but it is sensitive to noise. An information system S can be expressed as S= <U, R, V, f >, which U is a non-empty but finite object set, that U={x1,x2,...xn} is called universe of discourse; f:U*X→V is an information function, it specifies the U in the x property value of each object that VxЄU, VaЄR, have fa(x)ЄVa .For given information system S=<U, R, V, f >, for each subset XЄU and indiscernibility relation B, the upper approximate limit and lower approximate of X can be defined as:

$$B(x) = \{x: B(x) \subseteq X\} \text{ and } B(X) = \{x\in U: B(x) \cap X \neq \Phi\}$$

4   Proposed Algorithm

Input: Laboratory data for *n* diagnosed nephritis patients and *m* healthy Individuals.

Output: a) Diagnosis of disease of new un-diagnosed patients using multilayer perceptron and back propagation   algorithm.   b) Determination of significance factor of each chemical test element using rough set theory.   c) Determination of percentage accuracy of training during diagnosis.

We test our model on chemical test data for diagnosis of nephritis in a hospital. Seven kinds of minerals such as Na, K, Fe, Cu, Ca, Zn and Mg in human body were examined to help doctor make the diagnosis. Table 1 shows the test results of confirmed cases: No. 1-30 are the laboratory test results of 30 patients already diagnosed as nephritis; No. 31-60 are patients already diagnosed as healthy. Calculate the net inputs and outputs of the k output layer neurons are :

$$net^o_k= \sum_{j=1}^{j+1}V_{kj}y_j$$
$$Z_k= f(net^o_k)$$

Update the weights in the output layer (for all k, j pairs)

$$v_{kj} \leftarrow v_{kj} + c\lambda (d_k - Z_k) Z_k (1- Z_k) y_j$$

In our paper we have tried 30 new cases to be diagnosed as nephritis patients or healthy based on chemical test results in table 2, using the method of BP neural network. The first 20 data from the diagnosed nephritis patients and the first 20 data from healthy samples in No 1-60 were selected and input in the program in Matlab software.

Table 1. Dataset for chemical test data for diagnosed patients

Table 2. Dataset for chemical test data for undiagnosed patients

A gradient descent with acceleration was used to train the model. The parameter of learning speed was set as 0.05, and the error requirement was set as 0.001, the largest training recurrence was set as 100000 times. Then the best fitting result is shown in figure 3. The other 20 data in 60 diagnosed cases were input into the BP neural network model to predict the diagnosis result. Output matrix after calculation is shown as below:

[0.0078   0.0000   0.0027   0.4107   0.0738   0.1193   0.0007   0.0970
0.0000   0.9982   0.9998   0.0198   0.0005   0.4047   0.4047   0.9903
0.0809   0.2554   0.0970   0.3544   0.4107   0.3544   0.6227   0.4047
0.9957   0.3544   0.7035   0.9980   0.9980   0.9966   0.9778   0.9982
0.0095   0.6227   0.9767   0.9767   0.4047   0.9570   0.4031   0.9570]

The output matrix was transformed by 0 / 1 rule: make 0.5 as the center point, data closer to 0 are re-written as 0, and data closer to 1 are re-written as 1; data closer to 0.5 marked as undetermined.

Where 0 can be read as normal and 1 as nephritic. Results show that the conclusion from the model is entirely consistent with the real diagnosis result. 100% correct Prediction result of the model shows that this model works well. New cases of No 61-90 were input into the BP neural network model to predict the diagnosis. Process the output matrix by 0/1 rule as before. Updated data are:

[0   0   0   0   0   0   0   0   0   1   1   0   0   0   0   1   0   0   0   0
0   0   1   0   1   0   1   1 1   1   1   1   0   1   1   1   0   1   0   1]

In the next part, we have tried to find the significance factor of every key indicator i.e. the 7 chemical elements. Rough set theory has various terminologies for finding the significance factor . In order to find the significance factor of every element we use the formula for significance factor, as mentioned earlier. First of all, we do a further quantitative transformation on data of 60 diagnosed cases in table 1 by range transformation. All data after transformation are in range 0-1. Then we set rough value {content value 1、2、3、4} to represent the 4 categories. Here we consider the division into four levels: data in [0,0.25)are transformed as content value 1; data in [0.25,0.5)are transformed as content value 2; data in [0.5,0.75)are transformed as content value 3; data in [0.75,1) are transformed as content value 4. Therefore, after range transformation and the set of rough value, data of 60 diagnosed cases in table 1 are transformed as in table 3. According to rough set theory, we get

$$U/D = \{X1, X2\} \quad X1 = \{1...30\} \quad X2 = \{31...60\}$$

Reliance factor: Rd= (c1+c2)/card

Significance: sigf (1) =Rd-Rd1;

5   Implementation and Results

In our paper we tried to find the accuracy of diagnosis of Nephritis using ANN. After training and testing the network we got the accuracy to be 85% in 9 iterations. Following are the results obtained:

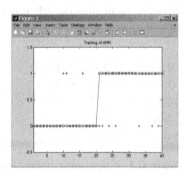
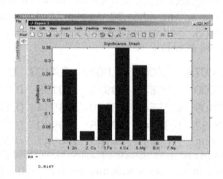

Fig. 3. Graph showing accuracy of elements   **Fig. 4.** Graph showing significance of elements

In later part we tried to find the impact factor or significance of every element used in diagnosis of nephritis. Our results show that, some of the key indicators play more important role as compared to others. This could help the patient to go for only few tests to inspect the elements with high impact factor inspite of having all laboratory tests.

Following are the results obtained:

sigf =   0.2667   0.0333   0.1333   0.3500   0.2833   0.1167   0.0167

We got the reliance factor to be 0.8167, with the significance factor of Ca = 0.3500, Mg = 0.2833, and Zn = 0.2667; Table 4 shows the graph for the significance factor of each element.Ca, Mg and Zn have more significance as compared to other elements. Table 5 shows the results of accuracy in diagnosis of nephritis after considering different combinations of elements for laboratory tests.

6   Conclusion and Future Scope

With the development of science and technology, it is more and more important to utilize mathematical modeling and quantitative research methods to deal with various types of practical problems. Neural networks are one of the recent areas of research. Neural networks have the strongly fault tolerance, self-organization, massively parallel processing and self-adapted. Hybrid systems can be used for optimization of problem. Many other data mining applications other that biomedical can be analyzed using different soft computing techniques. Other diseases can also be diagnosed using ANN. This proposed work can be compared with other soft computing technologies such as fuzzy logic to find better results. Combination of ANN with Genetic algorithm for optimization can also be done. The combination of neural network and rough set theory can also be used for other data mining applications also where large dataset is available and we have to find only the optimal data for our work. Thus it can be concluded that ANN have wide applications in various fields and the combination of rough set with other soft computing tool gives a more stronger mathematical tool for analysis of data or information.

References

[1] Nazari Kousarrizi, M.R., Seiti, F., Teshnehlab, M.: An Experimental Comparative Study on Thyroid Disease Diagnosis Based on Feature Subset Selection and classification. International Journal of Electrical & Computer Sciences IJECS-IJENS 12(1)

[2] Dai, J., Xu, Q., Wang, W.: A Comparative Study On Strategies of Rule Induction for Incomplete Data Based on Rough Set Approach. International Journal of Advancements in Computing Technology 3(3) (April 2011)

[3] Ding, S., Chen, J., Xu, X., Li, J.: Rough Neural Networks: A Review. Journal of Computational Information Systems 7, 7 (2011)

[4] Gharehchopoghi, F.S., Khalifelu, Z.A.: Neural Network Application in Diagnosis of Patient: A Case Study. IEEE (2011)

[5] Abbasi, M.M., Kashiyarndi, S.: Clinical Decision Support Systems: A discussion on different methodologies used in Health Care. IEEE (2011)

[6] Akgundogdu, A., Kurt, S., Kilic, N., Ucan, O.N., Akalin, N.: Diagnosis of Renal Failure Disease Using Adaptive Neuro-Fuzzy Inference System. Journal of Medical Systems 34(6), 1003–1009 (2010)

[7] Hannan, A., Mane, A.V., Manza, R.: Prediction of heart disease medical prescription using radial basis function. In: Computational Intelligence and Computing Research (ICCIC). IEEE (2010)

[8] Zhang, X., Yang, G., Xia, B., Wang, X., Zhang, B.: Application of the rough set theory and BP neural network model in disease diagnosis. In: 2010 Sixth International Conference on Natural Computation (ICNC). IEEE (2010)

[9] Paszek, P., Wakulicz–Deja, A.: Applying Rough Set Theory to Medical Diagnosing. In: Kryszkiewicz, M., Peters, J.F., Rybiński, H., Skowron, A. (eds.) RSEISP 2007. LNCS (LNAI), vol. 4585, pp. 427–435. Springer, Heidelberg (2007)

[10] Tsumoto, S.: Pawlak Rough Set Model, Medical Reasoning and Rule Mining. In: Greco, S., Hata, Y., Hirano, S., Inuiguchi, M., Miyamoto, S., Nguyen, H.S., Słowiński, R. (eds.) RSCTC 2006. LNCS (LNAI), vol. 4259, pp. 53–70. Springer, Heidelberg (2006)

[11] Ilczuk, G., Wakulicz-Deja, A.: Rough Sets Approach to Medical Diagnosis System. In: Szczepaniak, P.S., Kacprzyk, J., Niewiadomski, A. (eds.) AWIC 2005. LNCS (LNAI), vol. 3528, pp. 204–210. Springer, Heidelberg (2005)

[12] Ilczuk, G., Mlynarski, R., Wakulicz-Deja, A., Drzewiecka, A., Kargul, W.: Rough Set Techniques for Medical Diagnosis Systems. In: Computers in Cardiology. IEEE (2005)

[13] Brause, R.: Medical Analysis and Diagnosis by Neural Networks. In: Crespo, J.L., Maojo, V., Martin, F. (eds.) ISMDA 2001. LNCS, vol. 2199, pp. 1–13. Springer, Heidelberg (2001)

[14] Pawlak, Z.: Rough Sets and Data Mining. Springer, Heidelberg (2001)

[15] Tsumoto, S., Tanaka, H.: Extraction of Diagnostic Knowledge from Clinical Databases based on Rough Set Theory. IEEE (1996)

[16] Sharpe, P.K., et al.: ArtificialNeural Networks in Diagnosis of Thyroid Function. Clinical Chemistry (1993)

[17] Tan., K.C., et al.: Evolutionary computing for knowledge discovery in medical diagnosis. ACM

[18] Dakashata, et al.: An Expert System For Hepatitis B Diagnosis Using Artificial Neural Networks. In: International Conference & Workshop on Recent Trends in Technology (TCET) (2012)

[19] Mokhlessi, O., et al.: Utilization of 4 types of Artificial Neural Network on the diagnosis of valve-physiological heart disease from heart sounds. In: Proceedings of the 17th Iranian Conference of Biomedical Engineering (ICBME 2010), November 3-4 (2010)

Optimisation Using Levenberg-Marquardt Algorithm of Neural Networks for Iris

Asim Sayed[1], M. Sardeshmukh[1], and Suresh Limkar[2]

[1] Department of Electronics & Telecommunication, SAOE, Kondhwa, Pune, India
[2] Department of Computer Engineering, AISSMS IOIT, Pune, India
{asim27902,sureshlimkar}@gmail.com, manojsar@rediffmail.com

Abstract. This paper explores the optimisation technique of Damped Least Square Method also known as the Levenberg-Marquardt (LM) Algorithm for Iris recognition. The motive behind it is to show that even though there are many algorithms available which act as an alternative to the LM algorithm such as the simple gradient decent and other conjugate gradient methods be it the vanilla gradient decent or the Gauss Newton iteration, the LM algorithm outperforms these optimisation techniques due to the addressing of the problem by the algorithm as the Non-linear Least Square Minimisation. The results are promising and provide an insight into Iris recognition which are distinct pattern of individuals and are unique in case of every eye.

Keywords: Levenberg-Marquardt, LM algorithm, Iris, Biometrics of the eye, Optimisation of Iris Images, Least square method for Iris, Damped Least Square, Non Linear Least Square Method.

1 Introduction

The Iris is a very integral organ of our body, though it is very delicate it is highly protected by every individual. Unlike fingerprints used for biometrics which has chances for the finger getting mutilated or the finger facing defacement due to the use of the fingers for physical applications, the Iris serves as a more robust metric for identification. There are many reflexes of our body which help protect the eyes due to the delicate nature of the organ as compared to other biometric feature of the human body. To add to the advantages the Iris have a high bearing capacity to carry information with an in information density of 3.2 measurable bits per mm^2 [1]. Though our vision system processes up to 5 billion bits per second with great accuracy and precision [1] this speed is unmatched by any computer data processing unit. Hence Iris is used as a biometric measure to authenticate people since the Irides are unique to every individual and the fact that no two Irides of the same person match which means there is a vast diversity and distinction in every iris, which is an added advantage to use the Iris as a biometric measure.

The following figure gives an idea about the general structure of the Iris which contains the Pupil (in the centre), the Sclera (The white part) and the Iris which is between them, the Iris pattern is clearly shown in the image shown in fig. 1.

S.C. Satapathy, S.K. Udgata, and B.N. Biswal (eds.), *FICTA 2013*,
Advances in Intelligent Systems and Computing 247,
DOI: 10.1007/978-3-319-02931-3_12, © Springer International Publishing Switzerland 2014

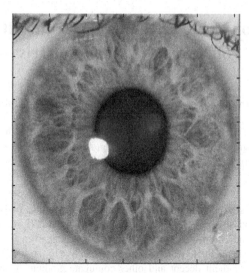

Fig. 1. Image of the Human Iris

With the pronounced need for reliable personal identification, iris recognition has become an important enabling technology in our society. Although an iris pattern is naturally an ideal identifier, the development of a high-performance iris recognition algorithm and transferring it from research lab to practical applications is still a challenging task. Automatic iris recognition has to face unpredictable variations of iris images in real-world applications [7]. For example, recognition of iris images of poor quality, nonlinearly deformed iris images, iris images at a distance, iris images on the move, and faked iris images all are open problems in iris recognition [8]. A basic work to solve the problems is to design and develop a high quality iris image database including all these variations. Moreover, a novel iris image database may help identify some frontier problems in iris recognition and leads to a new generation of iris recognition technology [7]. Hence the Database used is from Chinese Academy of Sciences Institute of Automation also dubbed as (CASIA – Iris).

A very reliable source of images it is a set of many subsets we use 200 images from these grey scale images since these images do not contain specular reflections due to the use of Near Infra-Red light for illumination.

2   The Back Propagation Theorem

When using neural networks for designing a solution to a problem we use layers as present in the human brain. Hence we simulate a machine called as Perceptron. The central idea behind this solution is that the errors for the units of the hidden layer are determined by back-propagating the errors of the units of the output layer. For this reason the method is often called the back-propagation learning rule [2]. Back-propagation can also be considered as a generalization of the delta rule for non-linear activation functions and multilayer networks as seen in Fig.2.

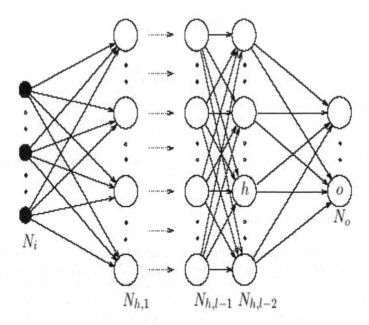

Fig. 2. A Neural Network with nodes [2]

When a learning pattern is clamped, the activation values are propagated to the output units, and the actual network output is compared with the desired output values, we usually end up with an error in each of the output units. Let's call this error e_o for a particular output unit o [26]. We have to bring e_o to zero The simplest method to do this is the greedy method: we strive to change the connections in the neural network in such a way that, next time around, the error e_o will be zero for this particular pattern. We know from the delta rule that, in order to reduce an error, we have to adapt its incoming weights accordingly [2]. The input units are merely 'fan-out' units; no processing takes place in these units. The activation of a hidden unit is a function Fi of the weighted inputs plus a bias, as given in in eq. 1.

$$y_k(t+1)=F_k(S_k(t))= F_k(\textstyle\sum_j(\omega_{jk}(t)\, y_j(t) + \theta_k(t))) \tag{1}$$

where, y_k (t+1) is the value for the next node, $\omega_{jk}(t)$ is the weight of the current node and $\theta_k(t)$ is the error which needs to be compensated. Since this mechanism only works as an feed forward network we have to look at providing the backward propagation using the Back Propagation Theorem [25], which is understood by the delta rule by using the activation function in eq. 2 which is the activation function

$$\left(y_k^{\,p}\right) = F\left(s_k^{\,p}\right) \tag{2}$$

In which y is a function of s which depends on weights between the nodes j and k with an addition of small deviation θ;

$$\left(s_k^P\right) = \sum_j \left(\omega_{jk}\, y_j^P\right) + \theta_k \tag{3}$$

To get the correct generalization of the delta rule as presented in the previous chapter, we must set the delta of weights by setting Error measure (E_p)

$$\Delta_p w_{jk} = -\gamma \frac{\partial E^P}{\partial w_{jk}}. \tag{4}$$

Hence this error is minimized and we get the least possible error and hence giving the weights of appropriate values[6].

3   The Proposed System

The Leverberg – Marquardt Algorithm is best suited for non linear optimisation of pseudo second order which implies it works with function evaluations and gradient information and by using the sum of outer products of gradient it estimates the Hessian Matrix [3]. Hence applying this method to find the best match to our iris database we get promising results. The technique invented by Levenberg was to 'blend' between the two extremes of minimum and maximum [4]. The Hessian matrix is approximated by

$$H = J^T J \tag{5}$$

Here J is Jacobian matrix containing the first order derivatives of network errors with respect to the weights and biases, and e is a vector of network errors. The Jacobian matrix can be computed through a standard back propagation technique that is much less complex than computing the Hessian matrix. Hence the LMA would try to get the solution by deducing itself to function as a Newton's method [5]. The Levenberg-Marquardt algorithm uses this approximation to the Hessian matrix in the following Newton-like update as shown in eq. 6

$$x_{k+1} = x_k - [J^T J + \mu I]^{-1} J^T e \tag{6}$$

Where x_{k+1} is the next value of x to be updated and x is the current value, J is the Jacobian and I is the identity matrix, the value of μ is a scalar and keeps decreasing, if μ is zero then it is the Newton's method.

4   Results

Giving the input image from the CASIA database we get the best matching result from the all the images available in the database. The input image is given in fig.3. The segmentation of Iris is derived by using the techniques such as Principal Component Analysis [19] [20] [21].

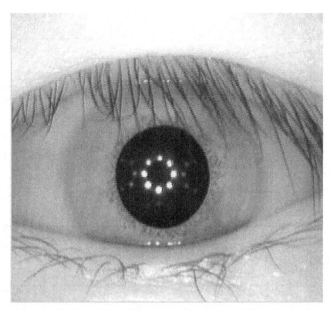

Fig. 3. Input image chosen as mask [12]

The result is found by the algorithm and the number of epochs used is derived along with the information that which epoch yielded the best results as shown in fig.4. Here a plot of epochs (MSE) has been plotted, the epochs gets the best validation performance at epoch no. 9, the MSE is the lowest at this point, and hereafter no significant change takes place and no further decrease takes place. Hence this is the best validation performance.

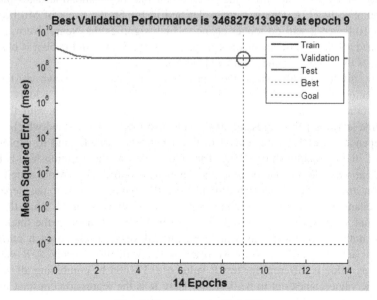

Fig. 4. Plot of Epochs Vs MSE

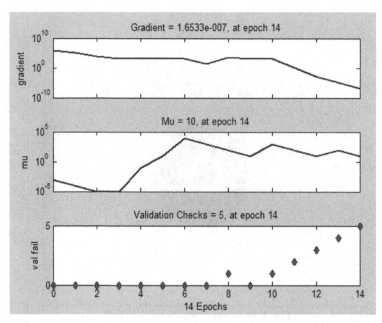

Fig. 5. Information about the training state [12] a) Gradient Vs Epochs b) Mu Vs Epochs c) Val Fails Vs Epochs

The parameters which are trained by the Neural Networks in the process to achieve the best validation are shown in Fig.5a [15]. It is observed that the Gradient is minimum at Epoch number 14 which plays an important part in deducing the conclusion along with the Mu value to ascertain that best validation takes place at Epoch number 14 [17]. This is further clarified in Fig.5b the value of Mu becomes stable at 10 hence it will no longer change which signifies the Neural Networks are well trained and no significant error is propagating backwards. In Fig.5c it is observed that 5 times the validation check failed [16] [18].

Fig. 6. gives an idea about the regression analysis also known as function approximation technique. Regression analysis [13] is widely used for prediction and forecasting, where its use has substantial overlap with the field of machine learning [9]. Regression analysis is also used to understand which among the independent variables are related to the dependent variable, and to explore the forms of these relationships [10]. The focus is on the relationship between a dependent variable and one or more independent variables [14]. More specifically, regression analysis helps one understand how the typical value of the dependent variable changes when any one of the independent variables is varied, while the other independent variables are held fixed. It is can be observed in Fig.5 the thick black lines are nothing but the points which are plotted very closely to each other, the training is achieved at R of 0.17151 whereas the validation occurs at R value of 0.17178this value is different for test R value of 0.171685 by combining all the data the final R value comes out to be 0.17234. Here R is the regression constant used in the Neural Network [11].

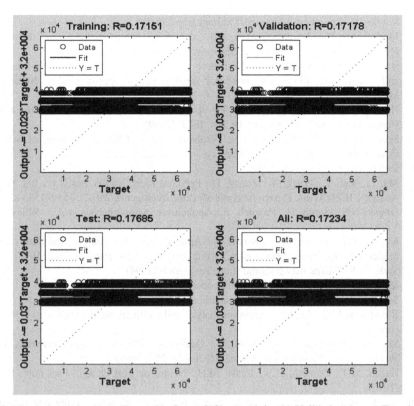

Fig. 6. Regressional Analysis- Target Vs Output[12] a) training b) Validation data c) Test data d) All of the data combined

5   Conclusion

We find that the training of the Neural networks using the Lavenberg Marquardt results in fast performance along with good accuracy as yielded in the proposed system, the efficiency of the algorithm we were able to find the correct matches each time with least no of epochs. We can conclude that the LMA is best suited for Neural Networks having hidden layers upto 100, after this count the LMA would more or less function as the Newton's method. A system of fuzzy logic and neural network combination is also available by various researchers and may or may not give the results as desired and may perform better or lower than the proposed system depending on the mixture of fuzzy logic with types of neural networks and how many hidden layers are involved [23] [24].

References

[1] Burghadt, T.: Inside Iris Recognition, Report on Identity, Assignment Information Security, COMS40213

[2] (2013), The NeuroAI website,

[3] http://www.learnartificialneuralnetworks.com/backpropagation.html

[4] Rowies, S.: Levenberg Marquardt Optimization (2005)
[5] Moré, J.J.: The Levenberg-Marquardt algorithm: implementation and theory. In: Numerical Analysis, pp. 105–116. Springer, Heidelberg (1978)
[6] Brown, K.M., Dennis Jr., J.E.: Derivative free analogues of the Levenberg-Marquardt and Gauss algorithms for nonlinear least squares approximation. Numerische Mathematik 18(4), 289–297 (1971)
[7] Hecht-Nielsen, R.: Theory of the back propagation neural network. In: International Joint Conference on Neural Networks, IJCNN. IEEE (1989)
[8] Specification of CASIA Iris Image Database(ver 1.0), Chinese Academy of Sciences (March 2007), http://www.nlpr.ia.ac.cn/english/
[9] irds/irisdatabase.htm
[10] Ma, L., Tan, T., Wang, Y., Zhang, D.: Personal Identification Based on Iris Texture Analysis. IEEE Trans. Pattern Analysis Machine Intelligence 25(12), 1519–1533 (2003)
[11] Draper, N.R., Smith, H., Pownell, E.: Applied regression analysis, vol. 3. Wiley, New York (1966)
[12] Kupper, L.L., Muller, K.E., Nizam, A.: Applied regression analysis and multivariable methods. In: Kleinbaum, D.G. (ed.), 3rd edn. Duxbury Press, Pacific Grove (1998)
[13] Sarle, W.S.: Neural networks and statistical models (1994)
[14] MATLAB version R2011b. The Mathworks Inc., Pune (2011)
[15] Seber, G.A.F., Lee, A.J.: Linear regression analysis, vol. 936. John Wiley & Sons (2012)
[16] Kleinbaum, D.G.: Applied regression analysis and multivariable methods. CengageBrain. com (2007)
[17] Michel, A.N., Farrell, J.A., Porod, W.: Qualitative analysis of neural networks. IEEE Transactions on Circuits and Systems 36(2), 229–243 (1989)
[18] Hagan, M.T., Demuth, H.B., Beale, M.H.: Neural network design. Pws Pub., Boston (1996)
[19] He, Y., et al.: Stability analysis for neural networks with time-varying interval delay. IEEE Transactions on Neural Networks 18(6), 1850–1854 (2007)
[20] Cochocki, A., Unbehauen, R.: Neural networks for optimization and signal processing. John Wiley & Sons, Inc. (1993)
[21] Hsieh, W.W.: Nonlinear principal component analysis by neural networks. Tellus A 53(5), 599–615 (2001)
[22] Baldi, P., Hornik, K.: Neural networks and principal component analysis: Learning from examples without local minima. Neural Networks 2(1), 53–58 (1989)
[23] Diamantaras, K.I., Kung, S.Y.: Principal component neural networks. Wiley, New York (1996)
[24] Haykin, S.S., et al.: Neural networks and learning machines, vol. 3. Prentice Hall, New York (2009)
[25] Ishibuchi, H., Tanaka, H.: Fuzzy regression analysis using neural networks. Fuzzy Sets and Systems 50(3), 257–265 (1992)
[26] Nauck, D., Klawonn, F., Kruse, R.: Foundations of neuro-fuzzy systems. John Wiley & Sons, Inc. (1997)
[27] Haykin, S.: Neural networks: A comprehensive foundation. Prentice Hall PTR (1994)
[28] Ham, F.M., Kostanic, I.: Principles of neurocomputing for science and engineering. McGraw-Hill Higher Education (2000)

Neural Networks – A Case Study in Gene Identification

V. Bhaskara Murthy[1] and G. Pardha Saradhi Varma[2]

[1] Padmasri Dr. BVRICE
Vishnupur, Bhimavaram, W.G.Dt. A.P.,
murthyvb@gmail.com
[2] Department of IT,
S.R.K.R. Engineering College,
Chinamiram, Bhimavaram., W.G.Dt. A.P.
gpsvarma@yahoo.com

Abstract. Gene prediction has been an interesting area of research in Bioinformatics. Many of the recent gene identification methods adopt different approaches which are more robust when dealing with uncertainty and ambiguity. In this paper details of Artificial Neural Networks and using them in study and analysis of Biological data are discussed. The types of neural networks in the area of bioinformatics are listed. The AI technique of simulated annealing is applied. Learning mechanism and evolution of neural networks in the field of bioinformatics are also listed.

Keywords: Gene Prediction Model, Artificial Neural Network, GRAIL System.

1   Introduction

A large amount of raw sequence data generated because of advancement in sequencing technology requires biological interpretation in an effective and optimal way. This is known as annotation. Although the Human Genome Project was completed in April, 2003, the exact number of genes encoded by the human genome is still unknown and estimated genes are in the range of 20,000-25,000. The steps involved in genome annotation are classified as three categories.

1. Gene Identification (Nucleotide level)
2. Structure determination of proteins. (Protein level)
3. Mechanism of biochemical reactions. (Process level)

Among three, nucleotide-level annotation is a primary step in molecular biology[1]. Only 80% of genes are accurately predicted at the nucleotide level , 45% are predicted at the exon level and nearly 20% at the whole genome level. Even if all human genes are experimentally determined, it would still be important to know how the structures of genes are organized and defined, and how they can be recognized.

2   Problem Definition

In this section, some basic terminology related to the problem of gene prediction is given. The problem is then stated formally.

S.C. Satapathy, S.K. Udgata, and B.N. Biswal (eds.), *FICTA 2013*,
Advances in Intelligent Systems and Computing 247,
DOI: 10.1007/978-3-319-02931-3_13, © Springer International Publishing Switzerland 2014

Basic Terminology

Gene: Gene is defined as a segment of DNA that contains the necessary information to produce a functional product, usually a protein.

Promoter: Promoter is the regulatory region of DNA located upstream of a gene. It provides a control point for regulated gene transcription.

Core Promoter: It is the minimal portion of the promoter required to initiate transcription properly. It serves as a binding site for RNA polymerase and general transcription factors.

Proximal Promoter: The proximal sequence upstream of the gene that tends to contain regulatory elements is known as proximal promoter. It serves as a binding site for specific transcription factors.

Open Reading Frame(ORF): An ORF is a sequence of DNA that starts with a start codon normally 'ATG' and ends with one of three codons TAA, TAG, TGA.

Coding sequence: Coding sequence(CDS) is the actual region of DNA that is translated to form proteins[2].

Gene Prediction: The characterization of genomic features using computational and experimental methods is called gene prediction.
   The following information is required:

1. Coding region for a protein
2. The DNA strand is used to encode the gene
3. The start and end positions of the gene
4. The exon-intron boundaries in eukaryotes
5. The regulatory sequences for that gene

2.1   Gene Prediction Problems

Prokaryotes: Easy to predict promoter region or start of coding region is able to determine a gene.

Eukaryotes: Hard to predict promoter, transcription or translation start region, splice sites, coding regions.
It is still difficult to predict genes accurately due to:

➤ DNA sequence have low information content
➤ Difficult to discriminate real signals.
➤ Contain sequencing errors.
➤ Short genes are found in prokaryotes that have little information
➤ The presence of alternate splicing mechanism in eukaryotic genes makes its detection difficult.

2.2   Gene Prediction Model

A model approach for Gene Prediction:

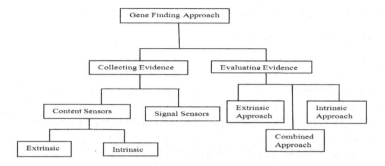

Gene Predicting approach has two factors
1. Collecting Evidence
2. Evidence Evaluation

Evidence collecting is through two different types, namely Content sensors and signal sensors to identify genes in genomic sequence. Content sensors are those measures try to divide DNA region into coding and noncoding. Extrinisic content senors consider a genomic sequence region and match to a protein or DNA sequence. Instrinisic content sensors used for prokaryotic genomes where only protein coding regions and intergenic regions are considered. Signal sensors are measures that try to detect the presence of the functional sites specific to a gene.

Evidence evaluation is through by combining all evidences to predict the complicated gene structure. In extrinisic approach is to combine similarity information with signal information obtained from signal sensors. This is used to refine the region boundary. Intrinisic gene finders is to locate all gene elements that occur in genomic sequence including partial gene structures at the border of the sequence. Combined approach of these two is used in recent gene prediction.

Computational gene finding is a process of the following:
➤   Identifying common phenomena in known genes.
➤   Building computational model that can accurately describe the common phenomena.
➤   Using the model to scan an uncharacterized sequence to identify regions that match the model.
➤   Test and validate the predictions.

3   Artificial Neural Network for Gene Prediction

The Objectives of Bioinformatics are:

1. To organize data in a way that provides users to access existing information and to submit new entries as they are produced

2. To invent tools and resources that helps in the analysis and organization of data.

3. To use this data to analyze and interpret the results using computers in a biologically meaningful way.

4. To help scientists in the pharmaceutical industry to understand and analyze the protein structures for design of drugs.

3.1   Gene Structure Prediction Methods:

- Homology Based Method[3]
- Markov Model[4]
- Hidden Markov Model[5]
- Neural Networks

Homology Based Methods

➢ Given a genomic sequence, search against cDNA or EST libraries
➢ GenomeScan
➢ Consensus-based programs: GeneComber

Markov Model

A Markov chain is a sequence of random variablesX1, X2, X3, ... with Markov property, namely that, given the present state, the future and past states are independent.

P(Xn+1=x|Xn=xn,...,X1=x1,X0=x0)=P(Xn+1=x|Xn=xn).   (first   order   Markov Model)

The possible values of Xi form a countable set S called the state space of the chain. A finite state machine is an example of a Markov chain. The probability of transition from one state to another state is called transition probability.

Markov Model for Gene Prediction

DNA sequences can be considered to be generated by two Markov Chains. One chain generates coding regions another chain generates non-coding regions. Each state in the chain can has four values: A, C, G, T.

Hidden Markov Model

HMM is Markov process where states are hidden (unseen), but the variables emitted from states can be observed. Challenge is to determine the hidden parameters from the observable parameters.

Three Problems in HMM

1. Prediction / Evaluation: Given parameters of the model, compute the probability of an output sequence (Forward / backward Algorithm)

2. Decoding: Given parameters of the model, find most likely sequence of hidden states. (Viterbi Algorithm)

3. Learning: Given a set of sequences generated by the model, learn the most likely model parameters (transition/emission probabilities) (Baum-Welch Algorithm)

In gene prediction, use coding and non-coding sequences to train a HMM and then use known HMM to make prediction for a new sequence.

3.2   Neural Networks (NN)

Bioinformatics research problems like protein structure prediction, multiple alignment of sequences, phylogenic inferences, etc are solved using Artificial Intelligence (AI) , which provides a powerful and efficient approach Artificial Neural Networks . It is one of the AI technique because of its capability to capture and represent complex nature input, output data. The objective of this paper is to provide an understanding role of ANN. Neural network is one of the most widely used methods in bioinformatics. It is used in gene structure prediction, protein structure prediction, and gene expression data analysis.

Neural network is a computational model[6] to follow the way how the brain works. Brain is made from small functional units called neurons. A neuron has a cell body, several short dendrites and single long axon. Each neuron connects to several other neurons by dendrites and axons. Dendrites receive signals from other neurons and act as the inputs to the neuron. These inputs increase or decrease the electrical potentials of the cell body and if it reaches a threshold, an electrical pulse is sent down the axon. This output becomes the input to several other neurons.

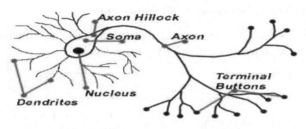

Fig. 1. Structure of Neuron

Novel Intelligence computational techniques like artificial intelligence techniques genetic algorithms, neural networks, which are top-to-bottom. These techniques differ from traditional and logic-based methods.

3.3   Relevance of Artificial Neural Network in Bioinformatics

An Artificial Neural Network (ANN) is an information processing model that is able to capture and represent complex input-output relationships. The development of the ANN technique came from a desire for an intelligent artificial system that could process information in the same way the human brain. Its novel structure is represented as multiple layers of simple processing elements, operating in parallel to solve specific problems. ANNs resemble human brain in two respects. Learning process and storing experiential knowledge. An artificial neural network learns and classifies a problem through repeated adjustments of the connecting weights between the elements. In other words, an ANN learns from examples and generalizes the learning beyond the examples supplied.

Artificial neural network applications have recently received considerable attention. The methodology of modeling, or estimation, is somewhat comparable to statistical modeling. Artificial neural networks (ANNs) [7] are computer algorithms based on modeling the neuronal structure of natural organisms. In general if given sufficient complexity, there exists an ANN that will map every input pattern to its appropriate output pattern, as mapping is not one-to-many.

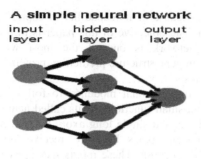

Fig. 2. A Simple Network

3.4   Designing Neural Networks for Bioinformatics

When designing NN for bioinformatics [8] applications, there are common designing issues that need to be addressed. Figure3 gives these details.

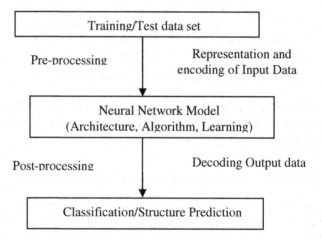

Fig. 3. Design Model of a Neural Network for bioinformatics

Preprocessing of data involves feature presentation and input encoding. This is an important element which determines the performance and the information entered to the NN. In order to get the full benefit of the NN, the designer has to represent prior knowledge about sequence structure and functions. This allows the extraction of salient features in the given sequence. After identifying the features needed to be represented in the NN application, there are several ways to represent these data to maximize information extraction. They are as follows:

- ➤ Real number measurements in a continuous scale.
- ➤ Vectors of distance or frequencies.
- ➤ Categorized into classes Using alternative alphabet to represent AA with similar features
- ➤ Hierarchical classes

In the NN application, data need to be encoded. Encoding can be local (involving single or neighboring residues in short sequence segments) or global (involving long range relationship in entire sequence). The sequence encoding method can be direct or indirect. Direct encoding converts each residue in to a vector and it preserves the positional information. Indirect encoding on the other hand provides overall information measures of the entire sequence.

Output decoding is not as complex as input encoding. It depends on the number of classification required in the application. However networks like self organizing maps automatically configure the number of output units. The value of the output units can be used qualitatively or quantitative measure of confidence level or activity level.

3.5   Coding Region Recognition and Gene Identification

Pattern recognition methods are mainly used in describing the location and significance of genes in a genome. In prokaryotes, the coding region is a continuous single reading frame. In eukaryotes, it consists of introns and exons. Therefore, the main task is to differentiate introns, exons and splice site detection. The GRAIL[9] system is an example of an application developed using NN for coding region recognition. GRAIL is a multiple sensor-neural network based system. It can locate genes in anonymous DNA sequence by recognizing features related to protein coding regions and the boundaries of coding regions. These recognized features are combined using a neural network system.

A typical neural network (shown in Figure 1) is composed of input units X1, X2,... corresponding to independent variables, a hidden layer known as the first layer, and an output layer (second layer) whose output units Y1,... correspond to dependent variables (expected number of accidents per time period).In between are hidden units H1, H2, … corresponding to intermediate variables.   These interact by means of weight

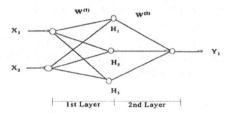

Fig. 4. A simplified Artificial Neural Network

Matrices W(1) and W(2) with adjustable weights. The values of the hidden units are obtained from the formulas:

$$H_j = f\left(\sum_k W_{jk}^{(1)} X_k \right)$$
$$Y_i = f\left(\sum_j W_{ij}^{(2)} H_j \right).$$

(1)

In One multiplies the first weight matrix by the input vector X = (X1, X2,...) and then applies an activation function f to each component of the result. Likewise the values of the output units are obtained by applying the second weight matrix to the vector H = (H1, H2,...) of hidden unit values, and then applying the activation function f to each component of the result. In this way one obtains an output vector Y= (Y1, Y2,....).The activation function f is typically of sigmoid form and may be a logistic function, hyperbolic tangent, etc.:

$$f(u) = \frac{1}{1 + e^{-u}}, \qquad f(u) = \frac{e^{u} - e^{-u}}{e^{u} + e^{-u}}.$$

(2)

Usually the activation function is taken to be the same for all components but it need not be. Values of W(1) and W(2) are assumed at the initial iteration. The accuracy of the estimated output is improved by an iterative learning process in which the outputs for various input vectors are compared with targets (observed frequency of accidents) and an average error term E is computed:

$$E = \frac{\sum_{n=1}^{N} (Y^{(n)} - T^{(n)})^2}{N}.$$

(3)

Here
N = Number of highway sites or observations
Y(n) = Estimated number of accidents at site n for n = 1, 2,..., N
T(n) = Observed number of accidents at site n for n = 1, 2,..., N.

After one pass through all observations (the training set), a gradient descent method may be used to calculate improved values of the weights W(1) and W(2), values that make E smaller. After reevaluation of the weights with the gradient descent method, successive passes can be made and the weights further adjusted until the error is reduced to a satisfactory level. The computation thus has two modes, the mapping mode, in which outputs are computed, and the learning mode, in which weights are adjusted to minimize E. Although the method may not necessarily converge to a global minimum, it generally gets quite close to one if an adequate number of hidden units are employed.

The most delicate part of neural network modeling is generalization, the development of a model that is reliable in predicting future accidents. Overfitting (i.e., getting weights for which E is so small on the training set that even random variation is accounted for) can be minimized by having two validation samples in addition to the training sample. According to Smith and Thakar (Smith and Thakar, 1993), the data set should be divided into three subsets: 40% for training, 30% to prevent overfitting, and 30% for testing. Training on the training set should stop at the epoch when the error E computed on the second set begins to rise (the second set is not used for training but merely to decide when to stop training). Then the third set is used to see how well the model performs. The cross validation helps to optimize the fit in three ways: by limiting/optimizing the number of hidden units, by limiting/optimizing the number of iterations, and by inhibiting network use of large weights.

The major advantages of neural networks in modeling applications are as follows:

Advantages

1. Adaptive learning: An ability to learn how to do tasks based on the data given for training origination experience.

2. Self-Organisation: An ANN can create its own organisation or representation of the information it receives during learning time.

3. Real Time Operation: ANN computations may be carried out in parallel, and special hardware devices are being designed and manufactured which take advantage of this capability.

4. Fault Tolerance via Redundant Information Coding: Partial destruction of a network leads to the corresponding degradation of performance. However, some network capabilities may be retained even with major network damage.

3.6   Learning Mechanism

Supervised Learning: The ANN architecture is called 'supervised' because the error between the expected output and actual output is used to back-propagate changes to the weights between the layers of the network.

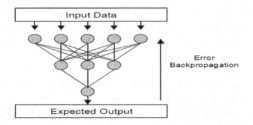

Fig. 5. Supervised Learning

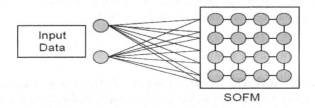

Fig. 6. Unsupervised Learning

Unsupervised Learning: Unsupervised neural networks have been frequently used in bioinformatics as they are a well tested method for clustering data. The most common technique used is the self-organizing-feature-map. This uses a map of interconnected neurons and an unsupervised learning algorithm which uses a neighborhood of units to find common features in the data and therefore develop clustering behavior.

3.7   Evolution of Neural Networks in Bioinformatics:

Neural networks have been widely used in biology since the early 1980s. They can be used to:

1. Predict the translation initiation sites in DNA sequences (Hatzigeorgious and Reckzo,2004).[10]

2. Explain the theory of neural networks using applications in biology (Baldi and Brunak,1998).

3. Predict immunologically interesting peptides by combining an evolutionary algorithm(Brusic et al., 1998).

4. Study human TAP transporter (Brusic et al., 1999).

5. Carry out pattern classification and signal processing successfully in bioinformatics; in fact, a large number of applications of neural network can be found in this area.

6. Perform protein sequence classification; neural networks are applied to protein sequence classification by extracting features from protein data and using them in combination with the Bayesian neural network (BNN) (Wu and Mclarty, 2000).

7. Predict protein secondary structure prediction (Chenand and Kurgan, 2007); Zhong et a.,2007).

8. Analyze the gene expression patterns as an alternative to hierarchical clusters (Toronen etal., 1999; Ma et al., 2000; Bicciato et al., 2001; Torkkola et al., 2001). Gene expression caneven be analyzed using a single layer neural network (Narayanan et al., 2003). Protein fold recognition using ANN and SVM (Ding and Dubchak, 2001).

In summary, a neural network is presented with a pattern on its input nodes, and an output pattern based on its learning algorithm during the training phase. Once trained, the neural network can be applied to classify new input patterns. This makes neural networks suitable for the analysis of gene expression patterns, prediction of protein structure, and other related processes in bioinformatics. If the input data are not linearly separable, a least means square solution is generated to minimize the means square error between the calculated output of the network and the actual desired output.

3.8   Neural Network for Gene Prediction

Given a sequence ACGGGGAATTCGTAGCT..., predict if it is an exon (coding region) or not. Extract features from the sequence and feed them into neural Network.

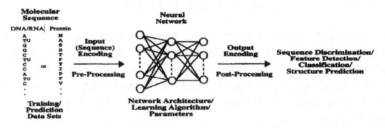

Fig. 7. Design issues of neural network applications for molecular sequence analysis

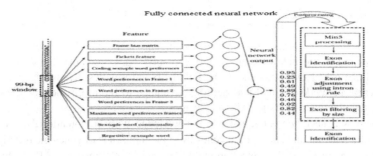

Fig. 8. Gene identification

A 99-base pair window is interrogated for a prediction of coding/noncoding on the central nucleotide. Input features are fed to the neural network, which provides an output value between -1 (noncoding) and +1(coding). Postprocessing features are used to make identification of a coding region (exon) boundary.

3.9   Grail System

Well known GRAIL software used an ANN to combine a number of coding indicators calculated within a fixed sequence window. The classification process using evolved ANNs proceeded as follows:

A sequence of DNA was interrogated using a window of 99 nucleotides. The ANN was used to classify the nucleotide in the center of the window as either coding or noncoding. For this analysis, the neural network architecture was fixed and consisted of 9 input nodes corresponding to 9 features as shown in fig: 8, 14 hidden nodes and one output node. The output decision was normalized on from -1 to +1 for each position in the sequence. If the output was less than -0.5 it was classified as coding, if it was less than +0.5 classified it as non coding. In evolved ANN, genetic algorithms have been used for determining the appropriate network architecture [11]. ANNs are combined with a rule-based system has been used for splice site prediction in human using a joint prediction scheme for local splice site assignment.

Table 1. A summary of NN methods for Protein coding Gene Prediction

Computing Technique	Organism/Data Source	Program/URL	Prediction Element
Back-propagation NN	Human, Mouse / GenBank	http://compbio.ornl.gov/ grailexp	Exons
Back-propagation NN	Human, Vertebrates /GenBank	http://beagle.colorado.ed u/~eesnyder/GeneParseer.ht ml	Exons, Introns
Back-propagation NN	Human, Mouse / GenBank	GRAIL-II	Exons
Back-propagation NN	Human, Mouse / GenBank	CODEX	Exons
Back-propagation NN	Vertebrates /GenBank	http://www.borkemblhei delberg.de/fmilpetz/GIN	Exons
Back-propagation NN	S. cerevisiae genome	MLAFANN (yeast genome)	Open reading frames

4   Conclusion

It is evident in the literature that artificial intelligence techniques like Neural Networks are heavily used in the field of bioinformatics to solve hard problems. These methods have proved and established its value in the field of bioinformatics. Knowledge and ability to use neural networks method add definite advantage to bioinformaticians to solve many types of problems in the field of bioinformatics.

There is a need to merge biological knowledge with computational techniques for extracting relevant and appropriate genes from the thousands of genes measured. There is growing interest in the application of artificial intelligence (AI) techniques in bioinformatics. In particular, there is an appreciation that many of the problems in bioinformatics need a new way of being addressed given either the intractability of current approaches or the lack of an informed and intelligent way to exploit biological data. On the other hand, artificial intelligence is an area of computer science that has been around since the 1950s, specializing in dealing with problems considered intractable by computer scientists through the use of heuristics and probabilistic approaches.

Ackknowledgement. The authors gratefully acknowledge the informative comments of the reviewers that will help in improving the quality and new sources of knowledge.

References

[1] Campbell, A.M., Hyer, L.J.: Discovering Genomics, Proteomics and Bioinformatics. Pearson Education (2004)

[2] Role of 5'- and 3'- untranslated regions of mRNAs in human diseases Sangeeta Chatterjee and Jayanta K. Pal, University of Pune, Pune 411007. India Biol. Cell 101, 251–262 (2009) (Printed in Great Britain), doi:10.1042/BC20080104

[3] Family-based Homology Detection via Pairwise Sequence Comparison, William Noble Grundy, department of Computer Science and Engineering, University of California, San Diego

[4] Gene Prediction with a Hidden Markov Model, Mario Stanke, University of Greifswald, Germany

[5] Enhancements to Hidden Markov Models for GeneFindings, Tomas Vinar, University of Waterloo, Canada

[6] Wu, C.H., McLarty, J.W.: Neural Networks and Genome Informatics, 1st edn. Methods in computational Biology and Biochemistry, vol. 1. Elsevier (2000)

[7] Haykin, S.: Neural Networks:A comprehensive Foundation Pearson Education (2002)

[8] Wu, C.H., McLarty, J.W.: Neural Networks and Genome Informatics, 1st edn. Methods in computational Biology and Biochemistry, vol. 1. Elsevier (2000)

[9] Uberacher, E.C., Xu, Y., Mural, R.J.: Discovering and understanding genes in human DNA sequence using GRAIL. Methods Enzymol. 266, 259–281 (1996)

[10] Molecular design and molecular docking (Rosin et al., 1997; Yang, Kao, 2000; Oshiro et al., 1995; Clark and Westhead, 1996; Venkata subramanian et al., 1994; Deaven and Ho,1995; Jones et al., 1995; Jones et al., 1999; cGarrah and Judson, 1993; Hou et al., 1999; Hatzigeorgiou and Reckzo, 2004)

[11] Fogel, G.B., Chellapilla, K., Corne, D.W.: Identification of coding regions in DNA sequences using evolved neural networks, pp. 195–218. Morgan Kaufmann (2002)

Breast Cancer Diagnosis: An Intelligent Detection System Using Wavelet Neural Network

V. Dheeba[1], N. Albert Singh[2], and J. Amar Pratap Singh[1]

[1] Department of Computer Science and Engineering,
Noorul Islam University, Kumaracoil, TN, India
{deeps_3u4,japsindia}@yahoo.com
[2] Bharat Sanchar Nigam Limited, Nagercoil, TN, India
mailalbertsingh@yahoo.co.in

Abstract. Breast cancer represents the leading cause of fatality among cancers for women and there is still no known way of preventing this pathology. Early detection is the only solution that allows treatment before the cancer spreads to other parts of the body. Diagnosis of breast cancer at the early stage is a very difficult task as the cancerous tumors are embedded in normal breast tissue structures. Aiming to model breast cancer prediction system, we propose a novel machine learning approach based on wavelets. The new model, called wavelet neural network (WNN), extends the existing artificial neural network by considering wavelets as activation function. The texture information in the area of interest provides important diagnostic information about the underlying biological process for the benign or malignant tissue and therefore should be included in the analysis. By exploiting the texture information, a computerized detection algorithm is developed that are not only accurate but also computationally efficient for cancer detection in mammograms. The texture features are fed to the WNN classifier for classification of malignant/benign cancers. An experimental analysis performed on a set of 216 mammograms from screening centres has shown the effectiveness of the proposed method.

Keywords: Breast Cancer, Computer Aided Diagnosis, Mammograms, Texture, Wavelet Neural Network.

1 Introduction

Breast cancer is the second frequently diagnosed cancer among women, especially in developed countries. In western countries about 53%-92% of the population has this disease. Though breast cancer leads to death, early detection of breast cancer can increase the survival rate. The current diagnostic method for early detection of breast cancer is mammography. Mammography still remains the key screening tool because it represents the most effective, low-cost and highly sensitive technique allowing the diagnosis of a breast cancer at a very early stage [1]. The mammograms are checked by the radiologist with the aim of detecting the abnormalities, but the complex

S.C. Satapathy, S.K. Udgata, and B.N. Biswal (eds.), *FICTA 2013*,
Advances in Intelligent Systems and Computing 247,
DOI: 10.1007/978-3-319-02931-3_14, © Springer International Publishing Switzerland 2014

structures and the signs of earlydisease are very small and subtle. Mammographies are low dose X-ray projections of the breast, and it is the best method for detecting cancer at the early stage.

Microcalcifications (MC) are quiet tiny bits of calcium, and may show up in clusters or in patterns and are associated with extra cell activity in breast tissue. Usually the extra cell growth is not cancerous, but sometimes tight clusters of microcalcification can indicate early breast cancer. Scattered microcalcifications are usually a sign of benign breast cancer. 80% of the MC is benign. MC in the breast shows up as white speckles on breast X-rays. The calcifications are small; usually varying from 100 micrometer to 300 micrometer, but in reality may be as large as 2mm. Though it is very difficult to detect the calcifications as such, when more than 10 calcifications are clustered together, it becomes possible to diagnose malignant disease. But the survival depends on how early the cancer is detected. So, any MC formation should be detected at the benign stage. Hence, a Computer Aided Diagnosis (CAD) system is used to detect MC clusters [6, 12, 13]. Many different algorithms have been proposed for automatic detection of breast cancer in mammograms. Features extracted from mammograms can be used for detection of cancers [2]. Studies reports that features are extracted from the individual MCs [3, 11] or from the ROI which contain MC clusters [4].

Matthew A. Kupinski et al. [9] presents a radial gradient index based algorithm and a probabilistic algorithm for detecting lesions in digital mammograms. Shape constraints are used to regularize the partitions analyzed, and simplifying the partition selection process by using utility functions based either on a single feature or probabilities. Nevine H. Eltonsy et al. [10] presents a method based on the presence of concentric layers surrounding a focal area with suspicious morphological characteristics and low relative incidence in the breast region. The suspicious focal area is localized using the morphological features and based on the minimum distance criterion the redundant focal areas are eliminated. The presence of concentric layers around the suspected focal regions is analyzed using multiple concentric layer criterions to detect the suspicious regions.

Berman Sahiner et al. used a Convolution Neural Network (CNN) classifier to classifier the masses and the normal breast tissue [5]. First, the Region of Interest (ROI) of the image is taken and it was subjected to averaging and subsampling. Second, gray level difference statistics (GLDS) and spatial gray level dependence (SGLD) features were computed from different subregions. The computed features were given as input to the CNN classifier.

Feature extraction phase is the key step in detecting abnormalities in mammograms [7]. Features are extracted from the ROI of the mammograms. Normal Texture measures includes mean, variance, etc which will be concatenated to a single feature vector and will be fed to a classifier to perform classification. The drawback in this method is much of the important information contained in the whole distribution of the feature values might be lost. MC clusters usually appear as a few pixels with brighter intensity embedded in a textured background breast tissue. By effectively extracting the texture information within any ROI of the mammogram, the region with abnormalities and the region without abnormalities can be differentiated. Laws texture energy measures (LTEM) has proven to be a successful method to highlight high energy points in the image. Anna et al. 2008 [8] suggests that LTEM has a best feature in analyzing texture of tissue for BC diagnosis. By considering the basic feature set like kurtosis, skewness, mean and Standard Deviation the accuracy achieved using LTEM is 90%.

The major objective of this paper is to take multiple texture features from the original image to discriminate between abnormal and the normal tissue in the breast. As a first stage, the original image is preprocessed and Region of Interest (ROI) is taken and features are extracted from the ROI image using Laws features. In the second stage, the extracted features are compared by means of their ability in detecting breast cancer using WNN. The database images have four different kinds of abnormalities namely: architectural distortions, stellate lesions, Circumscribed masses and calcifications. The proposed method is capable of detecting these abnormalities.

2   Image Preprocessing

All clinical mammograms used for this study were collected from screening clinics were positive for presence of microcalcifications. Mammograms were collected from 54 patients and all these patients have agreed to have their mammograms to be used in research studies. For each patient 4 mammograms were taken in two different views, one is the Craniocaudal (CC) and the other is the Mediolateral Oblique (MLO) view. The two projections of each breast (right and left) were taken for every case. For this study a total of 216 mammograms were taken, all the mammograms were digitized to a resolution of 290 x 290 Dots per Inch (DPI) which produces 24 bits/pixel. Each digitized mammograms was incorporated into a 2020 x 2708 pixel image (5.47 Mpixels).

The goal of preprocessing the image is to simplify recognition of abnormality without throwing away any important information. As a preprocessing step the breast area is separated from the background image. The breast area is chosen as ROI for the next stage of processing. This saves the processing rime and also the memory space. The block diagram of the proposed breast cancer detection methodology is shown in Fig. 1.

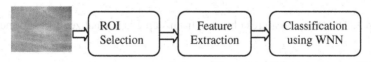

Fig. 1. Block Diagram of Proposed Methodology

3   Feature Extraction Methodology

3.1   Texture Feature Extraction

In image processing the texture of a region describes the pattern of spatial variation of gray tones (or in the different color bands in a color image) in a neighborhood that is small compared to the region. Texture features were extracted from the ROI in order to be used for classification of abnormalities. In image processing the texture of a region describes the pattern of spatial variation of gray tones in a neighborhood that is small compared to the region. By definition, texture classification is to identify the texture class in a region. The texture energy measures developed by Kenneth Ivan Laws [7] at the University of Southern California have been used for many diverse applications. These texture features are used to extract texture energy measures (TEM) from the ROI containing abnormalities and normal breast tissues [8]. These measures are computed by first applying small convolution kernels to the ROI and then performing a windowing operation. A set of 25, 5 x 5 convolution masks is used to compute texture energy, which is then represented by a vector of nine numbers for each pixel of the image being analyzed. The 2-D convolution kernels for texture discrimination are generated from the following set of 1-D convolution kernels of length five. The texture descriptions used are level, edge, spot, wave and ripple.

$$L5 = \begin{bmatrix} 1 & 4 & 6 & 4 & 1 \end{bmatrix}$$
$$E5 = \begin{bmatrix} -1 & -2 & 0 & 2 & 1 \end{bmatrix}$$
$$S5 = \begin{bmatrix} -1 & 0 & 2 & 0 & -1 \end{bmatrix}$$
$$W5 = \begin{bmatrix} -1 & 2 & 0 & -2 & 1 \end{bmatrix}$$
$$R5 = \begin{bmatrix} 1 & -4 & 6 & -4 & 1 \end{bmatrix}$$

From this above 1-D convolution kernels 25 different two dimensional convolution kernels are generated by convoluting a vertical 1-D kernel with a horizontal 1-D kernel. Texture energy measures are identified for each pixel in the ROI of a mammogram image by means of the following steps.

Step 1: Apply the two dimensional mask to the preprocessed image i.e. the ROI to get $F(i, j)$, where $F(i, j)$ is a set of 25 $N \times M$ features.

Step 2: To generate the TEM at the pixel, a non-linear filter is applied to $F(i, j)$. The local neighborhood of each pixel is taken and the absolute values of the neighborhood pixels are summed together. A 15×15 square matrix is taken for doing this operation to smooth over the gaps between the texture edges and other micro-features. The TEM features are obtained using equation (1)

$$E(x, y) = \sum_{j=-7}^{7} \sum_{i=-7}^{7} |F(x+i, y+j)| \tag{1}$$

Step 3: The texture features obtained from step 2 is normalized for zero-mean.

4   Proposed WNN Classifier

Recently, Wavelet Neural Network (WNN) have found many applications in function approximation, pattern recognition and signal processing. Wavelets have many desired properties like compact support, orthogonality, localization in time and frequency and fast algorithms. Wavelet networks are a class of neural networks that employ wavelets as activation functions [14, 15]. To identify the true cancer pixels (abnormality) in the mammograms, a good and optimized classification method has to be employed.Wavelets are functions that satisfy certain mathematical requirements and are used in representing data or other functions. These algorithms process data at different scales or resolutions. A wavelet is a real or complex valued function $\Psi(.)$ satisfying the following conditions,

$$\int_{-\infty}^{\infty} \Psi(u)du = 0 \text{ and } \int_{-\infty}^{\infty} |\Psi^2(u)|du = 1 \tag{2}$$

There are two functions in wavelet transform namely the scale function and the mother wavelet. Wavelets are powerful signal analysis tools. They can approximately realize the time-frequency analysis using a mother wavelet. The mother wavelet has a square window in the time- frequency space. Wavelet Neural Networks are designed by employing wavelet as hidden layer activation function. The proposed Wavelet Neural Network is a three-layer structure with input layer, hidden layer and output layer. Let Ψ be the activation function and is defined as the sum of the weighted inputs plus the bias and is represented as,

$$y_k^p = \Psi(s_k^p) \tag{3}$$

Where $s_k^p = \sum_j w_{j,k} y_j^p + \theta_k$, y_k^p is the output of the kth neuron when a pattern

p is fed, $w_{j,k}$ is the weight from the jth neuron and θ_k is the bias value of the kth neuron in the hidden layer and it is defined by wavelet activation function.

5   Experimental Results

The performance of the proposed methodology was tested on digitized mammograms collected from mammogram screening centers. The WNN algorithm classifies the input image into malignant and benign regions. For the classification experiments, the training dataset contain a total of 1064 patterns from the real database. These patterns contains pixels including true individual microcalcification clusters, circumscribed masses, ill defined masses and also pixels indicating normal tissues that includes blood vessels and dense breast tissues. If a mammogram is denser the margins are obscured and not clearly seen. Hence it is difficult for a radiologist to identify the mass in a denser tissue. Fig. 2 shows the detection of a mass in an obscured mammogram using the WNN classifier.

(a)   (b)   (c)

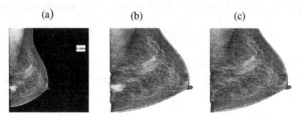

Fig. 2. Detection results of masses in mammogram (a) Original Image (b) ROI image (c) Classified Abnormality Region

Fig. 2 (a) shows the original image in the MLO view of the left breast mammogram and (b) shows the result after segmentation of ROI. Fig. 2 (c) shows the detected malignant masses in a denser mammogram. The real time mammograms are taken in varying intensities and hence detecting abnormalities in mammograms is a difficult task.

The main aim of the proposed system was that no case of malignancy-indicating microcalcification should escape radiological analysis. We therefore started from two basic assumptions: (i) the microcalcifications have an aspect that differentiates them from the other elements of the breast because of their different X-ray opacity; and (ii) since we are looking for microcalcifications that are in an incipient stage, they involve a very small proportion of the total area of the breast because they otherwise would be clearly visible to any radiologist and there would consequently be no point in using our system. Microcalcification clusters are tiny calcium deposits and these clusters that fail to demonstrate features characteristic of benignity have to be evaluated to determine for malignancy and their exact location in the breast. A mammographically significant cluster is usually considered to be 3-10 calcific particles within a volume of 1 cm^2.

(a)   (b)   (c)

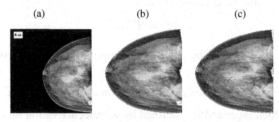

Fig. 3. Detection results of abnormality in real clinical Mammogram using WNN (a) Original Image (b) ROI image (c) Classified Abnormality Region

Fig. 3 (a) shows the ROI image of a digital mammogram in and Fig. 3 (b) shows the WNN classified microcalcification marked with red circles. The detected regions in a mammogram corresponding to tumor tissues have different texture patterns and gray levels than the normal ones and by employing WNN classifier it is possible to classify these regions. Several metrics can be determined for quantitative evaluations of the intelligent classifier. Receiver operating characteristic (ROC) curve is drawn with the sensitivity and specificity values. Fig. 4. shows the ROC curve for the proposed method. Sensitivity measures the proportion of actual positives which are correctly identified when the mammogram contains cancers tissues in it. Specificity measures the proportion of negatives which are correctly identified when cancer is not present in the mammogram. The following statistics can be defined,

$$sensitivity = \frac{TP}{(TP+FN)} \qquad specificity = \frac{TN}{(TN+FP)}$$

The overall performance of diagnostic systems has been measured and reported in terms of classification accuracy which is the percentage of diagnostic decisions that proved to be correct.

$$Accuracy = \frac{(TP + TN)}{(TP + FP + TN + FN)}$$

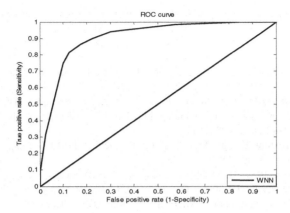

Fig. 4. ROC Curve for WNN Classifier

The WNN approach generates a sensitivity of 86.441 % with a specificity of 82.5 %. The misclassification rate is found to be 0.14557. When optimized learning is introduced, there is an improvement in classification accuracy which is 85.443%. Therefore, WNN have a great potential to be applied in automatic detection of abnormalities in mammograms by reducing the misclassification rate.

6   Conclusion

The novel approach presented in this paper demonstrated that the WNN classifier produces an improvement in classification accuracy to the problem of computer-aided analysis of digital mammograms for breast cancer detection. The algorithm developed here classifies mammograms into normal & abnormal. First, the ROI of the image is chosen then Laws features are extracted and classified using wavelet neural networks. Using the mammographic data from the real clinical centers a classification accuracy of 85.443% was achieved using the proposed approach. In our future work we would like to focus on expanding this research to find the appropriate rate for learning and momentum and test on a larger real time dataset using optimization technique applied to the WNN classifier. This might further improve classification accuracy.

References

1. Pisani: Outcome of screening by Clinical Examination of the Breast in a Trial in the Phillipines. Int. J. Cancer (2006)
2. Shen, L., Rangaan, R.M., Desautels, J.E.L.: Application of shape analysis to mammographic classifications. IEEE Trans. Med. Imag. 13(2), 263–274 (1994)
3. Lee, S.K., Chung, P., Chang, C.L., Lo, C.S., Lee, T., Hsu, G.C., Ang, C.W.: Classification of Clustered Microcalcifications using shape cognitron neural network. Neural Netw. 16(1), 121–132 (2003)

4. Dhawan, A.P., Chitre, C.K.-B., Moskoitz, M.: Analysis of mammographic microcalcification using gray-level image structure features. IEEE Trans. Med. Imag. 15(3), 11–150 (2005)
5. Sahiner, B., Chan, H.-P., Pretrick, N., Wei, D., Mark, A., Dorit, D., Mitchell, M.: Classification of Mass and Normal Breast Tissue A convolution Neural Network Classifier with Spatial Domain and Texture Images. IEEE Trans. Med. Imaging 15(5), 598–609 (1996)
6. Hu, K., Gao, X., Li, F.: Detection of Suspicious Lesions by Adaptive Threshoding Based on Multiresolution Analysis in Mammograms. IEEE Transactions on Instrumentation and Measurement 60(2), 462–472 (2011)
7. Laws, K.J.: Texture Energy Measures. In: Proceeding DARPA Image Understanding Workshop, pp. 47–51 (1979)
8. Karahaliou, A.N., Boniatis, I.S., Skiadopoulos, S.G., Sakellaropoulos, F.N., Arikidis, N.S., Likaki, E.A., Panayiotakis, G.S., Costaridou, L.I.: Breast Cancer Diagnosis: Analyzing Texture of Tissue Surrounding Microcalcifications. IEEE Transactions on Information Technology in Biomedicine 12(6), 731–738 (2008)
9. Kupinski, M.A., Giger, M.L.: Automated Seeded Lesion Segmentation on Digital Mammograms. IEEE Transactions on Medical Imaging 17(4), 510–517 (1998)
10. Eltonsy, N.H., Tourassi, G.D., Elmaghraby, A.S.: A Concentric Morphology Model for the Detection of Masses in Mammography. IEEE Transactions on Medical Imaging 26(6), 880–889 (2007)
11. Wong, A., Scharcanski, J.: Phase-Adaptive Superresolution of Mammographic Images Using Complex Wavelet. IEEE Transactions on Image Processing 18(5), 1140–1146 (2009)
12. Tsui, P.-H., Liao, Y.-Y., Chang, C.-C., Kuo, W.-H., Chang, K.-J., Yeh, C.-K.: Classification of Benign and Malignant Breast Tumors by 2-D Analysis Based on Contour Description and Scatterer Characterization. IEEE Transaction on Medical Imaging 29(2), 513–522 (2010)
13. Mencattini, A., Salmeri, M., Rabottino, G., Salicone, S.: Metrological Characterization of a CADx System for the Classification of Breast Masses in Mammograms. IEEE Transactions on Instrumentation And Measurement 59(11), 2792–2799 (2010)
14. Rying, E.A., Bilbro, G.L., Lu, J.-C.: Focused Local Learning with Wavelet Neural Networks. IEEE Transactions on Neural Networks 13, 304–319 (2002)
15. Zhang, Q., Benveniste, A.: Wavelet Networks. IEEE Transactions on Neural Networks 3, 889–898 (1992)

Efficient UMTS Location Update and Management Based on Wavelet Neural Networks

J. Amar Pratap Singh[1], J. Dheeba[1], and N. Albert Singh[2]

[1] Department of Computer Science and Engineering,
Noorul Islam University, Kumaracoil, TN, India
{japsindia,deeps_3u4}@yahoo.com
[2] Bharat Sanchar Nigam Limited, Nagercoil, TN, India
mailalbertsingh@yahoo.co.in

Abstract. Mobility management in wireless cellular networks is gaining more attention because of the increase in usage of mobile devices. As the number of mobile users increases rapidly, there is a need for efficient location management. Location management (LM) tracks mobile devices and locate them prior to establishing incoming calls. As the cell size becomes smaller the signalling cost increased in both location update and paging. In order to deal with the signalling cost issues, an efficient user activity based location management technique (UALM) is introduced in this paper. UALM is a profile based LM scheme, where the network takes the users past movement pattern and makes decisions on the future update and paging. This paper presents a novel intelligent Wavelet Neural Network (WNN) based UALM learning strategy to solve the LM problem in UMTS networks. A systematic comparative analysis is made with the existing location management schemes. The results show that the proposed WNN based UALM learning decreases the signalling cost. The proposed technique has the potential to reduce the network signalling cost and total LM cost that must be made to locate the users.

Keywords: Location Management, Wireless Networks, Wavelet Neural network, Location update, paging.

1 Introduction

Wireless mobile applications and usage of mobile devices have recently become more popular and has made a significant impact on our daily life. As the cell becomes smaller and the number of mobile user increases, the real challenge lies in providing reliable communication during movement of mobile terminals (MT). The protocol that has been adopted by the wireless systems is IS-41 [1] and GSM (Global System for Mobile Communication) [2]. Both these protocols deploy a two tier database architecture consisting of a single Home Location Register (HLR) and several Visitor Location Registers (VLR). The HLR will contain the network's subscriber profile for a network and the VLR stores the information of the user that are currently roaming within the Location Areas associated with that specific VLR.

S.C. Satapathy, S.K. Udgata, and B.N. Biswal (eds.), *FICTA 2013*,   119
Advances in Intelligent Systems and Computing 247,
DOI: 10.1007/978-3-319-02931-3_15, © Springer International Publishing Switzerland 2014

In wireless mobile networks the cells are grouped into location areas (LAs). LM enables the wireless mobile network to find the current point of attachment of a mobile terminal and directs call to it. When a MT moves from one location area (old LA) to a new location area (new LA), location update (LU) procedure is done to update the corresponding new location in the home location register (HLR). HLR can thus help the network to locate the users current location and deliver the messages to them upon call arrival. The user is simultaneously paged in all cells of its currently visited LA. This LA based mobility management scheme possesses two disadvantages. Firstly, the MT performs frequent LU when the MT resides in the boundary cell of the LA. This causes additional data traffic and unnecessary location updates particularly for the users located near the LA boundaries. Secondly, when an incoming call arrives the network is queried for the current position of a MT. When the LA has large number of cells, it may incur a large paging traffic to search the MT.

Recently many researches have been made on location management. LM is mainly categorized into two types: static and dynamic schemes. One major weakness with the static location update is that an MT's individual calling and mobility characteristics are not taken into account in location update decisions. Dynamic Location Management is a superior form of LM where the parameters of LM can be adapted to best fit individual users and conditions. Several dynamic schemes were proposed to overcome the shortcomings in static scheme. These mainly include the time-based scheme [1], [2], movement based scheme [3], [4], [5], [6], distance-based scheme [7], predictive distance based scheme [8] and state-based update scheme [9].

Javid and Albert (2007) [11], has proposed a dynamic location management scheme by combining both clustering and location area scheme. Alenjandro Quintero (2005) [10], used a User Profile Learning Strategy (UPL) to reduce the overhead in both location update and paging in UMTS networks. Wenchao Ma et al. (2007) [12], presented the User Mobility Pattern scheme to minimize the location update and paging traffic burden. The basic idea is to incorporate the time information in the mobility pattern profile; thus, different location update and paging schemes can be used based on the different user states.

In this proposed strategy the users are classified using the call to mobileity ratio. It is an alternative paging and location update algorithm to reduce the LM cost. Currently cellular system used geographic based registration in which registration occurs when the user crosses an LA. Reduction of signalling and database access traffic constitutes an important research challenge. Several strategies has been used to improve the 3G location management scheme [10,12]. The ultimate goal of this work is to predict the location area into which the mobile user will probably move. In this paper, a UALM learning strategy is used to reduce signalling cost of a location update by increasing the intelligence in updating and paging procedure in UMTS networks.

2   UMTS Architecture

Universal Mobile Telecommunications Service (UMTS) can be viewed as an evolution of GSM that supports 3G services. Generally, a UMTS network is divided in an access network and a core network as shown in Fig. 1. The former is dependent on the access technology, while the latter can theoretically handle different access networks. The access network is known as the UMTS Terrestrial Radio Access Network (UTRAN). The UTRAN is comprised of two types of nodes, the Radio Network Controller (RNC) and the Node B, which is a base station (BS). It controls the radio resources within the network and can interface with one or more stations (Node Bs). The air interface used between the user equipment (UE) and the UTRAN is WCDMA. The UTRAN communicates with the core network over the lu interface. The lu interface has two components: the lu-CS interface, supporting circuit-switched (CS) services, and the lu-PS interface, for packet-switched (PS) services.

The RNC that controls a given Node B is known as the Controlling RNC (CRNC). For a given connection between a UE and the core network, only one RNC can be in control, the Serving RNC (SRNC). The CRNC controls the management of radio resources for the Node B that it supports. The SRNC controls the radio resources that the UE is using. It is possible for a CRNC and a SRNC to coincide. UTRAN supports soft handovers (where an UE is communicating with a Node B whose CRNC is not the SRNC). The lur interface's purpose is to support this type of handover, that is, inter-RNC mobility.

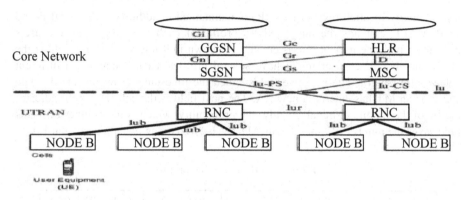

Fig. 1. UMTS Network Architecture

Location area (LA) - The current location area ID is available in home location register (HLR) and visitor locations register (VLR) because an HLR is a centralized database that maintains permanent information of mobile users, whereas VLR keeps the up-to-date locations of visiting mobile terminals.

Routing Area (RA) – It is composed of a group of cells that belong to only one RA. Several RAs can be included in the same LA, but an RA cannot span more than one LA. The PS domain to track the MT's location when in idle mode uses the RAs.

UMTS Registration Areas (URA) – They are an intermediate level between cells and RAs (or LAs). They are similar to RAs and LAs, but are used by the UTRAN to set trade-offs between the MT's location accuracy and signaling load. Furthermore, they are used to track the MT's location while it is in connected mode without using a logical channel. This concept is optional in UTRAN.

Cell ID - In order to maintain an active connection for a mobile user, it is most important to know in which cell the mobile user is located. The network knows in which cell an MT resides by sending polling messages, which can be acquired without additional cost during call origination/termination, or through location client service (LCS) management.

3   UALM Learning

UALM the network uses the past routine of the users to predict their current location. This basically involves two stages (i) acquiring users daily routine history. (ii) using intelligent system to identify the users based on the past history. The users are separated into classes as in [13]. Classes are divided as, (1) Deterministic users, follows a rigid daily routine that the system knows. (2) Quasi-deterministic users can be found in work during a certain time and at home during another time interval. The time intervals vary slightly from day to day and (3) Random users display no orderly behaviour whatsoever. The example users are given below,

C1: High speed car passengers (Quasi Deterministic Users)
C2: Office employers (Deterministic users)
C3: Sales and Business People (Quasi Deterministic Users)
C4: house wives and stable users (Random Users)

Users activity location register (UALR) is maintained in addition with the HLR and VLR. UALR contains the daily activity profile of each user. Based on the users movement path the user profile is created as shown in Table 1. The first field is the profile number. Each user might have several profiles, each containing a different behavioural pattern. The next two fields called the timestamp field indicate the date and time an MT has entered a new LA. The next field is the cell ID which indicates the ID of the mobile terminal. Finally, the probability, which is the number of Visits field saves the number of times that an MT has been roaming under a particular LA.

Table 1. An Example User profile

Examples	D_i(Day)	T_i(Time)	C_i(Cell Id)	Probability (%)
E1	Monday	02.15	5,1,6	90,50,10
E2	Monday	11.45	5,2,1	95,50,10
E3	Sunday	12.00	5,6,3	40,90,15
E4	Thursday	14.15	5,2,1	95,50,10
E5	Thursday	05.30	5,1,6	90,50,10

The time-normalized cost of the UPL is decomposed into three components.

- Location updating is assumed to occur on a geographic basis. If a user is inside the Location Area (LA), it notifies the system only if it leaves the area covered by the LA. When it is outside LA, it notifies the system whenever it crosses a location area boundary.
- Paging allows calls to be delivered to mobile terminals. When a user is inside LA, it is paged sequentially using the ordering within a list. If it is outside LA, it is paged in the location area where it last registered. In this work, we assume that if a terminal is present, it will be found when paged.
- List maintenance updates the information in the list of likely location areas.

4   Location Management using WNN

The users profile is learned using the WNN to predict the location of the user. Wavelet Neural Networks combine the theory of wavelets and neural networks into one. A wavelet neural network generally consists of a feed-forward neural network, with one hidden layer, whose activation functions are drawn from an orthonormal wavelet family. The hidden layer consists of neurons, whose activation functions are drawn from a wavelet basis. These wavelet neurons are usually referred to as wavelons.

The WNN consists of three layers: input layer, hidden layer and output layer. All the units in each layer are fully connected to the nodes in the next layer. The input layer receives the input variable $X = [x_1, x_2,..., x_n]^T$ and send it to the hidden layer. The nodes in this layer are given as the product of the jth multi-dimensional wavelet with N input dimensions as

$$\psi_j(x) = \prod_{i=1}^{N} f_{\lambda,t}(x_i) \tag{1}$$

$f_{\lambda,t}$ is the activation function of the hidden layer and λ and t are the dilations and translations respectively. The products of the hidden layer are then propagated to the output layer, where the output of the WNN will be the linear combination of the weighted sum of the hidden layer, which is represented in the form of,

$$y_j = \sum_i w_{ij} \psi_j(x) + b_j \tag{2}$$

where b_j the bias of node j between the hidden layer and the output layer $\psi_j(x)$ is taken as a Daubechies mother wavelet. The error is calculated by finding the difference between the target output (d_j) and the actual output (y_j).

$$E_j = \frac{1}{2}\sum_j (d_j - y_j)^2 \tag{3}$$

This error is then used mathematically to change the weights in such a way that the error will get smaller. The process is repeated again and again until the error is minimal. The localized wavelet activation function in the hidden layer of the WNN, the connection weight associated with the hidden nodes can be viewed as local piecewise constant models, which leads to learning efficiency and structure transparency.

5   Performance Evaluation and Results

The simulation models both the call delivery and mobility behavior of users offering the ability to consider different service types and different MT groups over a range of cell-layout scenarios. A medium sized cells (1 Km^2), is taken for simulation. In these experiments, 1,500 MTs are simulated and we generate 1,000 samples for each cell layout scenario assuming normal distributions using the statistics estimated from the real data. The simulations are made to run for the probability of a user being roaming within his associated list from 0.5 to 0.99. The user is assumed to be within that area covered by its list at least half of time. In the case of users whose position at a given moment is unpredictable and the past knowledge of their location cannot predict their future location, our strategy is not applicable. The call arrival rate (calls/MT/hour) is 3. The MTs are assumed to be moving at the same velocity V and their direction movement is uniformly distributed over [0, 2π]. MT is categorized based on normal distribution estimated from real data. The MT stays in a particular cell with a probability p and moves to one of its neighboring cells with the probability (1-p). A sample of 1500 Mobile terminals is considered roaming with a probability distribution for simulation period is one week. The mobility and the traffic conditions of the mobile users are characterized based on [13]. Several parameters that will be used for the simulation scenarios are given below:

C_{HLR} : Cost to access the HLR, C_{VLR} : Cost to access the VLR, C_{msg_f} : Cost of sending a message through the fixed network, $C_{msg_{wir}}$: Cost of sending a message through the wireless network , C_{LU} : Cost of Location Update, C_P : Cost of paging, C_{LM} : Total location management cost.

The location management cost involves the database access costs and signaling costs. The location update cost (C_{LU}) is defined as the cost of accessing the HLR, old VLR and new VLR and signaling cost to send authentication messages and location area identifier messages.

$$C_{LU} = C_{HLR} + 2*C_{VLR} + 4*(C_{msg_f} + C_{msg_{wir}})$$ (4)

The paging cost (C_P) is defines as the cost required for accessing the database at HLR and VLR to search the users in the specified location area and the signaling costs through the fixed and wireless network.

$$C_P = C_{HLR} + C_{VLR} + 2*(C_{msg_f} + C_{msg_{wir}})$$ (5)

The total cost (C) is calculated by adding the location update cost multiplied with average number of times the users changes the location area per time unit (μ) and paging cost multiplied with average number of calls to a target MT per time unit (λ). LU is the location update cost and P is the paging cost.

$$C = \mu LU + \lambda P$$

The concept of call to mobility ratio as described in [14] is considered for performance evaluation. CMR is the ratio of the average number of calls to a target MT per time unit (λ) to the average number of times the users changes the location area per time unit (μ). the value of CMR is varied from 1 to 5. We compare the proposed method with the standard UMTS procedure.

K : is the probability that the users adheres to the profile. for deterministic users the value will be greater than 0.8, for quasi-deterministic users the values will be between 0.5 and 0.8 and for the random users the values will be below 0.5.

The inputs to the WNN is the *Day(Di)* and the *Time (Ti)* resulting the Mobile Cell Identity and the probability (P1,P2 & P3) of the user's presence in the Cell Id. The number of neurons used in hidden layers is 100 and that of in the output layer is six. The hidden layer neuron is defined by wavelet activation function and the output layer neuron is defined by linear activation function. We use the Backpropagation algorithm to implement the learning process. In this problem, the learning process aims to derive a list with which we can find the cell in which the MT locates at any time of every day with high accuracy after observing the behaviour of the mobile user for a period, for example, a month. Fig. 2. shows the results using WNN technique for deterministic Users. When the value of K is 0.9, the relative cost of the proposed technique is lower. The results shows that in UPL strategy, when K=0.9 most of the time the profile identification is transferred rather than updates to the HLR. As the CMR increases the total cost also increases. When the CMR level is 1 is a drastic reduction in the cost of the location update and paging. This shows that when the users are not moving the LM cost is less and hence its clear that the proposed WNN algorithm avoids unnecessary location updates.

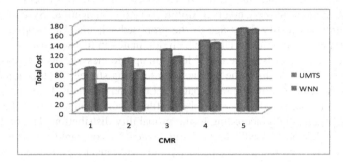

Fig. 2. Total Cost Comparison for Deterministic Users (K=0.9)

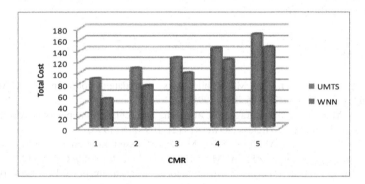

Fig. 3. Total Cost Comparison for Quasi Deterministic Users (K=0.7)

For each configuration, we observe that, when the location search cost dominates, the cost of our scheme increases as CMR values increase. This was to be expected since user pattern learning strategies tend to reduce location update costs at the expense of higher paging costs. The advantage of UALM is that the total cost is reduced by reducing the location update cost. In this strategy the search cost will be high if the user does not adhere to the profile. The reason is that paging is done sequentially according to the user's probability distribution; Although UALM scheme has a higher paging cost, when we examine the total cost in comparison with the conventional standard UMTS, the proposed method gives a better performance. The proposed algorithm outperforms the UMTS standard for every CMR between 1 and 5. The results obtained from our performance evaluation confirm the efficiency and the effectiveness of WNN obtaining UALM strategy in comparison with the UMTS standard and other well-known strategy. This improvement represents a large reduction in location update and paging signalling costs. This scheme is not applicable for random users because these users display no orderly behaviour whatsoever. Past knowledge of their location cannot predict their future location.

6   Conclusion

In this paper, a novel scheme using WNN based on UALM strategy is used for location management in the Universal Mobile Telecommunications System. We have used simulations to justify the usage of the proposed scheme. The propsoed method avoids the expensive queries to HLR, which dominates in this scenario. As more and more mobile users' increases, the users' location update message also increases which consumes more bandwidth. In the user activity learning strategy no location update is necessary if the users follow the same pattern. Location update is done only when the users goes to the location that is not in the profile. When a call arrives to the user, the user is paged sequentially according to the probability distribution. The performance of the proposed WNN based location management is compared with the standard UMTS procedure. The results obtained from our performance evaluation confirm the efficiency and the effectiveness of WNN based on UALM in comparison with the UMTS standard. This improvement represents a large reduction in location update and paging signalling costs.

References

1. Bar-Noy, A., Kessler, I., Sidi, M.: Mobile Users: To update or not to update? ACM Balter J. Wireless Networks 1(2), 175–186 (1995)
2. Rose, C.: Minimizing the average cost of paging and registration approach: A time based method. ACM-Baltzer. Wireless Networks 2(2), 109–116 (1996)
3. Akyildiz, I.F., Ho, J.S.M., Lin, Y.B.: Movement based location update and selective paging for PCS networks. IEEE/ACM Trans. Networking 4(4), 629–638 (1996)
4. Li, J., Pan, Y., Jia, X.: Analysis of dynamic location management for PCS network. IEEE Trans. Vehicular Technology 51(5), 1109–1119 (2002)
5. Mao, Z., Douligeris, C.: A location based mobility tracking scheme for PCS networks. Computer Communication 23(18), 1729–1739 (2000)
6. Xiao, Y.: Optimal Fractional movement based scheme for PCS location management. IEEE Comm. Letters 7(2), 67–69 (2003)
7. Wong, V., Leung, V.: An adaptive distance based location update algorithm for next generation PCS networks. IEEE J. Selected Areas in Comm. 19(10), 1942–1952 (2001)
8. Liang, B., Haas, Z.J.: Predictive distance-based mobility management for PCS networks. In: Proceedings of the IEEE Eighteenth Annual Joint Conference of the IEEE Computer and Communications Societies, vol. 3, pp. 1377–1384. IEEE (1999)
9. Wong, V.W.S., Leung, V.C.M.: Location management for next-generation personal communications network. IEEE Network 14(5), 18–24 (2000)
10. Quintero, A.: A User Pattern Learning Strategy for Managing Users' Mobility in UMTS Networks. IEEE Trans. on Mobile Comuting 4(6), 552–566 (2005)

11. Lyberopoulos, G.L.: Intelligent Paging Strategies for Third Generation Mobile Telecommunication Systems. IEEE Transactions on Vehicular Technology 44(3), 543–554 (1995)
12. Awduche, D.O., Ganz, A., Gaylord, A.: An optimal search strategy for mobile stations in wireless networks. In: 5th IEEE International Conference on Universal Personal Communications, vol. 2, pp. 946–950 (1996)
13. Pollini, G.P., Chih-Lin, I.: A profile-based location strategy and its performance. IEEE Journal on Selected Areas on Communication 15(8), 1415–1424 (1997)
14. Jain, R., et al.: A caching strategy to reduce networks impacts of PCS. IEEE Journal on Selected Areas on Communication 12, 1434–1444 (1994)

Discrimination between Alzheimer's Disease, Mild Cognitive Impairment and Normal Aging Using ANN Based MR Brain Image Segmentation

Tinu Varghese[1], R. Sheela Kumari[2], P.S. Mathuranath[3], and N. Albert Singh[1]

[1] Noorul Islam University, Kumara coil, Thuckalay, Tamilnadu
tinuannevarghese@gmail.com, albertsingh@rediffmail.com
[2] Sree Chitra Tirunal Institute for Medical Science and Technology,
Trivandrum, Kerala
sheela82nair@gmail.com
[3] Department of Neurology, Sree Chitra Tirunal Institute for Medical Science and Technology,
Trivandrum, Kerala
mathu@sctimst.ac.in

Abstract. Alzheimer's disease (AD) is the most common cause of dementia among people aged 60 years and older. Mild Cognitive impairment (MCI) is a pre-dementia condition that has been shown to have a high likelihood of progression to AD. In this prospective study evaluate the accuracy of the GM and CSF volumetry to help distinguish between patients with AD and MCI and subjects with elderly controls. This study we explored the ability of BP-ANN identify the structural changes of Grey Matter (GM), White Matter (WM) and Cerebrospinal fluid (CSF) in different groups using real MR images. The proposed approach employs morphological operations used for skull stripping and gabor filter for feature extraction. In these results we report a statistically significant trend towards accelerated GM volume loss in the MCI group compared to the NCI and AD from the MCI. We report the results of the classification accuracies on both training and test images are up to 96%.

Keywords: Alzheimer's disease, Mild Cognitive Impairment, No Cognitive impairment, Artificial Neural network, Magnetic Resonance Imaging.

1 Introduction

Alzheimer's disease (AD In) is a complex and progressive neuro degenerative disorder [1] .The main clinical feature of AD is increasing memory impairment followed by impairment of other cognitive domains, a characteristic pathological cortical and hippocampal atrophy, histological feature of senile plaques of amyloid deposits and neurofibrillary tangles consisting of intraneuronal tau fibrillary tangles. Neurofibrillary tangles and amyloid plaques are the histopathological hallmark of AD and are associated with neuronal loss and brain volume reductions [2]. The concept of MCI is a midway between normal aging and very early AD. It provides a window for intervention in the

S.C. Satapathy, S.K. Udgata, and B.N. Biswal (eds.), *FICTA 2013*,
Advances in Intelligent Systems and Computing 247,
DOI: 10.1007/978-3-319-02931-3_16, © Springer International Publishing Switzerland 2014

preclinical stage of dementia and thereby for possible prevention of dementia [3]. Early diagnosis of AD allows time to plan for the future and to treat patients before marked deterioration occurs. Structural imaging biomarkers can be used to predict those who are at risk or in preclinical stages of AD. It could possibly be useful even in predicting the conversion of Mild Cognitive Impairment (MCI) an early stage of AD to AD [4-5]. Three-dimensional MR imaging technique allows visualization of subtle anatomic changes and thus can help detect brain atrophy in the initial stages of the disease

The manual segmentation is both time consuming and subject to manual variations. The automatic segmentation techniques are become essentially important for morphometric analyses of the brain. In this prospective study, automated brain segmentation algorithms segment real MR images in to different tissue classes such as gray matter, white matter and cerebrospinal fluid. K-means and FCM are the unsupervised clustering techniques are widely used segmentation of brain MR images. But the main drawback of FCM is that the intensity values of background and CSF are almost same. Hence, the efficiency of the algorithm is considerably reduced in the case of noisy MR images and leads to some misclassification[6]. In such case, a supervised segmentation method utilizes artificial neural Network approaches are needed for tissue classification.

Neural network are the information processing paradigm inspired by biological nervous systems, such as our brain. These are configured for a specific application, such as pattern recognition or data classification, through a learning process. Neural networks are sometimes called machine learning algorithms, because changing of its connection weights (training) causes the network to learn the solution to a problem. The strength of connection between the neurons is stored as a weight-value for the specific connection. The system learns new knowledge by adjusting these connection weights. The learning ability of a neural network is determined by its architecture and by the algorithmic method chosen for training[7-8].

This paper presents current work a three layer back propagation neural network is employed for the classification task. This segmentation which results in the subdivision of an entire image into its constituent regions such as GM, WM and CSF. In research, different machine learning approaches such as neural network and fuzzy are used for the classification of different stages of AD. In the preclinical stage of AD, atrophy can be originated in the hippocampus, temporal lobe and entorhinal cortex[9]. Most of the brain volumetric measurements were significantly correlated with cognitive impairment, but in the group of NCI they were correlated with age. This study we explored the ability of BP-ANN identify the structural changes of Grey Matter (GM), White Matter (WM) and Cerebrospinal fluid (CSF) in AD patients using real MR images. Early recognition of AD allows time to plan for the future and to treat patients before marked deterioration occurs.Thus, the purpose of our study was to prospectively evaluate the accuracy of automated GM volumetry to distinguish patients with AD, MCI, and NCI, by using BP-NN with the reference standard.

2   Materials and Methods

2.1   Samples and Recruitment

All Participants in this study were selected in the Sree Chitra Tirunal Institute for Medical Science and Technology (SCTIMST), Trivandrum, Kerala dementia clinic. The subjects were underwent clinical examination, detailed neuropsychological and neuroimaging evaluation at baseline and after two year.25 MCI patients ,25 AD and 20 elderly controls were selected with a clinical diagnosis NINCDS/A-DRDA criteria. The MCI patients ranged in age from 52 to 75 years and the average score of Mini Mental State Examination (MMSE) was 23 ±3. 20 healthy volunteers group matched to patients on age and education and MMSE score of 28±2and 25 AD patients   was MMSE score on 20 ± 2.7.

2.2   MRI Acquisition and Preprocessing

Whole brain MRI scans were obtained on Siemens Magnetom-Avanto SQ engine, 1.5T MR Scanner. Whole brain volume was acquired by the 3D flash spoiled gradient echo sequence using standard parameters.TR=11msec, TE=4.95, flip angle=150, slice thickness=1mm, matrix=256x256, 112 axial plane images were made to cover the whole brain. The images were post processed in the fully equipped Brain mapping unit of Cognitive and Behavior Neurology Section (CBNS).

2.3   Flow Chart of the Proposed System

Our proposed system (fig:1) shows the different steps involved in the segmentation and volumetric analysis of real MR images in south indian population. Preprocessing plays a vital role in any image processing application. Preprocessing involves the major phase in medical image application and is known as skull stripping. Skull stripping removes the non cerebral tissues such as skull, scalp, vein etc from the original MR images. Most common techniques employed in skull stripping are region growing and mathematical morphology[10]. In this proposed study, the skull stripping process based on the use of mathematical morphology. Texture analysis have important role in the medical image segmentation. The purpose of texture analysis is to classify or segment different texture regions in an image. Specifically this work deals with 2D Gabor filter texture segmentation of an image [11]. In this paper, we propose to exploit the concept of Gabor filter texture segmentation to segment medical images containing various anatomical structures that are belonging to MR imaging modalities.

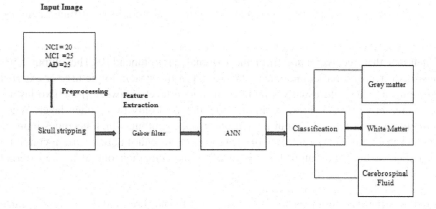

Fig. 1. Methodology of our proposed algorithm

In this study we have implemented supervised learning using feed forward network with back propagation (BP) algorithm(fig2) . The BP algorithm employed to train the ANN based on the number of hidden neurons, learning rate and momentum factor parameters to be set manually. As a result of the above steps we were able to extract the WM, GM, and CSF successfully. Fig. 3,4,5 shows segmented MR images of GM,WM and CSF in the NCI,MCI and AD study groups . The segmentation evaluation approach is based on the use of reference image as a ground truth .This allows us to compute the performance of our brain segmentation mainly the GM, WM and CSF compared with that obtained from real MR image manually segmented (ground truth) by two-expert neuro radiologist in this field. The volumes of GM, WM and CSF indicated important information, especially in the discrimination of AD. In contrast to the volume features, which are extracted from the whole three dimensional volume . Hence, a supervised clustering based segmentation algorithm is adopted to extract GM, WM and CSF probability maps from the source MRI data. The value of each pixel in the corresponding probability map denotes the posterior of the pixel belonging to the tissue by giving its gray intensity [12]. Finally Our proposed approach will be used for the early diagnosis of MCI in the NCI and AD in the MCI subjects.

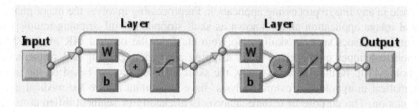

Fig. 2. Simple Architecture of ANN

3   Modeling Using Neural Network

Artificial neural networks are the simple clustering of the primitive artificial neurons. This clustering occurs by creating layers, which are then connected to one another as the figure above shows, the neurons are grouped into layers . The input layer consist of neurons that receive input from the external environment[10].The output layer consists of neurons that communicate the output of the system to the user or external environment. There are usually a number of hidden layers between these two layers; the figure above (fig:2) shows a simple structure with only two hidden layer. When the input layer receives the input its neurons produce output, which becomes input to the other layers of the system. [13-14] The process continues until a certain condition is satisfied or until the output layer is invoked and fires their output to the external environment.

The neural network is then trained using FFBP supervised learning algorithms which uses the data to adjust the networks weights and thresholds so as to minimize the error in its predictions on the training set. The classification performance by varying the hidden neurons, momentum and learning rate and showed that there is an increase in classification accuracy when there is a proper parameter setting. . The BP algorithm employed to train the ANN will require the to be set manually[15]. In this approach, a three-layer BP-ANN is employed for classification task . The input layer contains 24 neurons, and the output layer has 1 neuron. Hidden layer is composed of 100 neurons and the maximum iterations are set to 100 epochs. For the classification experiments, the training set consists of 1025 feature set and is used for training. The ANN classifier is performed with the learning rate and the momentum constant varied from 0 to 1 and the hidden neurons varied from 31 to 200.Using the proposed algorithm of ANN is achieved with Nh=100,Lr=0.01,and Mc=0.9.The parameters are optimally selected thereby adjusting the connection weights and analyze the classification performance depends on the Nh , Lr and Mc.

4   Experimental Results and Discussions

In the recent years human brain image segmentation in 3D MRI has gained a lot of importance in the area of Alzheimer's disease. In this work, we have employed supervised clustering algorithms(ANN) to segment MRI brain images and obtained the general segmented map.   The resultant segmentation map covered the entire image.. Two-experienced neuro radiologist validated ground truth images and the volume of GM, WM and CSF were calculated. Our study showed that automated segmentation was able to detect significant volume differences in both AD patients and patients with MCI.

Fig. 3(a), 4(a), 5(a) shows the skull stripped MR image of a control subject, MCI and AD patients . The skull stripped image contains only the three tissue classes GM, WM and CSF. Fig 3(b) represents the ANN segmented map. Fig. 3(c), 3(d) and 3(e) represents GM, WM and CSF extracted from the skull stripped images. Fig 4(a), 4(b), 4(c), 4(d) and 4(e) represents skull stripped, ANN segmented map and the GM, WM and CSF tissue maps. Fig(6) shows best training and regression plot performance results. Fig(7) represents the best training state results.

4.1   Structural Evaluation of Elderly Controls

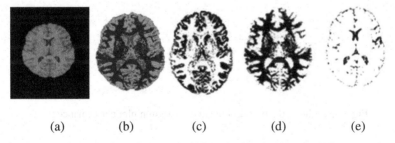

(a)   (b)   (c)   (d)   (e)

Fig. 3. Segmentation of MR brain images (NCI): (a) skull stripped (b)Segmentation map (c) GM (d) WM (e) CSF

4.2   Structural Evaluation of MCI Patients

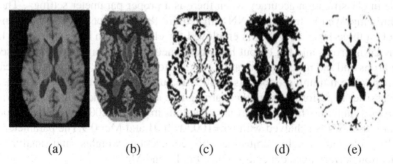

(a)   (b)   (c)   (d)   (e)

Fig. 4. Segmentation of MR brain images (MCI): (a) skull stripped (b)Segmentation map (c) GM (d) WM (e) CSF

4.3   Structural Evaluation of AD Patients

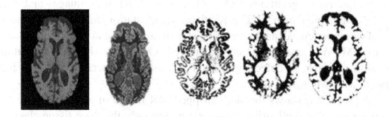

Fig. 5. Segmentation of MR brain images (AD): (a) skull stripped (b)Segmentation map (c) GM (d) WM (e) CSF

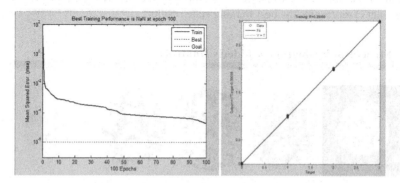

Fig. 6. Results of the best training and regression plot performance

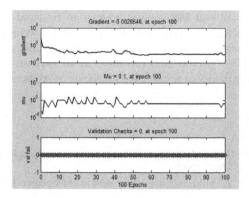

Fig. 7. Results of the best training state

Table 1. Classification results of different study groups

	Accuracy(%)				TPR (%)	FPR (%)	Sensititivity (%)	Specificity (%)	Yoden Index (%)
	GM	WM	CSF	Over all					
NCI	97.44	97.45	97.39	97.42	95.06	.007	95.47	97.45	94.24
MCI	95.10	96.21	96.01	95.66	92.81	.012	93.52	97.86	91.49
AD	95.62	96.08	95.99	95.9	93.23	.013	92.78	97.97	91.34

Table 2. Classification Brain Volumes of different study groups

	ANN		
	GM (mm)	WM((mm)	CSF((mm)
NCI	521.09	449.01	433.56
MCI	454.67	443.07	528.25
AD	258.54	440.21	916.45

From the validation results has shown good performance in improving ANN learning in terms of correct classification percentage. In this paper we have developed a novel classifier to distinguish normal and abnormal brain MRIs. The results show that our method obtained 97.42% on NCI, 95.66% on MCI and 95.9% on AD classifications accuracy on both training and test images.

5   Conclusions

In this paper, we have developed a computer aided diagnosis system for identifying the transition state and early diadnosis of AD.By using BP-ANN automated segmentation method on real MR images appears to be trend towards accelerated volume loss in grey matter and other regions of the brain were able to individually classify Alzheimer disease, mild cognitive impairment, and control participants with a high degree of accuracy. This method can serve as an alternative to manual tracing and may become a useful tool to assist in the diagnosis of Alzheimer disease.

References

1. Rahul, S., Desikan, H.J.C.: Automated MRI measures identify individuals with MCI and AD. Brain 132, 2048–2057 (2009)
2. Alistair, B., Michael, Z.: Mild cognitive impairment in older people. Lancet. 360, 1963–1965 (2002)
3. Wattamwar, P.R., Mathuranath, P.S.: An overview of biomarkers in Alzheimer's disease. Ann. Indian Acad. Neurol. 13, 116–123 (2010)
4. Mathuranath, P.S., Mathew, R.: Role of subjective memory complaints in defing MCI. Neurobiology of Aging 25, 74–79 (2004)
5. Barbra, R., Monica, N., Helle, W.: Investigating poor insight in AD: A survey research approaches. Dementia 6, 44–61 (2007)
6. Kannan, S.R., Sathya, A., Ramathilagam, S., Devi, R.: Novel segmentation algorithm in segmenting medical images. Journal of Systems and Software 8, 2487–2495 (2010)
7. Luca, M.D., Grossi, E., Borroni, B., Zimmermann, M., Marcello, E., Colciaghi, F., Gardoni, F., Intraligi, M., Padovani, A., Buscema, M.: Artificial neural networks allow the use of simultaneous measurements of Alzheimer Disease markers for early detection of the disease. Journal of Translational Medicine 3, 30–39 (2005)
8. Devanand, D.P., Liu, J., Hao, X., Pradhaban, G., Peterson, B.S.: MRI hippocampal and entorhinal cortex mapping in predicting conversion to AD. Neuroimage 60, 1622–1629 (2012)
9. Reed, R.T., du Buf, J.M.H.: A review of recent texture segmentation and feature extraction techniques. Comput. Vis. Graphics Image Processing 57, 359–372 (1993)
10. Yang, S.-T., Lee, J.-D., Huang, C.-H., Wang, J.-J., Hsu, W.-C., Wai, Y.-Y.: Computer-Aided Diagnosis of Alzheimer's Disease Using Multiple Features with Artificial Neural Network. In: Zhang, B.-T., Orgun, M.A. (eds.) PRICAI 2010. LNCS, vol. 6230, pp. 699–705. Springer, Heidelberg (2010)
11. Adams, R., Bischof, L.: Seeded region growing. IEEE Transactions on Pattern Analysis and Machine Intelligence 16, 641–646 (1994)
12. Varghese, T., Kumari, R.S., Mathuranath, P.S., Albert Singh, N.: Performance Evaluation of Bacterial Foraging Optimization Algorithm for the Early Diagonosis and Tracking of Alzheimer's Disease. In: Panigrahi, B.K., Das, S., Suganthan, P.N., Nanda, P.K. (eds.) SEMCCO 2012. LNCS, vol. 7677, pp. 41–48. Springer, Heidelberg (2012)
13. Deepa, S.N., Aruna Devi, B.: A survey on artificial intelligence approaches for medical image classification. Indian Journal of Science and Technology 4, 11 (2011)
14. Hojjatoleslami, S.A., Kittler, J.: Region Growing: A New Approach. IEEE Transactions on Image Processing 7, 7 (1998)
15. Shanthi, K.J., Sasi Kumar, M., Kesavadas, C.: Neural Network Model for Automatic Segmentation of Brain MRI. IEEE (2008)

Modified Mean Square Error Algorithm with Reduced Cost of Training and Simulation Time for Character Recognition in Backpropagation Neural Network

Sapna Singh[1], Daya Shankar Singh[1], and Shobhit Kumar[2]

[1] Computer Science and Engineering Department, Madan Mohan Malaviya Engineering College, Gorakhpur, India
[2] Information Technology Department, Manyawar Kanshiram Institute of Engineering & Information Technology, Akbarpur, India
{sapna2009singh,shobhit5786}@gmail.com, dss_mec@yahoo.co.in

Abstract. Neural Network concept is based on "Learn by example". Mean square error function is the basic performance function which affects the network directly. Reducing of such error will result in an efficient system. The paper proposes a modified mean squared error value while training Backpropagation (BP) neural networks. The new cost function is referred as *Arctan mean square error* (AMSE).The formula computed prove that the modification of MSE is optimal in the sense of reducing the value of error for an asymptotically large number of statistically independent training data patterns. The results shows improved network with reduced error value along with increment in performance consequently.

Keywords: Backpropagation algorithm, Mean square error algorithm, neural network.

1 Introduction

Backpropagation (BP) Neural Network classifiers show good performance in wide range of applications. Training this classifier proves to the minimization of a cost function over an available training set.

For pattern classification in general and for BP in particular, the cost function that is more efficiently being used than any other alternative is the standard mean square error(MSE).The standard mean square error is advantageous for having no prior knowledge for class distributions[1].This paper proposes training BP neural network using a new cost error function referred as the *arctan mean square error*(AMSE).The AMSE is optimal in yielding a network output with minimum variance for asymptotically large number of statistically independent training pattern, with respect to standard MSE. This paper also compares the performance of BP neural network trained by AMSE with respect to standard MSE. The data is compared with various other mathematical functions like finding the trigonometrically sin& cosine of the mean square error values. This paper is organized as follows:

Section 2 introduces the MSE and AMSE and other performance functions. Section 3 gives an overview of Backpropagation algorithm. Section 4describes the proposed

S.C. Satapathy, S.K. Udgata, and B.N. Biswal (eds.), *FICTA 2013*,
Advances in Intelligent Systems and Computing 247,
DOI: 10.1007/978-3-319-02931-3_17, © Springer International Publishing Switzerland 2014

concept by employing BP classifiers trained by AMSE through the deployment of arctan function by training a particular input binary data image of 20 pixel value.Section5 contains the methods being utilized in the research work with section 6 giving the comparative analysis of MSE with respect to the proposed concept i.e. AMSE. Finally, the conclusion of the work along with the results is being illustrated.

2   Performance Metrics

Artificial Neural Network is a representation of human brain that tries to learn and simulate its training input patterns by the predefined set of example patterns. The network is trained with particular specifications. The obtained output after training the network is compared with the desired target value and error is calculated based upon these values.

For training an input pattern and measuring its performance, a function must be defined. The various functions being included in neural network are:

2.1   Sum of Squared Error (SSE)

The first basic cost evaluation function. The Sum of Squared error is defined as

$$\text{SSE} = \sum_{p=1}^{p} \sum_{i=1}^{N} (t_{pi} - y_{pi})^2$$

Where, t_{pi} = Predicted value for data point i;

y_{pi} = Actual value for the data point i;

N = Total number of data points

2.2   Mean Squared Error (MSE)

This is most widely used and the effective performance function. The Mean Squared error is defined as

$$\text{MSE} = \frac{1}{N} \sum_{p=1}^{p} \sum_{i=1}^{N} (t_{pi} - y_{pi})^2$$

Where, t_{pi} = Predicted value for data point i;

y_{pi} = Actual value for the data point i;

N = Total number of data points

2.3   Root Mean Squared Error (RMSE)

The root Mean squared error is defined as

$$\text{RMSE} = \sqrt{\frac{1}{N} \sum_{p=1}^{p} \sum_{i=1}^{N} (t_{pi} - y_{pi})^2}$$

Where, t_{pi}= Predicted value for data point i;

y_{pi} =Actual value for the data point i;

N= Total number of data points

2.4   Mean Magnitude of Relative Error (MMRE) [4,6]

The mean magnitude relative error is defined as

$$\text{MMREE} = \frac{1}{N} \sum_{p=1}^{p} \sum_{i=1}^{N} \frac{t_{pi} - y_{pi}}{y_{pi}}$$

Where, t_{pi}= Predicted value for data point i;

y_{pi} =Actual value for the data point i;

N = Total number of data points

2.5   Relative Absolute Error (RAE)

The Relative absolute error is defined as the summation of the difference between predictive value and given value for the sample case j to that divide it by the summation of the difference between the given value and average of the given value. The relative absolute error of individual data set j is defined as

$$\text{RAE} = \frac{\sum_{i=1}^{N} t_{ij} - y_{ij}}{\sum_{i=1}^{N} y_i - y_m}$$

Where, t_{ij}= Predicted value by the individual dataset j for data point in i;

y_i=Actual value for the data point i;

N = Total number of data points

y_m = Mean of all y_{pi}

2.6   Root Relative Squared Error (RRSE)

The root relative squared error of individual data set j is defined as

$$\text{RRSE} = \sqrt{\frac{\sum_{i=1}^{N}(t_{ij} - y_{ij})2}{\sum_{i=1}^{N}(y_i - y_m)2}}$$

Where, t_{ij}= Predicted value by the individual dataset j for data point in i;

y_i =Actual value for the data point i;

$$N = \text{Total number of data points}$$
$$y_m = \text{Mean of all } y_{pi}$$

2.7   Mean Absolute Error (MAE)

The mean absolute error measures of how far the estimates are from actual values. It could be applied to any two pairs of numbers, where one set is "actual" and the other is an estimate prediction.

$$E = \frac{1}{N} \sum_{p=1}^{p} \sum_{i=1}^{N} t_{pi} - y_{pi}$$

Where, t_{pi}= Predicted value for data point i;
y_{pi}=Actual value for the data point i;
N = Total number of data points.

The above equation represents the output nodes, $\mathbf{t_{pi}}$ **and** $\mathbf{y_{pi}}$ which are target and actual network output unit on the pth pattern, respectively.

The network learns by adjusting weights. The process of adjusting the weights to make the neural network learn the relationship between the input and targets is known as learning or training. There are several techniques for training a network gradient descent method which is the most common.

3   The Backpropagation Approach

The backpropagation neural network is a neural network with a layered, feed-forward network structure and the generalised delta rule which updates its weight for each run of the network. It is a powerful mapping network which has been successfully applied to a wide variety of problems. Backpropagation is one of the finest algorithms that give better performance with relatively more accurate recognition ratio. Training is proficient by presenting the patterns to be classified to the network and determining its output. The actual output of the network is compared with the target output and an error measure is calculated. The higher the error value the less efficient the network is. Thus an effort has been made to decrease the error value for the Backpropagation network.

4   Proposed Mean Square Error Algorithm

In statistics, the **mean squared error** (MSE) of an estimator is one of many ways to quantify the difference between values implied by an estimator and the true values of the quantity being estimated. MSE is a risk function, corresponding to the expected value of the **squared error loss** or **quadratic loss**. MSE measures the average of the squares of the "errors." The error is the amount by which the value implied by the estimator differs from the quantity to be estimated. The difference occurs because of randomness or because the estimator doesn't account for information that could produce a more accurate estimate [wiki].

Minimizing squared error would increase the accuracy of a particular system with defined number of input training dataset. The decrease in the value of error is evaluated by a mathematical term known as *'arctan'*. *Arctan* is the term defined for *'inverse tangent'* of a particular error value. The formula is being described in the above section. The algorithm or the sequence of steps for the error calculation is as follows:

MSE Algorithm

Initialize 'n' number of input patterns (integer values)
Do

    For each training pattern 'n' train the network
    O=neural_net_output(network,n)
    T=neural_net_target(desired)
    Compute error e=(T-O) at the output units
    Square the error 'e'
    Calculate the summation of error 'e' for all input patterns
While(n!=0)
Divide the summation obtained by the 'n' number of patterns

The standard mean square error is advantageous for having no prior knowledge for class distributions. It is widely used as it results in the least error value as compared to other error values.

Minimizing MSE will further result in more accurate system for pattern recognition and other applications. Arctan Mean Square error (AMSE) has proved further minimization of the MSE. It is advantageous for a system with large dataset, where thousands of values are to be enumerated to result thousand of values. It has proved useful in calculation of error for Backpropagation NN.

The Arctan mean squared error can be estimated by the following formula

$$\text{AMSE} = \frac{1}{N} \sum_{p=1}^{p} \sum_{i=1}^{N} [(arctan(t_{pi} - y_{pi}))^2]$$

Where t_{pi}= Predicted value for data point i;

        y_{pi} =Actual value for the data point i;
        N = Total number of data points

AMSE Algorithm

Initialize 'n' number of input patterns (integer values)
Do

    For each training pattern 'n' train the network
    O=neural_net_output (network, n)
    T=neural_net_target (desired)
    Compute error e= (T-O) at the output units
    Compute inverse tangent of error 'e'
    Square the error e
    Calculate the summation of error 'e' for all input pattern
While (n!=0)
Divide the summation obtained by the 'n' number of patterns

The proposed Arctan Mean square error algorithm is a description of the way how the mean squared error value is being modified to improve its performance with reduced error cost. By applying the trigonometrical function i.e. inverse tangent of the MSE value to the network shows quite improvement in the performance. Thus, it proves that our mathematical modification in the formula of MSE shows better results with improvement in the recognition accuracy of the Backpropagation network.

5   Research Methodology

The Artificial Neural Network (ANN) and Gradient Descent learning techniques have been used for predicting the software effort using dataset of software projects in order to compare the performance results obtained from these models.

5.1   Empirical Data Collection

The data we have used is being evaluated from a binary image of 20 pixel values.The error value of a network is the mathematical value of target value difference the obtained value after training the network. The MSE is calculated for 20 pixel values. Then the AMSE is being calculated. A threshold value is set (let 0.5) so as to attain the number of training cases recognized. If the obtained output is more than the threshold value then the neural network has recognized the training pattern.

A feed-forward Backpropagation network is established using 900 input neurons, 12 hidden layer neurons and 2 output neurons. The output set is either 0 or 1.The 0 represent the pattern as unrecognized and 1 as recognized. Gradient descent algorithm with performance functions as mean square error evaluation algorithm. The network structure being evaluated is as shown in fig 1.

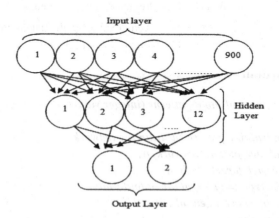

Fig. 1. Diagrammatical representation of the proposed network

5.2   Diagrammatical Representation of Error Calculation

Firstly initialize the training patterns and set a particular target value for the input training pattern. Set the maximum number of iterations up to which the input data's are to be iterated. Then train the input pattern by using Backpropagation training algorithm. Compare the obtained output with the target value set. If it is near the target value then the network has realized the training pattern else it has not. The flow of data in the whole recognition procedure is being shown in fig 2.

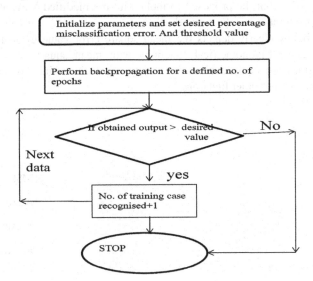

Fig. 2. Diagrammatical representation of number of recognition patterns

6   Performance Comparison

The mean square error is being calculated for the training input patterns .Then Mean square error value is calculated for various trigonometrically functions. The calculated error value is being shown through the table:

Table 1. Comparative analysis of various inverse Mean square errors (MSE)

	MSE	MSE using Tan^{-1}	MSE using Sin^{-1}	MSE using Cos^{-1}
Error	0.29091	0.2177	0.4198	0.4957

The readings are being obtained by training the Backpropagation neural network by the inverse of various trigonometrical functions. As being concluded from the readings above that the inverse of tangent functions shows positive results as compared to the other evaluated functions. The result is obtained from 20 pixel values through binarization of input data. The lesser the value of error signifies the more efficient and cost reduced the network is being devised.

7   Result

To show that the modified MSE formula shows better results, all the trigonometrically evaluated functions were being calculated for their respective MSE. The tabulated data's above proves that the proposed formula with the modified MSE algorithm has enhanced the performance of the network.

The graph below clearly depicts the mean square error value evaluated by using various inverses of trigonometrical functions with mean square error value. The different functions are being evaluated so as to check the affinity of our proposed function as compared to other functions.

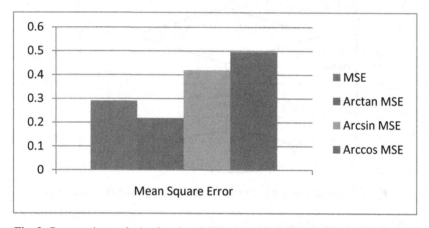

Fig. 3. Comparative analysis of various MSE and proposed Arctan Mean square error

The Arctan mean square error value shows the least error, from which it can be further concluded that Arctan MSE can be advantageous in increasing the accuracy of a large trained neural network along with the reduction in training time. Thus, our proposed concept of reducing the cost error function has shown positive and effective results.

8   Conclusion

The Backpropagation training algorithm has its advantage over training patterns as it increases the accuracy of the simulated data.The standard mean square error is beneficial for having no prior knowledge for class distributions. Through this paper it is being proposed that by reducing the error value, accuracy of a particular network can be increased. Further work is under process for reducing the training time and increasing the accuracy for a network using Backpropagation training algorithm with its application on hand written character recognition. This concept can be further applied to various applications related to classification and pattern recognition where larger dataset is being processed. Work can also be done to improve the accuracy and increasing recognition speed as its area of application is wide.

References

1. Wilamowski, B.M., Iplikci, S., Kaynak, O., ÖnderEfe, M.: An Algorithm for Fast Convergence in Training Neural, 0-7803-7044-9/01/$10.00 © (2001) IEEE
2. Rady, H.: Reyni's Entropy and Mean Square Error for Improving the Convergence of Multilayer Backpropagation Neural Networks: A Comparative. Study117905-8282 IJECS-IJENS 11(5) (October 2005)
3. Osman, H., Blostan, S.D.: New cost function for Backpropagation neural network with application to SAR imagery classification,K71396
4. Finnie, G.R., Wittig, G.E.: A Comparison of Software Effort Estimation Techniques: Using Function Points with Neural Networks, Case Based Reasoning and Regression Models. Journal of Systems and Software 39, 281–289 (1997)
5. Finnie, G.R., Wittig, G.E.: AI Tools for Software Development Effort Estimation. In: Proceedings of the International Conference on Software Engineering: Education and Practice, SEEP 1996 (1996)
6. Srinivasan, K., Fisher, D.: Machine Learning Approaches to Estimating Software Development Effort. IEEE Transactions on Software Engineering 21 (February 1995)
7. Arora, S., Bhattacharjee, D., Nasipuri, M., Malik, L., Kundu, M., Basu, D.K.: Performance Comparison of SVM and ANN for Handwritten Devnagari Character Recognition. IJCSI International Journal of Computer Science Issues 7(3) (May 2010)
8. Devireddy, S.K., Rao, S.A.: Hand written character recognition using backpropagation network. Journal of Theoretical and Applied Information Technology, JATIT (2005 - 2009)
9. Shahi, M., Ahlawat, A.K., Pandey, B.N.: Literature Survey on Offline Recognition of Handwritten Hindi Curve Script Using ANN Approach. International Journal of Scientific and Research Publications 2(5) (May 2012) ISSN 2250-3153
10. Garg, N., Kaur, S.: Improvement in Efficiency of Recognition of Handwritten Gurumukhi Script. IJCST 2(3) (September 2011)

A Novel Approach for Odia Part of Speech Tagging Using Artificial Neural Network

Bishwa Ranjan Das and Srikanta Patnaik

Department of Computer Science and Information Technology,
Institute of Technical Education and Research,
Siksha 'O' Anusandhan University, Khandagiri, Bhubaneswar, India
biswadas.bulu@gmail.com, patnaik_srikanta@yahoo.co.in

Abstract. This paper presents a challenging task for POS Tagging using Artificial Neural Network for Odia language. Neural Network is used for Odia POS Tagging. A Single Neural Network based POS Tagger with fixed length of context chosen empirically is presented first. Then a multiple neuron tagger which consists of multiple single-neuron taggers with fixed but different lengths of contexts is presented. Multi-neuron tagger performs tagging by voting on the output of all single neuron tagger. The experiments carried out are discussed, Neural Network for efficient recognition where the errors were corrected through forward propagation and rectified neuron values were transmitted by feed-forward method in the neural network of multiple layers, i.e. the input layer, the output layer and the middle layer or hidden layers. Neural networks are one of the most efficient techniques for identified the correct data. A small labeled training set is provided; a HMM based approach does not yield very good result. So in this work, morphological analyzer is used to improve the performance of the tagger. This tagger has an accuracy of about 81% on the test data provided.

Keywords: Single layer feed forward, Multi layer feed forward, Hidden layer, Panini grammar.

1   Introduction

POS Tagging is the process of assigning a part of speech, like noun, verb, pronoun, adverb, adverb or other lexical class marker to each word in a sentence. The input to a tagging algorithm is a string of words of a natural language sentence and a specific tagset the output is a single POS Tag for each word. There is different approach to the problem of assigning each word of a text with a parts of speech tag, which is known as POS tagging. In this paper the performance of a POS Tagging techniques are shown for Odia language. A supervised POS tagging approach requires a large amount of annotated training corpus. POS Tagging is the primary step in many applications, less accuracy at this stage affects latter stages. To build a POS tagger for new corpus or a new language, tuning of various parameters (Length of context) is a nontrivial task. Neural network is the one of the most efficient approaches for learning from a sparse data. Corpus based features play an important role to achieve

S.C. Satapathy, S.K. Udgata, and B.N. Biswal (eds.), *FICTA 2013*,
Advances in Intelligent Systems and Computing 247,
DOI: 10.1007/978-3-319-02931-3_18, © Springer International Publishing Switzerland 2014

adequate accuracy. This paper shows a neural network based approach which learns the parameters of POS tagger from a representative training dataset. Single-neuro tagger with fixed length of context chosen empirically is a neural network based POS tagger. Most of the Indian language in which length of context is fixed and chosen empirically, multi-neuron tagger is composed of multiple single-neuron taggers with fixed but different lengths of context and a voting based selection rule to obtain the final output. POS tagging is typically achieved by rule based systems, probabilistic data-driven systems, neural network systems or hybrid system. For languages like English or French, hybrid taggers have been able to achieve success percentage above 98% [Schulze et al, 1994].

Helmut Schmid [3], a new part of speech tagging method based on neural networks (net tagger) is presented and its performance is compared to that of a HMM-Tagger (Cutting et al 1992) and a trigram based tagger (Kempe, 1993). A part-of-speech tagger based on a multilayer perceptron network is presented. It similar to the network of Nakamura et al (1990) in so far as the same training procedure (Back propagation) is used; but it differs in the structure of network and also in its purpose (Disambiguation Vs Prediction). The performance of the tagger is measured and compared to that of two other taggers (Cutting et al. 1992; Kempe 1993).

2   POS Tagging in Odia

2.1   Various Terminology Uses in Odia

Various terminology uses in Odia language which are used for Odia Pos Tagging like Noun, Adjective, Verb, Pronoun, Adverb, Preposition etc. in Odia.

Noun -> Bisheshya, Adjective -> Bisheshana, Verb -> Kriya, Pronoun -> word use instead of Noun, Adverb -> Kriya bisheshana etc.

2.2   Morphological Analysis

To find the root or base word in Odia so many suffixes are there, these suffixes are used in verb as well as noun also. These noun suffixes are come from inflection list (Bivokti) and some suffix list are use in verb, from these suffix we find out no of nouns and verbs. Here suffix list are mentioned in which is they are used in noun and verb.

3   Artificial Neural Network

The functioning of neurons as information processing units in the human brain, so it is with neural networks made up of artificial neurons. The human nervous system may be viewed as a three-stage system,

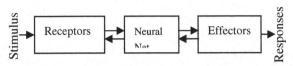

Fig. 1. Function of neuron

Central to the system is the brain represented by the neural net, which continually receives information perceives it, and makes appropriate decisions. Two sets of arrows are shown left to right arrow indicate the forward transmission of information bearing signals through the system. The arrows from right to left signify the presence of feed-back in the system. Different classes of neural network, i.e. Single layer feed forward & Multilayer feed forward network.

3.1   Single Layer Feed-Forward Network

Single layer Feed-forward Network consists of a single layer of weights, where the inputs are directly connected to the outputs, via a series of weights. The synaptic links carrying weights connect every input to every output, but not other way. This way it is considered a network of feed-forward type. The sum of the products of the weights and the inputs is calculated in each neuron node, and if the value is above some threshold the neuron fires and takes the activated value otherwise it takes the deactivated value.

$$y_j = f(net_j) = 1 \text{ if } net_j \geq 0$$

$$0 \text{ if } net_j \bullet 0 \text{ where } net_j = \bullet x_i w_{ij}$$

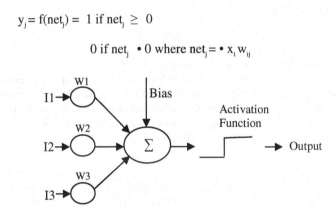

Fig. 2. Function of Single Layer Feed-Forward Network

3.2   Multi Layer Feed Forward Network

It consists of multiple layers. It has three layers, input, hidden layer & output layer. The computational units of the hidden layer are known as hidden neurons. A multi layer feed-forward network with l input neurons, m1 neurons in the first hidden layers, m2 neurons in the second hidden layers, and n output neurons in the output layers is written as (l-m1-m2-n).

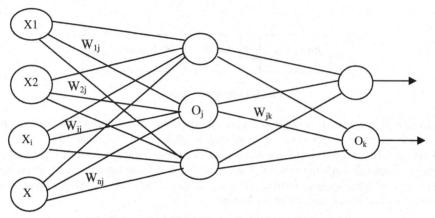

Fig. 3. Function of Multi Layer Feed Forward Network

$x_1, x_2, x_3, \ldots x_n$ are the input.
$w_{k1}, w_{k2}, w_{k3}, \ldots w_{kn}$ are synaptic weight.

$$U_k = \sum_{j=1}^{m} w_{kj} * x_j \qquad (1)$$

$$Y_k = \varphi(u_k + b_k) \qquad (2)$$

Where u_k= linear combiner output.

$\varphi()$ = is the activation function.

Y_k = Output signal of neuron apply an affine transformation to the output U_k of the linear combiner in the model

$$V_k = U_k + b_k \qquad (3)$$

Where b_k is bias value which may +ve or −ve

The relationship between the induced local field or activation potential v_k of neuron k and the linear combiner output u_k is modified. The graph of v_k versus u_k no longer passes through the origin. The Bias b_k is an external parameter of artificial neuron k. I may account for its presence as in equation 2, equivalently; we may formulate the combination of equation 1 & 3 as follows.

$$v_k = \sum w_{kj} x_j \qquad (4)$$

$$y_k = \varphi(v_k) \qquad (5)$$

4   Methodology

4.1   Training Data

It has been used our own training data set that was developed by ourselves. It gives the 100% correct result for our system.

4.2   Suffix & Prefix

Some suffix & prefix alphabets are used to identify NE which are mentioned in the features. Firstly a fixed length word suffix of the current and surrounding words are used as features.

4.3   Algorithm

The proposed algorithm used for finding the NE is as follows. Firstly the entire Odia Text is entered by user, and then it is divided into six steps which is described in the following algorithm.

Step 1. Enter a text.
Step 2. Convert entire text into token by tokenization.
Step 3. Finding root word using morphological analysis.
Step 3. Extract the suffix features from the each and every word.
Step 4. Compare each word with our valid suffix features.
Step 5. Identify Part of Speech.
The following flowchart to find POS tagging.

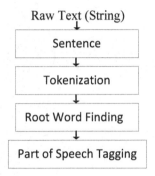

Fig. 4. Flowchart of POS tagger

Here the proposed model is based on ANN
For example:-
ମୁଁ ଘରକୁ ଯାଉଛି, (mu gharaku jauchhi/ I am going to home)
ମୁଁ – ବିଶେଷ୍ୟ, ଘରକୁ – ଘର+କୁ, ଯାଉଛି – ଯିବା+ଉଛି

In the above sentence "mu gharaku jauchhi". Mu/ମୁଁ represents as noun, gharaku/ ଘରକୁ represents as noun, ku/କୁ is used as inflection (Bivokti). In odia grammar, ku, nku, mananku used as Dittiya Bivokti(Inflection), Finding the root word is Ghara using morphological analysis. Like this the word jauchhi(ଯାଉଛି) used as verb, the root word ଯିବା/jiba, the word uchhi/ ଉଛି is used as suffix which is basically use in verb. In Odia, where the inflection suffix is used that word is known as Noun, and some suffix like uchhi/ଉଛି, ichhi/ଇଛି, ithili/ଇଥିଲି, uthila/ଉଥିଲା used as verb/Kriya. The Back propagation technique is used by iteratively processing a data set of training tuples, comparing the network's prediction for each tuple with the actual known target value may be the known class label of the training tuple or a continuous value. For each training tuple, the weights are modified so as to minimize the mean squared error between the network's prediction and the actual target value. These modifications are made in the "backward" direction, that is, from the output layer, through each hidden layer down to the first hidden layer. Although it is not guaranteed, in general the weights will eventually converge, and the learning processing stops.

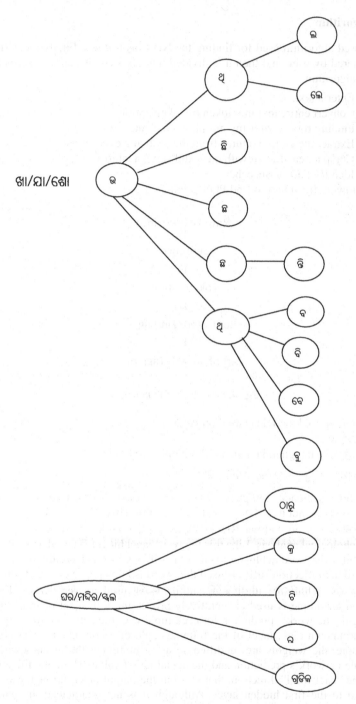

Fig. 5. Different form of Odia Verb and Noun

5   Experimental Result

Whenever a large amount of corpus is passed to the proposed system, using ANN technique to give the correct result with accuracy 81%. The result of this proposed model is compared with so many proposed models which are developed before.

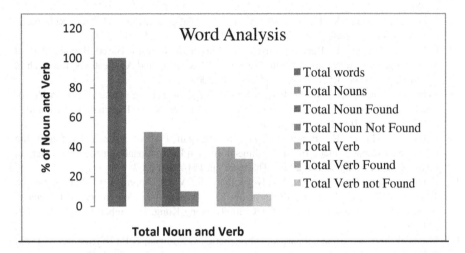

6   Conclusion

In this paper, the single-neuro tagger was presented, a part-of speech tagger which is based on single layer neural network (perceptrons) and uses multiple lengths of contexts to obtain final output. It gives the correct result without any error which accuracy is 81%. The ODI TAGGER gives the correct output without given any error. First it does the tokenization, and then morphological analysis is done successfully. After that it identifies correct noun & verb. It also helpful for implementing to design a Odia parser.

References

1. Parikh, A.: Part-of-speech Tagging using neural network. In: Proceedings of ICON 2009: 7th International Conference on Natural Language Processing, Report No: IIIT/TR/2009/232 (2009)
2. Ray, P.R., Harish, V., Sarkar, S., Basu, A.: Part of speech Tagging and Local word Grouping Techniques for Natural Language parsing in Hindi. In: Proceedings of the 1st International Conference on Natural Language Processing (ICON 2003), Mysore (2003)
3. Schmid, H.: Part-Of- Speech Tagging with Neural Networks. In: COLING 1994 Proceedings of the 15th Conference on Computational Linguistics, vol. 1, pp. 172–176 (1994)

4. Jena, I., Haudhury, S.C., Chaudhry, H., Sharma, D.M.: Developing Oriya Morphological Analyzer Using Lt-toolbox. Information Systems for Indian Languages Communications in Computer and Information Science 139, 124–129 (2011)
5. Singh, S., Gupta, K., Shrivastava, M., Bhattacharyya, P.: Morphological richness offsets resource demand – experiences in constructing a pos tagger for Hindi. In: Proceedings of the COLING/ACL 2006 Main Conference Poster Sessions, Sydney, Australia, pp. 779–786 (2006)
6. Ekbal, A., Mondal, S., Bandyopadhyay, S.: POS Tagging using HMM and Rule-based Chunking. In: Proceedings of SPSAL 2007, IJCAI 2007, pp. 25–28 (2007)
7. Ekbal, A., Haque, R., Bandyopadhyay, S.: Maximum Entropy Based Bengali Part of Speech Tagging. Advances in Natural Language Processing and Applications, Research in Computing Science (RCS) Journal (33), 67–78 (2008)
8. Ekbal, A., Bandyopadhyay, S.: Part of Speech Tagging in Bengali using Support Vector Machine. In: Proceedings of the International Conference on Information Technology (ICIT 2008), pp. 106–111. IEEE (2008)
9. Ekbal, A., Haque, R., Bandyopadhyay, S.: Bengali Part of Speech Tagging using Conditional Random Field. In: Proceedings of the 7th International Symposium on Natural Language Processing (SNLP 2007), Thailand, pp. 131–136 (2007)
10. Ekbal, A., Hasanuzzaman, M., Bandyopadhyay, S.: Voted Approach for Part of Speech Tagging in Bengali. In: Proceedings of the 23rd Pacific Asia Conference on Language, Information and Computation (PACLIC 2009), Hong Kong, December 3-5, pp. 120–129 (2009)
11. Dhanalakshmi, V., Anandkumar, M., Shivapratap, G., Soman, K.P., Rajendran, S.: Tamil POS Tagging using Linear Programming. International Journal of Recent Trendsin Engineering 1(2), 166–169 (2009)
12. Sreeganesh, T.: Telugu Parts of Speech Tagging in WSD. Language of India 6 (August 8, 2006)
13. Patel, C., Gali, K.: Part-Of-Speech Tagging for Gujarati Using Conditional Random Fields. In: Proceedings of the IJCNLP 2008 Workshop on NLP for Less Privileged Languages, Hyderabad, India, pp. 117–122 (2008)
14. Singh, T.D., Bandyopadhyay, S.: Morphology DrivenManipuri POS Tagger. In: Proceedings of the IJCNLP 2008 Workshop on NLP for Less Privileged Languages, Hyderabad, Hyderabad, India, pp. 91–98 (2008)
15. Saharia, N., Das, D., Sharma, U., Kalita, J.: Part of Speech Tagger for Assamese Text. In: Proceedings of the ACL IJCNLP 2009 Conference Short Papers, Suntec, Singapore, pp. 33–36 (2009)
16. Bharati, A., Chaitanya, V., Sangal, R.: Natural Language Processing A Paninian Perspective. In: Department of Computer Science and Engineering Indian Institute of Technology Kanpur, With contributions from K.V. Ramakrishnamacharyulu Rashtriya Sanskrit Vidyapeetha, Tirupati. Prentice-Hall, India (1994)
17. Data Mining, Concepts & Techniques, Jiawei Han and Micheline Kamber, 4th edn. Elsevier (2008)
18. Neural Networks, A.: Comprehensive Foundation, Simon Haykin. PHI (1998)
19. Machine Learning of Natural Language, Walter Daeleman, CNTS Language Technology Group Department of Linguistics, University of Antwerp, Belgium, http://walter.daelemans@ua.ac.be

Fractional Fourier Transform Based Features for Musical Instrument Recognition Using Machine Learning Techniques

D.G. Bhalke[1], C.B. Rama Rao[1], and D.S. Bormane[2]

[1] NIT Warangal (A.P.) India
[2] JSPM's RSCOE, Pune India
bhalkedg2000@yahoo.co.in, cbrr@nitw.ac.in,
bdattatraya@yahoo.com

Abstract. This paper reports the result of Musical instrument recognition using fractional fourier transform (FRFT) based features. The FRFT features are computed by replacing conventional Fourier transform in Mel Frequecny Cepstral coefficient (MFCC) with FRFT. The result of the system using FRFT is compared with the result of the system using Mel Frequency Cepstral Coefficients (MFCC), Wavelet and Timbrel features with different machine learning algorithms. The experimentation is performed on isolated musical sounds of 19 musical instruments covering four different instrument families. The system using FRFT features outperforms over MFCC, Wavelet and Timbrel features with 91.84% recognition accuracy for individual instruments. The system is tested on benchmarked McGill University musical sound database. The experimental result shows that musical sound signals can be better represented using FRFT.

Keywords: Musical instrument recognition, Mel Frequency Cepstral Coefficient (MFCC), Fractional Fourier transform (FRFT).

1   Introduction

Recognizing the objects in the environment from the sound they produce is primary function of auditory system. The aim of Musical instrument recognition is to identify the name and family of musical instrument from the sound they produce. So far many attempts were made for musical instrument recognition and classification. The statistical pattern-recognition technique for classification of 15 musical instrument tones with 31 features based on log-lag correlogram was discussed in Martin and Kin [2]. A study on pitch independent musical instrument recognition for 30 musical instruments with 43 features based on spectral, cepstral and temporal properties of sounds was described by Eronen and Klapuri [4].

Comparison of features for Musical Instrument recognition was discussed and used large set of features including MFCC, delta MFCC, Linear prediction cepstral coefficients, temporal feature, spectral features and modulation features for 16 orchestral instruments were used for experimentation by Eronen [1]. A work on

S.C. Satapathy, S.K. Udgata, and B.N. Biswal (eds.), *FICTA 2013*,
Advances in Intelligent Systems and Computing 247,
DOI: 10.1007/978-3-319-02931-3_19, © Springer International Publishing Switzerland 2014

instrument recognition for isolated monophonic notes using six features: cepstral coefficients, constant Q transform frequency spectrum, multidimensional scaling analysis trajectories, RMS amplitude envelope, spectral centroid and vibrato for 19 Instruments was described by Kaminskyj and Czaszejko [5]. Essid et al. performed study on use of hierarchical taxonomies for 20 musical instruments with wide set of 540 feature covering Temporal, Cepstral, spectral, wavelet and perceptual features [7]. Deng et al. [3] compared performance musical instrument recognition with work of [1], [2],[5],[6],[12], [13]. A study on feature analysis for recognition of classical instruments, using machine learning techniques to select and evaluate features extracted from a number of different feature schemes was described by Deng et al. [3]. The performance of Instrument recognition was analyzed using selected features with different feature selection and ranking algorithms.

Review of earlier work shows that, developing compact and efficient feature set for Musical Instrument Recognition has become challenging task and attracted the attention of researchers. Earlier work shows that number of features and recognition accuracy are main issues in Musical instrument recognition and classification.

This paper presents comparison of FRFT based features with MFCC, Timbrel and wavelet for Musical Instrument Recognition and classification using different machine learning algorithms.

The paper is organized as follows. Feature extraction and proposed features along with different machine learning algorithms is described in section 2. Database details and performance evolution is presented in section 3. Conclusion is summarized in section 4.

2   Feature Extraction and Proposed Features

Musical instrument recognition was tested using four different feature set as follows. 1. MFCC, 2.Waveletres, 3.Timbrel 4. FRFT (proposed).

The block schematic for musical instrument recognition system is shown in figure 1.

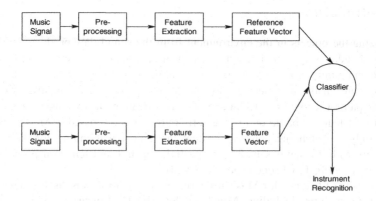

Fig. 1. Musical Instrument Recognition system

Musical Instrument recognition system consists of three stages as, i) pre-processing, ii) feature extraction, and iii) Classification. Initially, signal is given to pre-processing stage to make signals suitable for feature extraction. In pre-processing the silence part of the signal is removed. In preprocessing stage the signal is framed with 20 ms and windowed with hamming window. In feature extraction stage, acoustic features based on MFCC, Timbrel, Wavelet and FRFT have been extracted and applied to different classifiers. Table 1 shows features used for experimentation.

2.1   MFCC Features

Mel Frequency Cepstral Coefficients (MFCCs) are cepstral coefficients used for representing audio signal in a way that mimics the physiological properties of the human auditory system. MFCC feature extraction consists of pre emphasis, framing, windowing, FFT, Triangular band pass filter and DCT. Accordingly, 12 FCC features are extracted.

Table 1. Features Used

Sr. No.	Feature No.	No. of Features	Feature name	Feature Type
1	1-13	13	MFCC	MFCC
2	14-29	16	Wavelet sub-band energies (16 sub band energies using 'db1' wavelet)	Wavelet
3	30-45	16	Mean SC, STD of SC, Mean SR, STD SR, Mean SF, STD SF, Mean SS, STD SS , Mean ZCR, STD ZCR, Energy, Log attack time, Attack , Decay, Release, Sustain	Timbral
4	46-57	12	FRFT based MFCC features	FRFT

2.2   Wavelet Features

The wavelet analysis provides spectro-temporal information. The wavelet analysis decomposes the signal into "packets" by simultaneously passing the signal through a low decomposition filter (LDF) and a high decomposition filter (HDF) in a sequential tree like structure. In this work, we have used four level Daubenchies wavelet packet tree. The energy of each sub- band is computed and used as feature .

2.3   Timbrel Features

Timbre, also known as sound quality or tone color of music. It is defined as, when two sounds are heard that match for same pitch, same loudness, and same duration, and a difference can still be heard between the two sounds, that difference is called timbre. There are two physical correlates of timbre: spectrum envelope and amplitude envelope. Following feature based on spectral and temporal envelope are extracted.

2.4   Spectral Features

Spectral Centroid (SC): This is the amplitude-weighted average, or centroid, of the frequency spectrum, which can be related to a human perception or brightness of the instrument. The spectral centroid is given by

$$SC = (\frac{\sum_{k=1}^{K} P(f_k)f_k}{\sum_{k=1}^{k} P(f_k)}) \tag{1}$$

Where $P(f_k)$ is magnitude spectrum of kth sample and f_k is frequency corresponding to each magnitude element

Spectral flux (SF)

This is a measure of the amount of local spectral change. This is defined as the squared difference between the normalized magnitude spectra of successive frames and given by equation 2.

$$SF = (\sum_{k=2}^{k} |P(f_k) - P(f_{k-1})|) \tag{2}$$

Spectral spread (SS)

The spectral spread is a measure of variance (or spread) of the spectrum around the mean value μ .and is given by equation 3.

$$SS = \frac{\sqrt{\sum_{K=0}^{N/2}(P(f_k) - SC)^2}}{\sqrt{\sum_{k=0}^{N/2}(P(f_k)^2}} \tag{3}$$

where $P(f_k)$ = magnitude spectrum corresponding to each magnitude element and SC=spectral centroid.

Spectral skewness(SK)

The skewness is a measure of the asymmetry of the distribution around the mean value. The skewness is calculated from the 3rd order moment.

$$SK = \frac{\sum(freq - SC)^2 \times Mag}{\sum Mag} \tag{4}$$

where mag= magnitude spectrum, freq=frequency corresponding to each magnitude element and SC=spectral centroid.

The statistical values i.e. mean and standard deviation of SC,SF, SK , SS were computed and used as features.

2.5   Temporal Features

Temporal features are extracted in time-domain. Following temporal features are used for experimentation.

Energy: It is the sum of the amplitudes present in frame and is defined as:

$$Energy = \sum_{0}^{N-1} (x(n))^2 \tag{5}$$

where $x[n]$ is the amplitude of the sample.

Zero-Crossing Rate

This is the number of times the signal crosses zero amplitude during the frame, and can be used as a measure of the noisiness of the signal. It is defined as:

$$ZCR = \frac{1}{N} \sum_{0}^{N-1} |\,\text{sgn}[x(n)] - \text{sgn}[x(n-1)]\,| \tag{6}$$

where *sign* = 1 for positive arguments and 0 for negative arguments

Log-Attack Time

The log-attack time is the logarithm of time duration between the time the signal starts to the time it reaches its stable part. It can be estimated by taking the logarithm of the time from the start to the end of the attack.

ADSR envelope : Every musical sound are characterized by its temporal envelope which are characterized by Attack time, Decay time , Sustain time and release time. The ADSR values are computed and used as feature vector.

2.6   Fractional Fourier Transform Based Features

Fractional Fourier transform (FRFT):   The FRFT represents the signal in two orthogonal plane of time and frequency axis. It is a linear operator which corresponds to the rotation of the signal between time and frequency plane, where time axis corresponds alpha=0 and frequency axis corresponds to alpha= $\pi/2$ [8],[9]. FRFT is more flexible and suitable for non-stationary signal as compared to fourier transform because of its orthonormal basis of chirp signals and degree of freedom of rotation of time frequency axis, [8],[9],[10], [14].

The αth order fractional Fourier transform $F^{\alpha}(u)$ of f(t) is given by equation 7 to 9. Factional Fourier transform is general case of the Fourier transform with similar properties of Fourier transform such as, Linearity, additivity, commutatively, associatively, Time shift, Modulation, Multiplication, differentiation, Parseval's theorem etc.

$$
F^{\alpha}(u) =
\begin{cases}
\sqrt{\dfrac{1-j\cot\alpha}{2\pi}}\, e^{j\frac{u^2}{2}\cot\alpha} \displaystyle\int_{-\infty}^{\infty} e^{-j\frac{t^2}{2}\cot\alpha}\, e^{-jut\csc\alpha}\, f(t)dt \\[6pt]
\qquad\qquad\qquad\quad \text{if } \alpha \neq N\pi,\ \text{N is integer} \qquad (7)\\[4pt]
f(u) \qquad\quad\ \ \text{if } \alpha = 2N\pi,\ \text{N is integer} \qquad\qquad (8)\\[4pt]
f(-u) \qquad\ \ \text{if } \alpha = (2N+1)\pi,\ \text{N is integer} \qquad (9)
\end{cases}
$$

Since α varies from 0 to 1, the FRFT of f (t) changes from the time domain ($\alpha = 0$) to the frequency domain ($\alpha = 1$). Different value of α , provides additional flexibility and degree of freedom for processing of non-stationary signals [14]. The block schematic of FRFT based MFCC coefficients are shown in figure 2. The traditional FFT block in MFCC is substituted by FRFT and the coefficients are computed.

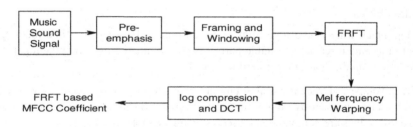

Fig. 2. Feature extraction using fractional Fourier transform

2.7 Machine Learning Techniques

Different machine learning algorithms are used to analyze the performance of the system. Decision tree, KNN, Naive Bayes, Support Vector Machine and Multilayer feed forward Neural Network have been used to test the performance of system. Waikato environment for knowledge analysis (WEKA) has been used to implement different classifiers [15]. These algorithms are briefly described as follows. 10-fold cross-validation technique with 70% notes for training and 30% notes for testing has been used to test the performance of machine learning algorithms. Confusion matrix are generated from these machine learning algorithms. In this paper, only overall average accuracy of the system is specified.

2.7.1 Decision Tree Algorithm
Decision Tree J48 is a supervised classification algorithm, which has been proposed in [16]. It is greedy algorithm and variant of ID3. Tree J48 is constructed top-down, where

parent nodes represent conditional attributes and leaf nodes represent decision outcomes. At each step tree finds most predictive attribute, and splits a node based on this attribute. Each node represents a decision point over the value of some attribute. J48 attempts to account for noise and missing data. It also deals with numeric attributes by determining where thresholds for decision splits should be placed. The main parameters that can be set for this algorithm are the confidence threshold, the minimum number of instances per leaf and the number of folds for reduced error pruning.

2.7.2 K Nearest Neighbours (KNN)

KNN is a simple algorithm that stores all available cases and classifies new cases based on a similarity measure (e.g., distance functions). It is a non-parametric technique used for pattern recognition and classification. A case is classified by a majority vote of its neighbours, with the case being assigned to the class most common amongst its K nearest neighbours measured by a distance function. Simple euclidian distance has been used for distance measure. If K = 1, then the case is simply assigned to the class of its nearest neighbour and so on.

2.7.3 Multilayer Feed Forward Neural Network

Multilayer Feed forward Neural network is trained with back propagation algorithm has been used as classifier. It is supervised classifier which adjusts its weight using back-propogation algorithm. Sigmodal threshold function has been used with the following parameters Learning rate 0.2, Momentum 0.3.

2.7.4 Support Vector Machine

Support Vector Machine is a machine learning method based on Statistical Learning theory and is used for classification and regression. Solution provided by SVM is theoretically elegant, computationally efficient and very effective in many large practical problems. The basic principle behind SVM is to find the optimal linear hyper plane which separates data from different categories with minimum error and maximum margin. John Pitt's Sequential Minimal Optimization (SMO) is a fast method to train SVM, performing well for large data sets. In this paper, SMO method has been employed to train SVM to achieve better classification accuracy. Training a SVM requires the solution of a very large quadratic programming (QP) optimization problem. SMO breaks this large QP problem into a series of smallest possible QP problems. For multiclass classification pair wise classification with normalized data has been used. Linear Kernel with binary SVM has been used for experimentation.

3   Database Details and Performance Evaluation

The dataset used for experimentation is taken from MUMS (McGill University Master Samples), which is set of 3-DVDs created by: Frank Opolko Joel Wapnick [11]. It is library of isolated music sound tones from a wide number of musical instruments, played with different articulation styles, covering entire pitch and recorded with 44.1 KHz sampling frequency as wave file. Experimentation is done on 760 monophonic isolated notes of 19 musical instruments covering string, Brass,

Woodwind and percussion families. For training the system, 70% notes are used for training and 30% notes are used for testing with cross 10-fold validation method and Instruments used for experimentation is listed in Table 2.

Table 2. Musical instruments used

Sr. No	Family	Instruments used
1	String	Guitar, Violin, Viola, Cello, Bass, Lute, Piano, harpsichord
2	Woodwind	Saxophone , Oboe classical , Oboe D, English Horn
3	Brass	Trumpet , Tuba, Cornet, Trombone , French Horn
4	Percussion	Steel drum, Tympani

Performance analysis:

In this section, the proposed method is analyzed and evaluated by performing various experiments on MUMS (McGill University Master Samples) database. Results in terms of Recognition accuracy is compared with proposed FRFT features with MFCC, Wavelet and Timberl features using different classifiers. Recognition accuracy is computed using KNN, Decision tree, Navie Bayes, SVM and Feed forward Neural Network algorithm. The result in percentage for different classifiers is given in table 3.

Table 3. Recognition accuracy in % for different classifiers

Features	No. Of features	KNN	Decision tree	Naive Bayes	SVM	NN
MFCC	13	76.06	65.42	76.32	58.51	75
Timbral	16	73.13	68.83	74.20	60.63	77.12
Wavelet	16	52.65	52.39	43.08	27.39	41.75
FRFT	12	88.94	76.84	78.94	75.78	91.84

4   Conclusion

In this paper, new features based on fractional Fourier transform have been proposed for Musical Instrument Recognition. The conventional Fourier transform is substituted by fractional fourier transform in MFCC and modified MFCC coefficients using FRFT have been proposed. These features represents the sub-band energies of mel filter bank of spectrum of music sound signal in fractional fourier domain. Recognition accuracy for 19 Muiscal instruments have been compared with proposed features with wavelet, MFCC and Timbral features using different machine learning algorithms. Result shows that performance of FRFT based features for musical instrument recognition system for all classifier is consistent and better. Further, it can be concluded that Music sound classes can be better represented in Fractional Fourier domain. In addition to this, proposed FRFT based features outperforms the conventional feature and only 12 features of FRFT based MFCC shows better performance as compared to MFCC, Wavelet, timbral features using any machine learning algorithms.

This work explores new tool of fractional fourier transform to the music signal research communities and Further research using FRFT may report significant achievements in this field.

References

1. Eronen, A.: Comparison of features for Musical instrument recognition. In: Proceeding of IEEE Workshop Applications of Signal Processing to Audio and Acoustic, pp. 19–22 (2001)
2. Martin, K.D., Kin: Musical Instrument recognition: A pattern recognition approach. Journal of Acoustical Society of America 109, 1068 (1998)
3. Deng, J.D., Simmermacher, C., Cranefield, S.: A study on Feature analysis for Musical Instrument Classification. IEEE Transaction on Systems, Man and Cybernetics 38(2), 429–438 (2008)
4. Eronen, A., Klapuri, A.: Musical Instrument Recognition using cepstral coefficients and Temporal features. In: ICASSP (2000)
5. Kaminskyj, I., Czaszejko, T.: Automatic Recognition of Isolated Monophonic Musical Instrument Sounds using k-NNC. Journal of Intelligent Information Systems 24(2/3), 199–221 (2005)
6. Agostini, G., Longari, M., Pollastri, E.: Content-Based Classification of Musical Instrument Timbres. IEEE Signal Processing Society (2003)
7. Essid, Richard, David: Hierarchical Classification of Musical Instruments on Solo Recordings. In: Proceedings of International Conference (2006)
8. Narayan, V.A., Prabhu, K.M.M.: The fractional Fourier transform: Theory, implementation and error analysis. Int. Journal of Microprocessors and Microsystems 27(10), 511–521
9. Ozaktas, H.M., Zalevsky, Z., Kutay, M.A.: The fractional Fourier transform with applications in optics and signal processing. Wiley, New York (2001)
10. Namias, V.: The fractional order Fourier transform and its application to quantum mechanics. IMA Journal of Appl. Math. 25(3), 241–265 (1980)
11. Mc gill University Master Sample, http://www.music.mcgill.ca/resources/mum/.html/mums.html
12. Agostini, G., Longari, M., Poolastri, E.: Musical instrument timbres classification with spectral features. EURASIP J. Appl. Signal Process. (1), 5–14 (2003), doi:10.1155/S1110865703210118
13. Kostek, B.: Musical instrument classification and duet analysis employing music information retrieval techniques. Proc. IEEE 92(4), 712–729 (2004)
14. Ajmera, P.K., Holambe, R.S.: Fractional Fourier transform based features for speaker recognition using support vector machine. Int. Journal of Computer and Electrical Engineering (2012)
15. Witten, H., Frank, E.: Data Mining: Practical Machine Learning Tools and Techniques, 2nd edn. Morgan Kaufmann, San Francisco (2005)
16. Quinlan, J.R.: C4.5: Programs for Machine Learning. Morgan Kaufmann, Springer, San Mateo, Appendix (1993)

Longitudinal Evaluation of Structural Changes in Frontotemporal Dementia Using Artificial Neural Networks

R. Sheela Kumari[1], Tinu Varghese[2], C. Kesavadas[3], N. Albert Singh[2]
and P.S. Mathuranath[4]

[1] Sree Chitra Tirunal Institute for Medical Science and Technology,
Trivandrum, Kerala
sheela82nair@gmail.com
[2] Noorul Islam University, Kumara coil, Thuckalay, Tamilnadu,
tinuannevarghese@gmail.com, albertsingh@rediffmail.com
[3] Department of Neurology, Sree Chitra Tirunal Institute
for Medical Science and Technology, Trivandrum, Kerala
chandkesav@yahoo.com
[4] Department of Neurology, National Institute of Medical Sciences
and Mental Health and Neurosciences, Banglore
psmathu@yahoo.com

Abstract. Automatic Segmentation of Magnetic Resonance (MR) Images plays an important role in medical image processing. Segmentation is the process of extracting the brain tissue components such as grey matter (GM), white matter (WM) and cerebrospinal fluids (CSF). The volumetric analysis of the segmented tissues helps in determining the amount of GM loss in specific disease pathology. Among the various segmentation techniques, fuzzy c means (FCM) is the most widely used one. The performance of traditional FCM is considerably reduces in noisy MR images. However, in the clinical analysis accurate segmentation of MR image is very important and crucial for the early diagnosis and prognosis. This paper put forward an Artificial Neural Network based segmentation to map the longitudinal structural changes overtime in Frontotemporal dementia (FTD) subjects that could be a better cue to impending behavioural changes. Our proposed approach has achieved an average classification accuracy of 96.7%, 96.4% and 97.96% for GM, WM and CSF respectively

Keywords: Magnetic Resonance (MRI), fuzzy C means (FCM), Artificial Neural Network (ANN), Frontotemporal Dementia (FTD).

1   Introduction

Magnetic Resonance Image (MRI) segmentation is one of the difficult and challenging tasks in the field of image processing. The main purpose of medical image segmentation is the production of a meaningful image that can be easily analyzed and

S.C. Satapathy, S.K. Udgata, and B.N. Biswal (eds.), *FICTA 2013*,
Advances in Intelligent Systems and Computing 247,
DOI: 10.1007/978-3-319-02931-3_20, © Springer International Publishing Switzerland 2014

implemented. Extraction of brain MRI into greymatter (GM), white matter (WM) and cerebrospinal fluids (CSF) [1,2] are widely used to measure the volumetric changes overtime in neurodegenerative diseases such as Alzheimer's Disease (AD), Fronto-temporal dementia (FTD), Epilepsy, Multiple Sclerosis (MS) and Schizophrenia. Traditional ROI- based manual segmentation is performed by expert radiologists. It has several disadvantages and may be inherently biased. In such situations computer based segmentation algorithms plays a vital role. The most widely used algorithms include thresholding, region growing and clustering. However, both region growing and thresholding are restrictive interms of precision of anatomical locations. Fuzzy clustering is an important unsupervised clustering technique for compensating intensity inhomogeneties [3]. Moreover, this algorithm only depends on the pixel intensity rather than their location or neighborhood properties. As a result, some stringent errors are occurred in the classification process.

In recent years, Artificial Neural Networks (ANN) has been used in MR image segmentation and promising results were obtained [4]. ANNs are the simplified version of human brain that utilizes connectionist perspectives. The ANN network is composed of a topology of neurons, which are capable of performing parallel computations. Multilayer Perceptron (MLP) is the simplest neural network architecture for the segmentation procedure. Some previous studies had successfully implemented a three-layered perceptron to segment the grey tones [5-7]. Later, Raff et al. used a Back Propagation Neural Network (BPNN) to segment GM, WM and CSF from Spin echo MR images [8]. However, BPANN for the quantification of GM, WM and CSF to map the progression of FTD have not been determined

2   Materials and Methods

2.1   Patients and MR imaging

Patients diagnosed with FTD at the decade old well-established Dementia Clinic at SCTIMST were included in this study. The subjects were undergone a detailed clinical, neuropsychological and neuroimaging analysis at baseline and a one year follow-up. 15 FTD patients with an age range of 56-79 yrs (mean ± SD, 65.4 ± 9.0) and total Frontal system Behavioural Scale (FrSBe) score (mean ± SD, 119.5± 27.9) were selected for this analysis.

MRI scanning was performed on a 1.5 Tesla Seimens Magnetom – Avanto, SQ MRI scanner. In all subjects, MR images of the entire brain were obtained using a three dimensional T1 weighted, spin echo sequence with the standard parameters (TR=11msec, TE=4.95, flip angle=150, slice thickness=1mm and matrix size =256x256). The post – processing was performed in the fully equipped brain mapping lab at SCTIMST.

2.2   Segmentation

Segmentation is the process of voxel level classification of predominant tissues of brain: GM, WM, and CSF. The various steps involved in the segmentation process are given below (Fig: 1)

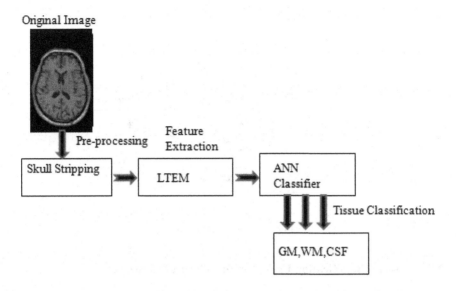

Fig. 1. Proposed Work Model

2.3 Preprocessing

Preprocessing involves the removal of noise and artifacts which are caused by inhomogenous magnetic fields, patient motion, thermal noise and noise due to metallic implants. Zadeh et al had done a comparitive analysis of image processing filters [13] and it was found that the implementation of median filter found to be very effective. Skull stripping is an important preprocessing task in MR image segmentation. It is the process of identifying brain and nonbrain voxels in the image. In this step, the nonbrain tissues such as skull, scalp, vein and dura are removed. Among the various skull-stripping techniques [14-16] mathematical morphological operation has showed a surpased rate of 95.5% than region growing [17]. Mathematical morphology involves basic operations like erosion and dilation to segment the brain image. Hence this analysis utilized mathematical morphology for skull stripping.

2.4 Feature Extraction

Feature extraction is the process in which texture features were extracted from the skull stripped MR image. The proposed system uses Laws of Texture Energy measure (LTEM) as feature vector [18]. The Laws of Texture energy measure is obtained by the convolution of an input image with 2-dimensional mask. The 2D convolution masks are generated from the following set of 1D convolution kernels of length five. The local averaging filter $L5 = (1, 4, 6, 4, 1)$, edge detector $E5 = (-1, -2, 0, 2, 1)$, spot detector $S5 = (-1, 0, 2, 0, -1)$, wave detector $W5 = (-1, 2, 0, -2, 1)$ and ripple detector $R5 = (1, -4, 6, -4, 1)$.

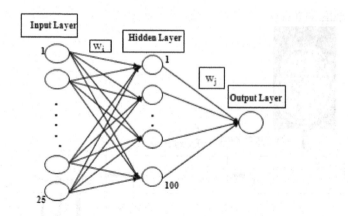

Fig. 2. Architecture of BPANN

2.5   Tissue Classification Using Artificial Neural Networks

ANNs are electronic models of nervous system that are capable of solving a variety of problems in pattern recognition and machine learning. Leaning algorithms are classified into supervised and unsupervised. The most commonly used are supervised learning. In ANN, learning is the process of updating network architecture and connection weights to perform a specific task. We have proposed a supervised ANN for the segmentation of MR image into GM, WM and CSF. The automatic segmentation depends only on the threshold values of brain tissues. Training was done for ANN using BP algorithm to classify GM, WM and CSF pixels itself [19]. As a result, we were able to classify GM, WM and CSF tissues. The process was made accurate by training the neural network based on its parameters like number of hidden layers, learning rate and momentum factor. The segmented results are given in Fig 3 and 4. The performance of our classification algorithm was evaluated with ground truth images. With the help of an expert radiologist we obtained the ground truth images. Previous studies in FTD have revealed significant GM volume reduction in medial frontal, orbito frontal and bilateral anterior temporal regions. Hence, we have done the longitudinal analysis of GM, WM and CSF volumes with the help of ground truth.

2.6   Model Learning

ANN is an adaptive system to model our real nervous system. The adaptive nature arises from the training phase. After the training phase, ANN recognizes the problem and set the parameter manually to get the desired output. As usual, the training of neural networks based on training algorithms. Among the various training algorithms, BPANN is found to be the most adaptable and robust technique for solving complex predicting problems. A simple architecture of a BPANN is shown in Fig2. In our analysis a three layer BPANN is used. The algorithm has been designed in a MATLAB framework, which aims at developing a CAD system for the longitudinal analysis of GM changes in FTD subjects. Our ANN contained 25 neurons in the input layer and 1 neuron in the output layer. For ANN training 100 hidden neurons, with a learning rate of 0.001 and momentum factor of 0.9, is involved with a maximum of 100 epochs iterations. The training set contained 1025 feature set for the brain tissue classification

3   Results and Discussions

The experiments were designed in MATLAB 7.1 package with neural network tool-box. The 15 MR images of FTD subjects obtained from SCTIMST Dementia Clinic were used for training the network and also were used as the ground truth image for testing the standard. In this analysis, better performance was obtained by varying the number of hidden neurons. As a result images were segmented into GM, WM and CSF. The volume of each of the tissue class was computed with the help of ground truth images. Our volumetric results revealed that patients with FTD had greater GM atrophy in visit compared to base. The performance evaluation was done on the basis of accuracy, sensitivity, specificity and youdens index parameters. Sensitivity is defined as the measure of the percentage of actual positive values that are correctly identified whereas specificity is the measure of percentage of negative values that are correctly identified. Accuracy deals with the quality of data and no of errors included in the data. Performance evaluation factor for the baseline and visit are included in Table1.

Fig. 3(a) shows the skull stripped MR image of an FTD patient in the baseline. The skull stripped image contains only the three tissue classes GM, WM and CSF. Fig 3(b) represents the ANN segmented map. Fig. 3(c), 3(d) and 3(e) represents GM, WM and CSF extracted from the skull stripped images. Fig 4(a), 4(b), 4(c), 4(d) and 4(e) represents skull stripped, ANN segmented map and the GM, WM and CSF tissue maps in the visit respectively.

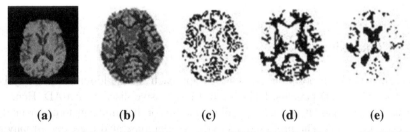

(a)   (b)   (c)   (d)   (e)

Fig. 3. Segmentation of a baseline MR brain image (FTD): (a) Original image (b) ANN Segmented map (c) GM (d) WM (e) CSF

Fig. 4. Segmentation of a follow-up MR brain image (FTD): (a) Original image (b) ANN Segmented map (c) GM (d) WM (e) CSF

Table 1. Average Volume of Segmented Tissues

Class	Volume using ANN		
	GM (mm^3)	WM (mm^3)	CSF (mm^3)
Base	358724	682494	279644
Visit	337108	680148	294428

Table 2. Quantitative Validation of Segmented Results

Class	Accuracy (%)			Sensitivity(%)	Specificity(%)	You(%)
	GM	WM	CSF			
Base	96.98	96.42	98.34	93.51	98.90	93.16
Visit	95.83	95	97.58	92.23	95.31	91.24

Our results are validated by an expert radiologist and have measured an increase in CSF volume in the visit. Table.1 compares brain tissue volumes in the baseline and visit. The tabulated results showed GM and WM atrophy in the follow-up compared to baseline. Table.2 indicates that the classification accuracy of GM in baseline is greater than the visit. Also the classification accuracy of GM and WM is more or less equal in base and followup after 1 year.

4   Conclusion

The current work proposed an ANN based approach for the longitudinal analysis of GM changes in FTD patients. FTD is a fast progressive disease than AD. Hence the quantitative and qualitative analysis will be helpful for the clinicians to understand the progression of FTD. The proposed approach has demonstrated a greater atrophy rate in follow-up than base. The obtained results have extended clinical applications such as to detect diagnostically relevant changes in FTD patients' overtime. Moreover, these automated methods for brain volumetry can be used to investigate disease pathology and volumetric changes. The average classification accuracy of our proposed approach is 96.7 %, 96.4% and 97.96% for GM, WM and CSF respectively. In future, an optimised ANN can be used to get more accurate results

Acknowledgments. The authors wish to express their thanks to Department of Science and Technology (DST) for funding.

References

1. Jobin Christ, M.C., Parvathi, R.M.S.: Segmentation of Medical Image using Fuzzy c-means Means Clustering and Marker Controlled Watershed Algorithm. European Journal of Scientific Research 71, 190–194 (2012)
2. Shah, B., Shah, S., Kosta, Y.P.: Novel Improved Fuzzy C-Mean Algorithm for MR-Image Segmentation. International Journal of Soft Computing and Engineering 2, 355–357 (2012)
3. Zanaty, E.A.: Determining the number of clusters for kernelized fuzzy C-means algorithms for automatic medical image segmentation. Egyptian Informatics Journal 13, 39–58 (2012)
4. Ashjaei, B., Soltanian-Zadeh, H.: A Comparative analysis of neural network methodologies for segmentation of magnetic resonance images". In: Proceedings of International Conference on Image Processing, vol. 2, pp. 257–260 (1996)
5. Blanz, W.E., Gish, S.L.: A connectionist classifier architecture applied to imagesegmentation. In: Proc. 10th ICPR, pp. 272–277 (1990)
6. Lawson, S.W., Parker, G.A.: Intelligent Segmentation of industrial radiographic images using neural networks. In: Proc. of SPIE, vol. 2347, pp. 245–255 (1994)
7. De Waard, W.P.: Neural techniques and postal code detection. Pattern Recognition Letters 15, 199–205 (1994)
8. Raff, U., Scherzinger, A.L., Vargas, P.F., Simon, J.H.: Quantitation of grey matter, white matter, and cerebrospinal fluid from spin-echo magnetic resonance images using an artificial neural network technique. Med. Phys. 21, 1933–1942 (1994)
9. Poletti, E., Veronese, E., Calabrese, M., Bertoldo, A., Grisan, E.: Supervised classification of brain tissues through local multi-scale texture analysis by coupling DIR and FLAIR MR sequences. In: SPIE 8314, Medical Imaging (2012)
10. Christodoulou, C.I., Michaelides, S.C., Pattichis, C.S.: Multifeature Texture Analysis for the Classification of Clouds in Satellite Imagery. IEEE Transactions on Geoscience And Remote Sensing 41, 2662–2668 (2003)
11. Grossman, M.: Frontotemporal dementia: A review. J. Int. Neuropsychol. Soc. 8, 566–583 (2002)
12. Zamboni, G., Huey, E.D., Krueger, F., Nichelli, P.F., Grafman, J.: Apathy and disinhibition in frontotemporal dementia: Insights into their neural correlates. Neurology 71, 736–742 (2008)
13. Zadeh, H.M., Windham, J.P., Peck, D.J., Yagle, A.E.: A Comparative Analysis of Several Transformations for Enhancement and Segmentation Magnetic Resonance images Scene sequences. IEEE Transaction on Medical Imaging 11, 302–318 (1992)
14. Smith, M.S.: Fast Robust Automated Extraction Human. Brain Mapping 17, 143–155 (2002)
15. Adams, R., Bischof, L.: Seeded region growing. IEEE Transactions on Pattern Analysis and Machine Intelligence 16, 641–646 (1994)

16. Hojjatoleslami, S.A., Kittler, J.: Region Growing: A New Approach. IEEE Transactions on Image Processing 7 (1998)
17. Roslan, R., Jamil, N., Mahmud, R.: Skull Stripping Magnetic Resonance Images Brain Images: Region Growing versus Mathematical Morphology. International Journal of Computer Information Systems and Industrial Management Applications 3, 150–158 (2011) ISSN 2150-7988
18. Suckling, J., Parker, J.: The Mammographic Images Analysis Society Digital Mammogram Database. In: Proc. of 2nd Int. Workshop Digital Mammography, U.K, pp. 375–378 (1994)
19. Jafari, M., Kasaei, S.: Neural network-based brain tissue segmentation in MR images using extracted features from intraframe coding in H.264. In: Proc. SPIE 8349, Fourth International Conference on Machine Vision (2012)

Content Based Image Retrieval Using Machine Learning Approach

Palepu Pavani and T. Sashi Prabha

Anil Neerukonda Institute of Technology and Sciences
Visakhapatnam, Andhra pradesh, India
{palepupavani22,tinkusashi}@gmail.com

Abstract. The rapid growth of computer technologies and the advent of the World Wide Web have increased the amount and the complexity of multimedia information. A content -based image retrieval (CBIR) system has been developed as an efficient image retrieval tool, whereby the user can provide their query to the system to allow it to retrieve the user's desired image from the image database. However, the traditional relevance feedback of CBIR has some limitations that will decrease the performance of the CBIR system, such as the imbalance of training-set problem, classification problem, limited information from user problem, and insufficient training set problem. Therefore, in this study, we proposed an enhanced relevance-feedback method to support the user query based on the representative image selection and weight ranking of the images retrieved. The support vector machine (SVM) has been used to support the learning process to reduce the semantic gap between the user and the CBIR system. From these experiments, the proposed learning method has enabled users to improve their search results based on the performance of CBIR system. In addition, the experiments also proved that by solving the imbalance training set issue, the performance of CBIR could be improved.

1   Introduction

Images are considered as the prime media type for the use of retrieving hidden information within data [6]. In general, an image retrieval system is a computer system for browsing, searching, and retrieving images from a large digital-images database [1]. In the early trend, image retrieval utilized certain methods of adding metadata, such as captioning, keywords, or descriptions to the images [1], so that retrieval can be performed over the annotation words. Obviously, annotating images manually is a time-consuming, laborious, and expensive task for large image databases, and is often subjective, context-sensitive, and in-complete [1]. Thus, content-based image retrieval (CBIR), which is an-other method of image retrieval, attempts to overcome the disadvantage of the keyword-annotation method. The CBIR aims to retrieve images based on their visual similarity to a user-supplied query image or user-specified image features. The visual contents of an image, such as color, shape, texture, and spatial layout, have been used to represent and index the image [1].

However, the performance of CBIR has been limited by several issues, such as subjectivity of human perception [6], similarity of visual feature [8], and semantic query-gap issues [3]. To solve these problems, interactive relevance feedback, which

S.C. Satapathy, S.K. Udgata, and B.N. Biswal (eds.), *FICTA 2013*,
Advances in Intelligent Systems and Computing 247,
DOI: 10.1007/978-3-319-02931-3_21, © Springer International Publishing Switzerland 2014

involves the interaction between human and system, was introduced. Relevance feedback is a supervised active learning technique, which uses the positive and negative feedback examples from the users to improve system performance. For a given query, the system first retrieves a list of ranked images according to the predefined similarity metrics. Subsequently, the user marks the retrieved images as relevant (positive examples) or irrelevant (negative examples) to the query. Thus, the system will refine the retrieval results based on the feedback and present a new list of images to the user. This process will go through several iterations until the user is satisfied with the retrieval result. Even though relevance feedback intends to reduce the problems of CBIR, its performance is still not satisfactory owing to the limitations of the traditional relevance feed-back of CBIR, such as query refinement band re-weighting (discussed in "Related Work" section). These problems include imbalance of training-set problem, classification problem, limited information from user problem, and in-sufficient training-set problem. Among these problems, imbalance training set issue will be the focus on this study.

In order to solve imbalance training set issue, we proposed, a relevance feedback that use the representative image selection and weight ranking approach. This approach will adapt support vector machines (SVMs) for the purpose of supervised learning and retrieve the image according to the users' preference. The related work on relevance feed-back based on CBIR is reviewed in Section 2 and the details of the proposed method are presented in Section 3. The experimental setup and its results are given in Section 4, while in Section 5; we have discussed the results of the experiments. In the last section, the conclusion and the direction for future work are presented.

2   Related Work

The traditional CBIR relevance-feedback techniques include query refinement [6] and re-weighting [9]. However, both techniques did not achieve a satisfactory performance for CBIR owing to several issues. One of the issue is how to incorporate positive and negative examples to refine the query or to adjust the similarity measure [1]. In fact, there has been a trend on CBIR researches focusing on the relevance-feedback issues. For instance, Cheng et al. [10] proposed a new relevance-feedback model suit-able for medical image retrieval; Qin et al. [11] proposed an active relevance-feedback framework to make the relevance feedback more informative, using unlabeled data in the training process; Das and Ray [5] provided a brief overview on feature re-weighting approach; Crucianu et al. [3] discussed the main issues related to relevance feedback for image retrieval as well as the recent developments in this domain, and Qi and Chang [4] introduced a composite relevance-feedback approach for image retrieval using transaction-based and SVM-based learning. According to earlier studies on SVM-based relevance feedback, the SVM is considered to normally treat the problem as a strict binary-classification problem, without noticing an important issue of relevance feedback , such as the imbalanced training-set problem in which the negative in-stances significantly overwhelm the positive ones [7]. In this study, relevant or positive image is the image that is similar to the query image in user perception, meanwhile, irrelevant or negative image is the image that is dissimilar to the query image. Consequently, this problem will degrade the performance of the CBIR system when the number of labeled positive-feedback samples is small [2][8][11]. As a result, the information we gain from the user is very limited. In addition, the training samples (user's labeled images) are insufficient because the users would only label a few images, and unable to label each feedback sample accurately all the time [2][11]. This issue had been mentioned previously as the limited information from user problem. Hence, a proper technique of relevance feedback that is adaptable with SVM is desired.

3   Proposed Method

In this study, the methodology consists of four main parts which are data collection, preprocessing, feature similarity measure and relevance feedback as shown in Fig. 1. How-ever, we focus on the relevance feedback part which has consists of five units. Among these units, we are highly emphasizing the use of representative image selection and weight ranking units. The benefits from these two units are:

1. Representative image selection unit is use to select a set of informative images from image database for the use of labeling process. According to Qin et al. [11], it is an effective way for the user to do labeling work which it will keep the mass of labeling task very little but meaningful. Besides, if there are more relevant images being retrieved, the possibility of imbalance training set problem to occur will be minimized.

2. Weight ranking unit is use to provide a better way for user to do labeling. User will rank the retrieved images in a sequence from most relevant to least relevant to the target image. According to Cheng et al. [10], it can help the system to quickly determine the user preference feature directly from the user feed-back.

Additionally, user ranking sequence can deter-mine which method is more concern by user and can adjust it accordingly. As a result, system will capable to increase the weight of important features and at the same time decrease the weight of insignificant features. This type of weight adjustment is adjusted according to the users' preference.

CBIR, user will provide a query image and it will go through preprocessing process to extract its features content such as color, texture and shape. In this study, the query image will be segmented to several significant regions by using MAP segmentation technique [13]. In each image region, the regions features such as color and texture features will be extracted to describe the particular region. Generally, the images features selection is a very fundamental issue in designing a content-based image retrieval system. The combination of two or more features are best representing the images content and there are no single feature can rep-resent the whole content of an image perfectly. Out of all the available features, the combination of color and texture features is chosen to represent the image regions.

In this study, the haar wavelet filter a[12][15] will be used to extract the texture features. However, the average RGB values of all pixels which belongs to the region will be de-fined as the color feature [14]. The extracted features will be compared and ranked through feature similarity measure process. In this study, Earth Mover Distance (EMD) will be used as the image similarity measure [17]. EMD is used to compute the dissimilarity between sets of regions and returns the correspondence between them [17]. After that, system will select a set of estimated possible positive image set (EPPIS), as an example, top 100 images from the ranked list will be classed as the EPPIS set. Its aim is to select a set of images that contain a subset of images from image database which are most likely relevant to the query [11]. In the representative image selection, system will select a set of representative images from the EPPIS set which have the minimum information loss when retrieve the images from EPPIS for user labeling process [11]. As a result, the selected images for user labeling process will fulfill two behaviors which the image has the similar behavior in training the classifier and do not contain redundancy. In the user labeling process, the selected representative images will be displayed and retrieved for the user to do labeling on it. In this study, user will give a sequence of images which is called the image ranking sequence and this sequence will be the feedback to the CBIR system. By using another word, user will rank a sequence of relevant images in order with respect to their similarity to the query image [10]. Thus, the retrieved images which are out of the image ranking sequence will be labeled as the irrelevant images. There are two types of user-defined preference relation which are > and = that will be used by the user to show the image ranking sequence.

Algorithm:
1. Browse an Input file(image) to the guide model
2. Display the path and Input image
3. Perform the basic query i.e.,,, extracting relevant Images using colour, texture and background and relevant data.
4. Perform SVM technique using wavelets and refined output images will be Displayed

4   Results

In this first we are selecting am image, it will display both the path and the image.

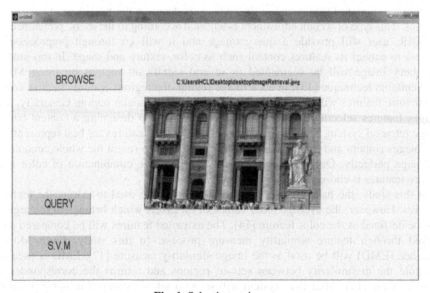

Fig. 1. Selecting an image

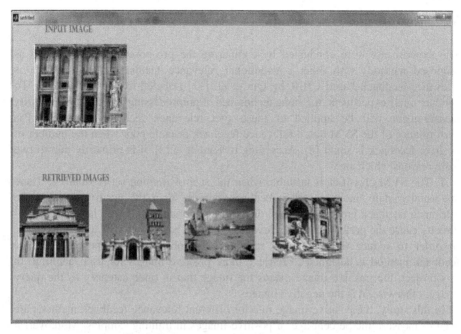

Fig. 2. Image retrieval using Query Technique

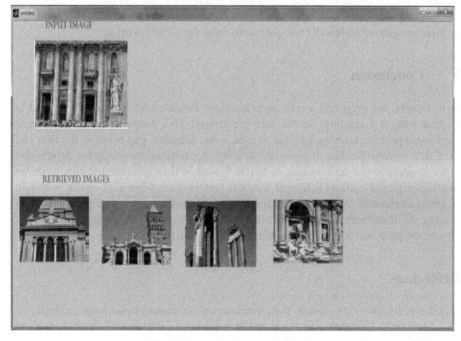

Fig. 3. Image Retrieval using SVM

5   Results and Discussion

The experiments were conducted by comparing the pro-posed method (labeled as proposed method) with other conventional relevance feedback method which is relevance-feedback based CBIR by Qin et al. [11] (labeled as conventional 2). To conduct a fair experiment, the same techniques of preprocessing and feature similarity measurement will be applied at these two relevance feedback methods. The performance of the SVM based relevance feedback became poor when the number of positive feedback is small [8]. According to Kim et al.[8], it is primarily due to two main reasons which are:

1. The SVM classifier is unstable when the size of training set is small. 2. There are usually many more negative feedback samples than the positive ones in the relevance feedback process. Besides, the poor performance of SVM classifier will indirectly cause the performance of relevance feedback based on CBIR to become poor. In order to reduce the imbalance training set problem, the proposed relevance feedback method is designed to retrieve more positive images for user labelling. In this project, the positive image means the image that is same category as the query image, otherwise, it is the negative image.

In this study, the positive image rate for different relevance feedback methods are used to measure the percent-age of positive images in training samples. Moreover, it will be used to show how the percentage of positive images in training samples will influence the performance of Support Vector Machine (SVM) and indirectly influence the performance of CBIR system. Figure 2 shows the relationship between the positive image rate and the F1 measurement value for SVM classifier.

6   Conclusion

In this study, we proposed a relevance feedback based on SVM learning method to retrieve images according to the user preference. This proposed method has been used to support the learning process to reduce the semantic gap between the user and the CBIR system. Besides, it also aims to solve the imbalance training set problem in order to improve the performance of CBIR. Based on the experiment results, it shows that the proposed method achieved the best performance when it compare with two others conventional methods. In addition, the experiments also been proven that by solving the imbalance dataset issue, the performance of CBIR could be improved. Hence, the proposed method is capable to solve the CBIR problems.

References

[1] Long, F., Zhang, H., David, D.F.: Fundamentals of content-based image retrieval. In: Multimedia Information Retrieval and Management - Technological Fundamentals and Applications. Springer (2003, 2010)

[2] Tao, D.C., Tang, X.O., Li, X.L., Wu, X.D.: Asymmetric bagging and random space for support vector machines-based relevance feedback in image retrieval. IEEE Trans. Pattern Analysis and Machine Intelligence 28(7) (July 2006)

[3] Crucianu, M., Ferecatu, M., Boujemaa, N.: Relevance feedback for image retrieval: A short survey. Report of the DELOS2 European Network of Excellence, 6th Framework Programme (October 10, 2004)

[4] Qi, X., Chang, R.: Image retrieval using transaction-based and SVM-based learning in relevance feedback sessions. In: Kamel, M.S., Campilho, A. (eds.) ICIAR 2007. LNCS, vol. 4633, pp. 638–649. Springer, Heidelberg (2007)

[5] Das, G., Ray, S.: A comparison of relevance feed-back strategies in CBIR. IEEE (2007)

[6] Rui, Y., Huang, T.S., Ortega, M., Methrotra, S.: Rel-evance feedback: A power tool in interactive content-based image retrieval. IEEE Trans. On Circuits and Systems for Video Technology 8(5), 644–655 (1998)

[7] Hoi, C.H., Chan, C.H., Huang, K.Z., Lyu, M.R., King, I.: Biased support vector machine for relevance feedback in image retrieval. In: Proceedings of Intl. Joint Conf. on Neural Networks (IJCNN 2004), Budapest, Hungary (2004)

[8] Kim, D.H., Song, J.W., Lee, J.H., Choi, B.G.: Sup-port vector machine learning for region-based image retrieval with relevance feedback. ETRI Journal 29(5) (October 2007)

[9] Rui, Y., Huang, T.S.: A novel relevance feedback techniques in image retrieval. In: Proc. 7th ACM Conf. on Multimedia, pp. 67–70 (1999)

[10] Cheng, P.C., Chien, B.C., Ke, H.R., Yang, W.P.: A two-level relevance feedback mechanism for image retrieval. Expert Systems with Applications 34(3), 2193–2200 (2008)

[11] Qin, T., Zhang, X.D., Liu, T.Y., Wang, D.S., Ma, W.Y., Zhang, H.J.: An active feedback framework for image retrieval. Pattern Recognition Letters 29(5), 637–6461 (2008)

[12] Smith, J.R., Chang, S.F.: Automated binary texture feature sets for image retrieval. In: Proc. ICASSP 1996, May 7-10 (1996)

[13] Blekas, K., Likas, A., Galatsanos, N., Lagaris, I.: A Spatially-Constrained Mixture Model for Image Segmentation. IEEE Transactions on Neural Networks 16(2), 494–498 (2005)

[14] Liu, Y., Zhang, D.S., Lu, G.: Region-Based Image Retrieval with High-Level Semantics using Decision Tree Learning. Pattern Recognition Ramakrishna Reddy.Eamani et al. / International Journal of Engineering Science and Technology (IJEST) 4(4), 1518 (2012); Recognition 41(8), 2554-2570 (August 2008)

[15] Manjunath, B., Wu, P., Newsam, S., Shin, H.: A texture descriptor for browsing and similarity retrieval. Signal Processing Image Communication (2001)

[16] Ohm, J., Bunjamin, F., Liebsch, W., Makai, B., Mller, K., Smolic, A., Zier, D.: A Set of Visual Feature Descriptors and their Combination in a Low-Level Description Scheme. Signal Processing: Image Communication 16, 157–179 (2000)

[17] Greenspan, H., Dvir, G., Rubner, Y.: Region Correspondence for Image Matching via EMD Flow. IEEE (2000)

[18] Chowdhury, G.G.: Introduction to modern information retrieval. Library Association Publishing, London (1999); Eamani, R.R., et al.: International Journal of Engineering Science and Technology (IJEST)

Hand Gesture Detection and Recognition Using Affine-Shift, Bag-of-Features and Extreme Learning Machine Techniques

M. Kranthi Kiran and T. ShyamVamsi

Department of Computer Science and Engineering,
Anil Neerukonda Institute of Technology and Sciences, Visakhapatnam, India
{kranthikiranm67,tshyamvamsi}@gmail.com

Abstract. This paper presents a real-time system for interaction with applications via hand gestures. Our system includes detection of bare hand from the video sequences by subtracting the background and including only the hand region. The system uses a machine learning approach. In the training stage the keypoints are extracted from the hand posture contour using the affine-scale invariance feature transform (ASIFT), and the keypoints are then clustered using the K-means clustering and mapped into the histogram vector (bag-of-features). Each vector is assigned a label which is treated as input to the Extreme Learning Machine (ELM) for training purpose. In the testing stage, for every frame captured using the webcam, the hand is detected, then, the keypoints are extracted from the hand segment only as described in our algorithm and fed into the cluster model to generate the vector and fed into the ELM training classifier to recognize the hand gesture.

Keywords: Affine scale invariance feature transform (ASIFT), Bag-of-Features, contour, Extreme Learning Machine (ELM), hand gesture, hand posture, K-means.

1   Introduction

HAND gesture recognition provides an intelligent, natural, and convenient way of human–computer interaction (HCI). Gesture recognition enables humans to communicate with the machine (HMI) and interact naturally without any mechanical devices. Using the concept of gesture recognition, it is possible to point a finger at the computer screen so that the cursor will move accordingly. This could potentially make conventional input devices such as mouse, keyboards and even touch-screens redundant.

Hand gestures are a collection of movements of the hand and arm that vary from the static posture of pointing at something to the dynamic ones used to communicate with others. The methods used for understanding these structures and movements are among the most classifying researches that still in progress. The gesture recognition methods are primarily divided into *Data-Glove Based* and *Vision Based methods* [1].

S.C. Satapathy, S.K. Udgata, and B.N. Biswal (eds.), *FICTA 2013*,
Advances in Intelligent Systems and Computing 247,
DOI: 10.1007/978-3-319-02931-3_22, © Springer International Publishing Switzerland 2014

The Data-Glove based approaches use mechanical or optical sensors connected to a glove that converts finger flexions into electrical signals for recognizing the hand posture. This method obstructs the ease and naturalness of the user interaction because it compels the user to carry a load of cables which are attached to the computer. However, the equipment is relatively expensive and awkward. On the contrary, the Vision Based techniques need only a camera, consequently achieving a natural interaction between humans and computers without the need for any extra devices. The Vision Based techniques are further classified into 3D hand model-based methods and Appearance-based methods. The paper describes the hand gesture recognition based on the appearance based models.

2   Related Work

In this paper, we will consider only appearance-based hand gestures captured with a camera. Appearance-based models can be can be further classified into two types: a *static and dynamic* gesture. Static hand gestures (hand postures/poses) are those in which the hand position does not change during the gesturing period. In dynamic hand gestures (hand gestures), the hand position is temporal and it changes continuously with respect to time. Dynamic gestures, which are actions composed of a sequence of static gestures, can be expressed as a temporal combination of static gestures [2].

The vision based hand gesture recognition can be classified into many categories of which lot of literature can be catogerised as here (1) Hidden Markov Model (HMM) based methods, (2) Neural network (NN) and learning based and (3) Other methods (Graph algorithm based methods), (4) 3D model based methods, (5) Statistical and syntactic methods, and (5) Eigen space based methods.

In [3], the efficiency of the keypoints to represent the images is presented. Keypoints are detected by robust feature detection methods like SIFT. In [4], ASIFT simulates three parameters: the scale, the camera longitude angle and the latitude angle (which is equivalent to the tilt) and normalizes the other three (translation and rotation). In [4], many experiments were conducted to evaluate performance of: SIFT, ASIFT.

Recently, bag-of-features (BoF) representations have shown outstanding performance for action and gesture recognition. However, despite recent developments, relatively few local descriptors in videos exist that benefit from combined spatial and temporal information. One limitations of BoF is that it has no explicit notion of objects or actors due to its order less representation, this lack of explicit object knowledge prevents modeling of spatial layout information which has been shown to increase performance [5]-[6]. Furthermore, BoF provides a global video representation which is inherently sensitive to background clutter [7].

In [3], [8], for training and classification purpose the variants of Support Vector Machines (SVM) is considered to give the best result and is a fast process. However, it is known that both neural networks and SVMs face some challenging issues such as: (1) slow learning speed, (2) trivial human intervene, and/or (3) poor computational scalability [9]. Extreme learning machine (ELM) as emergent technology which overcomes some challenges faced by other techniques has recently attracted the attention from more and more researchers.

3   System Overview

Our hand gesture system consists of two stages: the training and testing stages, much of the regonition framework is similar to [3] and is shown as in Figure-1.

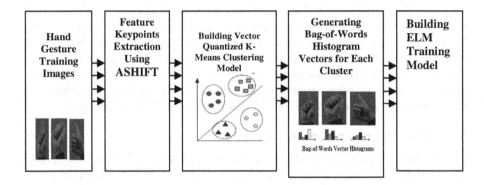

Fig. 1. Training Stage

A. Training Stage

The training stage is shown in Fig. 1. Initially the hand posture data sets are collected from Github, Sébastien Marcel - Hand Posture and Gesture Datasets [10]. We captured initially 572 images of hand gesture, which are the fist, index, palm, thumb, two finger gestures, for 5 different people, scales and rotations and under different illuminations conditions to increase the robustness of the ELM classifier and the cluster model.

The image database had many resolutions and when training all the finger gesture image size were reduced to 320 X 240 pixels and converted into PGM format to coincide with the size of images captured from the video file in the testing stage. The training stage was repeated with training images using different resolutions such as 320 X 240 pixels , 160 X 120 pixels , 50 X 50 pixels and new clusters were built for defined resolutions and for each gesture.

1) *Feature Extraction Using Affine Scale Invariant Feature Transform (ASIFT)*

ASIFT simulates with enough accuracy all distortions caused by a variation of the camera optical axis direction. Then it applies the SIFT method. In other words, ASIFT simulates three parameters: the scale, the camera longitude angle and the latitude angle (which is equivalent to the tilt) and normalizes the other three (translation and rotation). The mathematical proof that ASIFT is fully affine invariance will be given in [4].

B. K-Means Clustering

Among the different types of clustering, here k-means clustering algorithm defined in [11] is used. In the training stage, the images contain only the hand gestures on a white background, the keypoints that are extracted will represent the hand gesture only, and this depends on the number of extracted keypoints which range to 100- 200. Therefore, the training stage provides the minimum number of clusters that we can use. In the testing stage, the webcam will capture other objects besides the hand gesture such as the face and background. The keypoints will be around 800 keypoints of all the objects in the image. We chose the value 750 as the number of clusters (visual vocabularies or codebook) to build our cluster model.

Each keypoint, extracted from a training image, will be represented by one component in the generated bag-of-words vector with value equal to the index of the centroid in the cluster model with the nearest Euclidean distance. The generated bag-of-words vector, which represents the training image, will be grouped with all the generated vectors of other training images that have the same hand gesture and labeled with the same number, and this label will represent the class number.

1) *Building the Training Classifier Using ELM*

After mapping all the keypoints that represent every training image with its generated bag-of-words vector using k-means clustering, we fed every bag-of-words vector with its related class or label number into ELM classifier to build the ELM training classifier model.

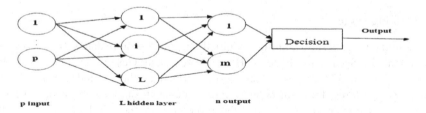

Fig. 2. Simplified Structure of ELM

Huang et al. [12] showed that single hidden layer feed forward neural network, also termed as ELM, can exactly learn N distinct observations for almost any nonlinear activation function with at most N hidden nodes as in Figure-2. ELM converts the learning problem into a simple linear system whose output weights can be analytically determined through a generalized inverse operation of the hidden layer weight matrices.

The profound use of multiple types of features is evident from improved detection results [13]-[14] since they provide complementary information for action recognition. However, trade-off between acquired accuracy and computational time poses a major bottleneck for real-time implementation of these schemes in various applications. In [15], much work of the ELM algorithm used here is described.

C. Testing Stage

The testing stage starts with capturing the frames from a video or a webcam and then the much similar skin types like the faces are detected and subtracted and only the hand region is captured and the hand posture contour algorithms are applied and the detected and the hand gesture image is saved in 50 X 50 pixel resolution images, as described in [3].

1) *Face Detection and Subtraction*

To get rid of other skin pixels majorly the face area, as described in [3], we detected the face using the most popular and well known Viola and Jones method [16]. Then the the face is subtracted and then the skin detection algorithm is applied to detect only the hand gesture.

2) *Hand Gesture Detection*

After the face is subtracted our described algorithm is applied by extracting the feature keypoints using ASIFT and the bag-of-features. The testing images are not be confined to the resolution of the captured images from the video or the webcam, because we always extract the keypoints in the 50 X 50 pixels of the bare hand only.

For our testing system we detected the hand gesture using the skin-color detection by converting Red, Green, Blue (RGB) color representation Hue, Saturation, Value color representation (HSV).If the contours of the skin area comply with any of the contours of the hand gestures templates, then, that area will be the region of interest by enclosing the detected hand gesture with a rectangle, which will be used in tracking the hand movements and saving hand gesture in a small image (50 X 50 pixels) for every frame captured.

3) *Hand Gesture Recognition*

We converted the small image (50 × 50 pixels) that contains the detected hand gesture only for every frame captured into a PGM format to reduce the time needed in extracting the keypoints. For every small PGM image, we extracted the keypoints using the ASIFT algorithm. The same training stage methodwas used. Each feature vector in the keypoints will be represented by one component in the generated vector with value equals to the index of centroid in the cluster model with nearest Euclidean distance. Finally, the generated bag-of-words histogram vector will be fed into the ELM training classifier model that was built in the training stage to classify and recognize the hand gesture.

4   Experimental Results

The experiments and the results are presented in this section. We tested five hand gestures, namely: the fist gesture, the index gesture, the palm gesture, the thumb gesture and the two gestures. The system described was tested on a Windows7 Operating system using the laptop webcam and a ordinary low resolution Logitech webcam. The low resolutions of the laptop 640 X 480, 800 X 600, were used at 19 frames per second.

The training was done using the benchmark hand gesture datasets downloaded from the github and [18]. 7 videos were recorded for 30 seconds each for the five gestures: fist, index, palm, thumb, and two. So total of 570 frames were available to test the system recorded under normal lightening conditions.

Table 1. Performance of the ELM Classifier Without any Objects (800 X 600 pixels)

Gesture Name	Number of frames	Correct	Incorrect	Correct rate	Recognition Time (second/ frame)
Fist	570	560	10	98.2 %	0.05320
Index	570	570	0	100 %	0.05332
Palm	570	565	5	99.1%	0.05313
Thumb	570	562	8	98.59%	0.05333
Two	570	560	10	98.2%	0.05331

Table-1 shows the performance of the ELM classifier for each of the gesture with testing against scale, rotation, illumination, change in longitude and latitude degrees. During the testing stage using the webcam there were not pause or latency to recognize the gestures, which shows that the system shows excellent results in terms of accuracy and speed because the features were extracted in real time and ASIFT was applied to get the keypoints which are invariant to scale and rotation. Also as the images were captured in different illumination conditions and trained, the ELM classifier is robust against illumination changes. The output is generated for each frame by giving the gesture name on the top left as shown in Figure-3.

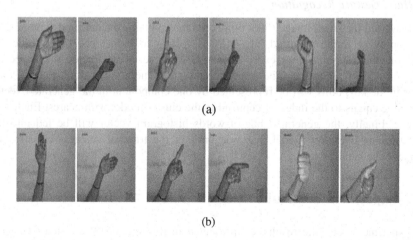

(a)

(b)

Fig. 3. Hand gesture recognition without any object and recognized gestures of palm, index, fist, for varying (a) scale (b) rotation

5   Conclusion

In this paper, we presented a real-time hand detection and posture recognition system using ASIFT features and ELM machine learning technique. Experiments show that the system can achieve satisfactory real-time performance regardless of the frame resolution size as well as high classification accuracy of 98.5% under variable scale, orientation and illumination conditions, change of different degrees and cluttered background.

References

1. Garg, P., Naveen, A., Sangeev, S.: Vision Based Hand Gesture Recognition. World Academy of Science, Engineering and Technology 49(173), 972–977 (2009)
2. Pisharady, P.K., Vadakkepat, P., Loh, A.P.: Attention Based Detection and Recognition of Hand PostureAgainst Complex Backgrounds. International Journal of Computer Vision 101, 403–419 (2013), doi:10.1007/s11263-012-0560-5
3. Dardas, N.H., Georganas, N.D.: Real-Time Hand Gesture Detection and Recognition Using Bag-of-Features and Support Vector Machine Techniques. IEEE Transactions on Instrumentation and Measurement 60(11) (November 2011)
4. Morel, J.-M., Yu, A.G.: Asift: A New Framework For Fully Affine Invariant Image Comparison. Image Processing On Line (2011)
5. Dalal, N., Triggs, B.: Histograms of oriented gradients for human detection. In: CVPR (2005)
6. Lazebnik, S., Schmid, C., Ponce, J.: Beyond bags of features: Spatial pyramid matching for recognizing natural scene categories. In: CVPR (2006)
7. Zhang, J., Marszalek, M., Lazebnik, S., Schmid, C.: Local features and kernels for classification of texture and object categories: A comprehensive study. International Journal of Computer Vision 73(2), 213–238 (2007)
8. Naik, G.R., Kumar, D.K., Jayadeva: Twin SVM for Gesture Classification Using the Surface Electromyogram. IEEE Transactions on Information Technology in Biomedicine 14(2) (March 2010)
9. Huang, G.-B., Lan, Y., Wang, D.H.: Extreme learning machines: A survey. International Journal of Machine Learning & Cybernetics 2, 107–122 (2011), doi:10.1007/s13042-011-0019-y
10. http://www.idiap.ch/resource/gestures/
11. MacKay, D.J.C.: Information Theory, Inference, and Learning Algorithms. Cambridge Univ. Press, Cambridge (2003)
12. Huang, G.B., Zhu, Q.Y., Siew, C.K.: Extreme learning machine: Theory and applications. Neurocomputing, 489–501 (2006)
13. Liu, J., Ali, S., Shah, M.: Recognizing human actions using multiple features. In: Proceedings of the International Conference on CVPR, pp. 1–8 (2008)
14. Liu, J., Luo, J., Shah, M.: Recognizing realistic actions from videos in the wild. In: Proceedings of the International Conference on CVPR (2009)
15. Minhas, R., Baradarani, A., Seifzadeh, S., Wu, Q.M.J.: Human action recognition using extreme learning machine based on visual vocabularies. Neurocomputing, 1906–1917 (2010)
16. Viola, P., Jones, M.: Robust real-time object detection. International Journal on Computer Vision 2(57), 137–154 (2004)

Comparative Study of Machine Learning Algorithm for Intrusion Detection System

K. Sravani and P. Srinivasu

Department of Computer Science and Engineering,
Anil Neerukonda Institute of Technology and Sciences, Visakhapatnam, India
{sravanikodukula,ursrinivasu}@gmail.com

Abstract. Now a day's, Intrusion detection is a very important research area in network security. Machine learning techniques have been applied to the field of intrusion detection. In this paper, we use KDD Cup 99' data set for taking samples. For these samples we use classification algorithms to classify the network traffic data. In this paper, we are going to compare our results with features selected using Naive Bayes, Neural Networks. We are trying to use standard measurements like detection rate, false positive, false negative, accuracy and Confusion Matrix.

Keywords: IDS, Machine learning algorithms, KDDCUP99 DATASET, Confusion Matrix.

1   Introduction

Now a day's Systems and Networks are prone to electronic attacks. These attacks are mostly happening in systems in order to crack system security. To secure the system, we have to trace them before they turn out to be severe. For this we can use Intrusion Detection System that helps computer to get aware of secure attacks.

The concept of intrusion detection system was introduced by James Anderson in the year 1980. James defined intrusion as an attempt or a threat which deliberates an unauthorized attempt to [4]

- Access information
- Manipulate information or
- Render a system unreliable or unusable.

Intrusion detection techniques are of two types:

1. *Anomaly detection* also known as behavior based detection, which detects the intrusions that occur in computer by scanning / monitoring the activities and then classify it as either normal or anomalous. It can identify novel attacks by analyzing the change of behavior from normal behavior. Here, anomaly detectors construct profiles which represent normal usage and then, use current behavior data to detect possible mismatch between profiles and recognize possible attack attempts[4]. It has a relatively high detection rate for new attack, but produces many false positives.

S.C. Satapathy, S.K. Udgata, and B.N. Biswal (eds.), *FICTA 2013*,
Advances in Intelligent Systems and Computing 247,
DOI: 10.1007/978-3-319-02931-3_23, © Springer International Publishing Switzerland 2014

2. *Misuse detection* is also known as Signature based IDS or Knowledge based IDS. It performs the simple process of matching patterns corresponding to a known attack type[4]. It has a relatively low rate of false alarms, which means it has relatively high precision.

Here we focus on developing an anomaly based intrusion detection system using machine learning techniques such as Naive Bayes, Neural Network, in order to reduce the number of false positives and false negatives and to improve the accuracy of the system in classifying the attacks.

As network based computer systems play vital roles in modern society, they can act as a intrusion detection systems provide following three essential security functions[4]:

Data confidentiality: Information that is being transferred through the network should be accessible only to those that are properly authorized.

Data integrity: Data has not been altered or destroyed in an unauthorized manner.

Data availability: The network or a system resource ensures that it is accessible and usable on demand, by an authorized system user.

2   Architecture

Figure 1 shows the Architecture of IDS based Machine Learning Techniques with different modules.

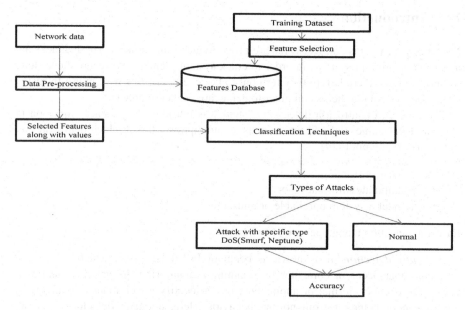

Fig. 1. Architecture of IDS based Machine Learning Techniques

3   KDDCUP'99 Dataset

In this paper we use KDDCUP99 data set. In KDDCUP99 there are 10% dataset for training and corrected dataset for testing. 10% data set contains 494022 records and corrected data set contains 311029 records. The data set contains, total 24 security attack types (Network) that fall into 4 major categories. They are Probe, Denial of Service (Dos), User to Root (U2R), Remote to local (R2L) as shown in Table 1. Each record is labeled either as a normal or as an attack, with exactly one specific type.

Table 1. Types of Attacks

Category	Attack types
Probe	ipsweep, mscan, nmap, portsweep, saint, satan
DoS	apache, back, land, mailbomb, neptune, pod,process table, smurf, teardrop, udpstorm
U2R	buffer overflow, load module, Perl, root kit, ps,sqlattack, xterm
R2L	ftp_write, guess password, imap, multihop

In this dataset, 41 attributes are used to describe the behavior of the access and grouped into three categories [6].

1. Basic features of   individual TCP connection, as in Table 2, S.No 1 - 9.
2. Content features within a connection suggested by domain knowledge, as in Table 2, S. No 10 - 22.
3. Traffic features computed using a two-second time window, as in Table 2, S.No. 23 – 31.

Table 2. Basic features of individual TCP connections

S. No.	Feature name	Description	Type
1	duration	length (number of seconds) of the connection	continuous
2	protocol_type	type of the protocol, e.g. tcp, udp, etc.	discrete
3	service	network service on the destination, e.g., http, telnet, etc.	Discrete
4	src_bytes	number of data bytes from source to destination	Continuous
5	dst_bytes	number of data bytes from destination to source	Continuous
6	flag	normal or error status of the connection	discrete
7	land	1 if connection is from/to the same host/port; 0 otherwise	discrete
8	wrong_fragment	number of ``wrong" fragments	Continuous
9	urgent	number of urgent packets	Continuous

Table 2. (*continued*)

S. No.	Feature name	Description	Type
10	host	number of ``host" indicators	Continuous
11	num_failed_logins	number of failed login attempts	Continuous
12	logged_in	1 if successfully logged in; 0 otherwise	Discrete
13	num_compromised	number of ``compromised" conditions	Continuous
14	root_shell	1 if root shell is obtained; 0 otherwise	Discrete
15	su_attempted	1 if ``su root" command attempted; 0 otherwise	discrete
16	num_root	number of ``root" accesses	continuous
17	num_file_creations	number of file creation operations	continuous
18	num_shells	number of shell prompts	continuous
19	num_access_files	number of operations on access control files	continuous
20	num_outbound_cmds	number of outbound commands in an ftp session	continuous
21	is_hot_login	1 if the login belongs to the ``hot" list; 0 otherwise	discrete
22	is_guest_login	1 if the login is a ``guest"login; 0 otherwise	discrete
23	count	Number of connections to the same host as the current connection in the past two seconds.	continuous
24	serror_rate	% of connections that have ``SYN" errors	Continuous
25	rerror_rate	% of connections that have ``REJ" errors	Continuous
26	same_srv_rate	% of connections to the same service	Continuous
27	diff_srv_rate	% of connections to different services	Continuous
28	srv_count	number of connections to the same service as the current connection in the past two seconds	Continuous
29	srv_serror_rate	% of connections that have ``SYN" errors	Continuous
30	srv_rerror_rate	% of connections that have ``REJ" errors	Continuous
31	srv_diff_host_rate	% of connections to different hosts	Continuous

10% data set and Corrected data set are divided into 4 categories in that DoS is one category. In these DoS we are using smurf and Neptune only. And other attack normal is also used in this[7].

Attack	No. of Samples	Class
Smurf	280790	DoS
Neptune	107201	DoS
Normal	97277	Normal

3.1   Feature Selection

Using Filter algorithm we get selected features as shown in Table 3, which are useful in detecting the intrusions[5]

Table 3. List of features for which the class is selected as most relevant

Class label	Relevant features
Neptune	3,4,5,23,26.29,30,31,32,34,36,37,38,39
Smurf	2,3,5,6,12,25,29,30,32,36,37,39
Normal	3,6,12,23,25,26,29,30,33,34,35,36,37,38,39

Normal, Smurf and Neptune are the most discriminative classes for major features. Based on the relevant features train and test the three algorithms.

4   Machine Learning Algorithms

4.1   Naive Bayes

A probabilistic classifier based on the Bayes theorem, Naive Bayes classifier which considers all attributes to independently contribute to the probability of a certain decision is used. The Naive Bayes classifier can be trained very efficiently [8] using

$$P(C \mid X) = \frac{P(X \mid C)P(C)}{P(C)}$$

4.2   Neural Networks

4.2.1   Back Propagation
Back propagation learns by iteratively processing a data set of training tuples. During the training session of the network, a pair of patterns is presented (T_i, T_o) where T_i is the input layer ant T_o is the output layer. The weights in the network are initialized to small random numbers (e.g., ranging from 0 to 1). At the output layer the difference

between the actual and target outputs yields an error signal. This error signal depends on the value of the weights of the neurons in each layer. This error is minimized, and during this process the new values of weights are obtained.

To start the Back Propagation process we need the following:

- The set of training patterns i.e. input and output.
- A value for the learning rate
- A criterion that terminates the Algorithm
- A methodology for updating the weights
- The non- linearity function
- Initial weights

The inputs to the nodes in the first layer in this case are the normalization values of the selected attributes. These are combined with the corresponding weights along with bias. For the first run these values are propagated to the next layer where outputs are generated from them according to the formula of sigmoid function which is as follows.

$$O_i = \frac{1}{(1+e^{-I_j})}$$

Where I_j represents the weighted sum of all inputs at that node along with respective bias.

$$I_j = \sum w_{ij} o_i + \Theta_j$$

Where w_{ij} is the weight of the ith node propagated to j^{th} node. O_i is the output of the ith node. θ_j is the bias at j^{th} node.

Now we calculate the error at the output node to be the difference between the expected and observed output values at the node

$$Err_j = O_j(1 - O_j)(T_j - O_j)$$

We propagate this error backwards from a node to the nodes connected to it from the previous layer and update the synaptic weights and the bias values.

$$Err_j = O_j(1 - O_j)\sum_k Err_j w_{jk}$$
$$\Delta W_{ij} = (l)Err_j O_i$$
$$w_{ij} = w_{ij} + \Delta W_{ij}$$

Once the error is adjusted among the different nodes, another run is required to check if the error has been properly adjusted.

Now the training is done with several sample inputs and once training is done, the neural network is to be validated by using the validation dataset to see that the network is under fitted not over fitted. Testing is to be done with some set of inputs and output is to be verified.

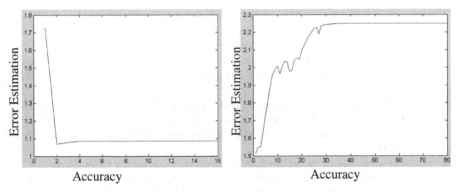

Fig. 2. Dataset containing smurf attack   **Fig. 3.** Dataset containing Neptune attack

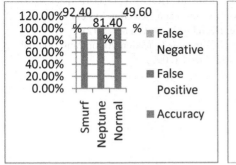

Fig. 4. Bar Chart for Neural Networks   **Fig. 5.** Bar Chart for Naive Bayes

5   Conclusion

In this paper we mainly focuses on classification of data to detect intruders trying to enter into a system. In this we used two algorithms Naive Bayes, Neural Network to detect such intrusions by applying them to KDD CUP 99 dataset with dimensionality reduced. The results obtained by implementing these algorithms are good with appreciable accuracy.

References

1. Farid, D., Harbi, N., Rahman, M.Z.: Combining Naïve Bayes And Decision Tree For Adaptive Intrusion Detection. International Journal of Network Security & Its Applications (IJNSA) 2(2) (April 2010)
2. Akbar, S., Nageswara Rao, K., Chandulal, J.A.: Intrusion Detection SystemMethodologies Based on Data Analysis. International Journal of Computer Applications 5(2), 0975–8887 (2010)

3. Chandolikar, N.S., Nandavadekar, V.D.: Selection of Relevant Feature for Intrusion Attack Classification by Analyzing KDD Cup 99. MIT International Journal of Computer Science & Information Technology 2(2), 85–90 (2012) ISSN No. 2230-7621
4. Srinivasu, P., Avadhani, P.S.: Approaches and Data Processing Techniques for Intrusion Detection System. International Journal of Computer Science and Network Security 9(12) (December 2009)
5. Olusola, A.A., Oladele, A.S., Abosede, D.O.: Analysis of KDD '99 Intrusion Detection Dataset for Selection of Relevance Features. In: Proceedings of the World Congress on Engineering and Computer Science, WCECS 2010, San Francisco, USA, October 20-22, vol. I (2010)
6. http://www.kdd.ics.uci.edu/databases/kddcup99/task.html
7. Singh, S., Silakari, S.: An ensemble approach for feature selection of Cyber Attack Dataset. International Journal of Computer Science and Information Security 6(2) (2009)
8. Jain, A., Sharma, S., Sisodia, M.S.: Network Intrusion Detection by using Supervised and Unsupervised Machine Learning Technique- A Survey. IJCTEE 1(3) (Deceember 2011)
9. Vinchurkar, D.P., Reshamwala, A.: A Review of Intrusion Detection System Using Neural Network and Machine Learning Technique. IJESIT 1(2) (November 2012)
10. Tang, H., Cad, Z.: Machine Learning-based Intrusion Detection Algorithms. Journal of Computational Information Systems 5(6), 1825–1831 (2009)

Grammatical Swarm Based-Adaptable Velocity Update Equations in Particle Swarm Optimizer

Tapas Si[1], Arunava De[2], and Anup Kumar Bhattacharjee[3]

[1] Department of Computer Science & Engineering
Bankura Unnayani Institute of Engineering, Bankura, W.B, India
c2.tapas@gmail.com
[2] Department of Information Technology
Dr. B.C Roy Engineering College, Durgapur, W.B, India
arunavade@yahoo.com
[3] Department of Electronics and Communication Engineering
National Institute of Technology, Durgapur, W.B, India
akbece12@yahoo.com

Abstract. In this work, a new method for creating diversity in Particle Swarm Optimization is devised. The key feature of this method is to derive velocity update equation for each particle in Particle Swarm Optimizer using Grammatical Swarm algorithm. Grammatical Swarm is a Grammatical Evolution algorithm based on Particle Swarm Optimizer. Each particle updates its position by updating velocity. In classical Particle Swarm Optimizer, same velocity update equation for all particles is responsible for creating diversity in the population. Particle Swarm Optimizer has quick convergence but suffers from premature convergence in local optima due to lack in diversity. In the proposed method, different velocity update equations are evolved using Grammatical Swarm for each particles to create the diversity in the population. The proposed method is applied on 8 well-known benchmark unconstrained optimization problems and compared with Comprehensive Learning Particle Swarm Optimizer. Experimental results show that the proposed method performed better than Comprehensive Learning Particle Swarm Optimizer.

Keywords: Particle Swarm Optimizer, Genetic Programming, Grammatical Evolution, Grammatical Swarm, Comprehensive Learning Particle Swarm Optimizer, Velocity update equations, Optimization.

1 Introduction

Particle Swarm Optimization(PSO) [1] was developed by Kennedy and Eberhart in 1995. PSO is a population based global optimization algorithm having stochastic nature. One advantage of PSO is its faster convergence speed. But it suffers from premature convergence in local optima due to lack in diversity. A lot of research has been already done in order to solve that local optima problem by creating diversity in the population. Different variants of PSO like FIPSO[2], CLPSO [3] are developed to enhance the performance of PSO. Different mutation strategies like Gausian

S.C. Satapathy, S.K. Udgata, and B.N. Biswal (eds.), *FICTA 2013*,
Advances in Intelligent Systems and Computing 247,
DOI: 10.1007/978-3-319-02931-3_24, © Springer International Publishing Switzerland 2014

mutation[4],cauchy mutation[5],adaptive mutation[6], polynomial mutation[7,8,9], differential mutation [10] are employed in PSO to overcome local optima problem. Here, description of related works which include PSO combined with Genetic Programming(GP) [12] is given.

M. Rashid [13] proposed GP based adaptable PSO (PSOGP) in which every particle used different velocity update equation to modify their position in the swarm space in order to achieve high exploration. Each equation is a GP expression. T. Si [14,15] proposed Grammatical Differential Evolution(GDE) based adaptation in PSO (GDE-APSO) in which each particles adopt different velocity update equation during the search process resulting in more exploration of the search space.

In this work, each particle uses its own velocity update equation evolved by grammatical swarm with the objective of creating diversity in the population so that the local optima problem can be avoided.

The remaining of the paper is structured as follows: In Section 2, classical PSO algorithm is described. Grammatical Swarm is described in Section 3. A detail description of the proposed method is given in Section 4. Experimental setup is given in Section 5. Section 6 comprises of results and discussions. Finally a conclusion with future work is given in Section 7.

2 Particle Swarm Optimization

Particle swarm optimization (PSO) [1] is a population based global optimization algorithm having stochastic nature. Each individual in PSO is called as particle and set of particles is called as swarm. Particle has its position X_i and velocity V_i where i is the index of particle. The position X_i is represented as $< X_{i1}, X_{i2}, X_{i3}, ..., X_{iD} >$ where D is the dimension of the problem to be optimized by the PSO. Each particle has its own memory to store its personal best X_i^{pbest} found so far. The best of all personal best solution is called the global best X^{gbest} of the swarm. Each particle is accelerated by its velocity and the velocity is updated by the following equation:

$$V_i(t+1) = W \times V_i(t) + C_1 R_1 (X_i^{pbest}(t) - X_i(t)) + C_2 R_2 (X^{gbest}(t) - X_i(t)) \quad (1)$$

and position is updated by following equation:

$$X_i(t+1) = X_i(t) + V_i(t+1) \quad (2)$$

In Eq. (1), W is the inertia weight in the range $(0,1)$, C_1 and C_2 are the personal cognizance and social cognizance respectively. R_1 and R_2 are two uniformly distributed random number in $(0,1)$ used for diversification.

Y. Shi and R.C Eberhart [11] introduced a linearly decreasing inertia weight with time in the range $(W_{min}, W_{max}) = (0.4, 0.9)$. The corresponding equation is given in below:

$$W = W_{max} - (W_{max} - W_{min}) \times (\frac{t}{t_{max}}) \quad (3)$$

3 Grammatical Swarm

Grammatical Swarm(**GS**)[16] algorithm is a Grammatical Evolution(**GE**)[17] based on PSO. GE is variant of Grammatical-based Genetic Programming that can write program in any arbitrary language. GE uses linear genome structure(variable-length bit string) instead of tree data structure in Genetic Programming(GP)[12]. The expressions in GS are evolved using PSO in swarm space. In Grammatical Swarm, each particle's position represents a set of integer values(codon) in the range [0,255]. The dimension of particle is the number of codons to be used to derive the expression from Backus-Naur Form(BNF) grammar. Particle's position represents the genotype which is mapped to phenotype(*fitness* corresponding derived expression).

4 Proposed Method

4.1 Algorithm

In the proposed method, GS adaptable PSO indicates the evolution of velocity update equations for each particle in PSO. Here, the search space of GS is denoted as *Grammatical Swarm Space* in which each particle represents a genome containing a number of codons. On the other hand, another *Swarm Space* is denoting the problem's search space where particles search the solution of the given problem. Therefore, in the proposed method **GSPSO**, a *Dual Swarm Space* is used. *Grammatical Swarm Space* is mapped to *Swarm Space*(i.e Genotype-Phenotype *mapping*). Therefore, the number of population in two swarm spaces are equal. The search space range in GS is $[0,255]$.

Table 1. GSPSO Algorithm

Algorithm:GSPSO
1. Initialize the population of PSO and GS
2. Calculate the fitness of particles
3. Calculate the pbest and gbest
4. While termination criteria
5. For each individual
6. Perform velocity and position update for GS
7. If derived expression from particle of GS is valid
8. Update the velocity using this new expression and update the position
9. Else
10. Update the velocity with pbest expression and update the position
11. End
12. Calculate new fitness
13. Update pbest and gbest of PSO
14. Update pbest expression if new expression is valid and gbest velocity updating equation in GS
15. End
16. End

And search space range of particles in other swarm space is $[Xmin, Xmax]$. The $Vmax$ for both GS and PSO are set to 50% of the search space range. The velocity in GS and PSO are strictly bounded in the range $[-Vmax, Vmax]$.

In the proposed algorithm, individuals in Grammatical Swarm share PSO's *fitness function* i.e *local fitness, pbest* and *gbest* of PSO in solution space.

The velocity update equation can be rewritten in the following form:

$$V_{(i)}(t+1) = f(a_j(t)), j = 1,2,3,4 \tag{4}$$

where $a_1 = V_i, a_2 = X_i, a_3 = X_i^{pbest}, a_4 = X_i^{gbest}$. The function set is $F = \{+,-,*,/\}$ and the terminal set is $T = \{a_1, a_2, a_3, a_4, r\}$ where r is random constant in $(0,1)$.

170	55	149	83	210

Fig. 1. Genotype

The Backus-Naur Form (BNF) Grammar is used in GE for genotype-phenotype mapping. BNF is a meta-syntax used to express Context-Free Grammar(CFG) by specifying production rules in simple, human and machine -understandable manner. An example of BNF grammar is described below:

```
1. <expr> := (<expr><op><expr>) (0)
            | <var> (1)
2. <op> :=   +       (0)
            |-       (1)
            |*       (2)
            |/       (3)
3. <var> := a1       (0)
            |a2       (1)
            |a3       (2)
            |a4       (3)
            |r        (4)

r represents a random number in the range (0,1).
```

A *mapping process* is used to map from integer-value to rule number in the derivation of expression using BNF grammar by the following ways:

rule=(codon integer value) MOD (number of rules for the current non-terminal)

In the derivation process,if the current non-terminal is $<expr>$, then, the rule number is generated by the following way:

rule number=(170 mod 2)=0

$<expr>$ will be replaced by $(<expr><op><expr>)$

```
<expr> :=(<expr><op><expr>)  (170 mod 2)=0
        :=(<var><op><expr>)   (55 mod 2)=1
        :=(r<op><expr>)       (149 mod 5)=4
        :=(r/<expr>)          (83 mod 4)=3
```

```
:=(r/<expr>)              (210 mod 2)=0
:=(r/<var>)               (175 mod 2)=1
:=(r/<var>)               (180 mod 2)=0
:=(r/a1)
```

The resultant derived expression will be (0.9345/a1) where r is replaced by a random number 0.9345.

The velocity update equation can be rewritten in the following form:

$$V_{(i)}(t+1) = f(a_j(t)), j = 1, 2, 3, 4 \qquad (5)$$

where $a_1 = V_i, a_2 = X_i, a_3 = X_i^{pbest}, a_4 = X_i^{gbest}$. The function set is $F = \{+, -, *, /\}$ and the terminal set is $T = \{a_1, a_2, a_3, a_4, r\}$ where r is random constant in (0, 1). The velocity update equations are evolved using above BNF grammar.

5 Experimental Setup

5.1 Benchmark Problems

There are 8 different global optimization problems, including 4 uni-modal functions $(f_1$-$f_4)$, 4 multi-modal functions$(f_5$-$f_8)$ are chosen in this experimental studies. These functions obtained from Ref. [18]. The function f_4 has platue like region. The funtion f_5 is highly multi-modal i.e it has too many local optima.The functions $(f_6$-$f_8)$ have few local optima.All functions are used in this work to be minimized.The description of these benchmark functions and their global optima are given in Table 2.

Table 2. The benchmark functions

Test Function	S	f_{min}
$f_1(x) = \sum_{i=1}^{D} x_i^2$	[-100,100]	0
$f_2(x) = \sum_{i=1}^{D} (\sum_{j=1}^{i} x_j)^2$	[-100,100]	0
$f_3(x) = \sum_{i=1}^{D} (10^6)^{\frac{i-1}{n-1}} x_i^2$	[-100,100]	0
$f_4(x) = \sum_{i=1}^{D} [100(x_{i+1} - x_i^2)^2 + (1 - x_i^2)^2]$	[-100,100]	0
$f_5(x) = \sum_{i=1}^{D} -x_i * sin(\sqrt{|x_i|})$	[-500,500]	-12569.5
$f_6(x) = \sum_{i=1}^{D} \frac{x_i^2}{4000} - \prod_{i=1}^{D} \cos(\frac{x_i}{\sqrt{i}}) + 1$	[-600,600]	0
$f_7(x) = -20 * exp(-0.2 * \sqrt{\frac{1}{D} \sum_{i=1}^{D} x_i^2})$		
$-exp(\frac{1}{D} \sum_{i=1}^{D} cos(2\pi x_i)) + 20 + e$	[-32,32]	0
$f_8(x) = \sum_{i=1}^{D} [x_i - 10\cos(2\pi x_i) + 10]$	[-5.12,5.12]	0

Table 3. Parameters Settings

Parameters	Values
Problem's Dimension(D)	30
Dimension in GS(i.e Length of Genome)	50
Number of Wrapping	2
Population Size(NP)	20
FEs (where FEs is the maximum number of function evaluations allowed for each run)	1,00,000
V_{max}	$0.5 \times (X_{max} - X_{min})$
$c_1 = c_2$	1.49445
$\omega_{max}, \omega_{min}$	0.9, 0.4
Threshold Error(e)	$1e-03$
Termination criteria	Maximum number of function evaluations or $E = |f(X) - f(X^*)| \le e$ where $f(X)$ is the current best and $f(X^*)$ is the global optimum. E is the best-error of a run of the algorithm. e is the threshold error.
Total number of runs for each problem	50

5.2 Parameters Settings

5.3 PC Configuration

1. System: Fedora 17
2. CPU: AMD FX -8150 Eight-Core 3.6 GHz
3. RAM: 16 GB
4. Software: Matlab 2010b

6   Results and Discussions

The devised method is applied on well-known benchmark unconstrained functions described in Section 5 for 30 dimension. The quality of solutions are measured in terms of mean and standard deviation of best-run-errors from 50 runs and are tabulated in Table 4. Best-run-error is the absolute difference between global optimum and best solution obtained from a single run. Success Rate(SR) is given in Table 5. Success Rate(SR) is calculated as follows:

$$SR = \frac{Number \ of \ achieved \ threshold \ errors}{Total \ runs} \tag{6}$$

Statistical t-test [19] has been carried out for sample size(number of runs)=50 and degrees of freedom=98 to compare the performance of CLPSO and GSPSO algorithms with statistical significance for each problem. The last column of Table 4 signifies whether the null hypothesis that the means of the two data are equal is accepted or rejected. The value "-" indicates that the approach will have a lower value with 95% of confidence, the value "+" represents that the approach will have a higher value with 95% of confidence and the value "≈" means that there is not statistically significant difference between the approaches.

Convergence speed is measured in terms of mean and standard deviation of number of function evaluations taken by the algorithms and it is given with average CPU time in Table 5. Better results in Table 4 and 5 are marked in bold face. Convergence graphs of GSPSO are given in Figure 2.

Table 4. Mean and Standard Deviation of best-run-error, success rate

Test#	GSPSO			CLPSO			Significance
	Mean	Std. Dev.	$SR(\%)$	Mean	Std. Dev.	$SR(\%)$	
f_1	**1.01E-04**	**2.40E-04**	**100.00**	9.34E-04	6.48E-05	**100.00**	+
f_2	**3.90E-05**	**1.35E-04**	**100.00**	6.11E-01	6.71E-01	0.00	+
f_3	**2.51E-05**	**1.31E-04**	**100.00**	9.28E-04	6.46E-05	**100.00**	+
f_4	**8.59E-05**	**2.08E-04**	**100.00**	34.33	23.46	0.00	+
f_5	**3023.498**	**1631.845**	0.00	4721.51	531.35	0.00	+
f_6	**1.15E-04**	**2.20E-04**	**100.00**	6.17E-03	7.90E-03	56.00	+
f_7	**6.84E-05**	**1.72E-04**	**100.00**	9.61E-04	3.86E-05	**100.00**	+
f_8	**4.54E-05**	**1.57E-04**	**100.00**	13.64	4.4465	0.00	+

Table 5. Mean,Standard Deviation of FES and mean CPU Time

Test#	GSPSO			CLPSO		
	Mean	Std. Dev.	Mean Time(Sec)	Mean	Std. Dev.	Mean Time(Sec)
f_1	**1212.8**	**847.4765**	32.7245	39605	645.5949	**1.0964**
f_2	**1516.8**	**998.7614**	41.9142	100000	0.00	**3.5592**
f_3	**1246.8**	**714.2584**	35.8272	45020	648.0047	**1.6755**
f_4	**5085.20**	**2749.30**	140.4408	100000	0.00	**2.7181**
f_5	100000	0.00	2646.648	100000	0.00	**3.763**
f_6	**996**	**701.718**	27.5345	68039	29385	**2.4653**
f_7	**1302.80**	**938.2399**	35.7485	43677	832.3986	**1.3273**
f_8	**1330**	**1381**	35.946	100000	0.00	**2.9752**

From Table 4, it is found that GSPSO outperformed over PSO in statistical significant way for all. GSPSO is more robust(always produce same result) than CLPSO. The success rate of GSPSO is 100% in finding out the target or threshold error for all except function f_5. GSPSO has also faster convergence speed than CLPSO in achieving target error value.

From Table 5, it is seen that average FEs of GSPSO is better than CLPSO. But GSPSO takes much more CPU time than CLPSO because extra CPU time is needed due to higher length of the genome and to derive the expressions from genome with additional multiple *wrapping*. If obtaining better solution has higher priority over computational time, then this experimental studies show that GSPSO outperformed over CLPSO and GSPSO is more efficient and effective than CLPSO in optimization of unconstrained functions.

Figure 2 depicts the convergence behaviour of GSPSO and it also shows that GSPSO maintains good diversity in the population(in solution space).

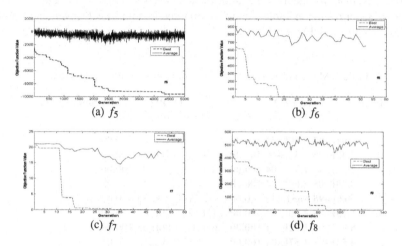

(a) f_5 (b) f_6

(c) f_7 (d) f_8

Fig. 2. Convergence graph of GSPSO for function $f_5 - f_8$

As particles update their velocity by different equations during the different runs, a set of best evolved equations in different runs in the optimization of function f_5 are given in Table 6.

Table 6. Evolved Equations

Sl. No	Evolved Equations	Simplified Equations
1	minus(minus(pdivide(x3,x3),x3),x2)	$1 - X_{pbest} - X$
2	minus(x1,minus(plus(x2,x4),x4))	$V - X$
3	minus(times(x4,x4),x3)	$X_{gbest}^2 - X_{pbest}$
4	minus(times(x3,minus(x4,0.44359)),x2)	$X_{pbest} \times (X_{gbest} - 0.44359) - X$
5	minus(pdivide(x4,x1),x2)	$X_{gbest}/V - X$
6	minus(plus(x3,x3),plus(plus(x4,x3),x2))	$-X - X_{gbest} + X_{pbest}$
7	minus(minus(x1,x3),x2)	$V - X - X_{pbest}$

7   Conclusions

In this paper, grammatical swarm based adaptable velocity update equations in particle swarm optimizer algorithm is devised. Each particles uses different velocity update equation to update their position in order to create diversity in the population where as particles in grammatical swarm use the classical velocity update equation. The GSPSO is applied to solve well-known benchmark unconstrained optimization problems. Experimental results established that GSPSO performed better than CLPSO in terms of quality of solutions, robustness and convergence behaviour. The analytical as well as experimental studies will be carried out as a future work in how different velocity equations in different times create diversity in the population. The future works of this study is also directed towards the training of artificial neural network and optimizing the more complex problems.

References

1. Kennedy, J., Eberhart, R.C.: Particle swarm optimization. In: IEEE International Conference on Neural Networks, Piscataway, NJ, pp. 1942–1948 (1995)
2. Mendes, R., Kennedy, J., Neves, J.: The Fully Informed Particle Swarm: Simpler, Maybe Better. IEEE Transactions on Evolutionary Computation 8(3), 204–210 (2004)
3. Liang, J.J., Qin, A.K., Suganthan, P.N., Baskar, S.: Comprehensive Particle Swarm Optimizer for Global Optimization of Multimodal Functions. IEEE Transactions on Evolutionary Computation 10(3), 281–295 (2006)
4. Higashi, N., Lba, H.: Particle Swarm Optimization with Gaussian Mutation. In: IEEE Swarm Intelligence Symposium, Indianapolis, pp. 72–79 (2003)
5. Li, C., Liu, Y., Zhou, A., Kang, L., Wang, H.: A Fast Particle Swarm Optimization Algorithm with Cauchy Mutation and Natural Selection Strategy. In: Kang, L., Liu, Y., Zeng, S. (eds.) ISICA 2007. LNCS, vol. 4683, pp. 334–343. Springer, Heidelberg (2007)
6. Tang, J., Zhao, X.: Particle Swarm Optimization with Adaptive Mutation. In: WASE International Conference on Information Engineering (2009)
7. Si, T., Jana, N.D., Sil, J.: Particle Swarm Optimization with Adaptive Polynomial Mutation. In: World Congress on Information and Communication Technologies (WICT 2011), Mumbai, India, pp. 143–147 (2011)
8. Si, T., Jana, N.D., Sil, J.: Constrained Function Optimization Using PSO with Polynomial Mutation. In: Panigrahi, B.K., Suganthan, P.N., Das, S., Satapathy, S.C. (eds.) SEMCCO 2011, Part I. LNCS, vol. 7076, pp. 209–216. Springer, Heidelberg (2011)
9. Jana, N.D., Si, T., Sil, J.: Particle Swarm Optimization with Adaptive Mutation in Local Best of Particles. In: 2012 International Congress on Informatics, Environment, Energy and Applications-IEEA 2012, IPCSIT, vol. 38. IACSIT Press, Singapore (2012)
10. Si, T., Jana, N.D.: Particle swarm optimisation with differential mutation. Int. J. Intelligent Systems Technologies and Applications 11(3/4), 212–251 (2012)
11. Shi, Y., Eberhart, R.C.: A modified particle swarm optimizer. In: Proceedings of the IEEE Congress on Evolutionary Computation (CEC 1998), Piscataway, NJ, pp. 69–73 (1998)
12. Koza, J.R.: Genetic Programming: On the Programming of Computers by Means of Natural Selection. MIT Press (1992)
13. Rashid, M.: Combining PSO algorithm and Honey Bee Food Foraging Behaviour for Solving Multimodal and Dynamic Optimization Problems, Ph.D Dissertation, Department of Computer Science, National University of Computer & Emerging Sciences, Islamabad, Pakistan (2010)
14. Si, T.: Grammatical Differential Evolution Adaptable Particle Swarm Optimization Algorithm. International Journal of Electronics Communications and Computer Engineering(IJECCE) 3(6), 1319–1324 (2012)

15. Si, T.: Grammatical Differential Evolution Adaptable Particle Swarm Optimizer for Artificial Neural Network Training. International Journal of Electronics Communications and Computer Engineering(IJECCE) 4(1), 239–243 (2013)

16. O'Neill, M., Brabazon, A.: Grammatical Swarm: The Generation of Programs by Social Programming. Natural Computing 5(4), 443–462

17. O'Neill, M., Ryan, C.: Grammatical Evolution. IEEE Trans. Evolutionary Computation 5(4), 349–358 (2001)

18. Yao, X., Liu, Y., Lin, G.: Evolutionary programming made faster. IEEE Transactions on Evolutionary Computation 3, 82–102 (1999)

19. Das, N.G.: Statistical Methods (Combined Vol). Hill Education Private Limited, Tata Mcgraw (2008)

Performance of Teaching Learning Based Optimization Algorithm with Various Teaching Factor Values for Solving Optimization Problems

M. Ramakrishna Murty[1], J.V.R. Murthy[2], P.V.G.D. Prasad Reddy[3],
Anima Naik[4], and Suresh Chandra Satapathy[5]

[1] Dept of CSE, GMRIT, Rajam, India
ramakrishna.malla@gmail.com
[2] Dept of CSE, JNTUK-Kakinada, India
mjonnalagedda@gmail.com
[3] Dept of CS&SE, Andhra University, Visakhapatnam, India
prasadreddy.vizag@gmail.com
[4] Dept of CSE, MITS, Rayagada, India
anmianaik@gmail.com
[5] Dept of CSE, ANITS, Visakhapatnam, India
sureshsapathy@ieee.org

Abstract. Teaching Learning Based Optimization (TLBO) is being used as a new, reliable, accurate and robust optimization technique scheme for global optimization over continuous spaces. This paper presents an effect of variation of a teaching factor T_F in traditional TLBO algorithm and then proposed a value for teaching factor T_F. The traditional TLBO algorithm with new teaching factor T_F value has been tested on several benchmark functions and shown to be statistically significantly better than other teaching factor values for performance measures in terms of faster convergence behavior.

Keywords: Convergence behavior, TF, TLBO, Performance.

1 Introduction

Constrained and unconstrained optimization problems are generally associated with many difficulties such as multi-modality, dimensionality and differentiability. Traditional optimization techniques generally fail to solve such problems, especially with nonlinear objective functions. To overcome these difficulties, there is a need to develop more powerful optimization techniques and research is continuing to find effective optimization techniques.

Rao et al. [1, 2] proposed a teaching-learning based optimization (TLBO) algorithm based on the natural phenomenon of teaching and learning. The implementation of TLBO does not require the determination of any algorithm specific controlling parameters which makes the algorithm robust and powerful. TLBO requires only common controlling parameters like population size and number of generations for its working. In this way TLBO can be said as an algorithm specific

S.C. Satapathy, S.K. Udgata, and B.N. Biswal (eds.), *FICTA 2013*,
Advances in Intelligent Systems and Computing 247,
DOI: 10.1007/978-3-319-02931-3_25, © Springer International Publishing Switzerland 2014

parameter-less algorithm. Rao and Patel [3] investigated the performance of TLBO algorithm for different elite sizes, population sizes and number of generations considering various constrained bench mark problems available in the literature to identify their effect on the exploration and exploitation capacity of the algorithm. Suresh Satapathy and Anima Naik [4] tried to propose a new approach to use TLBO to cluster data. They have proved that the TLBO algorithm can be used to cluster arbitrary data. Again they have shown another good characteristic of TLBO algorithm is the cost of computations are less in handling high dimensional problems in compare to other algorithms [5]. TLBO can able to show its application in the area of computer network by optimizing the multicast tree [6]. It can able to handle power system problem easily where optimal solution of the unit maintenance scheduling problem in which the cost reduction is as important as reliability [7]. TLBO algorithm has been adapted to solve multi-objective problems of an economic load dispatch problem with incommensurable objectives [8]. From literature we can see while clustering the data using fuzzy c-means (FCM) and hard c-means (HCM), the sensitivity to tune the initial clusters centers have captured the attention of the clustering communities for quite a long time. This problem has been addressed by TLBO in [9]. In the area of feature selection also TLBO is showing its performance in connection with rough set theory. From empirical results reveal that the Rough TLBO approach could be performed better in terms of finding optimal features and doing so in quick time in comparison with GA, PSO and DE [10]. Again there is some modification of TLBO have been done to improve its performance in optimization for global function optimization [11]. Again there is some modifications of TLBO have been done to improve its performance in optimization for global function optimization [11]. Like many other applications of TLBO can be seen from different papers [12]-[15]. This paper presents a variation on teaching factor T_F in traditional TLBO algorithms and proposed a new teaching factor T_F value for improving the convergence speed of algorithm. Due to this new teaching factor value the TLBO algorithm on several benchmark optimization problems shows a marked improvement in performance over the traditional TLBO.

The remaining of the paper is organized as follows: in Section 2, we give a brief description of TLBO. In Section 3, we describe the Teaching–Learning-Based Optimizer with proposed value for teaching factor. In Section 4, experimental settings and numerical results are given. The paper concludes with section 5.

2   Teaching-Learning-Based Optimization

This optimization method is based on the effect of the influence of a teacher on the output of learners in a class. It is a population based method and like other population based methods it uses a population of solutions to proceed to the global solution. A group of learners constitute the population in TLBO. In any optimization algorithms there are numbers of different design variables. The different design variables in TLBO are analogous to different subjects offered to learners and the learners' result is analogous to the 'fitness', as in other population-based optimization techniques.

As the teacher is considered the most learned person in the society, the best solution so far is analogous to Teacher in TLBO. The process of TLBO is divided into two parts. The first part consists of the 'Teacher Phase' and the second part consists of the 'Learner Phase'. The 'Teacher Phase' means learning from the teacher and the 'Learner Phase' means learning through the interaction between learners. In the sub-sections below we briefly discuss the implementation of TLBO.

A Initialization

Following are the notations used for describing the TLBO:

N: number of learners in a class i. e. "class size"

D: number of courses offered to the learners

$MAXIT$: maximum number of allowable iterations

The population X is randomly initialized by a search space bounded by matrix of N rows and D columns. The jth parameter of the ith learner is assigned values randomly using the equation

$$x_{(i,j)}^0 = x_j^{min} + rand \times (x_j^{max} - x_j^{min}) \tag{1}$$

where *rand* represents a uniformly distributed random variable within the range $(0, 1)$, x_j^{min} and x_j^{max} represent the minimum and maximum value for jth parameter. The parameters of ith learner for the generation g are given by

$$X_{(i)}^g = [x_{(i,1)}^g, x_{(i,2)}^g, x_{(i,3)}^g, \ldots \ldots, x_{(i,j)}^g, \ldots \ldots, x_{(i,D)}^g] \tag{2}$$

B Teacher Phase

The mean parameter M^g of each subject of the learners in the class at generation g is given as

$$M^g = [m_1^g, m_2^g, \ldots \ldots, m_j^g, \ldots \ldots, m_D^g] \tag{3}$$

The learner with the minimum objective function value is considered as the teacher $X_{Teacher}^g$ for respective iteration. The Teacher phase makes the algorithm proceed by shifting the mean of the learners towards its teacher. To obtain a new set of improved learners a random weighted differential vector is formed from the current mean and the desired mean parameters and added to the existing population of learners.

$$Xnew_{(i)}^g = X_{(i)}^g + rand \times (X_{Teacher}^g - T_F M^g) \tag{4}$$

T_F is the teaching factor which decides the value of mean to be changed. Value of T_F can be either 1 or 2. The value of T_F is decided randomly with equal probability as,

$$T_F = round[1 + rand(0,1)\{2 - 1\}] \tag{5}$$

If $Xnew_{(i)}^g$ is found to be a superior learner than $X_{(i)}^g$ in generation g , than it replaces inferior learner $X_{(i)}^g$ in the matrix.

C Learner Phase

In this phase the interaction of learners with one another takes place. The process of mutual interaction tends to increase the knowledge of the learner. The random interaction among learners improves his or her knowledge. For a given learner $X_{(i)}^g$, another learner $X_{(r)}^g$ is randomly selected($i \neq r$). The ith parameter of the matrix $Xnew$ in the learner phase is given as

$$Xnew_{(i)}^g = \begin{cases} X_{(i)}^g + rand \times \left(X_{(i)}^g - X_{(r)}^g\right) & if\ f\left(X_{(i)}^g\right) < f(X_{(r)}^g) \\ X_{(i)}^g + rand \times \left(X_{(r)}^g - X_{(i)}^g\right) & oterwise \end{cases} \tag{6}$$

D   Algorithm Termination

The algorithm is terminated after $MAXIT$ iterations are completed. Details of TLBO can be refereed in [1].

3   Proposed Value for Parameter TF

As it is known to obtain a new set of improved learners a random weighted differential vector is formed from the current mean and the desired mean parameters and added to the existing population of learners.

$$Xnew_{(i)}^g = X_{(i)}^g + rand \times (X_{Teacher}^g - T_F M^g)$$

where T_F is the teaching factor which decides the value of mean to be changed. Value of T_F can be varies from 2 to 4, 2 to 5 or 2 to 6. The value of T_F is decided randomly with equal probability as,

For T_F value 2 to 4

$\quad\quad F = 3;$
$\quad\quad T_F = floor(2 + rand * F)$

For T_F value 2 to 5

$\quad\quad F = 4;$
$\quad\quad T_F = floor(2 + rand * F)$

For T_F 2 to 6

$\quad\quad F = 5;$
$\quad\quad T_F = floor(2 + rand * F)$

4   Experimental Settings and Numerical Results

Settings

In all experiments in this section, the values of the common parameters used in each algorithm such as population size and total evaluation number were chosen to be the same. Population size was 20 and the maximum iteration was 50 for all functions.

We used 10 benchmark problems in order to test the performance of the TLBO algorithm with the variation of teaching factor. We have included many different kinds of problems such as unimodal, multimodal, regular, irregular, separable, non-separable and multidimensional. Initial range, formulation, characteristics and the dimensions of these problems are listed in Tables 1.

Table 1. Benchmark functions used in experiments 1. D: Dimension, C: Characteristic, U: Unimodal, M:Multimodal, S:Separable , N:Non-Separable

No.	Function	C	D	Range	Formulation	Value
f_2	Sphere	30	US	[-100,100]	$f(x) = \sum_{i=1}^{D} x_i^2$	$f_{min}=0$
f_4	Schwefel 1.2	30	UN	[-100,100]	$f(x) = \sum_{i=1}^{D} (\sum_{j=1}^{i} x_j)^2$	$f_{min}=0$
f_5	Schwefel 2.22	50	UN	[-10,10]	$f(x) = \sum_{i=1}^{D} |x_i| + \prod_{i=1}^{D} |x_i|$	$f_{min}=0$
f_6	Schwefel 2.21	30	UN	[-100,100]	$f(x) = \max_{i} \{|x_i|, 1 \le i \le D\}$	$f_{min}=0$
f_{10}	Rastrigin	30	MS	[5.12,5.12]	$f(x) = \sum_{i=1}^{D} [x_i^2 - 10\cos(2\pi x_i) + 10]$	$f_{min}=0$
f_{12}	Griewank	30	MN	[-600,600]	$f(x) = \frac{1}{4000} \sum_{i=1}^{D} x_i^2 - \prod_{i=1}^{D} \cos(\frac{x_i}{\sqrt{i}}) + 1$	$f_{min}=0$
f_{16}	Weierstrass	30		[-0.5, 0.5]	$f(x) = \sum_{i=1}^{D} (\sum_{k=0}^{kmax} [a^k \cos(2\pi b^k(x_i + 0.5))]) - D \sum_{k=0}^{kmax} [a^k \cos(2\pi b^k(x_i + 0.5))], where\ a = 0.5, b = 3, kmax = 20$	$f_{min}=0$
f_{15}	Multimod	30		[-10,10]	$f(x) = \sum_{i=1}^{D} |x_i| \prod_{i=1}^{D} |x_i|$	$f_{min}=0$
f_{16}	SumSquares	30	US	[-10,10]	$f(x) = \sum_{i=1}^{D} i x_i^2$	$f_{min}=0$
f_{17}	Zakharov	10	UN	[-5,10]	$f(x) = \sum_{i=1}^{D} x_i^2 + (\sum_{i=1}^{D} 0.5 i x_i)^2 + (\sum_{i=1}^{D} 0.5 i x_i)^4$	$f_{min}=0$

Table 2. Performance comparison with variation of T_F value

TF		Sphere	Schwefel 1.2	Schwefel2.21	Schwefel 2.22	Rastrigin
1	Mean	6.1918e+3	1.5106e+08	26.1892	35.8283	193.3782
	Std	1.8482e+3	5.0866e+07	4.3231	6.4906	23.8327
2	Mean	3.9495e-11	4.4854e-06	5.6993e-06	2.7612e-06	2.5557e-05
	Std	6.7983e-11	1.3586e-05	5.5087e-06	2.2764e-06	1.3286e-04
3	Mean	1.2841e-11	6.3967e-07	2.0874e-06	1.3191e-06	8.8072e-09
	Std	1.9833e-11	1.2402e-06	1.5209e-06	1.0359e-06	4.8237e-08
4	Mean	2.5434e-11	4.5480e-07	3.3016e-06	2.0515e-06	5.7969e-12
	Std	4.3724e-11	7.4897e-06	2.5699e-06	1.6888e-06	3.1737e-11
5	Mean	4.4062e-11	2.1694e-06	3.2156e-06	2.5127e-06	1.9571e-11
	Std	7.0214e-11	3.4808e-06	2.2464e-06	1.3373e-06	1.0818e-10
6	Mean	1.3902e-10	3.1982e-06	8.6444e-06	4.4488e-06	4.0068e-12
	Std	3.0173e-10	4.8522e-06	1.1199e-05	3.5504e-06	2.1855e-11
7	Mean	3.3690e-10	1.6031e-05	1.2174e-05	7.7374e-06	2.6705e-14
	Std	8.0350e-10	3.3457e-05	1.4185e-05	8.8219e-06	1.0210e-13
8	Mean	7.7322e-10	5.4849e-05	2.8029e-05	2.7892e-05	7.4368e-11
	Std	2.5454e-9	1.5900e-04	7.9706e-05	8.1715e-05	3.9416e-10
9	Mean	1.2827e-9	1.1482e-04	5.1738e-05	1.4464e-04	1.0666e-08
	Std	3.8219e-9	1.6911e-04	1.3722e-04	6.2482e-04	5.8238e08
10	Mean	0.0272	0.8982	2.0999e-04	0.0024	2.0599e-09
	Std	0.1440	4.9076	7.2461e-04	0.0095	1.0657e-08
11	Mean	0.0034	9.5859	0.2697	0.0361	3.8443e-05
	Std	0.0179	51.7678	1.1204	0.1761	2.1000e-04
12	Mean	0.0149	1.5277e+04	0.0641	2.9441e-04	5.8897e-05
	Std	0.0801	7.3757e+04	0.3438	0.0010	3.1832e-05
13	Mean	0.3212	2.3904e+03	0.2730	0.0335	1.0656e-05
	Std	1.7415	1.0923e+04	1.4865	0.1610	5.7922e-05
14	Mean	0.6920	1.8092e+03	0.0438	0.2050	0.0403

Table 2. (*continued*)

	Std	3.3305	7.9797e+03	0.2367	0.8765	0.2134
15	Mean	0.7780	1.1162e+5	0.2592	0.1840	0.1111
	Std	3.6029	3.5917e+05	1.0414	0.6527	0.4314
1-2	Mean	9.5800e-9	2.8747e-04	4.0419e-05	4.3453e-05	6.6863
	Std	3.2724e-8	8.9015e-04	1.1705e-04	7.5650e-05	36.6222
2-3	Mean	2.7982e-11	1.4634e-06	4.0030e-06	2.4219e-06	9.9056e-11
	Std	3.7543e-11	2.9744e-06	3.2655e-06	2.9230e-06	3.2677e-10
2-4	Mean	1.3699e-11	1.6038e-06	3.9789e-06	3.0433e-06	9.0132e-13
	Std	2.6261e-11	2.3010e-06	4.8483e-06	2.9201e-06	3.2556e-12
2-5	Mean	5.1806e-11	2.1595e-06	3.6632e-06	2.5207e-06	1.9889e-13
	Std	8.8652e-11	7.6822e-06	3.3493e-06	2.0677e-06	1.0894e-12
2-6	Mean	6.8947e-11	1.0063e-06	4.4519e-06	3.7744e-06	2.7328e-11
	Std	3.8598e-11	1.1537e-06	4.3339e-06	3.1426e-06	1.4776e-10
2-7	Mean	2.6792e-11	2.6833e-06	4.3684e-06	4.5288e-06	7.6611e-11
	Std	4.2245e-11	3.3367e-06	4.0909e-06	3.6845e-06	7.1902e-11
2-8	Mean	7.2269e-11	3.6868e-06	6.5355e-06	5.4211e-06	9.1129e-10
	Std	1.0859e-10	5.1759e-06	5.4311e-06	6.1327e-06	7.9112e-09
2-9	Mean	8.8615e-11	6.1796e-05	6.6697e-06	6.1414e-06	5.6123e-10
	Std	8.9516e-11	3.0022e-04	5.1717e-06	6.5258e-06	7.1123e-10
2-10	Mean	2.0658e-10	6.7516e-05	8.4150e-06	6.7788e-06	1.8982e-10
	Std	4.1065e-10	3.0180e-04	7.5606e-06	5.9072e-06	1.0396e-10
2-11	Mean	2.2192e-10	1.3717e-05	1.2042e-05	1.4282e-05	2.1162e-07
	Std	4.2913e-10	2.9127e-05	1.5148e-05	1.8476e-05	1.9226e-06
2-12	Mean	3.3484e-10	5.7227e-05	1.6533e-05	1.4225e-05	6.9121e-06
	Std	4.5657e-10	1.6126e-04	2.0678e-05	1.3402e-05	5.1123e-05

Table 3. Performance comparison with variation of T_F value

T_F		Griewank	weierstrass	Multimod	Sumsquares	Zakharov
1	Mean	55.5580	23.4720	2.3847e-09	953.5659	25.0354
	Std	14.1116	2.6380	1.0963e-08	264.1336	15.5119
2	Mean	4.3262e-10	0.0015	1.6286e-160	3.7443e-12	2.7258e-13
	Std	7.6586e-10	0.0011	8.9201e-160	7.1637e-12	6.3637e-13
3	Mean	2.4817e-11	7.1481e-04	4.3771e-186	1.8436e-12	6.9717e-14
	Std	3.0760e-11	7.6303e-04	0	3.1929e-12	1.0732e-13
4	Mean	4.7690e-10	6.9207e-04	1.1570e-177	3.5032e-12	7.2685e-14
	Std	2.0631e-09	3.1512e-04	0	4.6445e-12	1.2464e-13
5	Mean	1.0775e-10	5.1125e-04	9.2279e-180	6.1842e-12	2.0009e-13
	Std	1.3847e-10	5.3540e-04	0	1.3154e-11	2.4501e-13
6	Mean	5.1445e-10	0.0014	4.6011e185	4.3426e-11	4.286e-13
	Std	1.2672e-09	6.9462e-04	0	1.3842e-10	4.4554e-13
7	Mean	6.8956e-09	0.0019	1.5850e-170	5.8687e-11	3.9940e-12
	Std	3.6713e-09	0.0011	0	1.0357e-10	1.6901e-11
8	Mean	7.1735e-08	0.0026	5.9347e-144	4.1025e-09	5.8804e-12
	Std	2.3378e-07	0.0014	3.2506e-143	1.4930e-09	1.3532e-11

Table 3. (*continued*)

9	Mean	5.2794e-07	0.0177	3.5751e-171	2.1088e-09	1.6440e-10
	Std	2.0202e-06	0.0742	0	5.6267e-09	6.9280e-10
10	Mean	2.0686e-05	0.0102	1.2028e-151	2.2680e-05	1.0647e-07
	Std	8.2091e-05	0.0159	6.5080e-151	1.2421e-04	5.6175e-07
11	Mean	0.0767	0.0161	7.3971e-109	5.7555e-04	3.7316e-06
	Std	0.2922	0.0495	4.0515e-108	0.0030	2.0175e-05
12	Mean	0.0327	0.0583	6.5013e-92	0.0107	7.9374e-06
	Std	0.1776	0.1508	3.5609e-91	0.0569	6.6613e-05
13	Mean	0.1638	0.0840	1.4254e-130	0.0325	5.0717e-05
	Std	0.6864	0.1429	7.3450e-130	0.1778	6.1192e-05
14	Mean	0.1195	0.3166	1.4810e-73	0.0525	8.8330e-04
	Std	0.3520	1.0455	8.1119e-73	0.2876	0.0083
15	Mean	0.1272	0.2104	1.9486e-86	00463	0.0170
	Std	0.4774	0.6376	1.0571e-85	0.1642	0.0874
1-2	Mean	2.8223e-08	0.0052	2.1788e-130	5.8426e-07	6.7914e-10
	Std	1.1076e-07	0.0060	1.1934e-129	2.0360e-06	3.4773e-09
2-3	Mean	8.3134e-11	0.0010	1.4944e-182	3.6109e-12	1.3382e-13
	Std	1.7500e-10	5.7200e-04	0	4.2983e-12	2.6505e-13
2-4	Mean	8.8152e-11	0.0011	2.5461e-185	2.6774e-12	1.5962e-13
	Std	1.6391e-10	5.9164e-04	0	4.6347e-12	2.2701e-13
2-5	Mean	8.1400e-11	0.0012	5.3765e-183	6.2863e-12	3.6416e-13
	Std	9.8961e-11	6.0625e-04	0	9.3394e-12	6.4611e-13
2-6	Mean	3.8441e-09	0.0013	6.7905e-182	6.0734e-12	2.7093e-13
	Std	1.7467e-08	7.8478e-04	0	8.7874e-12	4.9394e-13
2-7	Mean	2.0784e-10	0.0012	1.6779e-183	1.6784e-11	8.4991e-13
	Std	4.0540e-10	6.0943e-04	0	3.3505e-11	2.7746e-12
2-8	Mean	4.4128e-10	0.0013	3.7751e-171	1.3785e-11	1.3110e-12
	Std	7.5992e-10	8.1316e-04	0	2.9825e-11	4.6599e-12
2-9	Mean	3.5070e-10	0.0017	3.7204e-180	6.1893e-11	5.9112e-12
	Std	4.2497e-10	9.6130e-04	0	1.2810e-10	6.1112e-12
2-10	Mean	1.3237e-09	0.0021	8.1606e-180	5.7582e-11	2.4294e-12
	Std	2.3003e-09	0.0010	0	1.2381e-10	4.6177e-12
2-11	Mean	7.4570e-08	0.0020	2.4436e-178	4.9985e-11	8.9038e-12
	Std	4.0116e-07	0.0012	0	7.9074e-11	2.2971e-12
2-12	Mean	6.1192e-07	0.0020	7.3362e-172	2.9286e-11	3.1675e-12
	Std	2.1126e-06	0.0011	0	4.7002e-11	6.7019e-12

In this experiment, we compared TLBO fitness values with the variation of T_F value by considering different unconstrained benchmark functions. Here we varies T_F value from 1 to 15 and the same time we have taken variation of T_F value in the range as 1-2, 2-3, 2-4, 2-5, 2-6, 2-7, 2-8, 2-9, 2-10, 2-11, 2-12. The result is given in table 2 and table 3. From the table it is clear that when T_F value varies from 2 to 4, 2 to 5 or 2 to 6 TLBO algorithm will give better value in compare to other variation in most of the cases.

5   Conclusion

In this paper, we have proposed variation of teaching factor in TLBO algorithm, to solve global numerical optimization problems by introducing a new search mechanism. We testify the performance of the proposed approach on a numbers of benchmark functions. The results show that the TLBO algorithm possesses superior performance in faster convergence with variation of teaching factor in algorithm. Hence, the TLBO algorithm with this variation of teaching factor may be a good alternative to deal with complex numerical optimization problems.

References

1. Rao, R.V., Savsani, V.J., Vakharia, D.P.: Teaching–learning-based optimization: A novel method for constrained mechanical design optimization problems. Computer-Aided Design 43, 303–315 (2011)
2. Rao, R.V., Savsani, V.J., Vakharia, D.P.: Teaching-learning-based optimization: A novel optimization method for continuous non-linear large scale problems. Inform. Sci. 183, 1–15 (2012)
3. Rao, R.V., Patel, V.: An elitist teaching-learning-based optimization algorithm for solving complex constrained optimization problems. Int. J. Ind. Eng. Comput. 3 (2012), http://dx.doi.org/10.5267/j.ijiec.2012.03.007
4. Satapathy, S.C., Naik, A.: Data clustering using teaching learning based optimization. In: Panigrahi, B.K., Suganthan, P.N., Das, S., Satapathy, S.C. (eds.) SEMCCO 2011, Part II. LNCS, vol. 7077, pp. 148–156. Springer, Heidelberg (2011)
5. Satapathy, S.C., Naik, A., Parvathi, K.: High dimensional real parameter optimization with teaching learning based optimization. International Journal of Industrial Engineering Computations, © 2012 Growing Science Ltd. All rights reserved (2012), doi:10.5267/j.ijiec.2012.06.001
6. Naik, A., Parvathi, K., Satapathy, S.C., Nayak, R., Panda, B.S.: QoS multicast routing using Teaching learning based Optimization. In: Kumar M., A., R., S., Kumar, T.V.S. (eds.) Proceedings of ICAdC. AISC, vol. 174, pp. 49–55. Springer, Heidelberg (2013)
7. Satapathy, S.C., Naik, A., Parvathi, K.: 0-1 integer programming for generation maintenance scheduling in power systems based on teaching learning based optimization (TLBO). In: Parashar, M., Kaushik, D., Rana, O.F., Samtaney, R., Yang, Y., Zomaya, A. (eds.) IC3 2012. CCIS, vol. 306, pp. 53–63. Springer, Heidelberg (2012)
8. Krishnanand, K.R., Panigrahi, B.K., Rout, P.K., Mohapatra, A.: Application of Multi-Objective Teaching Learning Based Algorithm to an Economic Load Dispatch Problem with Incommensurable Objectives. In: Panigrahi, B.K., Suganthan, P.N., Das, S., Satapathy, S.C. (eds.) SEMCCO 2011, Part I. LNCS, vol. 7076, pp. 697–705. Springer, Heidelberg (2011)
9. Naik, A., Satapathy, S.C., Parvathi, K.: Improvement of initial cluster center of c-means using Teaching learning based optimization. Accepted and will be published in Procedia Technology, Elsevier and indexed by Scopus
10. Naik, A., Satapathy, S.C.: Rough set and Teaching learning based optimization technique for Optimal Features Selection. Ref.: Ms. No. CEJCS-D-12-00042, Under Minor Review in Central European Journal of Computer Science

11. Satapathy, S.C., Naik, A.: Weighted Teaching-Learning-Based Optimization for global function optimization. Under Review in Applied Soft Computing Ms. Ref. No.: ASOC-D-12-00775
12. Rao, R.V., Patel, V.K.: Multi-objective optimization of combined Brayton and inverse Brayton cycles using advanced optimization algorithms. Engineering Optimization (2012), doi:10.1080/0305215X.2011.624183
13. Rao, R.V., Savsani, V.J.: Mechanical design optimization using advanced optimization techniques. Springer, London (2012)
14. Toğan, V.: Design of planar steel frames using Teaching–Learning Based Optimization. Engineering Structures 34, 225–232 (2012)
15. Rao, R.V., Kalyankar, V.D.: Parameter optimization of machining processes using a new optimization algorithm. Materials and Manufacturing Processes (2012), doi:10.1080/10426914.2011.602792
16. Potter, M.A., de Jong, K.A.: A cooperative coevolutionary approach to function optimization. In: Davidor, Y., Männer, R., Schwefel, H.-P. (eds.) PPSN 1994. LNCS, vol. 866, pp. 249–257. Springer, Heidelberg (1994)
17. Southwell, R.V.: Relaxation Methods in Theoretical Physics. Clarendon Press, Oxford (1946)
18. Friedman, M., Savage, L.S.: Planning experiments seeking minima. In: Eisenhart, C., Hastay, M.W., Wallis, W.A. (eds.) Selected Techniques of Statistical Analysis for Scientific and Industrial Research, and Production and Management Engineering, pp. 363–372. McGraw-Hill, New York (1947)
19. Das, S., Abraham, A., Konar, A.: Automatic Clustering Using an Improved Differential Evolution Algorithm. IEEE Transactions on Systems, Man, and Cybernetics—Part a: Systems and Humans 38(1) (January 2008)
20. Das, S., Abraham, A., Chakraborty, U.K., Konar, A.: Differential evolution using a neighborhood-based mutation operator. IEEE Trans. Evol. Comput. 13, 526–553 (2009)
21. Zhan, Z.H., Zhang, J., Li, Y., Chung, S.H.: Adaptive particle swarm optimization. IEEE Trans. Syst. Man Cybern. B Cybern. 39, 1362–1381 (2009)
22. Ratnaweera, A., Halgamuge, S., Watson, H.: Self-organizing hierarchical particle swarm optimizer with time-varying acceleration coefficients. IEEE Trans. Evol. Comput. 8, 240–255 (2004)
23. Zhang, J.Q., Sanderson, A.: JADE: Adaptive differential evolution with optional external archive. IEEE Trans. Evol. Comput. 13, 945–958 (2009)
24. Zhu, G.P., Kwong, S.: Gbest-guided artificial bee colony algorithm for numerical function optimization. Appl. Math. Comput. 217, 3166–3173 (2010)
25. Kang, F., Li, J.J., Ma, Z.Y.: Rosenbrock artificial bee colony algorithm for accurate global optimization of numerical functions. Inform. Sci. 12, 3508–3531 (2011)
26. Alatas, B.: Chaotic bee colony algorithms for global numerical optimization. Expert Syst. Appl. 37, 5682–5687 (2010)
27. Gao, W., Liu, S.: Improved artificial bee colony algorithm for global optimization. Information Processing Letters 111, 871–882 (2011)

Efficient Clustering of Dataset Based on Differential Evolution

Anima Naik[1] and Suresh Chandra Satapathy[2]

[1] MITS, India
[2] Dept of Computer Science and Engineering, ANITS, India
anmianaik@gmail.com, sureshsatapathy@ieee.org

Abstract. A novel approach to combining feature selection and clustering is presented. It uses selection of weighted Principal Components for features selection and automatic clustering based on Improved DE for clustering in order to reduce the complexity of high dimensional datasets and speed up the DE clustering process. We report significant improvements in total runtime. Moreover, the clustering accuracy of the dimensionality reduction DE clustering algorithm is comparable to the one that uses full dimensional datasets. The efficiency of this approach has been demonstrated with some real life datasets.

Keywords: Clustering, PCs, Dimension, DE.

1 Introduction

Selecting features in unsupervised learning scenarios is a much harder problem, due to the absence of class labels that would guide the search for relevant information. Problems of this kind have been rarely studied in the literature, for exceptions see e.g. [1, 2, 3]. The common strategy of most approaches is the use of an iterated stepwise procedure: in the first step a set of hypothetical partitions is extracted (the clustering step), and in the second step features are scored for relevance (the relevance determination step).

Feature selection has received considerable attention in various areas as a way to select informative features and to simplify the statistical model through dimensional reduction. One of the most widely used methods for dimensional reduction includes principal component analysis (PCA)[6]. Despite its popularity, PCA suffers from a lack of interpretability of the original feature because the reduced dimensions are linear combinations of a large number of original features. Traditionally, two or three dimensional loading plots provide information to identify important original features in the first few principal component dimensions. However, the interpretation of what constitutes a loading plot is frequently subjective, particularly when large numbers of features are involved. In this study, we use an unsupervised feature selection method that combines weighted principal components (PCs) with a thresholding algorithm. The weighted PC is obtained by the weighted sum of the first k PCs of interest. Each of the k loading values in the weighted PC reflects the contribution of each individual feature and the thresholding algorithm that identifies the significant features[4]. Differential evolution (DE) is well known as a simple and efficient scheme for global optimization over continuous spaces. However, DE does suffer from the problem of premature convergence to local optima. This problem of DE has been improved by using Improved DE[5]. Also, like most other stochastic optimization techniques DE is not free from the so-called "curse of dimensionality". In the improved DE, it is to try to avoid (or search and to facilitate convergence to the global optimum solution during the later stages of the search. We use some real life datasets both small as well as high dimensional datasets as our experimental data.

The rest of the paper is organized as follows: we introduce basic concepts of Dimensionality reduction technique Weighted PC's, Moving Range-Based thresholding Algorithm, Differential Evolution (DE), the automatic clustering DE algorithm which are taking directly from paper [4] and [5] respectively, in section 2. ACDRDDE algorithm (Automatic clustering on dimensional reduced data using Differential Evoluation) is discussed in section 3. Detailed simulation and results are presented in section 4. We conclude with a summary of the contributions of this paper in section 5.

2   Basic Concepts

A. Dimensionality Reduction Technique Weighted PC's

PCA is one of the most widely used multivariate data analysis techniques and is employed primarily for dimensional reduction and visualization [7]. PCA extracts a lower dimensional feature set that can explain most of the variability within the original data. The extracted features, PCi's (Yi) are each a linear combination of the original features with the loading values (α_{ij}, i, j=1,2,...,p). The Yi's can be represented as follows:

$$\left.\begin{array}{l} Y_1 = \alpha_{11}X_1 + \alpha_{12}X_2 + \cdots \ldots \ldots + \alpha_{1p}X_p \\ Y_2 = \alpha_{21}X_1 + \alpha_{22}X_2 + \cdots \ldots \ldots + \alpha_{2p}X_p \\ \ldots\ldots\ldots\ldots\ldots\ldots\ldots\ldots \\ Y_p = \alpha_{p1}X_1 + \alpha_{p2}X_2 + \cdots \ldots \ldots + \alpha_{pp}X_p \end{array}\right\} \tag{1}$$

The loading values represent the importance of each feature in the formation of a PC. For example, α_{ij} indicates the degree of importance of the jth feature in the ith PC. A two-dimensional loading plot (e.g., PC1 vs PC2 loading plot) may provide a graphical display for identification of important features in the first and second PC domains. However, the interpretation of a two-dimensional loading plot is frequently subjective, particularly in the presence of a large number of features. Moreover, in some situations, consideration of only the first few PCs may be insufficient to account for most of the variability in the data. Determination of the appropriate number of PCs (=k) to retain can be subjective. One can use a scree plot that visualizes the proportion of variability of each PC to determine the appropriate number of PCs .If a PCA loading value for the jth original feature can be computed from the first k PCs, the importance of the jth feature can be represented as follows:

$$w_j = \sum_{i=1}^{k} |\alpha_{ij}| \pi_i, \quad j=1,2\ldots\ldots\ldots\ldots,p \qquad (2)$$

where k is the total number of features of interest and π_i represents the weight of ith PC. The typical way to determine π_i is to compute the proportion of total variance explained by the ith PC. w_j can be called a weighted PC loading for the feature j. A feature with a large value of w_j indicates a significant feature. In the next section, we will present a systematic way to obtain a threshold that determines the significance of each feature.

B. Moving Range-Based ThresholdingAlgorithm

A moving range-based thresholding algorithm as a way to identify the significant features from the weighted PC loadings discussed in the above. The main idea of a moving range-based thresholding algorithm comes from a moving average control chart that has been widely used in quality control [8]. A control chart provides a comprehensive graphical display for monitoring the performance of a process over time so as to keep the process within control limits. A typical control chart comprises monitoring statistics and the control limit. When the monitoring statistics exceed (or fall below) the control limit, an alarm is generated so that proper remedial action can be taken. A moving range control chart is useful when the sample size used for process monitoring is one. Moreover, the average moving range control charts perform reasonably well when the observations deviate moderately from the normal distribution [8].

In this problem, it can consider the weighted PC loading values as the monitoring statistics. Thus, it plot these loading values on the moving range control chart and identify the significant features when the corresponding weighted PC loading exceeds the control limit (threshold). Given a set of the weighted PC loading values for individual features), $(\omega_1, \omega_2, \ldots\ldots\ldots\ldots\ldots\omega_p)$ the threshold γ can be calculated as follows [8].

$$\gamma = \bar{\omega} + \emptyset^{-1}(1-\alpha)\frac{\sqrt{\pi}}{2}\sigma \qquad (3)$$

Where, $\bar{\omega} = \frac{1}{p}\sum_{i=1}^{p}\omega_i$, $\emptyset^{-1}$is the inverse standard normal cumulative distribution function, and α is the Type I error rate that can be specified by the user. The range of α is between 0 and 1. In typical moving range control charts, σ can be estimated by $\overline{MR}$, calculated by the average of the moving ranges of two successive observations.

$$\overline{MR} = \frac{|\omega_1-\omega_2|+|\omega_2-\omega_3|+|\omega_3-\omega_4|+\cdots\ldots+|\omega_{p-1}-\omega_p|}{p-1} \qquad (4)$$

However, in our feature selection problems, because the weighted PC loading values for individual features $\omega_1, \omega_2, \ldots\ldots\ldots\ldots\ldots\omega_p$ are not ordered, we cannot simply use (4). To address this issue, we propose a different way of computing $\overline{MR}$ that can properly handle a set of unordered observations. Given the fact that there is no specific order of observations $\omega_1, \omega_2, \ldots\ldots\ldots\ldots\ldots\omega_p$, they are randomly reshuffled, $\overline{MRs}$ and are recalculated.

Therefore, we obtain a set of $\overline{MRs}$ $\overline{MR_{(1)}}$, $\overline{MR_{(1)}}$,........., $\overline{MR_{(N)}}$, where N is the number of features. The $\overline{MR^*}$ for unordered observations is calculated by

$$\overline{MR^*} = \frac{1}{N}\sum_{j=1}^{N} \overline{MR_{(j)}} \qquad (5)$$

Finally, the threshold of the feature selection method can be obtained by the following equation:

$$\gamma = \overline{\omega} + \emptyset^{-1}(1-\alpha)\frac{\sqrt{\pi}}{2}\ \overline{MR^*} \qquad (6)$$

A feature is reported as significant if the corresponding weighted PC loading exceeds the threshold γ.

C. Classical DE Algorithm and Its Modification

The classical DE is a population-based global optimization algorithm that uses a floating-point (real-coded) representation. The ith individual vector (chromosome) of the population at time-step (generation) t has d components (dimensions), i.e.

$$\overrightarrow{Z_i}(t)= [Z_{i,1}(t), Z_{i,2}(t), \ldots \ldots \ldots, Z_{i,d}(t)] \qquad (7)$$

For each individual vector $\overrightarrow{Z_k}(t)$ that belongs to the current population, DE randomly samples three other individuals, i.e., $\overrightarrow{Z_i}(t)$, $\overrightarrow{Z_j}(t)$ and $\overrightarrow{Z_m}(t)$, from the same generation (for distinct k, i, j, and m). It then calculates the (component wise) difference of $\overrightarrow{Z_i}(t)$ and $\overrightarrow{Z_j}(t)$, scales it by a scalar F (usually $\in [0,1]$), and creates a trial offspring $\overrightarrow{U_i}(t+1)$ by adding the result to $\overrightarrow{Z_m}(t)$. Thus, for the nth component of each vector.

$$U_{k,n}(t+1)=\begin{cases} Z_{m,n}(t) + F\left(Z_{i,n}(t) - Z_{j,n}(t)\right), \text{if } rand_n(0,1) < C_r \\ Z_{k,n}(t), \qquad\qquad\qquad\qquad \text{otherwise} \end{cases} \qquad (8) \qquad C_r$$

$\in [0,1]$ is a scalar parameter of the algorithm, called the crossover rate. If the new offspring yields a better value of the objective function, it replaces its parent in the next generation; otherwise the parent is retained in the population, i.e.,

$$\overrightarrow{Z_i}(t+1) = \begin{cases} \overrightarrow{U_i}(t+1), \text{ if } f\left(\overrightarrow{U_i}(t+1)\right) > f(\overrightarrow{Z_i}(t)) \\ \overrightarrow{Z_i}(t) \quad \text{if } f\left(\overrightarrow{U_i}(t+1)\right) \le f(\overrightarrow{Z_i}(t)) \end{cases} \qquad (9)$$

where $f(\cdot)$ is the objective function to be maximized.

To improve the convergence properties of DE, we have tuned its parameters in two different ways here. In the original DE, the difference vector $(\overrightarrow{Z_i}(t) - \overrightarrow{Z_j}(t))$ is scaled by a constant factor F. The usual choice for this control parameter is a number between 0.4 and 1. We propose to vary this scale factor in a random manner in the range (0.5, 1) by using the relation

$$F = 0.5 * (1 + rand(0, 1)) \qquad (10)$$

where rand(0, 1) is a uniformly distributed random number within the range [0, 1]. The mean value of the scale factor is 0.75. This allows for stochastic variations in the amplification of the difference vector and thus helps retain population diversity as the search progresses. The DE with random scale factor (DERANDSF) can meet or beat the classical DE. In addition to that, here we also linearly decrease the crossover rate Cr with time from $Cr_{max} = 1.0$ to $Cr_{min} = 0.5$. If $Cr = 1.0$, it means that all components of the parent vector are replaced by the difference vector operator However, at the later stages of the optimizing process, if Cr is decreased, more components of the parent vector are then inherited by the offspring. Such a tuning of Cr helps exhaustively explore the search space at the beginning but finely adjust the movements of trial solutions during the later stages of search, so that they can explore the interior of a relatively small space in which the suspected global optimum lies. The time variation of Cr may be expressed in the form of the following equation:

$$Cr = (Cr_{max} - Cr_{min}) * (MAXIT - iter)/MAXIT \qquad (11)$$

where Cr_{max} and Cr_{min} are the maximum and minimum values of crossover rate Cr, respectively; iter is the current iteration number; and MAXIT is the maximum number of allowable iterations.

D. Auto-DE Clustering Algorithm

In this algorithm for n data points, each d dimensional, and for a user-specified maximum number of clusters K_{max}, a chromosome is a vector of real numbers of dimension $K_{max} + K_{max} \times d$. The first K_{max} entries are positive floating point numbers in [0, 1], each of which controls whether the corresponding cluster is to be activated (i.e., to be really used for classifying the data) or not. The remaining entries are reserved for K_{max} cluster centers, each d dimensional.

The jth cluster center in the ith chromosome is active or selected for partitioning the associated data set if $T_{i,j} > 0.5$. On the other hand, if $T_{i,j} < 0.5$, the particular jth cluster is inactive in the ith chromosome. Thus, the $T_{i,j}$'s behave like control genes (we call them activation thresholds) in the chromosome governing the selection of the active cluster centers. The rule for selecting the actual number of clusters specified by one chromosome is

$$\textbf{IF } T_{i,j} > 0.5, \textbf{ THEN} \text{ the } j\text{th cluster center } \overrightarrow{m_{i,j}} \text{ is } \textbf{ACTIVE}$$
$$\textbf{ELSE } \overrightarrow{m_{i,j}} \quad \text{is } \textbf{INACTIVE} \qquad (12)$$

The quality of the partition yielded by such a chromosome can be judged by an appropriate cluster validity index.

When a new offspring chromosome is created according to (8) and (9), at first, the T values are used to select [using (12)] the active cluster centroids. If due to mutation some threshold $T_{i,j}$ in an offspring exceeds 1 or becomes negative, it is forcefully fixed to 1 or 0, respectively. However, if it is found that no flag could be set to 1 in a chromosome (all activation thresholds are smaller than 0.5), we randomly select two thresholds and reinitialize them to a random value between 0.5 and 1.0. Thus, the minimum number of possible clusters is 2.

The fitness of a particle is computed with the CS measure. The CS measure is defined as

$$CS(K) = \frac{\frac{1}{K}\sum_{i=1}^{K}\{\frac{1}{N_i}\sum_{x_j \in C_i} \max_{x_k \in C_i}\{d(x_j, x_k)\}\}}{\frac{1}{K}\sum_{i=1}^{K}\{\min_{j \in K, j \neq i}\{d(z_i, z_j)\}\}}$$

$$= \frac{\sum_{i=1}^{K}\{\frac{1}{N_i}\sum_{x_j \in C_i} \max_{x_k \in C_i}\{d(x_j, x_k)\}\}}{\sum_{i=1}^{K}\{\min_{j \in K, j \neq i}\{d(z_i, z_j)\}\}} \tag{13}$$

$$z_i = \frac{1}{N_i}\sum_{x_j \in C_i} x_j \text{ , i=1,2,}\ldots\ldots\text{,K} \tag{14}$$

Where z_i is the cluster center of C_i, and C_i is the set whose elements are the data points assigned to the ith cluster, and N_i is the number of elements in C_j, d denotes a distance function. This measure is a function of the ratio of the sum of within-cluster scatter to between-cluster separation.

The objective of the ACDE is to minimize the CS measure for achieving proper clustering results. The fitness function for each individual particle is computed by

$$F = \frac{1}{CS_i + eps} \tag{15}$$

Where CS_i is the CS measure computed for the ith particle, and eps is a very small-valued constant.

There is a possibility that, in our scheme, during computation of the CS measures, a division by zero may be encountered. This may occur when one of the selected cluster centers is outside the boundary of distributions of the data set. To avoid this problem, we first check to see if any cluster has fewer than two data points in it. If so, the cluster center positions of this special chromosome are reinitialized by an average computation. We put n/K data points for every individual cluster center, such that a data point goes with a center that is nearest to it.

3   Proposed ACDRDDE Algorithm (Automatic Clustering on Dimensional Reduced Data Using DE)

We first need to introduce the basic principles:
  The proposed algorithm works in two stages:

- First stage: As we are using dimensionality reduction techniques as data preprocessing, the first stage will be the dimensional reduction technique. Here weighted PC's based on moving ranged base thresholding algorithm is the dimensional reduction technique.
- Second stage: In this stage Auto_DE clustering algorithm is used for clustering reduced datasets.

Then the Pseudocode for the complete **ACDRDDE algorithm** is given here

(A) Pseudocode for first stage
1.   PCA extracts a lower dimensional feature set that can explain most of the variability   within the original data. The extracted features, PCi's (Yi) are each a linear combination of the original features with the loading values (α_{ij}, i, j=1,2,...,p). The Yi's can be represented as equation (1). The loading value represents the importance of each feature in the formation of PC.
2.   Determination of actual number of PCs (=k) to retain, use screen plot.

3. PCA loading value for the jth original feature can be computed from the first k PCs, the importance of the jth feature can be represented as follows:

$$w_j = \sum_{i=1}^{k} |\alpha_{ij}| \pi_i \quad ,j=1,2\ldots\ldots\ldots\ldots\ldots,p$$

    where k is the total number of features of interest
    and π_i represents the weight of ith PC .
    The π_i is to compute the proportion of total variance explained by the ith PC. w_j can be called a weighted PC loading for the feature j.

4. Given a set of the weighted PC loading values for individual features, $(\omega_1, \omega_2, \ldots\ldots\ldots\ldots\ldots \omega_p)$ the threshold γ can be calculated using equation (6), where $\overline{MR^*}$ can calculated using equation (5).
5. Identify the significant features when the corresponding weighted PC loading exceeds the control limit (threshold). Using those significant features makes a reduced data set.

At the end of this stage we will get reduced data set, which will be input for second stage of our proposed algorithm.

(B) Pseudocode for second stage
1. Initialize each chromosome to contain K number of randomly selected cluster centers and K (randomly chosen) activation thresholds in [0, 1].
2. Find out the active cluster centers in each chromosome with the help of the rule described in (12).
3. For t = 1 to t_{max} do
    a) For each data vector $\overrightarrow{X_p}$, calculate its distance metric $d(\overrightarrow{X_p}, \overrightarrow{m_{1,j}}$) from all active cluster centers of the ith chromosome $\overrightarrow{V_1}$.
    b) Assign $\overrightarrow{X_p}$, to that particular cluster center $\overrightarrow{m_{1,j}}$,where

    $$d(\overrightarrow{X_p}, \overrightarrow{m_{1,j}}) = \min_{\forall b \in \{1,2,\ldots,K\}} \{d(\overrightarrow{X_p}, \overrightarrow{m_{1,b}})\}$$

    c) Check if the number of data points that belong to any cluster center, $\overrightarrow{m_{1,j}}$ is less   than 2. If so, update the cluster centers of the chromosome using the concept of average computation.
    d) Change the population members according to the DE algorithm outlined in (8)–(11). Use the fitness of the chromosomes to guide the evolution of the population.
4. Report as the final solution the cluster centers and the partition obtained by the globally best chromosome (one yielding the highest value of the fitness function) at time $t = t_{max}$.

4   Experiment and Results

We run dimensionality reduction technique to reduce the dataset as lower dimensional dataset. We apply clustering on the data before and after the dimensionality reduction to verify and compare the results. The accuracy of the clustering results and the runtime of the algorithms will be compared. The runtime for the dimensionality reduction based AUTO-DE algorithm will include the dimensionality reduction time and the AUTO-DE clustering time.

A. Experimental Setup

The parameters of the AUTO-DE algorithm for all examples are defined as follows:
Pop_size = 20.

B. Datasets used

The following real-life data sets are used in this paper which are taken from UCI Machine Repository. Here, n is the number of data points, d is the number of features, and K is the number of clusters.

1) **Iris** plants database ($n = 150$, $d = 4$, $K = 3$): This is a well-known database with 4 inputs, 3 classes, and 150 data vectors. The data set consists of three different species of iris flower: *Iris setosa*, *Iris virginica*, and *Iris versicolour*. For each species, 50 samples with four features each (sepal length, sepal width, petal length, and petal width) were collected. The number of objects that belong to each cluster is 50.

2) **Glass** ($n = 214$, $d = 9$, $K = 6$): The data were sampled from six different types of glass: 1) building windows float processed (70 objects); 2) building windows nonfloat processed (76 objects); 3) vehicle windows float processed (17 objects); 4) containers (13 objects); 5) tableware (9 objects); and 6) headlamps (29 objects). Each type has nine features: 1) refractive index; 2) sodium; 3) magnesium; 4) aluminum; 5) silicon; 6) potassium; 7) calcium; 8) barium; and 9) iron.

3) **Wisconsin breast cancer data set** ($n = 683$, $d=9$, $K=2$): The Wisconsin breast cancer database contains nine relevant features: 1) clump thickness; 2) cell size uniformity; 3) cell shape uniformity; 4) marginal adhesion; 5) single epithelial cell size; 6) bare nuclei; 7) bland chromatin; 8) normal nucleoli; and 9) mitoses. The data set has two classes. The objective is to classify each data vector into benign (239 objects) or malignant tumors (444 objects).

4) **Wine** ($n = 178$, $d = 13$, $K = 3$): This is a classification problem with "well-behaved" class structures. There are 13 features, three classes, and 178 data vectors.

5) **Pima Diabates data** ($n=768$, $d=8$, $K=2$): This data set consists of 768 data vectors, 8 features and 2 classes.

6) **Haberman's Survival Data Set** ($n=306$, $d=3$, $K=2$): This data set consists of 306 data vectors, 3 features and 2 classes.

C. Population Initialization

For the AUTO-DE algorithm, we randomly initialize the activation thresholds (control genes) within [0, 1]. The cluster centroids are also randomly fixed between X_{max} and X_{min}, which denote the maximum and minimum numerical values of any feature of the data set under test, respectively. To make the comparison fair, the populations for both the AUTO-DE clustering algorithm and proposed **ACDRDDE algorithm** (for all problems tested) were initialized using the same random seeds.

D. Simulation Strategy

In this paper, while comparing the performance of our proposed **ACDRDDE** algorithm with AUTO-DE clustering techniques, we focus on two major issues: as 1) ability to find the optimal number of clusters; and 2) computational time required to find the solution. For comparing the speed of the algorithms, the first thing we require is a fair time measurement. The number of iterations or generations cannot be accepted as a time measure since the algorithms perform different amount of works in their inner loops, and they have different population sizes. Hence, we choose the number of *fitness function evaluations (FEs)* as a measure of computation time instead of generations or iterations. Since the algorithms are stochastic in nature, the results of two successive runs usually do not match. Hence, we have taken 30 independent runs (with different seeds of the random number generator) of each algorithm. The results have been stated in terms of the mean values and standard deviations over the 30 runs in each case.

Finally, we would like to point out that all the experiment codes are implemented in MATLAB. The experiments are conducted on a Pentium 4, 1GB memory desktop in Windows XP 2002 environment.

E. Experimental Results

To judge the accuracy of the AUTO-DE and **ACDRDDE**, we let each of them run for a very long time over every benchmark data set, until the number of FEs exceeded 10^4. Then, we note the number of clusters found.

Table 1. Final Solution (Mean and Standard Deviation Over 30 Independent Runs) After Each Algorithm Was Terminated After Running For 10^4 FEs, with the Cs-Measure-Based Fitness Function

Data sets name	Algorithm Used	Average Number of cluster found	Actual number of clusters
wine data	AUTO-DE	3.0250±0.6973	3
	ACDRDDE	3.0000±0.6417	
Iris data	AUTO- DE	3.0000±0.6831	3
	ACDRDDE	3.1290±0.7634	
Breast cancer data	AUTO- DE	2.333±0.5164	2
	ACDRDDE	2.3636±0.5045	
Pima Diabates data	AUTO- DE	2.2000 ±0.0000	2
	ACDRDDE	2.0000±0.0000	
Haberman's Survival Data	AUTO- DE	2.5000± 0.7071	2
	ACDRDDE	2.3000±0.4321	
Glass data	AUTO- DE	5.900±0.1972	6
	ACDRDDE	6.1250±0.7428	

To compare the speeds of different algorithms, we selected a threshold value of CS measure for each of the data sets. This cutoff CS value is somewhat larger than the minimum CS value found by each algorithm in TABLE I. Now, we run a clustering algorithm on each data set and stop as soon as the algorithm achieves the proper number of clusters, as well as the CS cutoff value. We then note down the number of fitness FEs that the algorithm takes to yield the cutoff CS value. A lower number of FEs corresponds to a faster algorithm.

Table 2. Mean and standard deviations of the number of fitness FEs (over 30 independent runs) required By each algorithm to reach a predefined cutoff value of the CS validity index

Data sets name	Algorithm Used	Mean number of function evaluation required	CS Cutoff Value
wine data	AUTO- DE	745±34.1245	1.90
	ACDRDDE	128.45±78.5289	
Iris data	AUTO- DE	600.87±80.3288	0.95
	ACDRD E	90.38±70.1902	
Breast cancer data	AUTO- DE	190.72±50.4512	1.10
	ACDRDDE	40.12±11.6536	
Pima Diabates data	AUTO- DE	588.49±87.9001	1.00
	ACDRD E	60.56±10.8711	
Haberman's Survival Data Set	AUTO- DE	490.23±361.4228	0.90
	ACDRDDE	23±10.1456	
Glass data	AUTO- DE	3901±256.3401	1.80
	ACDRDDE	665.42±38.4216	

F.   Discussion on Results

The Tables II reveals the fact that the ACDRDDE algorithm is faster algorithm in compare to AUTO-DE where as ACDRDDE run on reduce data and AUTO-DE run on original high dimensional data.

5   Conclusion and Future Scope

We have presented a new method of clustering in high-dimensional datasets. The proposed method combines PCA techniques and a moving range-based thresholding algorithm with automatic clustering of DE. We first obtained the weighted PC, which can be calculated by the weighted sum of the first k PCs of interest. Each of the k loading values i the weighted PC reflects the contribution of each individual feature. To identify the significant features, we proposed a moving-range thresholding algorithm. Features are considered to be significant if the corresponding weighted PC loadings exceed the threshold obtained by a moving-range thresholding algorithm. Using this technique we get reduced dataset of original high dimensional data set. On that reduced data we applied automatic clustering DE to get cluster. Our experimental results with real datasets will demonstrate that the proposed method could successfully find the true clusters of high dimensional data set on less computation.

Our study extends the application scope of **ACDRDDE algorithm**. We hope that the procedure discussed here stimulates further investigation into development of better procedures for clustering of high dimensional datasets.

References

[1] Ben-Dor, A., Friedman, N., Yakhini, Z.: Class discovery in gene expression data. In: Procs. RECOMB, pp. 31–38 (2001)

[2] Law, M.H., Jain, A.K., Figueiredo, M.A.T.: Feature selection in mixture-based clustering. In: Advances in Neural Information Processing Systems, vol. 15 (2003) (to appear)

[3] Heydebreck, A.V., Huber, W., Poustka, A., Vingron, M.: Identifying splits with clear separation: A new class discovery method for gene expression data. Bioinformatics 17 (2001)

[4] Kim, S.B., Rattakorn, P.: Unsupervised Feature Selection Using Weighted Principal Components (2010)

[5] Das, S., Konar, A., Braham, A.: Automatic Clustering Using an Improved Differential Evolution Algorithm. IEEE Transactions on Systems, Man, and Cybernetics—Part a: Systems and Humans 38(1) (January 2008)

[6] Boutsidis, C., Mahoney, M.W., Drineas: Unsupervised Feature Selection for Principal Components Analysis. In: KDD 2008, Las Vegas, Nevada, USA, August 24-27 (2008)

[7] Jolliffe, I.T.: Principal Component Analysis. Springer, New York (2002)

[8] Vermaat, M.B., Ion, R.A., Does, R.J.M.M., Klaassen, C.A.J.: A comparison of Shewhart individuals control charts based on normal, non-parametric, and extreme-value theory. Quality and Reliability Engineering International 19, 337–353 (2003)

Numerical Optimization of Novel Functions Using vTLBO Algorithm

S. Mohankrishna[1], Anima Naik[2], Suresh Chandra Satapathy[3],
K. Raja Sekhara Rao[4], and B.N. Biswal[5]

[1] IT Dept, Gitam University and K.L University
[2] MITS, India
[3] Dept of Computer Science and Engineering, ANITS, India
[4] Dept of Computer Science and Engineering, K.L University, India
[5] Bhubaneswar Engineering College, Bhubaneswar
{smkrishna,sureshsatapathy}@ieee.org, anmianaik@gmail.com,
rajasekhar.kurra@klce.ac.in, bhabendra_biswal@yahoo.co.in

Abstract. Teaching–Learning-Based Optimization (TLBO) is recently being used as a new, reliable, accurate and robust optimization technique for global optimization. It outperforms some of the well-known metaheuristics regarding constrained benchmark functions, constrained mechanical design, and continuous non-linear numerical optimization problems. However, the success of TLBO in solving some specific types of problems such as shifted function goes down. In this paper we have modified little bit in code of TLBO to improve its performance while solving shifted type of functions. The modified code of TLBO is named as vTLBO (variant TLBO). The performance of vTLBO algorithm is extensively evaluated on 9 shifted and 9 shifted rotated numerical optimization problems and compares favorably with the DE, PSO and conventional TLBO. The results show the better performance of the vTLBO algorithm. Also we have shown that whenever the performance of vTLBO compare with TLBO by taking simple benchmark function, its performance has been degraded.

Keywords: metaheuristics, TLBO, shifted function, shifted rotated function, numerical optimization.

1 Introduction

Constrained and unconstrained optimization problems are generally associated with many difficulties such as multi-modality, dimensionality and differentiability. Traditional optimization techniques generally fail to solve such problems, especially with nonlinear objective functions. To overcome these difficulties, there is a need to develop more powerful optimization techniques and research is continuing to find effective optimization techniques.

Rao et al. [1, 2] proposed a teaching-learning based optimization (TLBO) algorithm based on the natural phenomenon of teaching and learning. The implementation of TLBO does not require the determination of any algorithm specific

S.C. Satapathy, S.K. Udgata, and B.N. Biswal (eds.), *FICTA 2013*,
Advances in Intelligent Systems and Computing 247,
DOI: 10.1007/978-3-319-02931-3_27, © Springer International Publishing Switzerland 2014

controlling parameters which makes the algorithm robust and powerful. TLBO requires only common controlling parameters like population size and number of generations for its working. In this way TLBO can be said as an algorithm specific parameter-less algorithm. Rao and Patel [3] investigated the performance of TLBO algorithm for different elite sizes, population sizes and number of generations considering various constrained bench mark problems available in the literature to identify their effect on the exploration and exploitation capacity of the algorithm. Suresh satapathy and Anima Naik [4] tried to propose a new approach to use TLBO to cluster data. They have proved that the TLBO algorithm can be used to cluster arbitrary data. Again they have shown another good characteristic of TLBO algorithm is the cost of computations are less in handling high dimensional problems in compare to other algorithms [5]. TLBO can able to show its application in the area of computer network by optimizing the multicast tree [6]. It can able to handle power system problem easily where optimal solution of the unit maintenance scheduling problem in which the cost reduction is as important as reliability [7]. TLBO algorithm has been adapted to solve multiobjective problems of an economic load dispatch problem with incommensurable objectives [8]. From literature we can see while clustering the data using fuzzy c-means (FCM) and hard c-means (HCM), the sensitivity to tune the initial clusters centers have captured the attention of the clustering communities for quite a long time. This problem has been addressed by TLBO in [9]. In the area of feature selection also TLBO is showing its performance in connection with roughset theory. From empirical results reveal that the RoughTLBO approach could be performed better in terms of finding optimal features and doing so in quick time in comparison with GA, PSO and DE [10]. Again there is some modification of TLBO have been done for improve its performance in optimization for global function optimization [11]-[12]. Like many others applications of TLBO can be seen from different papers [13 [16]

In this paper we are going to prove that although TLBO can able to solve large number of optimization problems still the performance of TLBO degrade in solving some specific kind of problems such as shifted function. In this paper we have tried to modify little bit in code of TLBO to improve its performance while solving shifted type of functions whereas vTLBO degrade its performance while solving simple benchmark functions. The modified code of TLBO is named a vTLBO. The performance of vTLBO algorithm is extensively evaluated on shifted and shifted rotated numerical optimization problems (using codes available from http://www.ntu.edu.sg/home/epnsugan for shifted and shifted rotated function) and compares favorably with the DE, PSO and conventional TLBO.

The remainder of this paper is organized as follows. The conventional TLBO is explained in detail in Sections 2 and it has been copied directly from paper [1] to give proper comparison, the modified code of TLBO i.e. vTLBO is explained in detail in section 3. Comparison of code of vTLBO and TLBO in implementation level is explained in section 4. Numerical Experiment and results demonstrating the performance of vTLBO in comparison with DE, PSO, and conventional TLBO over a suite of 9 shifted and 9 shifted rotated numerical optimization problems and comparison on performance of TLBO and vTLBO by taking 12 simple benchmark functions are presented in Section 5. Section 6 concludes this paper.

2   Teaching Learning Based Optimization

This optimization method is based on the effect of the influence of a teacher on the output of learners in a class. Like other nature-inspired algorithms, TLBO [1, 2] is also a population based method that uses a population of solutions to proceed to the global solution. A group of learners are considered as the population. In TLBO, different subjects offered to learners are considered as different design variables for the TLBO. The learning results of a learner is analogous to the 'fitness', as in other population-based optimization techniques. The teacher is considered as the best solution obtained so far.

There are two parts in TLBO: 'Teacher Phase' and 'Learner Phase'. The 'Teacher Phase' means learning from the teacher and the 'Learner Phase' means learning through the interaction between learners.

Teacher Phase

In our society the best learner is mimicked as a teacher. The teacher tries to disseminate knowledge among learners, which will in turn increase the knowledge level of the whole class and help learners to get good marks or grades. So a teacher increases the mean learning value of the class according to his or her capability i.e. say the teacher T_1 will try to move mean M_1 towards their own level according to his or her capability, thereby increasing the learners' level to a new mean M_2. Teacher T_1 will put maximum effort into teaching his or her students, but students will gain knowledge according to the quality of teaching delivered by a teacher and the quality of students present in the class. The quality of the students is judged from the mean value of the population. Teacher T_1 puts effort in so as to increase the quality of the students from M_1 to M_2, at which stage the students require a new teacher, of superior quality than themselves, i.e. in this case the new teacher is T_2.

Let M_i be the mean and T_i be the teacher at any iteration i. T_i will try to move mean M_i towards its own level, so now the new mean will be T_i designated as M_{new}. The solution is updated according to the difference between the existing and the new mean given by

$$\text{Difference\_mean}_i = r_i(M_{new} - T_F M_i) \quad\quad\dots\dots\dots\dots\dots\dots\dots\dots\dots \quad (1)$$

where T_F is a teaching factor that decides the value of mean to be changed, and r_i is a random number in the range [0, 1]. The value of T_F can be either 1 or 2, which is again a heuristic step and decided randomly with equal probability as

$$T_F = \text{round}[1 + \text{rand}(0,1) * (2 - 1)]. \quad\quad\dots\dots\dots\dots\dots\dots\dots\dots \quad (2)$$

This difference modifies the existing solution according to the following expression

$$X_{new,i} = X_{old,i} + \text{Difference\_mean}_i \quad\quad\dots\dots\dots\dots\dots\dots\dots\dots \quad (3)$$

Learner Phase

Learners increase their knowledge by two different means: one through input from the teacher and the other through interaction between themselves. A learner interacts randomly with other learners with the help of group discussions, presentations, formal communications, etc. A learner learns something new if the other learner has more knowledge than him or her. Learner modification is expressed as

For $i = 1: P_n$

Randomly select two learners X_i and X_j, where $i \neq j$

If $f(X_i) < f(X_j)$ $X_{new,i} = X_{old,i} + r_i (X_i - X_j)$

Else $X_{new,i} = X_{old,i} + r_i (X_j - X_i)$

End If

End For

Accept X_{new} if it gives a better function value.

3   Proposed vTLBO

The proposed vTLBO is same with original TLBO. There is little bit difference in coding parts of TLBO both in Teacher Phase' and 'Learner Phase'.

Teacher phase

In the teacher phase the eqn (1) is written as

$$\text{Difference\_mean}_i = (M_{new} - T_F M_i) \qquad \dots\dots\dots\dots\dots\dots\dots\dots (4)$$

where T_F is a teaching factor that decides the value of mean to be changed. The value of T_F can be either 1 or 2, which is again a heuristic step and decided randomly with equal probability as

$$T_F = \text{round}[1 + \text{rand}(0,1) * (2 - 1)].$$

So equation (2) of TLBO remain same.

This difference modifies the existing solution according to the following expression

for k=1: dn

$$X_{new,i,k} = X_{old,i,k} + r_k * \text{Difference\_mean}_i \qquad \dots\dots\dots\dots\dots (5)$$

End

Where dn is the number of variable or number of subject assigned to learners and r_k is random number varies with k.

Learner phase

In the learner phase the Learner modification is expressed as

For $i = 1: P_n$

Randomly select two learners X_i and X_j, where $i \neq j$

If $f(X_i) < f(X_j)$

```
          For k=1:dn
              Xnew,i,k = Xold,i,k+ rk (Xi − Xj )
          end
      Else
          For k=1:dn
              Xnew,i,k = Xold,i,k+ rk (Xj − Xi )
          end
      End If
  End For
```

Where r_k is the random number varies with k.
Accept X_{new} if it gives a better function value.

4 Comparison of Code of vTLBO and TLBO in Implementation Level

From conventional TLBO and vTLBO it clear that in case of conventional TLBO, in teacher phase the random value varies with iteration and in learner phase random value varies with learner, whereas in vTLBO the random value varies with dimension of learner in both teacher phase and learner phase. The step by step implementation of conventional TLBO code has been explained in detail by taking an example in [2].

5 Numerical Experiments and Results

Experiments 1:

A. Test functions

We used 12 benchmark problems in order to test the performance of vTLBO and TLBO algorithms. This set is sufficient to include many different kinds of problems such as unimodal, multimodal, regular, irregular, separable, non-separable and multidimensional. Initial range, formulation, characteristics and the dimensions of these problems are listed in Tables 1.

Table 1. Benchmark functions used in experiments 1. D: Dimension, C: Characteristic, U: Unimodal, M: Multimodal, S:Separable , N:Non-Separable

Function	D	C	Range	Formulation	Value
Sphere	30	US	[-100,100]	$f(x) = \sum_{i=1}^{D} x_i^2$	$f_{min} = 0$
Schwefel 1.2	30	UN	[-100,100]	$f(x) = \sum_{i=1}^{D} (\sum_{j=1}^{i} x_j)^2$	$f_{min} = 0$

Table 1. (*continued*)

Schwefel 2.22	50	UN	[-10,10]	$$f(x) = \sum_{i=1}^{D} |x_i| + \prod_{i=1}^{D} |x_i|$$	$f_{min} = 0$
Schwefel 2.21	30	UN	[-100,100]	$$f(x) = \max_i \{|x_i|, 1 \le i \le D\}$$	$f_{min} = 0$
Rastrigin	30	MS	[-5.12,5.12]	$$f(x) = \sum_{i=1}^{D} [x_i^2 - 10\cos(2\pi x_i) + 10]$$	$f_{min} = 0$
Griewank	30	MN	[-600,600]	$$f(x) = \frac{1}{4000} \sum_{i=1}^{D} x_i^2 - \prod_{i=1}^{D} \cos\left(\frac{x_i}{\sqrt{i}}\right) + 1$$	$f_{min} = 0$
Weierstrass	30		[-0.5, 0.5]	$$f(x) = \sum_{i=1}^{D} \left(\sum_{k=0}^{kmax} [a^k \cos(2\pi b^k(x_i + 0.5))] \right) - D \sum_{k=0}^{kmax} [a^k \cos(2\pi b^k(x_i + 0.5))],$$ where $a = 0.5, b = 3, kmax = 20$	$f_{min} = 0$
Multimod	30		[-10,10]	$$f(x) = \sum_{i=1}^{D} |x_i| \prod_{i=1}^{D} |x_i|$$	$f_{min} = 0$
SumSquares	30	US	[-10,10]	$$f(x) = \sum_{i=1}^{D} ix_i^2$$	$f_{min} = 0$
Zakharov	10	UN	[−5,10]	$$f(x) = \sum_{i=1}^{D} x_i^2 + \left(\sum_{i=1}^{D} 0.5 i x_i \right)^2 + \left(\sum_{i=1}^{D} 0.5 i x_i \right)^4$$	$f_{min} = 0$

Table 1. (*continued*)

Noncontinuous Rastrigin	30	MS	[-5.12,5.12]	$f(x)$ $$= \sum_{i=1}^{D} [y_i^2 - 10\cos(2\pi y_i) + 10]$$ Where $y_i =$ $\begin{cases} x_i & |x_i| < 0.5 \\ \frac{round(2x_i)}{2} & |x_i| \geq 0.5 \end{cases}$	$f_{min} = 0$
Ackley	30	MN	[-32,32]	$f(x)$ $$= -20\exp\left(-0.2\sqrt{\frac{1}{D}\sum_{i=1}^{D} x_i^2}\right)$$ $$- \exp\left(\frac{1}{n}\sum_{i=1}^{D} \cos(2*pi *x_i)\right) + 20 + e$$	$f_{min} = 0$

B. Settings

In all experiments in this section, the values of the common parameters used in each algorithm such as population size and total evaluation number are chosen to be the same. Population size is 20 and the maximum number generation for all functions is 100.

C. Experimental Results and Discussions

Since the algorithms are stochastic in nature, the results of two successive runs usually do not match. Hence, we have taken 30 independent runs (with different seeds of the random number generator) of each algorithm. The results have been stated in terms of the mean values and standard deviations over the 30 runs in each case.

Table 2 report the mean and standard deviation of function values by applying vTLBO and TLBO algorithms to optimize all functions of table 1. From table 2 it is clear that TLBO algorithm is always best then vTLBO algorithm while solving simple benchmark functions.

Table 2. Performance comparision of vTLBO and TLBO

Name of the function		Vtlbo	TLBO
Sphere	Mean	3.6391e-15	5.7087e-23
	Std	8.2767e-15	7.9737e-23
Schwefel 1.2	Mean	4.4392e-11	1.8212e-17
	Std	4.8832e-11	8.2011e-17
Schwefel 2.22	Mean	2.7420e-08	2.6401e-13
	Std	1.9328e-08	3.1162e-13
Schwefel 2.21	Mean	1.4571e-06	4.4544e-12
	Std	1.2370e-06	5.8088e-12
Rastrigin	Mean	65.2122	0
	Std	28.5556	0
Griewank	Mean	2.0188e-14	0
	Std	5.3469e-14	0
Weierstrass	Mean	9.2893e-05	0
	Std	4.9760e-05	0
Multimod	Mean	4.1113e-192	0
	Std	0	0
SumSquares	Mean	5.4496e-16	8.2117e-24
	Std	1.2108e-15	3.8844e-23
Zakharov	Mean	5.0741e-08	1.0871e-24
	Std	9.4480e-08	3.3443e-24
Noncontinuous Rastrigin	Mean	101.7492	0
	Std	29.1472	0
Ackley	Mean	1.5139e-08	1.2822e-12
	Std	1.2092e-08	2.9046e-12

Experiments 2:

A. Test Functions

As discussed in [17], many benchmark numerical functions commonly used to evaluate and compare optimization algorithms may suffer from two problems. First, global optimum lies at the center of the search range. Second, local optima lie along the coordinate axes or no linkage among the variables/dimensions exists. To solve these problems, we can shift or rotate the conventional benchmark functions. For benchmark functions suffering from the first problem, we may shift the global optimum to a random position so that the global optimum position has different numerical values for different dimensions, i.e., $F(x) = f(x - o_{new} + o_{old})$, where $F(x)$ is the new function, $f(x)$ is the old function, o_{old} is the old global optimum, and o_{new} is the new global optimum with different values for different dimensions and not lying at the center of the search range. For the second problem, we can rotate the function $F(x) = f(M * x)$, where M is an orthogonal rotation matrix obtained using Salmon's method [18], to avoid local optima lying along the coordinate axes while retaining the properties of the test function. We hereby shift nine commonly used benchmark functions, and further rotate them. So in this way we are getting 18 different type of novel functions. These are explained below

1) Shifted sphere function

$$f(x) = \sum_{i=1}^{D} z_i^2$$

$$z = x - o$$

$O = [\,^\backprime O_1, O_2, \ldots\ldots, O_D\quad]$: the shifted global optimum.

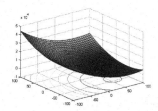

2) Shifted Schwefel's 1.2

$$f(x) = \sum_{i=1}^{D} (\sum_{j=1}^{i} z_j)^2$$

$$z = x - o$$

$O = [\,^\backprime O_1, O_2, \ldots\ldots, O_D\quad]$: the shifted global optimum.

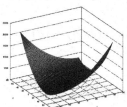

3) Shifted Schwefel's 2.21

$$f(x) = \max_{i}\{z_i,\ 1 \le i \le D\}$$

$$z = x - o$$

$O = [\,^\backprime O_1, O_2, \ldots\ldots, O_D\quad]$: the shifted global optimum.

4) Shifted Schwefel's 2.22

$$f(x) = \sum_{i=1}^{D}|z_i| + \prod_{i=1}^{D}|z_i|$$

$$z = x - o$$

O=[$O_1, O_2, ..., O_D$]: the shifted global optimum.

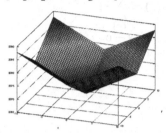

5) Shifted Elliptic function

$$f(x) = \sum_{i=1}^{D}(10^6)^{\frac{i-1}{D-1}} z_i^2$$

$$z = x - o$$

O=[$O_1, O_2, ..., O_D$]: the shifted global optimum.

6) Shifted Rastrigin's function

$$f(x) = \sum_{i=1}^{D}(z_i^2 - 10cos(2\pi z_i) + 10)$$

$$z = x - o$$

O=[$O_1, O_2, ..., O_D$]: the shifted global optimum

.

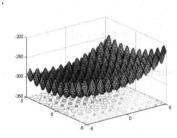

7) Shifted Griewank's function

$$f(x) = \sum_{i=1}^{D}\frac{z_i^2}{4000}$$

$$z = x - o$$

O=[$O_1, O_2, ..., O_D$]: the shifted global optimum.

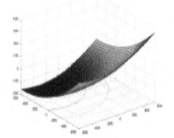

8) Shifted Ackley's function

$$f(x) = -20exp(-0.2\sqrt{\frac{1}{D}\sum_{i=1}^{D}z_i^2})\text{-}exp(\frac{1}{D}\sum_{i=1}^{D}\cos{(2\pi z_i)})+20+e$$

$$z = x - o$$

O=[$`O_1, O_2, \dots \dots, O_D$]: the shifted global optimum.

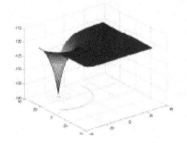

9) Shifted Weierstrass function

$$f(x) =$$
$$\sum_{i=1}^{D}\left(\sum_{k=0}^{kmax}\left[a^k\cos{\left(2\pi b^k(z_i + 0.5)\right)}\right]\right) - D\sum_{k=0}^{kmax}[a^k\cos{(2\pi b^k(z_i + 0.5))}]$$

$$z = x - o$$

O=[$`O_1, O_2, \dots \dots, O_D$]: the shifted global optimum.

Similarly for shifted rotated function we can find as
10) Shifted rotated sphere function

$$f(x) = \sum_{i=1}^{D} z_i^2$$

$$z = M(x - o)$$

M is the orthogonal rotation matrix. O=[$`O_1, O_2, \dots \dots, O_D$]: the shifted global optimum.

11) Shifted rotated Schwefel's 1.2

$$f(x) = \sum_{i=1}^{D} \left(\sum_{j=1}^{i} z_j \right)^2$$

$$z = M(x - o)$$

M is the orthogonal rotation matrix, O=[$^{\backprime}O_1, O_2, \dots \dots, O_D$]: the shifted global optimum.

12) Shifted rotated Schwefel's 2.21

$$f(x) = \max_i \{z_i, \ 1 \le i \le D\}$$

$z = M(x - o)$

M is the orthogonal rotation matrix ,O=[$^{\backprime}O_1, O_2, \dots \dots, O_D$]: the shifted global optimum.

13) Shifted rotated Schwefel's 2.22

$$f(x) = \sum_{i=1}^{D} |z_i| + \prod_{i=1}^{D} |z_i|$$

$$z = M(x - o)$$

M is the orthogonal rotation matrix,O=[$^{\backprime}O_1, O_2, \dots \dots, O_D$]: the shifted global optimum.

14) Shifted rotated Elliptic function

$$f(x) = \sum_{i=1}^{D} (10^6)^{\frac{i-1}{D-1}} z_i^2$$

$$z = M(x - o)$$

M is the orthogonal rotation matrix, O=[$^{\backprime}O_1, O_2, \dots \dots, O_D$]: the shifted global optimum.

15) Shifted rotated Rastrigin's function

$$f(x) = \sum_{i=1}^{D} \left(z_i^2 - 10\cos(2\pi z_i) + 10 \right)$$

$$z = M(x - o)$$

M is the orthogonal rotation matrix,O=[$^{\backprime}O_1, O_2, \dots \dots, O_D$]: the shifted global optimum.

16) Shifted rotated Griewank's function

$$f(x) = \sum_{i=1}^{D} \frac{z_i^2}{4000}$$

$$z = M(x - o)$$

M is the orthogonal rotation matrix,O=[$^{\backprime}O_1, O_2, \dots \dots, O_D$]: the shifted global optimum.

17) Shifted rotated Ackley's function

$$f(x) = -20exp(-0.2\sqrt{\frac{1}{D}\sum_{i=1}^{D}z_i^2})-\exp(\frac{1}{D}\sum_{i=1}^{D}\cos{(2\pi z_i)})+20+e$$

$$z = M(x-o)$$

M is the orthogonal rotation matrix ,O=[`$O_1, O_2, ..., O_D$]: the shifted global optimum.

18) Shifted rotated Weierstrass function

$$f(x) =$$
$$\sum_{i=1}^{D}\left(\sum_{k=0}^{kmax}\left[a^k\cos\left(2\pi b^k(z_i+0.5)\right)\right]\right) - D\sum_{k=0}^{kmax}[a^k\cos{(2\pi b^k(z_i+0.5))}]$$

$$,z = M(x-o)$$

M is the orthogonal rotation matrix, O=[`$O_1, O_2, ..., O_D$]: the shifted global optimum.

B.   Settings

In all experiments in this section, the values of the common parameters used in each algorithm such as population size and total evaluation number were chosen to be the same. Population size was 20 and the maximum number fitness function evaluation was 40,000 for all functions. The other specific parameters of algorithms are given below:

PSO Settings: Cognitive and social components c_1, c_2 are constants that can be used to change the weighting between personal and population experience, respectively. In our experiments cognitive and social components were both set to 2. Inertia weight, which determines how the previous velocity of the particle influences the velocity in the next iteration, was 0.5 [19]

DE Settings: In DE [20], F is a real constant which affects the differential variation between two Solutions and set to F = 0.9. Value of crossover rate, which controls the change of the diversity of the population, was chosen to be Cr = 0.1.

TLBO Settings: For TLBO there is no such constant to set.
vTLBO Settings: For vTLBO there is no such constant to set.

Table 3. Global optimum , search ranges, and Initialization ranges of the test functions

Functions	Dimension	Global optimum x^*	$f(x^*)$	Search range	Initialization range
Shifted sphere	10 and 30	O	0	[-100,100]	[-100,100]
Shifted schwefel 1.2		O	0	[-100,100]	[-100,100]
Shifted schwefel 2.21		O	0	[-100,100]	[-100,100]
Shifted schwefel 2.22		O	0	[-100,100]	[-100,100]
Shifted Elliptic		O	0	[-100,100]	[-100,100]
Shifted Rastrigin		O	0	[-5.12,5.12]	[-5.12,5.12]
Shifted Griewank		O	0	$\Re$	[-600,600]
Shifted Ackley		O	0	[-32,32]	[-32,32]
Shifted Weiestrass		O	0	[-0.5,0.5]	[-0.5,0.5]
Shifted rotated sphere		O	0	[-100,100]	[-100,100]
Shifted rotated schwefel 1.2		O	0	[-100,100]	[-100,100]
Shifted rotated schwefel 2.21		O	0	[-100,100]	[-100,100]
Shifted rotated schwefel 2.22		O	0	[-100,100]	[-100,100]
Shifted rotated Elliptic		O	0	[-100,100]	[-100,100]
Shifted rotated Rastrigin		O	0	[-5.12,5.12]	[-5.12,5.12]
Shifted rotated Griewank		O	0	$\Re$	[-600,600]
Shifted rotated Ackley		O	0	[-32,32]	[-32,32]
Shifted rotated Weiestrass		O	0	[-0.5,0.5]	[-0.5,0.5]
Shifted rotated sphere		O	0	[-100,100]	[-100,100]
Shifted rotated schwefel 1.2		O	0	[-100,100]	[-100,100]
Shifted rotated schwefel 2.21		O	0	[-100,100]	[-100,100]

O is the shifted vector.
O is the shifted vector.

C. Experimental Results and Discussions

In this paper, while comparing the performance of algorithms, we focus on computational time required to find the solution. For comparing the speed of the algorithms, the first thing we require is a fair time measurement. The number of iterations or generations cannot be accepted as a time measure since the different algorithms perform different amount of works in their inner loops. Hence, we choose the number of *fitness function evaluations (FEs)* as a measure of computation time instead of generations or iterations. Since the algorithms are stochastic in nature, the results of two successive runs usually do not match. Hence, we have taken 30 independent runs (with different seeds of the random number generator) of each algorithm. The results have been stated in terms of the mean values and standard deviations over the 30 runs in each case.

Tables 4 and 5 report the mean and standard deviation of function values by applying four algorithms to optimize 10-D and 30-D numerical shifted functions respectively. The best results are typed in bold. Similarly tables 6 and 7 report the mean and standard deviation of function values by applying four algorithms to optimize 10-D and 30-D numerical shifted rotated functions respectively. The best results are typed in bold.

a) *Comparing vTLBO with Conventional TLBO:* In this section, we intend to show how well the proposed vTLBO algorithm performs when compared to the conventional TLBO. It is clear from table 4, 5, 6 and 7 that in almost all the cases vTLBO is shown best result compare to conventional TLBO.

b) *Comparing vTLBO with other algorithms:* The performance of the vTLBO is compared with DE and PSO algorithm. In 10-D shifted function cases the performance of vTLBO either better than or equivalent with PSO. In 4 cases vTLBO equivalent with DE, in 3 cases DE is better and 2 cases vTLBO is better. In 30-D shifted function cases the performance of vTLBO is always better than PSO. In 1 cases vTLBO equivalent with DE, in 6 cases DE is better and 2 cases vTLBO is better. In 10-D shifted rotated function cases the performance of vTLBO is always better than PSO and DE. In 30-D shifted rotated function cases the performance of vTLBO is always better than PSO. In 4 cases vTLBO is better than DE and 5 cases DE is better than vTLBO.

Finally, we would like to point out that all the experiment codes are implemented in MATLAB. The experiments are conducted on a Pentium 4, 1GB memory laptop in Windows 7 environment.

Table 4. Shifted function with dim 10

Name of the function		PSO	DE	TLBO	vTLBO
shifted Sphere	Mean	0	0	2.7571e+004	0
	std	0	0	1.0317e+004	0
shifted Schwefel 1.2	Mean	0	0	2.6875e+06	0
	std	0	0	8.5178e+5	0
shifted Schwefel 2.21	Mean	11.7277	6.8368e-007	48.3534	**6.1212e-7**
	std	1.5981	1.4626e-007	12.8919	1.1311e-7
shifted Schwefel 2.22	Mean	379.6616	. **378.8248**	428.1518	**378.8248**
	std	20.2319	5.8593e-014	25.8912	5.8593e-014
shifted Elliptic	Mean	0	0	4.4969e+009	0
	std	0	0	1.2312e+9	0
shifted Rastrigin	Mean	23.8789	**0**	90.7058	0.0012
	std	11.8945	0	40.3419	0.0009
shifted Griewank	Mean	0.3052	**0**	70.2341	5.9020e-5
	std	0.2319	0	27.1039	2.8912e-5
shifted Ackley	Mean	5.2312	2.6645e-14	14.2481	**1.2341e-014**
	std	2.9012	1.2824e-29	2.8821	1.6022e-030
shifted Weierstrass	Mean	1.6057	**2.5422e-14**	5.0572	2.1298e-5
	std	0.8912	0	0.9003	1.5612e-5

Table 5. all are shifted function with dim 30

Name of the function		PSO	DE	TLBO	vTLBO
shifted Sphere	Mean	216.3035	**6.6258e-011**	7.1563e+04	2.6624e-05
	std	58.2221	4.7898e-011	2.5612e+04	1.4512e-05
shifted Schwefel 1.2	Mean	0.0137	**9.0692e-10**	1.2965e+09	2.7823e-04
	std	0.0041	5.3278e-10	8.5408e+08	2.1123e-04
shifted Schwefel 2.21	Mean	90.5421	3.4201	76.8660	**3.2131**
	std	8.9231	1.8917	3.1325	1.0023
shifted Schwefel 2.22	Mean	1.3050e+03	**1.2107e+03**	1.4113e+003	**1.2107e+03**
	std	452.4297	2.3328e-013	234.8912	2.3128e-013
shifted Elliptic	Mean	2.8945	**3.1331e-08**	5.6755e+9	2.2216e-05
	std	0.7548	2.8934e-08	1.7811e+9	1.2012e-05
shifted Rastrigin	Mean	150.2483	**1.0745e-07**	411.5017	1.9012
	std	35.7812	9.8393e-08	23.9012	1.3491
shifted Griewank	Mean	0.0123	**1.5870e-08**	765.1969	5.9034e-05
	std	0.0067	3.4449e-08	68.6788	8.9008e-05
shifted Ackley	Mean	18.0540	**9.5623e-06**	20.1890	1.0089e-5
	std	3.8912	5.8912e-06	1.6723	1.0002e-5
shifted Weierstrass	Mean	13.2977	8.9362e-05	38.4033	**7.9927e-06**
	std	2.0056	7.3826e-05	3.9089	4.4612e-06

Table 6. All are shifted rotated function with dim 10

Name of the function		PSO	DE	TLBO	vTLBO
shifted rotated Sphere	Mean	**0**	**0**	4.4573e+03	**0**
	std	0	0	378.7812	0
shifted rotated Schwefel 1.2	Mean	4.0323e+05	1.5611e+03	7.5864e+05	**3.9954e-04**
	std	2.8069e+05	902.5042	2.7026e+05	3.1125e-04
shifted rotated Schwefel 2.21	Mean	17.0754	1.4126e-04	23.8646	**1.3415e-007**
	std	5.6299	1.3126e-04	10.3737	1.2902e-007
shifted rotated Schwefel 2.22	Mean	1.0081e+08	4.6446e+07	3.7700e+06	**2.6056e+05**
	std	6.7431e+07	2.5208e+07	1.1394e+006	2.1192e+05
shifted rotated Elliptic	Mean	2.5924e+07	6.0012e+05	1.0159e+007	**1.8862e+05**
	std	9.0094e+06	1.6239e+05	7.9017e+006	1.6669e+05
shifted rotated Rastrigin	Mean	65.5063	32.9917	56.2292	**11.9023**
	std	12.8135	1.5093	4.2900	5.9112
shifted rotated Griewank	Mean	1.0423	0.5612	57.4703	**0.4518**
	std	0.8629	0.2149	19.3044	0.0452
shifted rotated Ackley	Mean	3.4722	2.6645e-14	13.6046	**2.0428e-14**
	std	1.7872	1.2824e-29	2.9322	1.3323e-30
shifted rotated Weierstrass	Mean	9.5539	10.0922	7.1355	**6.4512**
	std	0.9812	2.5542	2.4394	2.0718

Table 7. Shifted rotated function with dimension 30

Name of the function		PSO	DE	TLBO	vTLBO
shifted rotated	Mean	579.6219	**5.6258e-11**	8.1563e+04	2.0012e-07
Sphere	std	63.3035	4.7898e-11	2.5745e+04	1.6229e-07
shifted rotated	Mean	3.8610e+008	9.0692e+08	1.2965e+09	**8.0332e+08**
Schwefel 1.2	std	1.0041e+008	5.3278e+08	8.5408e+08	1.0002e+08
shifted rotated	Mean	85.9214	**32.4201**	65.8667	58.6102
Schwefwl 2.21	std	9.5421	6.1636	5.0064	21.7232
shifted rotated	Mean	7.3150e+036	1.2107e+035	1.3583e+035	**6.4618e+34**
Schwefel 2.22	std	3.8911e+035	1.9231e+035	1.4443e+035	1.2511e+34
shifted rotated	Mean	5.8284e+08	3.1331e+08	5.6555e+09	**2.2012e+08**
Elliptic	std	1.7548e+08	1.1013e+08	1.6961e+09	1.1012e+08
shifted rotated	Mean	250.1125	**0.0745**	378.9141	237.9012
Rastrigin	std	26.2483	223.0095	69.5017	21.4321
shifted rotated	Mean	1.8923	**1.5870e-08**	889.1969	1.0045
Griewank	std	0.0298	0.4499	99.8912	0.0012
shifted rotated	Mean	17.6388	**2.5416**	20.1890	7.9012
Ackley	std	3.8813	1.3931	2.6734	2.9012
shifted rotated	Mean	43.2977	41.9361	44.5475	**34.9575**
Weierstrass	std	2.4982	0.3826	6.6879	5.1125

6   Conclusion

This paper presents a modified code of TLBO for solving shifted type of functions such as shifted function and shifted rotated function. From the analysis and experiments, we observe that based on the results of the four algorithms on the 18 chosen test problems belonging to two classes, we can conclude that vTLBO significantly improves the TLBO's performance and gives the best performance on shifted and shifted rotated function when compared with four other optimization algorithms. Again the performance of vTLBO algorithm compare with traditional TLBO algorithm and shown that the performance of vTLBO is degraded while solving simple benchmark functions.

So here we conclude that although the vTLBO may not the best choice for solving simple real-world unconstrained optimization problems, but it is the good method for solving shifted type of functions. Another attractive property of the vTLBO is that it does not introduce any complex operations to the original simple TLBO framework. The only difference from the original TLBO, the random value is update in every dimension updating. The vTLBO is also simple and easy to implement like the original TLBO.

Acknowledgments. We sincerely thank to P. N. Suganthan, Associate Professor in the School of Electrical and Electronic Engineering, Nanyang Technological University, Singapore for his code on shifted and shifted rotated function.

References

1. Rao, R.V., Savsani, V.J., Vakharia, D.P.: Teaching-learning-based optimization: A novel method for constrained mechanical design optimization problems. Comput. Aided Des. 43(3), 303–315 (2011)
2. Rao, R.V., Savsani, V.J., Vakharia, D.P.: Teaching-learning-based optimization: A novel optimization method for continuous non-linear large scale problems. Inform. Sci. 183, 1–15 (2012)
3. Rao, R.V., Patel, V.: An elitist teaching-learning-based optimization algorithm for solving complex constrained optimization problems. Int. J. Ind. Eng. Comput. 3 (2012), http://dx.doi.org/10.5267/j.ijiec.2012.03.007
4. Satapathy, S.C., Naik, A.: Data clustering using teaching learning based optimization. In: Panigrahi, B.K., Suganthan, P.N., Das, S., Satapathy, S.C. (eds.) SEMCCO 2011, Part II. LNCS, vol. 7077, pp. 148–156. Springer, Heidelberg (2011)
5. Satapathy, S.C., Naik, A., Parvathi, K.: High dimensional real parameter optimization with teaching learning based optimization. In: International Journal of Industrial Engineering Computations, © 2012 Growing Science Ltd. All rights reserved (2012), doi:10.5267/j.ijiec.2012.06.001
6. Naik, A., Parvathi, K., Satapathy, S.C., Nayak, R., Pandap, B.S.: QoS multicast routing using Teaching learning based Optimization. In: Aswatha Kumar, M., Selvarani, R., Suresh Kumar, T.V. (eds.) ICAdC 2012. AISC, vol. 174, pp. 49–55. Springer, Heidelberg (2012)
7. Satapathy, S.C., Naik, A., Parvathi, K.: 0-1 Integer Programming For Generation maintenance Scheduling in Power Systems based on Teaching Learning Based Optimization (TLBO). In: Parashar, M., Kaushik, D., Rana, O.F., Samtaney, R., Yang, Y., Zomaya, A. (eds.) IC3 2012. CCIS, vol. 306, pp. 53–63. Springer, Heidelberg (2012)
8. Krishnanand, K.R., Panigrahi, B.K., Rout, P.K., Mohapatra, A.: Application of Multi-Objective Teaching Learning Based Algorithm to an Economic Load Dispatch Problem with Incommensurable Objectives. In: Panigrahi, B.K., Suganthan, P.N., Das, S., Satapathy, S.C. (eds.) SEMCCO 2011, Part I. LNCS, vol. 7076, pp. 697–705. Springer, Heidelberg (2011)
9. Naik, A., Satapathy, S.C., Parvathi, K.: Improvement of initial cluster center of c-means using Teaching learning based optimization, Accepted and will be published in Procedia Technology, Elsevier and indexed by Scopus
10. Naik, A., Satapathy, S.C.: Rough set and Teaching learning based optimization technique for Optimal Features Selection. Ref.: Ms. No. CEJCS-D-12-00042, Under Minor Review in Central European Journal of Computer Science
11. Satapathy, S.C., Naik, A.: Weighted Teaching-Learning-Based Optimization for global function optimization. Under Review in Applied Soft Computing Ms. Ref. No.: ASOC-D-12-00775
12. Satapathy, S.C., Naik, A.: A Modified Teaching-Learning-Based Optimization (mTLBO) for Global Search. Under Review in Swarm and Evolutionary Computation
13. Rao, R.V., Patel, V.K.: Multi-objective optimization of combined Brayton and inverse Brayton cycles using advanced optimization algorithms. Engineering Optimization (2012), doi:10.1080/0305215X.2011.624183
14. Rao, R.V., Savsani, V.J.: Mechanical design optimization using advanced optimization techniques. Springer, London (2012)
15. Toğan, V.: Design of planar steel frames using Teaching–Learning Based Optimization. Engineering Structures 34, 225–232 (2012)

16. Rao, R.V., Kalyankar, V.D.: Parameter optimization of machining processes using a new optimization algorithm. Materials and Manufacturing Processes (2012), doi:10.1080/10426914.2011.602792

17. Liang, J.J., Suganthan, P.N., Deb, K.: Novel composition test functions for numerical global optimization. In: Proc. IEEE Swarm Intell.Symp., Pasadena, CA, pp. 68–75 (June 2005)

18. Salomon, R.: Reevaluating genetic algorithm performance under coordinate rotation of benchmark functions. BioSystems 39, 263–278 (1996)

19. Shi, Y., Eberhart, R.C.: A modified particle swarm optimizer. In: Proc. IEEE Congr. Evol. Comput., pp. 69–73 (1998)

20. Storn, R., Price, K.: Differential evolution-a simple and efficient adaptive scheme for global optimization over continuous spaces. TR-95-012 (1995),
http://http.icsi.berkeley.edu/

Sensitivity Analysis of Load-Frequency Control of Power System Using Gravitational Search Algorithm

Rabindra Kumar Sahu, Umesh Kumar Rout, and Sidhartha Panda

Department of Electrical Engineering
Veer Surendra Sai University of Technology (VSSUT), Burla-768018, Odisha, India
rksahu123@gmail.com, umesh6400@gmail.com,
panda_sidhartha@rediffmail.com

Abstract. This paper investigates the sensitivity analysis of load-frequency control for multi area power system based Proportional Integral Derivative controller with derivative Filter (PIDF) by Gravitational Search algorithm (GSA). At first, a two area non reheat thermal system without physical constraints is considered. A modified objective function which includes ITAE, damping ratio of dominant eigenvalues, settling times of frequency and peak overshoots with appropriate weight coefficients is proposed. Further, the proposed objective function is extended to a more realistic power system model by considering the physical constraints such as reheat turbine, Generation Rate Constraint (GRC) and Governor Dead Band nonlinearity. Finally the robustness of the system is verified, with operating load condition and time constants of speed governor, turbine, tie-line power are changed from their nominal values in the range of +50% to -50% in steps of 25%. It is observed that the proposed controllers are robust and perform satisfactorily for a wide range of the system parameters and operating load conditions.

Keywords: Load Frequency Control (LFC), Proportional Integral Derivative controller with derivative Filter (PIDF), Gravitational Search Algorithm (GSA), Sensitivity Analysis.

1 Introduction

The frequency of a power system is a very important performance signal to the system operator for stability and security considerations. The desired power system frequency should stay within a very small, acceptable interval around its nominal value. Otherwise, the operator needs to take relevant actions immediately. In the past decades, there are many research works on power system frequency regulation [1]-[3]. The main objective of a power system utility is to maintain continuous supply of power with an acceptable quality to all the consumers in the system. The system will be in equilibrium, when there is a balance between the power demand and the power generated [4]. There are two basic control mechanisms used to achieve power balance; reactive power balance (acceptable voltage profile) and real power balance (acceptable frequency values). The former is called the Automatic Voltage Regulator

S.C. Satapathy, S.K. Udgata, and B.N. Biswal (eds.), *FICTA 2013*,
Advances in Intelligent Systems and Computing 247,
DOI: 10.1007/978-3-319-02931-3_28, © Springer International Publishing Switzerland 2014

(AVR) and the latter is called the Automatic Load Frequency Control (ALFC) or Automatic Generation Control (AGC). For multi area power systems, which normally consist of interconnected control area, AGC is an important aspect to keep the system frequency and the interconnected area tie-line power as close as possible to the intended values. The mechanical input power to the generators is used to control the system as it is affected by the output electrical power demand and to maintain the power exchange between the areas as planned. AGC monitors the system frequency and tie-line flows, calculates the net change in the generation required according to the change in demand and changes the set position of the generators within the area so as to keep the time average of the ACE (Area Control Error) at a low value. ACE is generally treated as controlled output of AGC. As the ACE is adjusted to zero by the AGC, both frequency and tie-line power errors will become zero [4],[5]. In most of the LFC studies, the effect of the Governor dead band (GDB) nonlinearity and, Generation Rate Constraint (GRC) is neglected for simplicity. But for a realistic analysis of the performance of the system, this should be included as it has a considerable effect on the amplitude and settling time of the oscillations [6]. The effect of GDB nonlinearity is considered in this study, by describing function approach in the state space model.

The aim of the present work is:

❖ to study the sensitivity analysis of the power system, by wide range of the system parameters and operating load conditions
❖ to study the effect of physical constraint (GDB and GRC) is considered
❖ in order to obtained the optimum parameter values of the PIDF controller by GSA technique based on modified objective function
❖ to show advantages of using a controller structure and objective function to further increase the performance of the power system

2   System under Study

System under investigation consists of two area interconnected power system of non reheat thermal plant as shown in Fig. 1. Each area has a rating of 2000 MW with a nominal load of 1000 MW. The system is widely used in literature is for the design and analysis of automatic load frequency control of interconnected areas [7], [8]. In Fig. 1, B_1 and B_2 are the frequency bias parameters; ACE_1 and ACE_2 are area control errors; u_1 and u_2 are the control outputs form the controller; R_1 and R_2 are the governor speed regulation parameters in pu Hz; T_{G1} and T_{G2} are the speed governor time constants in sec; ΔP_{V1} and ΔP_{V2} are the change in governor valve positions (pu); ΔP_{G1} and ΔP_{G2} are the governor output command (pu); T_{T1} and T_{T2} are the turbine time constant in sec; ΔP_{T1} and ΔP_{T2} are the change in turbine output powers; ΔP_{D1} and ΔP_{D2} are the load demand changes; ΔP_{Tie} is the incremental change in tie line power (p.u); K_{PS1} and K_{PS2} are the power system gains; T_{PS1} and T_{PS2} are the power system time constant in sec; T_{12} is the synchronizing coefficient and ΔF_1 and ΔF_2 are the system frequency deviations in Hz. The relevant parameters are taken from [7]. Each area of the power system consists of speed governing system, turbine and generator as shown in Fig.1.

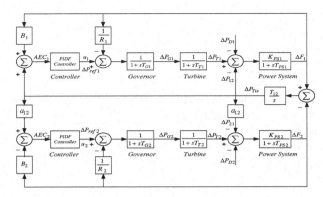

Fig. 1. Transfer function model of two-area non reheat thermal system

3   Overview of Gravitational Search Algorithm

Gravitational Search Algorithm (GSA) is one of the newest heuristic algorithms inspired by the Newtonian laws of gravity and motion [9]. In GSA, agents are considered as objects and their performance is measured by their masses. All these objects attract each other by the force of gravity and this force causes a global movement of all objects towards the objects with a heavier mass. Hence masses co-operate using a direct form of communication through gravitational force.

3.1   Law of Gravity

Each particle attracts every other particle and the gravitational force between the two particle is directly proportional to the product of their masses and inversely proportional to the distance between them R. It has been reported in literature that R provides better results than R^2 in all experiment cases [10].

3.2   Law of Motion

The current velocity of any mass is equal the sum of the fraction of its previous velocity and the variation in the velocity. Variation in the velocity or acceleration of any mass is equal to the force acted on the system divided by mass of inertia.

For a system with 'n' agent (masses), the *i-th* position of an agent X_i is defined by:

$$X_i = (x_i^1,........,x_i^d,........x_i^n) \tag{1}$$

for $i = 1,2,.....n$

Where

x_i^d represents the position of *i-th* agent in the *d-th* dimension.

At a specific time '*t*', the force acting on mass '*i*' from mass '*j*' is defined as:

$$F_{ij}^d(t) = G(t)\frac{M_{pi}(t)*M_{aj}(t)}{R_{ij}(t)+\in}(x_j^d(t)-x_i^d(t)) \tag{2}$$

Where

$M_{aj}$is the active gravitational mass related to agent j

$M_{pi}$is the passive gravitational mass related to agent i

$G(t)$is the gravitational constant at time t

$\in$ is small constant

$R_{ij}(t)$is the Euclidian distance between two agents i and j given by:

$$R_{ij}(t) = \left\| X_i(t), X_j(t) \right\|_2 \tag{3}$$

The stochastic characteristic in GSA algorithm is incorporated by assuming that the total forces that act on agent 'i' in a dimension'd' be a randomly weight sum of d-th components of the forces exerted from other agents as:

$$F_i^d(t) = \sum_{j=1, j\neq i}^{n} rand_j F_{ij}^d(t) \tag{4}$$

Where $rand_j$ is a random number in the interval [0, 1]

The acceleration of the agent 'i' at the time t and in the direction d-th, $a_i^d(t)$is given by the law of the motion as:

$$a_i^d(t) = \frac{F_i^d(t)}{M_{ii}(t)} \tag{5}$$

Where $M_{ii}(t)$ is the inertia mass of i-th agent.

The velocity of an agent is updated depending on the current velocity and acceleration. The velocity and position are updated as:

$$v_i^d(t+1) = rand_i*v_i^d(t) + a_i^d(t) \tag{6}$$

$$x_i^d(t+1) = x_i^d(t) + v_i^d(t+1) \tag{7}$$

Where $rand_i$ is a uniform random variable in the interval (0, 1). The random number is used to give a randomized characteristic to the search process.

The gravitational constant G is initialized at the beginning. To control the search accuracy it is reduced with time and expressed as function of the initial value (G_0) and time t as:

$$G(t) = G_0 e^{(-\alpha t/T)} \tag{8}$$

Where α is a constant and T is the number of iteration.

To achieve a good compromise between exploration and exploitation, the number of agents is reduced with lapse of Eq. (5) and therefore a set of agents with bigger mass are used for applying their force to the other. The performance of GSA is improved by controlling exploration and exploitation. To avoid trapping in a local optimum GSA must use the exploration at beginning. By lapse of iterations, exploration must fade out and exploitation must fade in.

4   Proposed Approach

4.1   Controller Structurer

In the present paper, identical controllers have been considered for the two areas as the two areas are identical. The structure of PID controller with derivative filter is shown in Fig. 2 where K_P, K_I and K_D are the proportional, integral and derivative gains respectively, and N is the derivative filter coefficient. When used as PI controller, the derivative path along with the filter is removed from Fig. 2.

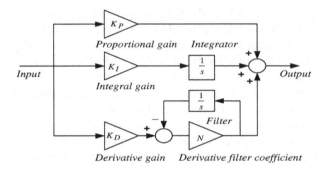

Fig. 2. Structure of PID controller with derivative filter

4.2   Objective Function

To determining the optimum values of controller parameters, modified objective functions are considered. The objective functions considered as given below:

$$(9)$$

$$J = \int_0^{t_{sim}} \omega_1 \cdot \int_0^{t_{sim}} (|\Delta F_1| + |\Delta F_2| + |\Delta P_{Tie}|) \cdot t \cdot dt + \omega_2 \cdot \frac{1}{\min(\sum_{i=1}^{n}(1 - \zeta_i))} + \omega_3 (ST)$$

Where, ΔF_1 and ΔF_2 are the system frequency deviations; ΔP_{Tie} is the incremental change in tie line power; t_{sim} is the time range of simulation; ζ_i is the damping ratio and n is the total number of the dominant eigen values; ST is the sum of the settling times of frequency and tie line power deviations respectively; ω_1 to ω_3 are weighting factors. The problem constraints are the PIDF controller parameter bounds. Therefore, the design problem can be formulated as the following optimization problem.

Minimize J (10)

Subject to

For PIDF controller:

$$K_{P\min} \leq K_P \leq K_{P\max}, K_{I\min} \leq K_I \leq K_{I\max} \quad K_{D\min} \leq K_D \leq K_{D\max} \quad (11)$$

The minimum and maximum values of PID controller parameters are chosen as -2.0 and 2.0 respectively. The range for filter coefficient N is selected as 1 and 300.

5   Results and Discussions

5.1   Application of GSA

The model of the system under study has been developed in MATLAB/SIMULINK environment and GSA program has been written (in .mfile). The developed model is simulated in a separate program (by .m file using initial population/controller parameters) considering a 10% step load change in area 1. The objective function is calculated in the .m file and used in the optimization algorithm. The process is repeated for each individual in the population. Using the objective function values, the population is modified by GSA for the next iteration. Following parameters are chosen for the application of GSA: population size NP=20; maximum iteration =100; gravitational constants G_0=100 and α =20; K_0= total number of agents and decreases linearly to 1 with time. Optimization is terminated by the prespecified number of iterations. Simulations were conducted on an Intel, core 2 Duo CPU of 2.4 GHz and 2 GB RAM computer in the MATLAB 7.10.0.499 (R2010a) environment. The optimization was repeated 50 times and the best final solution among the 50 runs is chosen as proposed controller parameters. The flowchart of proposed optimization is shown in Fig. 3.

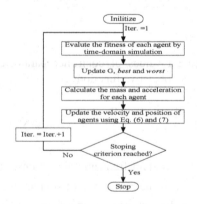

Fig. 3. Flow chart of proposed GSA optimization approach

5.2   Simulation Results

Case-I: Without Physical Constraints

A step load increase of 10 % is considered for area-1 at $t = 0$ sec.The optimum values of PIDF controller parameters with proposed objective function are

$$K_P = 1.4011, \ K_I = 1.9981, \ K_D = 0.7102, \ N = 93.2760$$

Sensitivity analysis is carried out to study the robustness the system to wide changes in the operating conditions and system parameters [11]. Taking one at a time, the operating load condition and time constants of speed governor, turbine, tie-line power are changed from their nominal values [8] in the range of +50% to -50% in steps of 25%. PIDF controller optimized employing GSA using modified objective function J_5 is considered. The optimum values of controller parameters, at changed loading conditions and changed system parameters (for a step increase in demand of 10% at $t = 0$ sec in area-1) are provided in Table 1. The corresponding performance indexes (ITAE values, settling times and minimum damping ratios) with the above varied system conditions are given in Table 1. Critical examination of Table 1 clearly reveals that the performance indexes are more or less same. The frequency deviation response of area 1 with above varied conditions is shown in Fig. 4. It can be observed from Fig. 4 that the effect of the variation of operating loading conditions on the system responses is negligible. So it can be concluded that, the proposed control strategy provides a robust control and the controller parameters obtained at the nominal loading with nominal parameters, need not be reset for wide changes in the system loading or system parameters.

Table 1. Sensitivity analysis of system without physical constraints

Parameter variation	% Change	Tuned controller Parameter				Settling time T_s(Sec)			ζ	ITAE
		K_P	K_I	K_D	N	ΔF_1	ΔF_2	ΔP_{tie}		
Nominal	0	1.4011	1.9981	0.7102	93.2760	1.92	3.19	2.86	0.4470	0.1362
Loading condition	+50	1.4103	1.9945	0.7257	89.0014	1.96	3.21	2.88	0.4501	0.1386
	+25	1.4081	1.9983	0.7255	92.7390	1.95	3.20	2.87	0.4515	0.1379
	-25	1.4067	1.9909	0.7137	86.4434	1.94	3.20	2.87	0.4424	0.1372
	-50	1.4178	1.9901	0.7052	93.9838	1.93	3.22	2.88	0.4409	0.1362
T_G	+50	1.5001	1.9923	0.7746	95.0011	1.99	3.29	2.92	0.2906	0.1373
	+25	1.4386	1.9942	0.7212	93.7709	1.89	3.23	2.88	0.3518	0.1348
	-25	1.3293	1.9991	0.6013	94.9823	1.83	3.14	2.79	0.5067	0.1315
	-50	1.3112	1.9986	0.6021	90.4946	1.87	3.11	2.79	0.5949	0.1313
T_T	+50	1.5792	1.9338	0.8224	91.9358	1.82	3.37	2.99	0.3368	0.1349
	+25	1.4975	1.9972	0.7406	94.9587	1.80	3.24	2.88	0.3728	0.1322
	-25	1.3002	1.9992	0.6004	95.0012	1.93	3.16	2.81	0.5032	0.1360
	-50	1.3009	1.9911	0.6009	94.8310	2.09	3.25	2.86	0.5112	0.1483

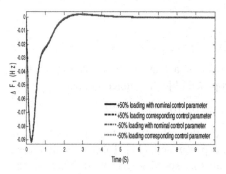

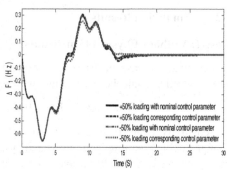

Fig. 4. Frequency deviation of area-1 for change in nominal load

Fig. 5. Frequency deviation of area- with physical constraints

Case-II: With Physical Constraints

The study is further extended to a more realistic power system by considering the effect of reheat turbine, GRC, and GBD. GRC of 3%/ min and GBD of 0.036 Hz are considered in the present work [6]. The optimum values of PIDF controller parameters are:

$$K_P= 0.8589, K_I= 0.0791, K_D= 1.992, N= 44.5678$$

Sensitivity analysis is done to study the robustness the system to wide changes in the operating conditions and system parameters as before. The various performance indexes (ITAE values, settling times and minimum damping ratios) under normal and parameter variation cases are given in Table 2.

Table 2. Sensitivity analysis of system with physical constraints

Parameter variation	% Change	Tuned controller Parameter				Settling time T_s(Sec)			ζ	ITAE
		K_P	K_I	K_D	N	ΔF_1	ΔF_2	ΔP_{tie}		
Nominal	0	0.8589	0.0791	1.9920	44.5678	18.21	19.50	37.44	0.4400	40.4612
Loading condition	+50	0.8630	0.0753	1.9057	43.3623	14.34	13.96	42.37	0.4492	31.4246
	+25	0.8397	0.0758	1.8949	49.1008	16.83	18.98	41.29	0.4670	35.6197
	-25	0.7508	0.0831	1.7503	43.2105	20.56	21.60	34.95	0.4743	49.8424
	-50	0.7500	0.0765	1.9995	48.4272	20.62	21.64	37.78	0.4511	50.4765
T_G	+50	0.8000	0.0800	1.6614	48.8137	19.73	20.51	36.12	0.3802	49.2855
	+25	0.8000	0.0800	1.9168	44.4899	19.57	20.52	36.57	0.3926	43.8609
	-25	0.8811	0.0841	1.8552	44.2036	18.51	19.79	35.09	0.5445	40.4841
	-50	0.9560	0.0855	1.8971	44.4794	17.13	18.43	34.84	0.6684	38.1144
T_T	+50	0.7757	0.0787	1.8821	42.9658	18.03	19.42	37.09	0.4944	40.1139
	+25	0.7908	0.0720	1.9924	47.5757	15.17	19.06	43.01	0.4840	38.3354
	-25	0.7036	0.0749	1.8137	46.2931	16.25	19.15	39.76	0.4297	38.3024
	-50	0.7003	0.0761	1.7980	49.4480	15.85	18.85	38.84	0.3897	38.2674

It can be noticed from Table 2 that when physical constraints are introduced, the variations in performance index are more prominent. So it can be concluded that in the presence of GBD, GRC and reheat turbine, the system becomes highly non-linear (even for small load perturbation) and hence the performance of the designed controller is degraded. To complete the analysis, a 10 % step load increase in area-1 at t = 0 sec is considered and the frequency deviation response of area-1 for the above varied conditions are shown in Fig.5. From Table 2 and Fig. 5 it can be concluded that once the controller parameters are tunned in the presence of physical constraints, the controller need not be reset for wide changes in the system loading or system parameters.

6   Conclusion

Firstly, the system without any physical constraint is optimized PIDF controller with modified objective functions by GSA. Sensitivity analysis reveals that the optimum PIDF controller tuned at the nominal condition are quite robust and need not be reset for wide changes in system loading conditions or in system parameters. Finally, the proposed approach is extended to a more realistic power system model by considering the physical constraints such as reheat turbine, GRC and governor dead band nonlinearity and it is observed that the when physical constraints are introduced, the variations in performance index are more prominent as evident from the sensitivity analysis.

References

[1] Elgerd, O.I.: Electric energy systems theory. An introduction. Tata McGraw-Hill, New Delhi (1983)
[2] Willems, J.L.: Sensitivity Analysis of the Optimum Performance of Conventional Load-Frequency Control. IEEE Transactions on Power Apparatus and Systems PAS-93, 51287–51291 (1974)
[3] Ibraheem, P., Kothari, D.P.: Recent philosophies of automatic generation control strategies in power systems. IEEE Trans. Power Syst. 20(1), 346–357 (2005)
[4] Kundur, P.: Power System Stability and control, 8th reprint. TMH, New Delhi (2009)
[5] Shoults, R.R., Jativa Ibarra, J.A.: Multi area adaptive LFC developed for a comprehensive AGC simulation. IEEE Trans. Power Syst. 8(2), 541–547 (1993)
[6] Golpîra, H., Bevrani, H., Golpîra, H.: Application of GA optimization for automatic generation control design in an interconnected power system. Energy Conversion and Management 52, 2247–2255 (2011)
[7] Ali, E.S., Abd-Elazim, S.M.: Bacteria foraging optimization algorithm based load frequency controller for interconnected power system. Elect. Power and Energy Syst. 33, 633–638 (2011)
[8] Rout, U.K., Sahu, R.K., Panda, S.: Design and analysis of differential evolution algorithm based automatic generation control for interconnected power system. AinShamsEngJ (2012), http://dx.doi.org/10.1016/j.asej.2012.10.010

[9] Rashedi, E., Nezamabadi-pour, H., SaryazdiJ, S.: GSA: A Gravitational Search Algorithm. Information Sciences 179, 2232–2248 (2009)

[10] Rashedi, E., Nezamabadi-pour, H., SaryazdiJ, S.: Filter modeling using gravitational search algorithm. Engineering Applications of Artificial Intelligence 24, 117–122 (2011)

[11] Chaoshun, L., Jianzhong, Z.: Parameters identification of hydraulic turbine governing system using improved gravitational search algorithm. Energy Conversion and Management 52, 374–381 (2011)

Privacy Preserving Distributed Data Mining with Evolutionary Computing

Lambodar Jena[1,2], Narendra Ku. Kamila[3], and Sushruta Mishra[1]

Department of Computer Science & Engineering,
[1] Gandhi Engineering College, Bhubaneswar, India
[2] Utkal University, Bhubaneswar, India
[3] C.V.Raman College of Engineering, Bhubaneswar, India
lmjena@yahoo.com

Abstract. Publishing data about individuals without revealing sensitive information about them is an important problem. Distributed data mining applications use sensitive data from distributed databases held by different parties. This comes into direct conflict with an individual's need and right to privacy. It is thus of great importance to develop adequate security techniques for protecting privacy of individual values used for data mining. Here, we study how to maintain privacy in distributed data mining. That is, we study how two (or more) parties can find frequent itemsets in a distributed database without revealing each party's portion of the data to the other. In this paper, we consider privacy-preserving naïve-Bayes classifier for horizontally partitioned distributed data and propose data mining privacy by decomposition (DMPD) method that uses genetic algorithm to search for optimal feature set partitioning by classification accuracy and k-anonymity constraints.

Keywords: Distributed database, privacy, data mining, classification, k-anonymity.

1   Introduction

Information sharing is a vital building block for today's business world. Data mining techniques have been developed successfully to extract knowledge in order to support a variety of domains. But it is still a challenge to mine certain kinds of data without violating the data owners' privacy. As data mining become more pervasive, privacy concerns are increasing[1]. Distributed data mining is a process to extract globally interesting associations, classifiers, clusters, and other patterns from distributed data [1], where data can be partitioned into many parts either vertically or horizontally [2]. Till now a number of privacy- preserving data mining algorithms and protocols have been proposed, such as those for association rule mining [3–7], clustering [8,9], naive Bayes classifiers [10–13], etc. So far, there have been two main approaches for privacy-preserving data mining as follows: One is the randomization approach, another is the cryptographic approach. Several researchers have proposed methods for incorporating privacy-preserving requirements in various data mining tasks, such as classification [27]; frequent itemsets [29]; and sequential patterns [23].

S.C. Satapathy, S.K. Udgata, and B.N. Biswal (eds.), *FICTA 2013*,
Advances in Intelligent Systems and Computing 247,
DOI: 10.1007/978-3-319-02931-3_29, © Springer International Publishing Switzerland 2014

We propose a new method data mining privacy by decomposition (DMPD) for achieving k-anonymity. The basic idea is to divide the original dataset into several disjoint projections such that each one of them adheres to k-anonymity. It is easier to make a projection comply with k-anonymity if the projection does not contain all quasi identifier features. The k-anonymity is still preserved even if the attacker attempts to rejoin the projections. A classifier is trained on each projection and subsequently, an unlabelled instance is classified by combining the classifications of all classifiers. Because DMPD preserves the original values and only partitions the dataset. DMPD supports classification, but can be extended to support other data mining tasks by incorporating various types of decomposition [15].

DMPD employs a genetic algorithm for searching for optimal feature set partitioning. The search is guided by k-anonymity level constraint and classification accuracy. Both are incorporated into the fitness function. Our new approach significantly outperforms existing suppression-based and generalization-based methods that require manually defined generalization trees. In addition, DMPD can assist the data owner in choosing the appropriate anonymity level[26].

2   Backgrounds

2.1   Naive Bayes Classification

The naive Bayes classifier, or simple Bayes classifier, works as follows:

Each data sample is represented by an m+1 dimensional feature vector (a1, a2, . . . , am, c), depicting m+1 measurements made on the sample from m+1 attributes, respectively, A1,A2, . . . ,Am, C, where C is the class attribute and c is the class label.

Suppose that the domain of C is $(C_1, C_2, \ldots, C_l)$ where $C_i \neq C_j$ for $i \neq j$, and thus there exist λ classes. Given an unknown data sample, X =(a1, a2, . . . , am) (i.e., having no class label),the classifier will predict that X belongs to the class having the highest posterior probability, conditioned on X. That is, the naive Bayes classifier assigns an unknown sample X to class Ci if and only if $P(C_i|X) > P(C_j|X)$, for $1 \leq j \leq \lambda$, $j \neq i$. Thus we maximize $P(C_i|X)$. The class Ci for which $P(C_i|X)$ is maximized is called the maximum posteriori hypothesis.

By Bayes theorem, we have $P(C_i|X) = [P(X|C_i)P(C_i)] /[P(X)]$. As P(X) is constant for all classes, only $P(X|C_i)P(C_i)$ need to be maximized. The class prior probabilities may be estimated by $P(C_i) = S_i/S$, where S_i is the number of training samples of class Ci and S is the total number of training samples.

In order to reduce computation in evaluating $P(X|C_i)$, the naive assumption of class conditional independence is made. Thus $P(X|C_i) = \prod_{k=1}^{m} P(a_k|C_i)$. The probabilities $P(a1|C_i)$, $P(a2|C_i)$,...., $P(am|C_i)$ can be estimated from the training sample, namely, $P(a_k|C_i) = S_{ik}/S_i$,where s_{ik} is the number of training samples of class C_i having the value a_k for A_k. In order to classify an unknown sample X, $P(X|C_i)P(C_i)$ is evaluated for each class C_i . Sample X is   then assigned to the class C_i if and only if $P(X|C_i)P(C_i) > P(X|C_j)P(C_j)$, for $1 \leq j \leq \lambda$, $j \neq i$.

In other word, it is assigned to the class Ci for which $P(X|C_i)P(C_i)$ is maximum.

2.2   Genetic Algorithm Framework for Multiobjective Optimization

Genetic algorithms (GA), a type of evolutionary algorithm (EA), are computational abstractions, derived from biological evolution, for solving optimization problems through a series of genetic operations [21]. A GA requires a fitness function that assigns a score (fitness) to each candidate in the current population sample (generation). Being a population-based approach, genetic algorithms are well suited to solve multiobjective optimization problems. A generic single-objective GA can be modified to find a set of multiple, non-dominated solutions in a single run. The ability of the GA to simultaneously search different regions of a solution space makes it possible to find a diverse set of solutions for difficult problems.

The DMPD algorithm presented here is extended in a natural way to perform a multiobjective optimization to assist data owners in deciding about an appropriate anonymity level in released datasets. The multiobjective genetic algorithm (MOGA) was designed to solve multiobjective problems where the objectives are generally conflicting thus preventing simultaneous optimization of each objective. The final choice of the solution depends on the user characterizing a subjective approach. User participation in this process is important for obtaining useful results [16].

Optimizing competing objective functions is different from single function optimization in that it seldom accepts one perfect solution, especially for real-world applications [24]. The most successful approach for multiobjective optimization is to determine an entire Pareto optimal solution set or its representative subset [17,21]. A Pareto optimal set is a set of solutions that are non-dominated with respect to each other. While moving from one Pareto solution to another, there is always a certain amount of sacrifice in one objective(s) vs. a certain amount of gain in the other (s). In [22], the authors reported that 90% of the approaches to multiobjective optimization aimed to approximate the true Pareto front for the underlying problem. The ultimate goal of a multiobjective optimization algorithm is to identify solutions in the Pareto optimal set. A practical approach to multiobjective optimization is to investigate a set of solutions (the best-known Pareto set) that represent the Pareto optimal set as best as possible.

3   Privacy-Preserving Naive Bayes Classifier On Distributed Data

3.1   Two-Party Privacy-Preserving Naive Bayes Classifier

Here we consider as scenario in which two semi-honest users U1 and U2 owning their confidential databases DB1 and DB2, respectively, wish to learn a naive Bayes classifier on the union DB = DB1 U DB2, without revealing privacy of their databases. We assume that the two databases have the same attributes (A1, A2, . . . , Am, C), where C is the class attribute with a domain of {C1, C2, . . ., Cl}. A user is semi-honest in terms that he provides correct inputs to the naive Bayes classifier, but may want to learn something that violates the privacy of another database. The two users jointly classify a new instance X =(a1, a2, . . . , am).

3.2   Multi-party Privacy-preserving Naive Bayes Classifier

In this section, we consider as scenario in which n(n≥2) semi-honest users U1,U2, . . . , Un owning their confidential databases DB1, DB2, . . . , DBn, respectively, wish to learn a naive Bayes classifier on the union DB = $\bigcup_{i=1}^{n}$ DBi of their databases with the help of two semi-trusted mixers, without revealing privacy of their databases. We assume that all databases have the same attributes (A1,A2, . . . , Am, C), where C is the class attribute with a domain of {C1, C2, . . . , Cl}. The semi-trusted mixer model is used in multi-party protocol, in which each data site sends messages to two semi-trusted mixers.

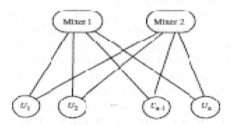

Fig. 1. Semi-trusted mixer model

4   Experimental Evaluation

4.1   Experimental Process

Fig.2 graphically represents the experimental process that was conducted. The main aim of this process was to estimate the generalized accuracy (i.e. the probability that an instance was classified correctly). First, the dataset (box 1) was divided into a train dataset (box 3) and test dataset (Step 4) using five iterations of a two-fold cross vali-dation (Step 2 – known as the 5x 2 CV procedure). The 5x 2 CV is known to be better than the commonly used 10-fold cross-validation because of the acceptable Type-1 error [14]. At each iteration, the dataset is randomly partitioned into two equal-sized sets, S1 and S2, such that the algorithm is evaluated twice. During the first evaluation S1 is the train dataset and S2 the test dataset, and vice versa during the second evalua-tion. We apply (Step 5) the k-anonymity method on the train dataset and obtain a new anonymous train dataset (Step 6). Additionally, we obtain a set of anonymity rules (Step 7) that is used to transform the test dataset into a new anonymous test dataset (Step 8). In the case of DMPD, the rule of partitioning must be applied on an original feature set. For generalization and suppression-based techniques, generalization or suppression rules are applied to different original values in the dataset. An inducer is trained (Step 9) over the anonymous train dataset to generate a classifier (Step 10). Finally the classifier is used to estimate the performance of the algorithm over the anonymous test dataset (Step 11).

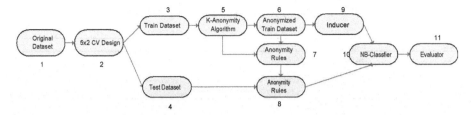

Fig. 2. The Experimental Process

4.2   Datasets

Privacy-preserving classification algorithms are usually evaluated only on the Adult dataset which has become a commonly used benchmark for k-anonymity [28,20,18]. Recently Fung et al. [19] used a German credit dataset to evaluate the TDR algorithm. In our experimental study we used an additional seven datasets that were also selected from the UCI Machine Learning Repository [25] and which are widely used by the machine-learning community for evaluating learning algorithms. An additional dataset, drawn from a real-world case study performed on commercial banks, is described below. The datasets vary across such dimensions as the number of target feature classes, instances, input features and their type (nominal, numeric).

4.3   Effect of k-anonymity Constraints on Classification Accuracy

In this section we analyze the effect of the value of k (anonymity level) on classification accuracy. Table 1 shows the accuracy results obtained by the proposed algorithm for four different values of k for various datasets using different inducers. In this section we take the top nine features as quasi-identifiers (top eight for nursery and pima datasets). The column with k = 1 represents the DMPD algorithm result (i.e. when no k-anonymity constraints were implied) enabling us to examine the effect of anonymity on the accuracy of the results. The superscript ''*'' indicates that the degree of accuracy of the original dataset was significantly higher than the corresponding result with a confidence level of 95%.

As expected, the results indicate that there is a tradeoff between accuracy performance and the anonymity level for most datasets. Usually, increasing the anonymity level decreases accuracy. For some datasets, the feature set partitioning approach improves baseline accuracy, even despite applying k-anonymity constraints such as for vote, cmc, pima, wisconsine or nursery. Supervised discretization, as a part of DMPD, also contributes to classification accuracy, for example, in the heart dataset. These above results are marked out for both inducer types.

Table 1. Accuracy vs. anonymity for DMPD algorithm

Dataset	Inducer	k-Anonymity level					
Adult	K	Baseline	1	50	100	200	500
	C4.5	85.58 ± 0.51	86.59 ± 0.22	83.03 ± 0.10*	83.00 ± 0.10*	82.63 ± 1.15*	82.99 ± 0.08*
	Naïve Bayes	82.68 ± 0.27	84.87 ± 0.20	81.81 ± 0.13*	81.41 ± 0.79*	81.47 ± 0.77*	81.19 ± 1.11*
	K	Baseline	1	10	20	30	50
Credit	C4.5	85.53 ± 2.45	85.82 ± 1.54	86.71 ± 1.15	85.08 ± 1.75	85.29 ± 1.99	85.66 ± 2.19
	Naïve Bayes	76.82 ± 1.87	85.66 ± 1.89	85.94 ± 1.42	86.46 ± 1.46	85.66 ± 2.45	85.45 ± 2.98
Vote	C4.5	96.23 ± 2.41	96.35 ± 1.95	96.26 ± 2.09	96.35 ± 2.38	95.04 ± 4.06	96.70 ± 1.91
	Naïve Bayes	90.78 ± 1.98	96.26 ± 2.05	96.43 ± 1.85	96.52 ± 2.09	96.61 ± 1.85	96.52 ± 1.97
Wisconsine	C4.5	93.28 ± 1.83	97.00 ± 0.90	95.97 ± 1.40*	94.88 ± 1.00*	95.09 ± 1.71	95.23 ± 1.77
	Naïve Bayes	93.28 ± 1.78	96.11 ± 1.01	95.27 ± 1.57	94.77 ± 2.09	93.89 ± 2.03	93.04 ± 1.09*
German	C4.5	69.88 ± 2.12	73.23 ± 7.44	71.86 ± 2.35	70.78 ± 2.22	71.12 ± 2.51	63.01 ± 16.49
	Naïve Bayes	73.81 ± 1.13	75.03 ± 2.03	73.79 ± 2.37	73.71 ± 2.30	71.26 ± 2.59	69.84 ± 1.87*
Heart	C4.5	74.89 ± 2.97	84.03 ± 6.71	79.18 ± 4.14	78.96 ± 3.53	77.31 ± 3.76	74.78 ± 3.83
	Naïve Bayes	84.89 ± 3.44	82.61 ± 4.37	79.93 ± 4.43	77.69 ± 4.84	78.06 ± 4.75	74.93 ± 3.38*
Portfolio	C4.5	74.82 ± 0.98	76.41 ± 0.56	72.04 ± 1.34*	71.61 ± 2.37*	72.07 ± 1.13*	71.76 ± 1.57*
	Naïve Bayes	58.96 ± 1.81	75.46 ± 0.71	68.74 ± 2.67	69.11 ± 1.84*	68.58 ± 2.72*	68.46 ± 2.47*
Cmc	C4.5	50.90 ± 1.61	56.59 ± 2.73	52.53 ± 4.64	51.51 ± 4.53	51.02 ± 4.19	48.30 ± 2.30
	Naïve Bayes	48.78 ± 1.37	54.04 ± 2.57	54.04 ± 3.33	52.73 ± 4.69	52.12 ± 3.49	48.07 ± 2.00*
Pima diabetes	C4.5	72.09 ± 2.20	78.20 ± 2.05	77.75 ± 1.53	77.62 ± 2.66	76.79 ± 3.01	75.85 ± 1.87
	Naïve Bayes	75.20 ± 2.67	77.75 ± 1.93	77.75 ± 2.27	78.02 ± 2.34	78.41 ± 2.45	78.30 ± 2.26
	K	Baseline	1	50	100	150	200
Nursery	C4.5	96.16 ± 0.31	97.13 ± 0.52	92.64 ± 1.15*	91.15 ± 0.81*	90.86 ± 1.27*	88.91 ± 4.79
	Naïve Bayes	90.08 ± 0.40	90.25 ± 0.51	90.24 ± 0.57	90.27 ± 0.67	90.00 ± 0.77	89.46 ± 0.94

4.4   Scalability Analysis

Here we examine the DMPD's ability to handle expanding datasets in an elegant manner. The scalability is tested using the procedure that Fung et al. [19] proposed to measure the runtime costs of algorithms on large datasets. Our scalability test was based on a German dataset. For this purpose, the original dataset containing 1000 records was expanded as follows: for every original instance q, we added r -1 variations where r is a scale factor. Together with all original instances, the enlarged dataset has r x 1000 instances. Each instance variation was generated by randomly drawing appropriate feature values (xqi, yq) from the feature domain (dom(ai), dom(y)).

The experiments have been conducted on a hardware configuration that included a desktop computer implementing a Windows XP operating system with Intel Pentium 4, 2.8 GHz, and 1 GB of physical memory. Table 2 presents the average time (in minutes) measured on 10 runs for various values of r, a quasi-identifier that includes top 5 features and a k-anonymity level = 10. The time, including model generation time, reflects the runtime cost of a J4.8 inducer in a WEKA package and is beyond our control. Fig.3 shows that the execution time is almost linear in the number of records. This confirms the DMPD's ability to handle a growing search space since an increase in train dataset size leads to more partitions that the algorithm must consider.

4.5   Genetic Algorithm(GA) Settings

Our default settings for DMPD were 100 generations and 50 individuals in a population. Here we present some experiments from the German and cmc datasets. We present two of 12 experiments that were performed on the dataset with a C4.5 inducer

with a minimum of k-anonymity constraints (top5 and k = 10). The number of generations considered is 50, 100, 150 and 200. Test were carried out on two populations
numbering 50 and 100 individuals, respectively.

It is observed that, increasing GA settings beyond 100 generations and 50 parti-
tionings does not significantly improve classification accuracy. Similar behavior is
true for other datasets used in the experimental study. The evidence points to the
DMPD's ability to achieve good feature set partitioning with a relatively low number
of generations and population size.

Table 2. Scalability analysis for DMPD

Scalability factor	Algorithm runtime (min)	Dataset size
1	1.41	1000
5	6.55	5000
10	13.90	10,000
15	24.48	15,000
20	32.76	20,000

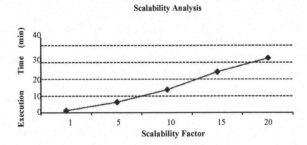

Scalability Analysis

Fig. 3. Scalability trend in the dataset

5 Conclusions

In this paper we presented a new method for preserving privacy in classification tasks
using a naive Bayes classifier and k-anonymity framework with genetic algorithm.
The proposed method was designed to work with no prior knowledge and with any
inducer. Compared to existing state-of-the-art methods, the new method also shows a
higher predictive performance on a wide range of datasets from different domains.
Additional issues to be further studied include: Examining DMPD with other classification algorithms, Examining the possibility of using DMPD along with known gene-
ralization/suppression-based methods that could result in improved data mining
results in terms of classification accuracy and discovered patterns, Extending
k-anonymity to l-diversity framework. Extending DMPD to handle multiobjective
optimization with different quasi-identifier sets. The main idea is to eliminate the
k-anonymity model assumption; the quasi-identifier set is determined prior to
performing data anonymization.

References

1. Kargupta, H., Chan, P.: Advances in Distributed and Parallel Knowledge Discovery. MIT, AAAI Press, Cambridge, New York (2000)
2. Vaidya, J., Clifton, C.: Privacy-preserving data mining: Why, how and when. IEEE Security and Privacy, 19–27 (November/December 2004)
3. Evfimievski, A., Ramakrishnan, S., Agrawal, R., Gehrke, J.: Privacy- preserving mining of association rules. In: Proceedings of the 8th ACM SIGKDD International Conference on Knowledge Discovery and Data Mining, Edmonton, Alberta, Canada (July 2002)
4. Kantarcioglu, M., Vaidya, J.: Privacy preserving naive Bayes classifier for horizontally partitioned data. In: Proceedings of IEEE Workshop on Privacy Preserving Data Mining (2003)
5. Vaidya, J., Clifton, C.: Privacy-preserving association rule mining in vertically partitioned data. In: Proceedings of the 8th ACM SIGKDD International Conference on Knowledge Discovery and Data Mining, pp. 639–644. ACM Press, New York (2002)
6. Verykios, V.S., Elmagarmid, A.K., Bertino, E., Saygin, Y., Dasseni, E.: Association rule hiding. IEEE Transactions on Knowledge and Data Engineering 16(4), 434–447 (2004)
7. Rizvi, S.J., Haritsa, J.R.: Maintaining data privacy in association rule mining. In: Proceedings of the 28th International Conference on Very Large Data Bases, pp. 682–693 (2002)
8. Clifton, C., Kantarcioglou, M., Lin, X., Zhu, M.Y.: Tools for privacy preserving distributed data mining. SIGKDD Exploration 4(2), 1–7 (2002)
9. Vaidya, J., Clifton, C.: Privacy-preserving k-means clustering over vertically partitioned data. In: Proceedings of 9th ACM SIGKDD International Conference on Knowledge Discovery and Data Mining, pp. 206–215. ACM Press, New York (2003)
10. Kantarcioglu, M., Vaidya, J.: Privacy-preserving naive Bayes classifier for horizontally partitioned data. In: IEEE Workshop on Privacy Preserving Data Mining (2003)
11. Vaidya, J., Clifton, C.: Privacy preserving naive Bayes classifier on vertically partitioned data. In: 2004 SIAM International Conference on Data Mining (2004)
12. Wright, R., Yang, Z.: Privacy-preserving Bayesian network structure computation on distributed heterogeneous data. In: KDD 2004, Seattle, Washington, USA (August 2004)
13. Yang, Z., Zhong, S., Wright, R.: Privacy-preserving classification of customer data without loss of accuracy. In: Proceedings of the 5th SIAM International Conference on Data Mining, Newport Beach, CA (April 2005)
14. Alpaydin, E.: Combined 5 _ 2 CV F-test for comparing supervised classification learning classifiers. Neural Computation 11, 1975–1982 (1999)
15. Cohen, S., Rokach, L., Maimon, O.: Decision-tree instance-space decomposition with grouped gain-ratio. Information Sciences 177(17), 3592–3612 (2007)
16. Fayyad, U., Piatetsky-Shapiro, G., Smyth, P.: From data mining to knowledge discovery: An overview. In: Advances in Knowledge Discovery and Data Mining, pp. 1–31. AAAI Press, Menlo Park (1996)
17. Fonseca, C.M., Fleming, P.J.: Genetic algorithms for multiobjective optimization: Formulation, discussion and generalization. In: Forrest, S. (ed.) Proc. of the Fifth International Conference on Genetic Algorithms, pp. 416–423. Morgan Kaufmann, San Mateo (1993)
18. Friedman, A., Schuster, R.W.: Providing k-anonymity in data mining. VLDB 17(4), 789–804 (2008)
19. Fung, B.C.M., Wang, K., Yu, P.S.: Anonymizing classification data for privacy preservation. IEEE Transactions on Knowledge and Data Engineering 19(5), 711–725 (2007)

20. Fung, B.C.M., Wang, K., Yu, P.S.: Top-down specialization for information and privacy preservation. In: Proc. of the 21st IEEE International Conference on Data Engineering, ICDE 2005, pp. 205–216. IEEE Computer Society, Washington, DC (2005)
21. Goldberg, D.E.: Genetic Algorithms in Search, Optimization, and Machine Learning. Addison-Wesley, Boston (1989)
22. Jones, D.F., Mirrazavi, S.K., Tamiz, M.: Multiobjective meta-heuristics: An overview of the current state-of-the-art. European Journal of Operational Research 137(1), 1–9 (2002)
23. Kim, S.W., Park, S., Won, J.I., Kim, A.W.: Privacy preserving data mining of sequential patterns for network traffic data. Information Sciences 178(3), 694–713 (2008)
24. Konaka, D.W., Coitb, A.E.: Smithc, Multi-objective optimization using genetic algorithms: A tutorial. Reliability Engineering and System Safety 91, 992–1007 (2006)
25. Mitchell, M.: An Introduction to Genetic Algorithms. MIT Press, Cambridge (1996)
26. Meints, M., Moller, J.: Privacy preserving data mining – a process centric view from a European perspective (2004), http://www.fidis.net
27. Sharpe, P.K., Glover, R.P.: Efficient GA based techniques for classification. Applied Intelligence 11, 277–284 (1999)
28. Zhang, J., Zhuang, J., Du, H., Wang, S.: Self-organizing genetic algorithm based tuning of PID controllers. Information Sciences 179(7), 1007–1018 (2009)
29. Zitzler, E., Deb, K., Thiele, L.: Comparison of multiobjective evolutionary algorithms: Empirical results. Evolutionary Computation 8(2), 173–195 (2000)

Boundary Searching Genetic Algorithm: A Multi-objective Approach for Constrained Problems

Shubham J. Metkar and Anand J. Kulkarni

Optimization and Agent Technology (OAT) Research Lab
Maharashtra Institute of Technology, 124 Paud Road, Pune 411038, India
{Sjmetkar,ajkulkarni}@oatresearch.org, kulk0003@ntu.edu.sg

Abstract. The Performance of most of the Nature-/bio-inspired optimization algorithms is severely affected when applied for solving constrained problems. The approach of Genetic Algorithm (GA) is one of the most popular techniques; however, similar to other contemporary algorithms, its performance may also degenerate when applied for solving constrained problems. There are several constraint handling techniques proposed so far; however, developing efficient constraint handling technique still remains a challenge for the researchers. This paper presents a multi-objective optimization approach referred to as Boundary Searching GA (BSGA). It considers every constraint as an objective function and focuses on locating boundary of the feasible region and further search for the optimum solution. The approach is validated by solving four test problems. The solutions obtained are very competent in terms of the best and mean solutions in comparison with contemporary algorithms. The results also demonstrated its robustness solving these problems. The advantages, limitations and future directions are also discussed.

Keywords: Boundary Searching Genetic Algorithm, Constrained Optimization, Multi-Objective Optimization.

1 Introduction

Recently, a variety of Nature-/bio-inspired optimization techniques have been developed such as Evolutionary Algorithms (EAs) Genetic Algorithms (GAs), Particle Swarm Optimization (PSO), Ant Colony Optimization (ACO), Simulated Annealing (SA), Probability Collectives (PC), etc. and have been successfully applied in a variegated application domain [1]. However, being basically unconstrained problem solution techniques, their performance may degenerate when dealing with the real world problems which are inherently constrained ones. This necessitated a strong and generic constraint handling techniques to be developed and incorporated into them.

The approach of handling constraints using multi-objective optimization techniques redefines the problem by considering m constraints involved in the problem as objective functions, i.e. the problems becomes unconstrained with $m+1$ objectives [2-6]. The approach of Vector Evaluated GA (VEGA) [4] assigns fitness

S.C. Satapathy, S.K. Udgata, and B.N. Biswal (eds.), *FICTA 2013*,
Advances in Intelligent Systems and Computing 247,
DOI: 10.1007/978-3-319-02931-3_30, © Springer International Publishing Switzerland 2014

values to every individual using constraints and objective function values; however, as the number of constraints increases the associated number of subpopulations also increases. The approach in [3] used two objectives, first was to optimize the objective function and second was to minimize the collection of constraints and instead of using crossover operator, line search was performed which was computationally expensive and diversity was not preserved [2].

A constraint handling method in [5] used evolutionary algorithm in which ranking of solutions was done separately based on the values of objective function and constraints and then intelligent mating restrictions were applied based on available feasibility information. This approach was computationally less expensive and less robust. In an approach based on stochastic ranking [6], individuals were compared with their adjacent neighbors through a certain number of sweeps which aimed to maintain balance between objective function and a penalty function so that no value from them can dominate the search. The approach was found to be dependent on the associated probability factor of using only objective function for comparing individuals while assigning corresponding rank [2].

This paper proposes a multi-objective GA approach referred to as Boundary Searching Genetic Algorithm (BSGA) based on the assumption [7] that at least one constraint is active at the global optimum. The algorithm searches the boundary of the feasible region to find the global optimum. Similar to the Strength Pareto Evolutionary Algorithm II (SPEA-II) [8], the proposed BSGA uses fitness assignment procedure in which individuals in the population were compared based on their strengths which were decided based on the objective function value and the distance of feasible constraint from the boundary of the feasible region. This strength value was further used to compute the raw fitness which focused on the infeasible solutions by considering the amount of constraint violation and aimed to assign minimum raw fitness value to the fitter individual. Unlike SPEA-II density information was not used and the solutions with unequal fitness values were accepted to avoid the probable repetition of the individuals within the population. In addition, the best fit individuals referred to as pure elitist individuals were selected from the combination of parent and offspring population. This selection approach was similar to the Non-dominated Sorting GA II (NSGA-II) [9]; however, differs in the fitness assignment and selection procedure. The approach has been validated by solving four well studied benchmark test problems.

The remainder of this paper is organized as follows. Section 2 discusses BSGA algorithm in detail including the fitness assignment procedure, selection procedure, mating selection for crossover, perturbation and elitism approach. Section 3 provides validation of the approach by solving four benchmark test problems and further comparing the results with other contemporary approaches. The concluding remarks and future work are discussed at the end of the paper.

2   BSGA Algorithm

The proposed BSGA is described in the context of multi-objective optimization problems (in minimization sense) as follows:

$$\text{Minimize} \quad f(\mathbf{x})$$

$$
\begin{aligned}
\text{Subject to} \quad & g_i(\mathbf{x}) \le 0, & i = 1, 2, \dots, q \\
& h_i(\mathbf{x}) = 0, & i = 1, 2, \dots, r \\
& \psi_i^{lower} \le \mathbf{x}_i \le \psi_i^{upper} & i = 1, 2, \dots, n
\end{aligned}
\tag{1}
$$

According to [10], the equality constraints $h_i(x) = 0$, $i = 1, 2, \dots, r$ can be transformed into inequality constraints using a tolerance value δ as $h_i = 0 \Rightarrow |h_j(\mathbf{x})| - \delta \le 0$ $i = 1, 2, \dots, r$. Thus, the total number of constraints becomes $s = q + r$. The modified representation of the problem can be represented as follows :

$$\text{Minimize} \quad f(\mathbf{x})$$

$$
\begin{aligned}
\text{Subject to} \quad & g_i(\mathbf{x}) \le 0, & i = 1, 2, \dots, s \\
& \psi_i^{lower} \le \mathbf{x}_i \le \psi_i^{upper} & i = 1, 2, \dots, n
\end{aligned}
\tag{2}
$$

For solving the problem using multi-objective approach, the constraints are considered as objective functions [2-6]. Thus, the total number of functions to be optimized becomes $s + 1$. The function matrix $\mathbf{f}$ for population size p can be represented as follows:

$$
\mathbf{f} =
\begin{bmatrix}
f_{11} & g_{11} & \cdots & g_{1s} \\
f_{21} & g_{21} & \cdots & g_{2s} \\
\vdots & \vdots & \ddots & \vdots \\
f_{p1} & g_{p1} & \cdots & g_{ps}
\end{bmatrix}
\tag{3}
$$

Step 1. Compute the function matrix $(\mathbf{f})$, correspond to the population $(\mathbf{u})$. Initialization of $\mathbf{u}$ is based on the random generation. It is represented as:

$$
\mathbf{u} =
\begin{bmatrix}
u_{11} & u_{21} & u_{31} & \cdots & u_{n1} \\
\vdots & \vdots & \vdots & \ddots & \vdots \\
u_{1p} & u_{2p} & u_{3p} & \cdots & u_{np}
\end{bmatrix}
\tag{4}
$$

$$u_i = \psi_i^{lower} + (\psi_i^{upper} - \psi_i^{lower}) \times R \qquad i = 1, 2, \dots n$$

where R is any random number $R \in [0, 1]$.

Step 2. For determining the fitter solutions amongst the available p individual solutions, every individual solution i is assigned a strength value $S(i)$ representing the number of solutions it dominates [8]. In addition it is important to note that the strength of an individual is based on its values of objective function and constraints as well. The strength $S(i)$ of every individual i is computed as pseudo code shown in Fig.1 (a).

(a) column:

```
REPEAT
    ADD 1 to j
        IF f(i) < f(j)
            S(i) = S(i) + 1
        ENDIF
    b = 0
    REPEAT
        ADD 1 to b
        IF g_b(i) ≥ g_b(j) AND g_b(i) ≤ 0
            S(i) = S(i) + 1
        ENDIF
    UNTIL b = s
UNTIL j = p
```

(a) Strength evaluation

(b) column:

```
j = 0
REPEAT
    ADD 1 to j
        IF f(i) > f(j) OR any one constraints violates
            R(i) = R(i) + S(j)
        ENDIF
    b = 0
    REPEAT
        ADD 1 to b
        IF g_b(i) > 0 AND g_b(i) > g_b(j)
            R(i) = R(i) + S(j)
        ENDIF
        IF g_b(i) < 0 AND g_b(i) < g_b(j)
            R(i) = R(i) + S(j)
        ENDIF
    UNTIL b = s
UNTIL j = p
fitness(i) = R(i)
```

(b) Raw Fitness Assignment

Fig. 1. Pseudo code for Strength and Raw fitness assignment

On the basis of $S(i)$ value associated raw fitness $R(i)$ of the individual i is calculated. The raw fitness of every individual i is deter-mined by the strength of number of solutions dominate it. The raw fitness of an individual i is decided based on the observation that very often the global solution lies on the boundary of feasible region [7]. For deciding the fitness $fitness(i)$ of every individual i, the individuals are compared with one another as pseudo code shown in Fig.1 (b).

It is important to note that $fitness(i)$ should be minimum for the fittest individual solution i, i.e. the higher value of $fitness$ for the individual i means more number of solutions dominate that solution.

Step 3. Sort the function matrix $\mathbf{f}$ in ascending order based on the fitness values and select the first N solutions to form an archive set $\mathbf{n}$, where N is size of the archive set given by: $N = p / m$, m is archive factor.

Step 4. Each individual solution $i = 1, 2, ..., N$ in the archive set forms m crossover pairs. The pairing individuals with solution i are selected randomly from the archive set, provided that all the m individuals forming pair should differ from one another. Crossover between the integer and the fraction part is performed separately. Single point crossover is used and crossover point is generated randomly. Each solution gives m solutions as after crossover only the changes in i^{th} individual are considered. Thus, N solutions generate $p = m \times N$ solutions to form crossover population $(\mathbf{u}')$.

Step 5. Compute the function matrix $(\mathbf{f}')$, correspond to the population $(\mathbf{u}')$. Combine both function set $(\mathbf{f})$ and the crossover function set $(\mathbf{f}')$ to form a combined function set $(\mathbf{cf})$. Compute the fitness value for each solution as mentioned is step 2, the only difference here being the size of set is $2 \times p$. Now sort this fitness set in the ascending order and select the first p solutions having unequal fitness values, this is to preserve the diversity within the population. If less than p solutions are available with different fitness values generate the remaining solutions randomly to complete the population. This gives population for the next generation $\mathbf{u}_{t+1}$, t is the generation number. Also select first N solutions to form an archive set $\mathbf{n}_{t+1}$.

Step 6. Search the fittest solution (elite solution) in the current generation population $(\mathbf{u})$ and store this solution in the minimum function solution set.

Step 7. If the elite solution does not update for the η iterations it might be possible that whole population get stuck into local minima. For providing an opportunity to this population to come out from this possible local minima hoping for better solution perturb the solutions to its nearby values as $x_i \Rightarrow x_i \pm \alpha$, $i = 1, 2, \ldots, n$. Perturbation iterations η depend on the efficiency of crossover algorithm to generate fitter solutions. Perturbation range (α) is decided based on the preliminary trials and its value changes in two steps (i.e. α_1 changes to α_2 such that $\alpha_1 > \alpha_2$) once elite solution is found to be constant for $\lambda/3$ iterations, where λ iterations represent stopping criterion.

Step 8. Carry τ elite solutions into the next generation population $\mathbf{u}_{t+1}$ to provide better opportunity during crossover to generate fitter solutions.

Step 9. The terminating conditions being 1) If elite solution does not improve for λ iterations, where λ is decided based on preliminary trials or 2) If iterations reach their maximum limit whichever comes first, accept the current elite solution set as a best solution and terminate the program.

Step 10. If terminating condition is not satisfied go back to step 4.

3   Test Problems and Numerical Results

The performance of BSGA discussed in Section 2 was validated by solving four test problems G03, G04, G06 and G08. These have been well studied in the literature and used to compare the performance of various optimization algorithms [6, 13-16].

Table 1. Results of proposed BSGA

Pro- blem	Best	Mean	Worst	Std. Dev.	Avg. Comp. Time (sec.)	Avg. Func. Evalua- tions	Closeness to Best Sol. (%)
G03	1.0002401	1.0001860	1.0000582	4.2e-5	344.24	687125	2e-3
G04	-30665.5267	-30665.49113	-30664.4019	0.041	145.73	269003	4e-7
G06	-6961.79290	-6961.78347	-6961.69858	0.029	209.84	611050	3e-4
G08	0.095825041	0.095825039	0.095825023	5.4e-9	3.88	8250	1e-5

Table 2. Set of parameters for BSGA

Problem	G03	G04	G06	G08
α_1, α_2	$\alpha_1 = 0.1, \alpha_2 = 0.02$	$\alpha_1 = 1, \alpha_2 = 0.1$	$\alpha_1 = 1, \alpha_2 = 0.1$	$\alpha_1 = 0.1, \alpha_2 = 0.01$

Table 3. Comparison of Best and Mean solutions

Methods		G03	G04	G06	G08
HM [14]	Best	0.9997	-30664.5	-6952.1	0.0958250
	Mean	0.9989	-30655.3	-6342.6	0.0891568
SR [6]	Best	1	-30665.539	-6961.814	0.095825
	Mean	1	-30665.539	-6875.940	0.095825
FSA [16]	Best	1.00000	-30665.5380	-6961.81388	0.095825
	Mean	0.99918	-30665.4665	-6961.81388	0.095825
Bacerra et al. [15]	Best	0.99541	-30665.53867	-6961.81387	0.095825
	Mean	0.78863	-30665.53867	-6961.81387	0.095825
Farmani et al. [12]	Best	1.00000	-30665.5000	-6961.8000	0.0958250
	Mean	0.99990	-30665.2000	-6961.8000	0.0958250
Ray et al. [5]	Best	N.A.	-30651.662	-6852.5630	N.A.
	Mean	N.A.	-30647.105	-6739.0479	N.A.
Proposed BSGA	**Best**	**1.00024**	**-30665.5267**	**-6961.7929**	**0.095825041**
	Mean	**1.00018**	**-30665.4019**	**-6961.6985**	**0.095825039**

The BSGA was coded in MATLAB 7.7.0 (R2008b) and simulations were run on a Windows platform using Intel Core i5-2400, 3.10GHz processor speed with 3.16 GB RAM. Every problem was solved 20 times. The best, mean and worst solutions, associated standard deviation, computational time, number of function evaluations and closeness to the best reported solution is presented in Table 1. The associated set of parameters such as archive size N and elitism τ is kept twenty (20%) and two (2%) percent of population size p, perturbation iterations η was chosen $1/10^{th}$ of stopping criterion λ iterations. Perturbation range α_1 and α_2, are given in Table 2. Table 3 summarizes the BSGA results and comparison with the various methods.

The solutions to the test problems using BSGA approach presented here are compared with several contemporary constraint handling techniques. Some of the approaches such as self adaptive penalty approach [11] and a dynamic penalty scheme [12] produced premature convergence and were quite sensitive to the additional set of parameters. A stochastic bubble sort algorithm in [6] attempted the direct and explicit balance between the objective function and the penalty function; however, the approach was sensitive to the associated probability parameter and needed several preliminary trials as well. The Homomorphous Mapping (HM) [14] needed initial feasible solutions along with the problem dependent parameters to be set. The lack of diversity of the population was noticed in the cultural differential evolution (CDE) [15]. The multi-objective approach in Filter SA (FSA) [16] required numerous problem dependent parameters to be tuned along with the associated preliminary trials. The multi-criteria approach in [10] could not solve the fundamental problem of balancing the objective function and the constraint violations.

The approach of BSGA was found to be producing consistently optimum results irrelevant of the size of the feasible region, In addition, BSGA did not require any initial feasible solution which was essential for PSO and HM methods, and no premature convergence was observed which was unlike the self adaptive and a dynamic penalty approach. From the comparison with other contemporary using BSGA were very competitive, sufficiently robust with associated reasonable computational time and number of function evaluations. Importantly, the algorithm did not use any diversity preserving technique such as in [8]; however, the solutions with unequal fitness values were selected to avoid repeated individuals. This facilitated the inclusion of diverse solutions at the beginning and allowed solutions which were very close to one another providing efficient local search to improve the solution further, this avoids the need of any additional local search technique such as niching required in ASCHEA approach.

4   Conclusions and Future Work

The BSGA approach locating the active constraint boundaries of the constraints by assigning the fitness based on closeness of the constraint to boundary region was found to be very effective. The solutions obtained were very competitive with the contemporary approaches. In addition, the approach was found to be robust with reasonable computational cost. In all the four problems, it was found that the approach was effective to locate the feasible region irrespective of the size of its feasible region as the amount of constraint violation was also considered while deciding the fitness.

In some problems it might be possible that no constraint is active at the global optimum which may require a self adaptive method that can be implemented to redefine the boundary using the available information. In addition, the parameters such as τ elite individuals, perturbation iterations η and perturbation range α required to be fine tuned and needs preliminary trials as well. In the future, a self adaptive scheme could be developed to choose these parameters which may help to reduce the preliminary trials and associated computational cost as well. A variety of class of problems can be solved to further strengthen the approach to make it capable solving real world problems.

References

1. Kulkarni, A.J., Tai, K.: Solving Constrained Optimization Problems Using Probability Collectives and a Penalty Function Approach. International Journal of Computational Intelligence and Applications (10), 445–470 (2011)
2. Coello Coello, C.A.: Theorotical and Numerical Constraint-Handling Techniques Used with Evolutionary Algorithms: A Survey of the State of the Art. Computer Methods in Applied Mechanics and Engineering 191(11-12), 1245–1287 (2002)
3. Camponogara, E., Talukdar, S.: A Genetic Algorithm for Constrained and Multiobjective Optimization. In: Alander, J.T. (ed.) 3rd Nordic Workshop on Genetic Algorithm and their Applications, pp. 49–62 (1997)
4. Coello Coello, C.A.: Treating Constraints as Objective for Single-Objective Evolutionary Optimization. Engineering Optimization 32(3), 275–308 (2000)
5. Ray, T., Tai, K., Chye, S.K.: An Evolutionary Algorithm for Constrained Optimization. In: Proceedings of the Genetic and Evolutionary Computation Conference, San Francisco, California, pp. 771–777 (2000)
6. Runarson, T.P., Yao, X.: Stochastic Ranking for Constrained Evolutionary Optimization. IEEE Transactions on Evolutionary Computation 4(3), 284–294 (2000)
7. Schoenauer, M., Michalewicz, Z.: Evolutionari Computation at the Edge of Feasibility. In: Ebeling, W., Rechenberg, I., Voigt, H.-M., Schwefel, H.-P. (eds.) PPSN 1996. LNCS, vol. 1141, pp. 245–254. Springer, Heidelberg (1996)
8. Zitzler, E., Laumanns, M., Thiele, L.: SPEA2: Improving the Strength Pareto Evolutionary Algorithm. In: Giannakoglou, K.C. (ed.) Evolutionary Methods for Design, Optimisation and Control with Application to Industrial Problems (EUROGEN 2001), pp. 95–100 (2002)
9. Deb, K., Agrawal, S., Pratap, A., Meyarivan, T.: A Fast Elitist Non-Dominated Sorting Genetic Algorithm for Multi-Objective Optimization: NSGA-II. Evolutionary Computation 6(2), 182–197 (2002)
10. Tai, K., Prasad, J.: Target-Matching Test Problem for Multiobjective Topology Optimization Using Genetic Algorithms. Structural and Multidisciplinary Optimization 34, 333–345 (2007)
11. Coello Coello, C.A.: Use of a Self-Adaptive Penalty Approach for Engineering Optimization Problems. Computers in Industry 41(2), 113–127 (2000)
12. Farmani, R., Wright, J.A.: Self-Adaptive Fitness Formulation for Constrained Optimization. IEEE Transactions on Evolutionary Computation 7(5), 445–455 (2003)
13. Lampinen, J.: A Constraint Handling Approach for the Differential Evolution Algorithm. In: Proceedings of the IEEE Congress on Evolutionary Computation, vol. 2, pp. 1468–1473 (2002)
14. Koziel, S., Michalewicz, Z.: Evolutionary Algorithms, Homomorphous Mappings, and Constrained Parameter Optimization. Evolutionary Computation 7(1), 19–44 (1999)
15. Becerra, R.L., Coello Coello, C.A.: Cultured Differential Evolution for Constrained Optimization. Computer Methods in Applied Mechanics and Engineering 195(33-36), 4303–4322 (2006)
16. Hedar, A.R., Fukushima, M.: Derivative-Free Simulated Annealing Method for Constrained Continuous Global Optimization. Journal of Global Optimization 35(4), 521–549 (2006)

Security Analysis of Digital Stegno Images Using Genetic Algorithm

G. Praneeta[1] and B. Pradeep[2]

Department of Computer Science and Engineering,
Anil Neerukonda Institute of Technology and Sciences, Visakhapatnam, India
{pranita513,pradeep.inspire}@gmail.com

Abstract. Data hiding is a technique that can embed the secret data in digital images for transmitting messages confidentially through internet. These carriers before transmitting are called cover images and after transmission are called stepno images. An effective stenography algorithm should have lass embedding distortion and capable of evading visual and statistical detection. So there is scope for finding the strength of data hiding technique by considering as optimization problem. Genetic algorithm can solve optimization problems effectively then simulated annealing technique. This paper mainly describes the framework that can analyze security strength of data hiding technique called adaptive pixel pair matching technique and finds penalization parameter and kernel parameter. This paper follows subtractive pixel adjacency matrix model and Experimental results depicted that Genetic algorithm gives better performance than the simulated annealing Technique.

Keywords: Optimal pixel adjustment process, Genetic algorithm, simulated annealing, SPAM.

1   Introduction

In this Internet Century, The people in various locations can transmit various messages in different parts of world. The main requirement of transmission is security in long distance communication. So the need for development of schemes which can capable of transmitting secret data with carrier and receiver extracts the secret data from carrier. If images are used for communicating hidden data, that will called as Image steganography. It is art of science in which it can hide to (or) extract from the images. The steganography is used to transmit message in side of a image such that status of message is secured and it is not early to extract when third person found that image.

The steganalysis means that it is science of discrimination between Stego Objects and cover objects. This process needs to be done without any knowledge of secret key used for embedding. First goal in design of data hiding algorithm is to get less entropy between cover image and Stegno image, next it should be difficult detect by using techniques like visual detection and statistical detection.

S.C. Satapathy, S.K. Udgata, and B.N. Biswal (eds.), *FICTA 2013*,   277
Advances in Intelligent Systems and Computing 247,
DOI: 10.1007/978-3-319-02931-3_31, © Springer International Publishing Switzerland 2014

The Subtractive Pixel Adjacency Matrix (SPAM) is statistical steganalysis scheme. To obtain the detect ability of diamond encoding using SPAM, it trains the steganalyzer. The genetic algorithm is used in the optimization process to find parameters such that error rate is minimum.

The contributions in this paper are as follows First we implement data hiding technique adaptive pixel pair matching. Second we analysis the diamond encoding technique by SPAM Method. Third calculation of parameter used in SPAM can treated as optimization problem by taking objective as minimum error rate. This problem can be solved by using Genetic algorithms. Finally we solve this problem by implementing Genetic algorithms.

Rest of paper is structural as follows in section 2 we summarize the background work related to Steganography, Genetic algorithms, Steganalysis. Problem description in various aspects like adaptive pixel pair matching, SPAM security analysis and Genetic algorithms and simulated annealing are presented in section 3. Section 4 describes security analysis of digital Stegno images. The experimental results printed in section 5. Finally we conclude the paper in section 6.

2   Related Work

This section describes the back ground work related to the steganalysis of digital images. This can be decomposed into the following sub sections.

2.1   Background Work on Steganography

Now day's world the art of embedding and extracting the hidden text has attracted the researcher's attention and it has many challenges. Shashikala Channalli et al [1] proposed a new form of steganography, on-line hiding of secret text on the output screens. Their method could be used for announcing a secret message in public place. Their data hiding method is very similar to image steganography and video steganography.

3   Problem Description

Definition1: The data hiding algorithm is to be e % secured if the relative entropy between the cover image and Stegno image is at most e

$$\int P_e.\log(\frac{P_c}{P_s}) < e$$

Definition2: The data hiding algorithm is to be perfectly secured the relative entropy is zero. i.e. the pixel values of both cover image and Stegno images are same . But in reality it is not possible.

3.1   Adaptive Pixel Pair Matching

The key functionality of the Pixel Pair Marching based steganography method is to use pixel pair as the coordinate (x ,y), and searching a coordinate (x1,y1) within a predefined neighborhood set C(x,y) such that D(x1,y1)= s ,where D is the extraction function and s is the message digit in a B-ary notational system to be concealed. Data encryption is done by replacing (x, y) with(x1,y1) . For a PPM-based method, suppose a digit s is to be concealed. The range of is s between 0 and B-1, and a coordinate C(x1, y1) belongs to X has to be found such that D(x1, y1) =s. Therefore, the range of D(x, y) must be integers between 0 and B-1, and each integer must occur at least once. In addition to reduce the distortion, the number of coordinates in should be as small as possible. The best PPM method shall satisfy the following three requirements: 1) there are exactly B coordinates in C(x, y). 2) The values of extraction function in these coordinates are mutually exclusive. 3) The design of and should be capable of embedding digits in any notational system so that the best can be selected to achieve lower embedding distortion.

3.2   Genetic Algorithms

Genetic algorithms are stochastic search algorithms which are able to solve optimization problems. Genetic Algorithms are the heuristic search and optimization techniques that mimic the and optimization techniques that mimic the process of natural evolution. The process of representing a solution in the form of a string that conveys the necessary information. Just as in a chromosome, each gene controls a particular characteristic of the individual; similarly, each bit in the string represents a characteristic of the solution. Most common method of encoding is binary coded. Chromosomes are strings of 1 and 0 and each position in the chromosome represents particular characteristics of the problem A fitness function value quantifies the optimality of a solution. The value is used to rank a particular solution against all the other solutions. A fitness value is assigned to each solution depending on how close it is actually to the optimal solution of the problem.

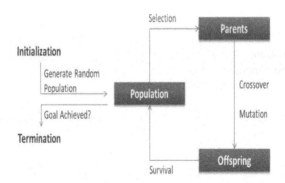

Fig. 1. Genetic operations

4   Security Analysis for Digital Stegno Images

4.1   Architecture

The above architecture describes the sender give secret message and image cover along with the sender key it will generate Stegnos as an output. The receiver receives the message threw transmission channel and the he decrypt the message by using receiver key for getting original data. (SPAM) is statistical steganalysis scheme, to obtain the detect ability of diamond encoding using SPAM, it transis the steganalyzer. The genetic algorithm is used in the optimization process to find parameters such that error rate is minimum.

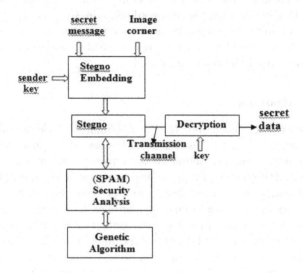

Fig. 2. Architecture

4.2   Algorithms

Function Genetic ()

```
{
Initialize population;
Calculate fitness function;
While (fitness value! = termination criteria)
{
Selection;
Crossover;
Mutation;
Calculate fitness function;
}
}
```

5 Results

The following represent output screens generated. Dotnet frame work used implements the paper. Here c# can be used as coding language. Genetic algorithms (G.A) produce less error rate when compare to simulated annealing (S.A)in the calculation of parameters.

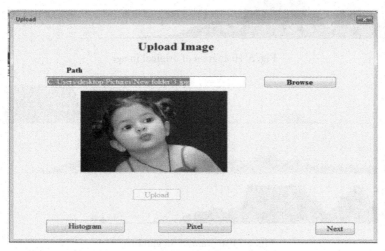

Upload
Upload Image
Path
Browse
Upload
Histogram
Pixel
Next

Fig. 3. Upload image

DisplayPixels
Display Pixel Data
Sno X Y Red
49570 100 191 0
49571 101 191 0
49572 102 191 0
49573 103 191 0
49574 104 191 1
49575 105 191 1
49576 106 191 3
49577 107 191 4
49578 108 191 5
Close

Fig. 4. Pixel level data

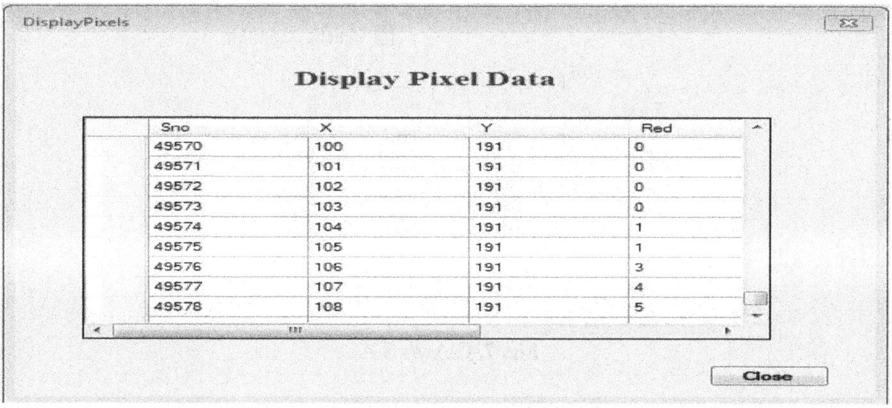

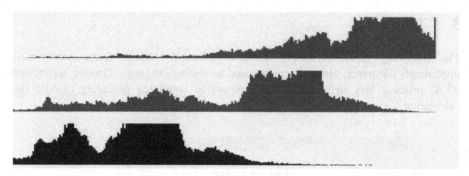

Fig. 5. Histogram of original image

Fig. 6. Histogram of Stegno image

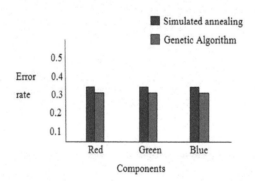

Fig. 7. G.A v/s S.A

6   Conclusion

This paper proposed a simple and efficient data embedding method based on PPM. Two pixels are scanned as an embedding unit and a specially designed neighborhood set is employed to embed message digits with a smallest notational system.   This paper mainly described the framework that could analyze security strength of data hiding technique called adaptive pixel pair matching technique and found penalization parameter and kernel parameter. This paper focused subtractive pixel adjacency matrix model an. Experimental results depicted that Genetic algorithm gave better performance than the simulated annealing Technique.

References

1. Channalli, S., et al.: International Journal on Computer Science and Engineering 1(3), 137–141 (2009)
2. Goel, M.K., Jain, N.: A Novel Visual Cryptographic Steganography Technique. International Journal of Computer, Electronics & Electrical Engineering 2(2) (1999) ISSN: 2249 - 9997
3. Hamid, N., Yahya, A., Badlishah Ahmad, R., Al-Qershi, O.M.: Image Steganography Techniques: An Overview. International Journal of Computer Science and Security (IJCSS) 6(3) (2012)
4. Provos, N., Honeyman, P.: Hide and seek: An introduction to steganography. IEEE Security Privacy 3(3), 32–44 (2003)
5. Zhang, W., Zhang, X., Wang, S.: A double layered plus-minus one data embedding scheme. IEEE Signal Process. Lett. 14(11), 848–851 (2007)
6. Chao, R.M., Wu, H.C., Lee, C.C., Chu, Y.P.: A novel image data hiding scheme with diamond encoding. EURASIP J. Inf. Security 2009, Article ID 658047 (2009), doi:10.1155/2009
7. Wang, J., Sun, Y., Xu, H., Chen, K., Kim, H.J., Joo, S.H.: An improved section-wise exploiting modification direction method. Signal Process. 90(11), 2954–2964 (2010)
8. Cheddad, A., Condell, J., Curran, K., McKevitt, P.: Digital image steganography: Survey and analysis of current methods. Signal Process. 90, 727–752 (2010)
9. Ker, A.D.: Steganalysis of LSB matching in grayscale images. IEEE Signal Process. Lett. 12(6), 441–444 (2005)
10. Chan, C.K., Cheng, L.M.: Hiding data in images by simple LSB substitution. Pattern Recognit. 37(3), 469–474 (2004)

Probability and Priority Based Routing Approach for Opportunistic Networks

Kiran Avhad[1], Suresh Limkar[2], and Anagha Kulkarni[1]

[1] Department of Computer Engineering, GHRCOEM, Ahmednagar, India
[2] Department of Computer Engineering, AISSMS IOIT, Pune, India
{kiranavhad87,sureshlimkar,anaghak313}@gmail.com

Abstract. Opportunistic networks are wireless networks that have sporadic network connectivity, thus rendering the existence of instantaneous end-to-end paths from a source to a destination difficult. Hence, in such networks, message delivery relies heavily on the store-and-forward paradigm to route messages. However, limited knowledge of the contact times between the nodes poses a big challenge to effective forwarding of messages. This paper describes probability and priority based approach for routing in opportunistic networks, rendering traditional routing protocols unable to deliver messages between hosts. This approach minimizes the copies of the messages in the network and ensures successful delivery of messages.

Keywords: Epidemic routing, Opportunistic Networks, Probabilistic Routing, Prioritized routing.

1 Introduction

The pervasive nature of wireless communications in today's communication scenario has seen a rapid rise in heterogeneous networks. A heterogeneous network is defined as a network that connects computers and other devices which run on different operating systems and uses diverse communication protocols or access technologies. For example, a wireless network that provides a service through wireless LAN and is able to maintain the service when network switches to a cellular network. Some heterogeneous networks which operate in mobile or extreme terrestrial environments lack continuous network connectivity. Opportunistic Networking is an approach that seeks to address the issues that rendered with communication in heterogeneous networks difficult. Opportunistic Networking [1], [2], [3] has a wide scope in all the rising communication paradigms in wireless mobile communications. Opportunistic network may be a sort of delay tolerant networks. Once a node want to communicate with another node however there doesn't exists the direct connection between them, then information packets will be forwarded to intermediate collaborating nodes that carries data packets from   source to destination [4],[5],[6],[7]. In some cases, intermediate nodes buffer the packets received for a long time. In opportunistic networks successful delivery of messages is major issue, because of this numerous

S.C. Satapathy, S.K. Udgata, and B.N. Biswal (eds.), *FICTA 2013*,
Advances in Intelligent Systems and Computing 247,
DOI: 10.1007/978-3-319-02931-3_32, © Springer International Publishing Switzerland 2014

routing protocols were planned to maximize message delivery rate. One of the foremost documented routing protocol for opportunistic network is, PRoPHET. PRoPHET uses predictability value, calculated using the history of encounters between nodes to evaluate the packet forwarding preference, because of the FIFO queuing nature of PRoPHET, packets may be dropped consistently when forwarded to a couple of focused nodes. Packets may also be lost because of node failures or incomplete transmissions [1].

This paper presents probability and priority based routing protocol for successful delivery of messages in the opportunistic networks. This routing protocol effectively use probabilistic routing [1],[2],[3],with priority of messages in the buffers of the nodes. The rest of the paper is organized as follows. Section 2 deals with related work. The probabilistic model is discussed in section 3. Section 4, presents probability and priority based model and section 5 and 6 discuss simulation setup and results respectively. And section 7 concludes the paper.

2   Related Work

PROPHET [1], a Probabilistic ROuting Protocol using History of Encounters and Transitivity makes use of observations that real users largely move in a very predictable fashion. If a user has visited a location many times before, there is additional likelihood to go to that location once more. PROPHET uses this info to boost routing performance.

PRoPHET+ [2] is improved version of PRoPHET. PRoPHET+ computes a deliverability value to determine the routing path for packets. Deliverability is calculated using a weighted function consisting of evaluations of nodes' buffer size, power, location, popularity, and the predictability value from PRoPHET.

In Epidemic routing [3], [4] the sender node replicates packets and forwards them to all nodes that come in contact with it, therefore essentially flooding the network with identical data. PREP [5] is an example of epidemic routing protocol that prioritizes flooding based on buffer. While Epidemic routing effectively minimizes the chance of data loss due to node failure situations, it also produces unnecessary buffer and power overhead [6].

SpatioTempo [7] exploits both the spatial (context) and the temporal (periodical behavior) characteristics of people to better route and optimize the message retransmissions.

Probabilistic Erasure Coding [8], is a forwarding algorithm based on erasure coding. Erasure coding [9] generates large amounts of code blocks with fixed overhead. According to this algorithm they provide a sophisticated but realistic method to allocate the generated code blocks to nodes that relies on a probabilistic metric for evaluating node potential.

The protocol proposed in [10] known as disconnected transitive Communication (DTC). It utilizes an application-tunable utility perform to find the node within the cluster of presently connected nodes that it's best to forward the message to base on the needs of the application. In each step, a node searches the cluster of currently connected nodes for a node that's "closer" to the destination, wherever the closeness is given by a utility function that may be tuned by the appliance to give appropriate results.

Work presented in [11] deals with the same problem of communication in disconnected networks. They propose a solution wherever nodes actively change their trajectories to create connected paths to accommodate the data transmission. Whereas this may add military applications and in some robotic device networks, in most situations it's not likely that nodes can move simply to accommodate communication of alternative nodes (if it's even attainable to communicate the need for it).

The routing protocol presented in [12] works for forwarding of messages from a mobile source to a stationary destination that exploits movement structure by learning the motion patterns of the peers. It maintains information about *meetings* between participants and their *visits* to locations and uses this information for routing and buffer allocation.[13]

3   Probability Based Approach

The probability based approach presented in [1], focuses on the movement property of mobile users, and takes forwarding decision, i.e. the mobile users move in specific pattern or move randomly. If any user has visited a location many times before, there's additional chance to visit that location once again. In this, forwarding decision depends on delivery predictability information. Delivery predictability information is nothing but the contact probability values between the nodes, which is calculated by using the number of encounter value.

To accomplish this, every node must know the contact probabilities of all the nodes presently available within the network. Each node maintains a probability matrix same as described in [1]. Every cell represents contact probability between two nodes for e.g. A and B, which shows the number of times node A and B visits before. The nodes exchange their contact probability matrices, when they meet with each other. Nodes compare their own contact matrix with all other nodes. A node updates its matrix with another nodes matrix if another node has more recent updated time attribute. In this way, two nodes can have identical contact probability matrices once communication takes place.

3.1   Delivery Predictability Calculation

The calculation of the delivery predictabilities described in [1] has three elements. The primary factor is updating the metric whenever a node is encountered; the nodes that are often encountered have high message delivery predictability. Once node 'A' meets node 'B', the delivery probability of node 'A' for 'B' is updated by "Eq. 1". Where P_{init} [0, 1], is an initialization constant.

$$P(A,B) = P(A,B)\text{old} + (1-P(A,B)\text{old}) \times P_{init} \qquad (1)$$

If node A doesn't meet with node B for a few predefined time slots, they're less likely to be good forwarders of messages to each other, therefore the delivery predictability values are aged, which are calculated in ageing equation. Refer to "Eq.2"where $\gamma \ \varepsilon \ [0,1]$ is the ageing constant, and k is the number of your time units that have elapsed since the last time the metric was aged. The time unit used can differ, and should be defined based on the application and the expected delays within the targeted network.

$$P(A,B) = P(A,B)\text{old} \times \gamma^{k.} \qquad (2)$$

The delivery predictability also has a transitive property, that is if node A often encounters with node B, and node B often encounters with node C, then node C most likely could be a sensible node to forward messages destined for node A. refer to "Eq. 3", which shows how this transitivity affects the delivery predictability, where $\beta \ \varepsilon \ [0, 1]$ is a scaling constant that decides how large impact the transitivity should have on the delivery predictability.

$$P(A,C) = P(A,C)\text{old} + (1-P(A,C)\text{old}) \times P(A,B) \times P(B,C) \times \beta \qquad (3)$$

4   Probability and Priority Based Routing Strategy

Probability and priority based routing strategy used priority of messages with probabilistic routing. The priority factor is the number of packets with higher priority in the queues of the neighboring nodes and the position of the current packet in the queue of the neighbor[14].

In opportunistic networks, a path from source to destination won't offered therefore the nodes buffer messages that it has originated as well as messages that it's buffering on behalf of other hosts. A hash table indexes this list of messages, keyed by a novel symbol associated with every message.[3],[4] Every host stores a small vector known as the summary vector that indicates that entries in their local hash tables are set and list of the priorities of the message. When two nodes meet, they exchange summary vectors and delivery predictability data and priorities of messages hold at the nodes. This data is used to update the inner delivery predictability vector, and then the information within the summary vector is used to make a decision of which messages to request from the opposite node.

In probability and priority based routing strategy, at the start a source node search all nodes in his communication range, then all these nodes exchange delivery predictability, summary vector information and priorities of the messages in the buffer. Now source compares all predictability values including its own and decides the neighbors having greater delivery predictability value than its own. Then out of that selected nodes, source compares message priorities in the queue of each node. The number of messages that have a priority greater than the packet to be forwarded is counted. The packet is forwarded to the two nodes that have the least number of packets (or no packets) with a priority greater than that of the packet that needs to be forwarded. After that source perform logical 'AND' operation between summary vectors of self and selected two nodes vectors, through which it finds the missing messages in buffers of selected nodes, then source sends missing messages to them, which reduce congestion in network. Similarly, these two nodes that get messages forward it to successive two's in their range, with on every occasion checking for the destination.

Algorithm

```
If (node having message to forward)
    Broadcast (Request message);
    Get reply (Delivery predictability, Summary vector, Priority);
    Array A[i]=delivery predictability;
    Sort (A[i]);
While (List is not fully traversed)
    Temp=own delivery predictability value;
    Searchposition (Temp, A[i]);
    Return(position);
endwhile
    Selectnodes(position, A[i])
    List[j]=Position+1 to i;
Do
    Priority = Priority of message to be forwarded;
    count = 0 ;
while (Current node has an unchecked node)
Do
while (List[j] is not fully traversed do)
    temp = Next Message in the List;
if (temp: priority > priority) then
    increment count
else
    continue traverse
end if
end while
endwhile
end do
end if
    Forward message to two neighbors with lowest count;
```

Algorithm 1. Probability and priority based routing

Increment priority()
{
prioritynew = priorityold + delayinqueue(time);
}

Algorithm 2. Avoiding unnecessary delay in queue of the low priority messages

Descriptively, every time a node has a message that needs to be forwarded, the following actions happen at the node:

- Broadcast the "Hello" message in his range.
- Get delivery predictability, summary vector and priority from all the nodes that get message.
- Compare delivery predictability values with its own and list all the nodes that have more than its own value.
- The priority of the packet that needs to be forwarded is obtained from the packet header.
- In the queue of these selected nodes, the number of packets that have a priority greater than the packet to be forwarded is counted.
- The packet is forwarded to the two nodes that have the least number of packets (or no packets) with a priority greater than that of the packet that needs to be forwarded.

5   Simulation Setup

The performance of the proposed protocol is studied and evaluated through ONE (Opportunistic Network Environment) simulator. The simulator contains a model of the wireless nodes with 50 no of nodes. Map based movement model is used for nodes movement in simulator. Each node has Bluetooth interface with range 10 meter. Each node generates message at 25-35 sec interval with size varying between 500kb-1Mb.The performance of the proposed protocol is compared with regard to the metrics such as message delivery delay, and number of messages transferred successfully among nodes. The simulation runs more than 30 times for 1800seconds each time and results are captured.

6   Results

Figure 1, shows that the graph for delivery rate of message. This is intuitive, since a larger queue size means that more messages can be buffered, and the risk of throwing away a message decreases. In this scenario, the performance is little bit similar for all three protocols; even though Probability &Priority based protocol seem to perform slightly better. Proposed protocol performs well when buffer size is less, but increasing buffer size fails to perform like PROPHET. Figure 2, shows delivery delay graph, here it seems like increasing the queue size, also increases the delay for messages. Large buffers might lead to problems in being able to exchange all messages between two nodes, leading to a higher delay. Proposed protocol performs better than PROPHET and Epidemic Routing for shorter buffers and slightly outperforms than PROPHET on bigger buffers.

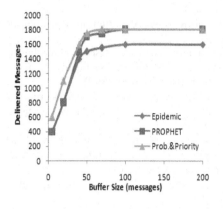
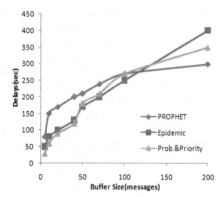

Fig. 1. Delivery rate graph   **Fig. 2.** Delivery delay graph

7   Conclusion

Opportunistic network should be a rising system that's getting growing interest in networking research community. Proposed system describes a probability and priority based routing approach for opportunistic network, This system focuses to use functionality of epidemic and probabilistic routing techniques and advanced it with adding new technique i.e. selection of two intermediate nodes that have higher priority to the forwarding message on every forwarding decision. This protocol results in improved message delivery, low overhead on resources than epidemic routing and drastically reduce the black hole attack from probabilistic approach. Proposed algorithm works better than epidemic routing in terms of the number of copies of a message that are in the network, i.e. very little number of copies in the network, which results in low overhead on resources.

References

1. Lindgren, A., Doria, A., Schelén, O.: Probabilistic Routing in Intermittently Connected Networks. In: Dini, P., Lorenz, P., de Souza, J.N. (eds.) SAPIR 2004. LNCS, vol. 3126, pp. 239–254. Springer, Heidelberg (2004)
2. Huang, T.-K., Lee, C.-K., Chen, L.-J.: PRoPHET+: An Adaptive PRoPHET- Based Routing Protocol for Opportunistic Network. In: IEEE Proceedings (2011)
3. Vahdat, A., Becker, D.: Epidemic routing for partially connected ad hoc networks. Technical Report CS-200006, Duke University (April 2000)

4. Vogels, W., van Renesse, R., Birman, K.: The power of epidemics: robust communication for large-scale distributed systems. In: Proceedings of First Workshop on Hot Topics in Networks (HotNets-I), Princeton, New Jersey, USA, October 28-29 (2002)
5. Ramanathan, R., Hansen, R., Basu, P.: Prioritized epidemic routing for opportunistic networks. In: MobiOpp 2007, San Juan, Puerto Rico, USA, pp. 978–971 (June 11, 2007), Copyright 2007 ACM 978-1-59593-688-2/07/0006
6. Avhad, K., Limkar, S., Kulkarni, A.: An Efficient Hybrid Approach for Opportunistic Networks. In: CSA 2013. LNCS, pp. 217–224 (2013)
7. Nguyen, H.A., Giordano, S.: Spatio temporal Routing Algorithm in Opportunistic Networks. 978-1-4244-2100-8/08/2008 IEEE
8. Tsapeli, F., Tsaoussidis, V.: Routing for Opportunistic Networks Based on Probabilistic Erasure Coding. In: Koucheryavy, Y., Mamatas, L., Matta, I., Tsaoussidis, V. (eds.) WWIC 2012. LNCS, vol. 7277, pp. 257–268. Springer, Heidelberg (2012)
9. Wang, Y., Jain, S., Martonosi, M., Fall, K.: Erasure coding based routing for opportunistic networks. In: WDTN 2005: Proceedings of the 2005 ACM SIGCOMM Workshop on Delay-Tolerant Networking, pp. 229–236. ACM, New York (2005)
10. Chen, X., Murphy, A.L.: Enabling disconnected transitive communication in mobile ad hoc networks. In: Proc. of Workshop on Principles of Mobile Computing, collocated with PODC 2001, Newport, RI (USA), 21 to 27, pp. 21–27 (August 2001)
11. Li, Q., Rus, D.: Communication in disconnected ad-hoc networks using message relay. Journal of Parallel and Distributed Computing (2003)
12. Marasigan, D., Rommel, P.: Mv routing and capacity building in disruption tolerant networks. In: Proceedings of the IEEE 24th Annual Joint Conference of the IEEE Computer and Communications Societies, INFOCOM 2005, vol. 1, pp. 398–408 (March 2005)
13. Avhad, K., Limkar, S., Patil, A.: Survey of Routing Techniques in Opportunistic Networks. IJSRP 2(10) (October 2012) ISSN 2250-3153
14. Rotti, P.G.: Opportunistic lookahead routing protocol for delay tolerant networks.Thesis, Submitted to the Faculty of the Louisiana State University (December 2012)
15. Lilien, L., Kamal, Z.H., Bhuse, V., Gupta, A.: Opportunistic networks: The concept and research challenges in privacy and security. In: Proc. of the WSPWN. cs.wmich.edu (2006)
16. Lindgren, A., Doria, A., Schelén, O.: Poster: Probabilistic routing in intermittently connected networks. In: Proceedings of The Fourth ACM International Symposium on Mobile Ad Hoc Networking and Computing, MobiHoc 2003 (June 2003)
17. Dvir, Vasilakos, A.V.: Backpressure-based routing protocol for dtns. In: Proceedings of the ACM SIGCOMM 2010 Conference, SIGCOMM 2010, pp. 405–406. ACM, NewYork (2010)
18. Jacquet, P., Muhlethaler, P., Clausen, T., Laouiti, A., Qayyum, A., Viennot, L.: Optimized link state routing protocol for ad hoc networks. In: Proceedings of IEEE INMIC 2001, pp. 62–68 (2001)
19. Ryu, J., Ying, L., Shakkottai, S.: Back-pressure routing for intermittently connected networks. In: Proceedings of IEEE INFOCOM 2010 (2010)
20. Glance, N., Snowdon, D., Meunier, J.-L.: Pollen: Using people as a communication medium. Computer Networks 35(4), 429–442 (2001)

D&PMV: New Approach for Detection and Prevention of Misbehave/Malicious Vehicles from VANET

Megha Kadam[1] and Suresh Limkar[2]

[1] Department of Computer Engineering, GHRCEM, Pune, India
[2] Department of Computer Engineering, AISSMS IOIT, Pune, India
{megha.desai1,sureshlimkar}@gmail.com

Abstract. VANET means vehicular ad hoc network is nothing but the group of independent vehicles nodes which are moving throughout the wireless network freely. Such kind of networks are temporary as the vehicles and their positions are not fixed and hence the all the routing paths which are established in order to make the communication in between the source and destination are on demand and depends on the nodes movement into the network. The architecture is not at all needed for such kind of networks. Role of routing protocols is most important for the VANET which is used to route the data from source to destination, but they are also vulnerable to the many of the security attacks in the VANET. Due to the unprotected nature of the VANET networks routing protocols, such networks also unprotected from the malicious vehicles in the network itself. This paper presents new approach for not only the detection of malicious vehicles attack but also their prevention from the VANET. Proposed algorithm is referred as Detection and Prevention of Malicious Vehicles (D&PMV). The malicious vehicles detected using the monitoring process over the VANET, once they are detected, proposed algorithm is applied for the prevention of the same. The detection of malicious vehicles is based on DMV algorithm presented earlier.

Keywords: Abnormal behavior, Vehicular Ad Hoc Networks, Honest vehicle, Secure communication, malicious vehicle, Detection, Prevention, MANET.

1 Introduction

Vehicular Ad Hoc Networks (VANETs) are appropriate networks that can be applied to intelligent transportation systems [1]. VANET is based on short-range wireless communication between vehicle- to- vehicle and some roadside infrastructure. Moreover, a large number of Certification Authorities (CAs) will exist, where each CA is responsible for the identity management of all vehicles registered in its region (e.g., national territory, district, country) [14].

Several types of messages are exchanged among vehicles such as traffic information, emergency incident notifications, and road conditions. It is important to forward correctly message in VANET; however, attacker nodes may damage the messages. Attackers or malicious vehicles perform in several ways and have different objectives such as attackers eavesdrop the communication between vehicles, drop, and change or inject packets into the network.

To provide network services under the presence of misbehaving nodes, it is necessary to consider "fault tolerance" as a main objective at the design level of routing protocols. To address this concern, several secure routing protocols have been proposed recently. Some of these protocols handle attacks by malicious nodes but not the selfish nodes and some handle selfish nodes but not malicious nodes. At the best of our knowledge, there is no solution that handles all misbehaving nodes actions. So it's necessary to provide a simulation study that measures the impact of misbehaving nodes in order to provide protocol designers with new guidelines that help in the design of fault / attack tolerant routing protocols for VANET.

1.1   VANET Communications

With an immense improvement in technological innovations, we find Vehicular Communication (VC) as a solution to many problems of our modern day communication system in roads. VC involves the use of short range radios in each vehicle, which would allow various vehicles to communicate with each other which is also known as (V-V) communication and with road side infrastructure(V-I) communication. These vehicles would then form an instantiation of ad hoc networks in vehicles, popularly known as Vehicular Ad Hoc Networks (VANET). It is a subset of Mobile Ad Hoc Networks (MANET). The similarity between these two networks is characterized by the movement and self organization of nodes. Also the difference between these ad hoc networks is that MANET nodes cannot recharge their battery power where as VANET nodes are able to recharge them frequently. We can understand VANETs as subset of MANET and best example of VANET is Bus System of any University which is connected. These buses are moving in different parts of city to pick or drop students if they are connected, make an Ad hoc Network.

VANET is mainly designed to provide safety related information, traffic management, and infotainment services. Safety and traffic management require real time information and this conveyed information can affect life or death decisions. Simple and effective security mechanism is the major problem of deploying VANET in public. Without security, a Vehicular Ad Hoc Network (VANET) system is wide open to a number of attacks such as propagation of false warning messages as well as suppression of actual warning messages, thereby causing accidents. This makes security a factor of major concern in building such networks. VANET are of prime importance, as they are likely to be amongst the first commercial application of ad hoc network technology. Vehicles are the majority of all the nodes, which are capable of forming self organizing networks with no prior knowledge of each other whose security level is very low and they are the most vulnerable part of the network which can be attacked easily. The capacity of VANET technology is high with a wide range of applications being deployed in aid of consumers, commercial establishments such as toll plazas, entertainment companies as well as law enforcement authorities. However, without securing these networks, damage to life and property can be done at a greater extent.

1.2   Technology

In VANET defines an intelligent way of using Vehicular Networking. In VANET integrates on multiple ad-hoc networking technologies [13] such as WiFi IEEE 802.11p, WAVE IEEE 1609, WiMAX IEEE 802.16, Bluetooth, IRA, ZigBee for easy, accurate, effective and simple communication between vehicles on dynamic mobility. Effective measures such as media communication between vehicles can be enabled as a well method to track the automotive vehicles is also preferred.

Vehicular Networks are envisioned of the Intelligent Transportation Systems (ITS). Vehicles communicate with each other via Inter-Vehicle Communication (IVC) as well as with roadside base stations via Roadside-to-Vehicle Communication (RVC). The optimal goal is that vehicular networks will contribute to safer and more efficient roads in the future by providing timely information to drivers and concerned authorities.

2   VANET Applications

Major applications of VANET include providing safety information, traffic management, toll services, location based services and infotainment. One of the major applications of VANET include providing safety related information to avoid collisions, reducing pile up of vehicles after an accident and offering warnings related to state of roads and intersections. Affixed with the safety related information are the liability related messages, which would determine which vehicles are present at the site of the accident and later help in fixing responsibility for the accident. There are several applications of VANET like Location-based services [12], Traffic optimization [15] and collision avoidance [5].

3   Security of VANET

To become a real technology that can guarantee public safety on the roads, vehicular networks need an appropriate security architecture that will protect them from different types of security attacks such as Denial of Service [17], Message Suppression Attack [17], Fabrication Attack [17], Alteration Attack [17] and Replay Attack [17]. VANET security should satisfy goals like Information authenticity, message integrity, source authentication, privacy and robustness of system. VANET should held Greedy drivers, Pranksters, industrial insiders and malicious attackers that affects on security of VANET.

4   Proposed Work

Our main aim with this is to simulate the misbehaving vehicles attack in the VANET by modifying existing routing protocol DSR (Dynamic routing protocol) in order to detect and prevent such attack in the network. Our research methods are implemented over the DSR for presenting the impact of presence of malicious vehicle in network. Proposed DSR algorithm also addresses the all kinds of misbehaving nodes such as selfish as well as malicious nodes. Simulation study presents how proposed DSR detect the misbehaving nodes and how to prevent them from dropping the data. Following are the main objectives:

> To present the various significance of the VANET networks.
> To present the detailed study over VANET
> To analyze the malicious nodes attack in the VANET.
> To present approaches to provide security to the mobile ad hoc networks from malicious vehicle attacks.

In this proposed research work, we are presenting the design of a novel misbehavior detection scheme at the application layer, called DMV (Detection of Malicious Vehicles), and tag vehicles using their distrust values [4]. In DMV, verifiers operate independently from each other. In addition, DMV can improve the performance of verifier selection at high speeds. In DMV [4], a number of vehicles are located in a cluster. Each cluster has one main cluster-head and one spare cluster-head, where the spare cluster-head is the trustiest vehicle after the main cluster-head. Each vehicle is monitored by some of its trustier neighbors which are defined as verifier nodes. If verifier nodes observe an abnormal behavior from vehicle V (a node that drops or duplicates packets), they increase the distrust value of vehicle V. The identification code of vehicle V is reported to its Certificate Authority (CA) as a malicious node when its distrust value becomes higher than a threshold value.

Algorithm Design
In this section, we introduce a monitoring algorithm to detect malicious vehicles and prevent them from honest ones. Following algorithm shows the monitoring process for vehicle V when it joins to a cluster.

Algorithm 1: Malicious vehicle detection

Step 1: Vehicle V joins the network
Step 2: Get the cluster keys
Step 3: Assigning the verifiers to newly joined vehicle V.
Step 4: Start monitoring behavior of vehicleV.
Step 5: If (verifier detecting the abnormalbehavior of vehicle V)
          Report to cluster_head (CH)
          goto step 6;
     else
          goto step 4;
Step 6: CH modifies the disturb value (d)
Step 7: if disturb value less than or equal to threshold (t) value (which is set for each
          new vehicle once that join the network) then update the white list of
          network
          else generation of warning message to cluster agents of CH.

```
if (d <= t)
    goto step 4
else
    goto step 8
```

Step 8: Alarm generation in order to provide the warning message among all the
other vehicles under the same cluster head.

Algorithm 2: Prevention of Malicious Vehicle

Step 1: As once in algorithm 1 and step 8, alarm is generated for the detection of
malicious vehicle in VANET, our second algorithm is executed for its
prevention. Information retrieval of malicious vehicle is done at first.

Step 2: Parsing all communication paths in the network.

Step 3: Identify the presence of malicious nodes, if it's presented their in path,
simply discard that path and use alternate communication path.

5   Research Methodology

Research methodologies for the project work are related to the analysis of the
misbehaving vehicles which are responsible for the black hole attacks in the
VANET. Mainly in the Blackhole attack, traffics for the network are redirected to
the mobile node in the network which not at all exists in the network. Thus in this
case the network traffic is disappear into the one of the special mobile node such
node is called as Blackhole node. The Blackhole attack has two characteristics: first
one, the misbehave node advertise itself regarding to the information that it has
shortest route to the destination with the intention of dropping the packets or
intercepting packets in the routing protocols like DSR, AODV. Intercepted packets
then consumed by the node: this is the second characteristics. For the simulation of
the Blackhole attack, our research works on the misbehaving nodes simulation on the
DSR protocol and prevention of it. We used Qualitative and Quantitative research
methods.

6   Mathematical Model

6.1   NP-Hard Problem

As per from given algorithm, we are finding the malicious vehicle using the
monitoring process. Here the process of monitoring is NP-Hard problem. We have
following is the mathematical model of this problem:

The purpose of monitoring is to collect information about the behavior of all
vehicles in the network. Vehicles that perform the monitoring process are called
"verifier" vehicles. A verifier is a vehicle which has a smaller or equal Td compared

with the Td of vehicle V, and is located inside the region z (V, CH). This region shows the intersection area of both vehicle V and its CH (see Fig. 1).

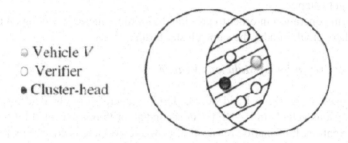

Fig. 1. Illustration of region z (V,CH)

Using this way, we are sure that all verifiers which monitor vehicle V are able to send their reports to CH [18]. Note that the area of CH is equal to its transmission range but the area of V is obtained from Eq. (1):

$$Area(V) = TR(V) - T_f(S_{max} - S_{min}) \quad . (1)$$

Definition of general variables

β	Number of transmitted packets per second
CA	Certificate Authority
CH	Cluster head
I_v	Abnormal behavior rate
L	Number of verifiers
P_c	Probability of missing or duplicating a packet
P_{CA}	Radius of the CA
S_{max}	Legal maximum speed of the vehicle
S_{min}	Legal minimum speed of the vehicle
T_d	Distrust value
T_f	Packet Latency
TR	Transmission range
t_{rem}	Maximum time that malicious vehicle requires to pass through the CA region
U	Verifier
V	Certain vehicle
X	Number of missed or duplicated packets
Y	Number of packets received to vehicle V
γ	Legal number of drops or duplicates packets
λ	Average number of dropped or duplicated packets by each vehicle before detection
σ	Threshold value on distrust

6.2   NP-Complete Problem

Determining abnormal behavior of a vehicle by its neighbors:
From a source vehicle point of view in VANET, another vehicle can play either destination role or relaying role. When vehicle V plays the relaying role, some of its verifier vehicles verify the behavior of V via monitoring. When u_i operates as verifier of V, it counts the number of packets that V receives (denoted by parameter y) and those missed or duplicated by V that verifier u_i detects (denoted by parameter x). For this purpose verifier u_i sends its report to its CH. If after elapsing T_f, vehicle V does not forward a received packet or sends a packet twice, u_i considers it as an abnormal behavior and then increases the parameter x by one unit.

The parameter T_d is automatically mapped to each vehicle and can change when the vehicle performs either as a relaying vehicle or as a source vehicle. When T_d of vehicle V changes, the new T_d is broadcast to neighbors, and the neighbors update their white lists.

Vehicles cooperate with vehicle V if its T_d is lower than a threshold value σ. When T_d of vehicle V is larger than σ, its ID should be reported to the relevant CA as a malicious vehicle. Then, CA broadcasts the ID of malicious vehicle V to all vehicles and roadside infrastructure units. In order to update T_d, the CH calculates parameter I_V for vehicle V using Eq. (2):

$$Iv = \sum_{j=0}^{L} \frac{(1-Pc)^{yj-xj}(Pc)^{xj}}{Tdj} \quad (2)$$

Where L is the number of verifier vehicles of malicious vehicle V. In Eq. (2), we assume that vehicle V forwards a received packet or misses or duplicates it. The parameter Pc is the probability of missing or duplicating a packet by a malicious vehicle. The parameters y_j and x_j show the number of packets received to vehicle V as well as those missed or duplicated by V, respectively, counted by the jth verifier. According to Eq. (2), the CH computes a normalized I_V using coefficient in order to reduce the effect of verifiers with high distrust values. Using I_V, the CH computes new T_d of V using Eq. (2):

$$T_d(new) = T_d(old) + I_v \quad (3)$$

7   Performance Evaluation

1. Throughput

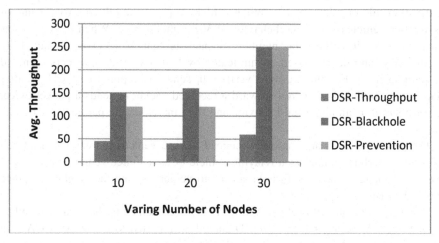

Fig. 2. Average Throughput

Table 1. Table of reading for Avg Throughput

Avg. Throughput/ Varying No of Nodes	DSR- Throughput	DSR-Blackhole	DSR-Prevention
10	45	150	120
20	40	160	120
30	60	250	250

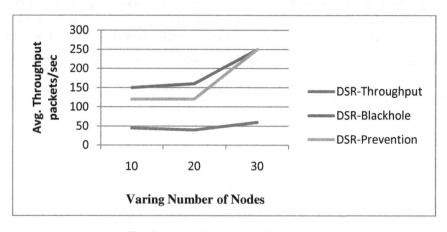

Fig. 3. Average Throughput (Packet/Sec)

2. Delay

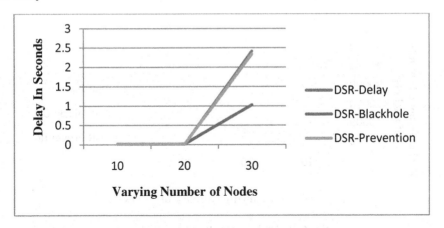

Fig. 4. Delay in Seconds

3. Jitter

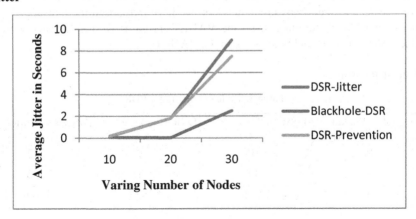

Fig. 5. Average Jitter in Seconds

Table 2. Table of reading for Avg. Jitter

Avg. Jitter in Sec / Varying No of Nodes	DSR-Jitter	Blackhole-DSR	DSR-Prevention
10	0.16	0.07	0.18
20	1.8	0.02	1.8
30	9	2.5	7.5

4. Packet Dropped Ratio

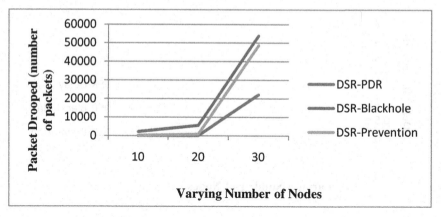

Fig. 6. Packet Drooped (Number of Packets)

From above graphs, it's obvious that at the presence ofblackhole attack in VANET network makes the extra packet drops as compared to normal working of DSR. Hence to reduce such packet drops in the network, we have introduced the new approach to mitigate the blackhole attack from the VANET. From above results, this approach reduced the effect of blackhole attack in the VANET.

Table of Reading

Table 3. Table of reading for PDR

No of Packet / Varying No Of Nodes	DSR-PDR	DSR-Blackhole	DSR-Prevention
10	10	2180	242
20	44	5565	856
30	22126	53853	48604

8   Conclusion

In this research work, we have proposed and evaluated a misbehavior detection scheme for detecting vehicles that drop or duplicate received packets. The simulation results show that DMV can detect malicious vehicles before the number of dropped or duplicated packets becomes more than a given number of drops or duplicates (γ). In addition, DMV can improve the performance of verifier selection at high speeds. In this algorithm, a distrust value has been introduced that can be used to determine trustiness value of each vehicle when it forwards messages. Therefore, vehicles can select trustier vehicles to forward messages based on the vehicles distrust value.

References

1. Abdulhamid, H., Tepe, K.E., Abdel-Raheem, E.: Performance of DSRC systems using conventional channel estimation at high velocities. Int. J. Electron. Commun., 556–561 (2007)
2. Artimy, M.: Local density estimation and dynamic transmission-range assignment in Vehicular Ad Hoc Networks. IEEE Trans. Intell. Transp. Syst. 8(3), 400–412 (2007)
3. Bettstetter, C.: Smooth is better than sharp: A random mobility model for simulation of wireless networks. In: 4th ACM International Work-shop on Modeling, Analysis, and Simulation of Wireless and Mobile Systems (MSWiM 2001), Rome, Italy (2001)
4. Detection of malicious vehicles (DMV) through monitoring in Vehicular Ad-Hoc Networks. AmenehDaeinabi & Akbar GhaffarpourRahbar Springer Science+Business Media, LLC (2011)
5. Raya, M., Hubaux, J.P.: The security of vehicular ad hoc networks. J. Comput. Secur. Spec. Issue Secur. Ad Hoc Sensor Netw. 15(1), 39–68 (2007)
6. Wang, N.W., Hauang, Y.M., Chen, W.M.: A novel secure communication scheme in vehicular ad hoc networks. Comput. Commun. 31(12), 2827–2837 (2008)
7. Fan, P., Haran, J.G., Dillenburg, J., Nelson, P.C.: Cluster-Based Framework in Vehicular Ad-Hoc Networks. In: Syrotiuk, V.R., Chávez, E. (eds.) ADHOC-NOW 2005. LNCS, vol. 3738, pp. 32–42. Springer, Heidelberg (2005)
8. Hu, Y.-C., Perrig, A.: Survey of Secure Wireless Ad Hoc Routing. IEEE Security & Privacy 2(3), 28–39 (2004), doi:10.1109/MSP.2004.1
9. Yan, G., Olariu, S., Weigle, M.C.: Providing VANET security through active position. Comput. Commun. 31(12), 2883–2897 (2008)
10. Raya, M., Hubaux, J.P.: The Security of Vehicular Ad Hoc Networks. In: SASN 2005, Alexandria, Virginia, USA (November 7, 2005), Copyright 2005 ACM 95932275/05/0011
11. Raya, M., Papadimitratos, P., Hubaux, J.P.: Securing Vehicular Communications. IEEE Wireless Communications 13 (October 2006)
12. Golle, P., Greene, D., Staddon Detecting, J.: Staddon Detecting and Correcting Malicious Data in VANETs. In: VANET 2004, Philadelphia, Pennsylvania, USA (October 1, 2004), Copyright 2004 ACM 1581139225/04/0010
13. Abdalla, G.M.T., Senouci, S.M.: Current Trends in Vehicular Ad Hoc Networks. In: Proceedings of UBIROADS Workshop (2007)
14. Raya, M., Jungels, D., Papadimitratos, P., Aad, I., Hubaux, J.P.: Certificate Revocation in Vehicular Networks, Laboratory for Computer Communications and Applications (LCA) School of Computer and Communication Sciences. EPFL, Switzerland (2006)
15. Nadeem, T., Shankar, P.: A comparative study of data dissemination models for VANETs. IEEE Mob. Ubiquitous Syst. Netw. Serv., 1–10 (2006)
16. Picconi, F., Ravi, N., Gruteser, M., Iftode, L.: Probabilistic validation of aggregated data in Vehicular Ad Hoc Networks. In: International Conference on Mobile Computing and Networking, Los Angeles, CA, USA, pp. 76–85 (2006)
17. Parno, B., Perrig, A.: Challenges in Securing Vehicular Networks
18. Raya, M., Papadimitratos, P., Aad, I., Jungels, D., Hubaux, J.P.: Eviction of Misbehaving and FaultyNodes in Vehicular Networks. IEEE J. Sel. Areas Commun. 25(8), 1557–1568 (2007)

On the Use of MFCC Feature Vector Clustering for Efficient Text Dependent Speaker Recognition

Ankit Samal, Deebyadeep Parida, Mihir Ranjan Satapathy,
and Mihir Narayan Mohanty

Siksha 'O' Anusandhan University, Bhubaneswar, Odisha, India
ankitsamal@rediffmail.com, ddeebya@gmail.com,
mrsatapathy@sify.com, mihir.n.mohanty@gmail.com

Abstract. The paper describes an experimental study and the development of a computer agent for Speaker recognition. It presents an efficient method to verify authorised speakers and identify them using MFCC Feature vector clustering. For clustering of the MFCC features, Vector Quantisation using *Linde-Buzo-Gray* (LBG) algorithm has been presented. This approach proves to be an efficient ASR technique.

Keywords: ASR, MFCC, LBG, Clustering.

1 Introduction

Automatic Speaker recognition has an important role in applications of forensics. Lot of research work has been done in the past few decades; still the field remains undiscovered. Although difficult to extract, human voice is a mine of information. Certain voice features pertain to the biologically unique structure of the nasal cavity, vocal tract, teeth and the tongue. Some features even attribute to cultural and social aspects, like the accent.

Security systems can be used to identify wanted criminals from their respective voices and have a great demand on accuracy. These systems can even be installed in public places to capture voices of miscreants. It can also be implemented as a password for internet banking, e-commerce websites and to secure confidential information. The best advantage is that this password can never be forgotten because Biometrics refers to 'What you are' rather than 'what you carry' [1].

In [2] the authors have described a system built around the likelihood ratio test for verification, using simple and effective Gaussian mixture models (GMMs) for likelihood functions whereas Chiyomi Miyajima *et. al.,* in [3] shows the modeling of speech spectra and pitch for text-independent speaker identification using Gaussian mixture models based on multi-space probability distribution (MSD-GMM).

Of late research has revealed several techniques for Automatic Speaker Recognition. Here, an Automatic Speaker Recognition (ASR) system using Mel Frequency Cepstral Coefficients (MFCC) as features of the speech signal has been implemented. Further we clustered the features through Vector Quantization method using

S.C. Satapathy, S.K. Udgata, and B.N. Biswal (eds.), *FICTA 2013*,
Advances in Intelligent Systems and Computing 247,
DOI: 10.1007/978-3-319-02931-3_34, © Springer International Publishing Switzerland 2014

Linde-Buzo-Gray Algorithm. The minimum Euclidean distance between the input clustered features and the trained data is evaluated to recognise the speaker.

Section 2 describes a brief introduction to Speaker Recognition and its types. The proposed Speaker recognition algorithm is detailed in section 3. This includes the algorithm for VQ-LBG. Feature extraction using MFCC is illustrated in section 4. Section 5 documents the results. Finally section 6 concludes this piece of work.

2   Speaker Recognition Basics

The technique of Speaker recognition involves the authentication or identification of a speaker from a large ensemble of possible speakers [4]. Speaker recognition can be classified into two types: Speaker verification & Speaker Identification [5]. A *Speaker Verification* system is used decide if a speaker is the person he claims to be. On the other hand, a *Speaker identification* is used to decide which speaker among an ensemble of speakers produced a given speech utterance [4]. In a sense speaker verification is a 1:1 match where one speaker's voice is matched to one template (also called a "voice print" or "voice model") whereas speaker identification is a 1:N match where the voice is compared against N templates. Speaker verification is usually employed as a "gatekeeper" in order to provide access to a secure system (e.g.: telephone banking).

Speaker recognition methods can also be divided into text-independent and text-dependent methods. A text-independent Speaker recognition system identifies irrespective of what one is saying. On the other hand, in a text-dependent system, the recognition of the speaker's identity is based on his or her speaking one or more specific phrases, like passwords, card numbers, PIN codes, etc.

3   Proposed Algorithm for Recognition

Speaker recognition algorithms generally consist of two phases: Training & Testing. In the *Training* phase, the system needs to be trained to recognise authorised users. Figure 1 shows the flow diagram for training.

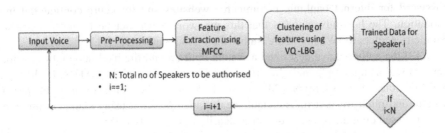

Fig. 1. Flow Diagram for Training

3.1   Clustering Using Vector Quantization Technique through *Linde-Buzo-Gray* Algorithm

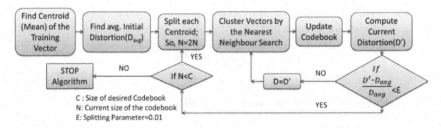

Fig. 2. Flow Diagram for VQ-LBG

The algorithm is described as follows:

1. Given T, the training vector obtained from feature extraction phase (MFCC). Design a 1 vector codebook. It is basically the centroid of the entire set of training vectors. (By centroid we refer to finding the mean). Let $N = 1$ and

$$c_1^* = \frac{1}{M} \sum_{m=1}^{M} x_m$$

where M is half the frame size used in MFCC feature extraction and x_m is the source vector which is K-dimensional, where k is the number of filters used in the *Mel* filter bank.

2. Fix $\in$, the splitting parameter (error value) to be a small number, assume it as $\in = 0.01$.

3. Calculate D_{ave}^*, the average initial distortion by using

$$D_{ave}^* = \frac{1}{Mk} \sum_{m=1}^{M} \|x_m - c_1^*\|^2$$

4. Double the size of the codebook by splitting it till we reach the desired codebook size. If size of codebook is greater than the desired size we stop.

   For $i = 1,2,...,N$ set

$$c_i = (1+\in)c_i^*, \ c_{N+i} = (1-\in)c_i^*, \text{ Set } N = 2N.$$

5. Find the nearest neighbours of each codeword. For $m = 1,2,...,M$, find the minimum value of

$$\|x_m - c_n\|^2 n = 1,2,...,N$$

Let n^* be the index which achieves the minimum. Set $Q(x_m) = c_{n^*}$. In the end one must ensure that each code vector c_n is closest to atleast one source vector x_m.

6. Update the codeword in each cell using the centroid of the training vectors assigned to that cell. After all the source vectors closest to each code vector have been found, update the codebook by updating each codevector c_n by finding the centroid of each   cell, resulting in a corresponding updated code vector c_n . For   = $1, 2, ..., N$, update the codevector.

$$c_n = \frac{\sum_{Q(x_m)=c_n} x_m}{\sum_{Q(x_m)=c_n} 1}$$

7. Next calculate the average distortion after updating the codebook ,

$$D_{ave} = \frac{1}{Mk} \sum_{m=1}^{M} \|x_m - Q(x_m)\|^2$$

Check if $(D_{ave}^* - D_{ave}/D_{ave}^*) > \in$, Set $D_{ave}^* = D_{ave}$ and go back to Step 6. Where D_{ave}^* is the initial distortion and D_{ave} is the current distortion. The initial distortion is assigned with the current distortion. Else go to Step 5 and continue until the desired codebook size is obtained.

Figure 3 shows the codebooks and respective centroids of 4 speakers. After the system is trained properly, it is ready to verify authorised users and identify them.

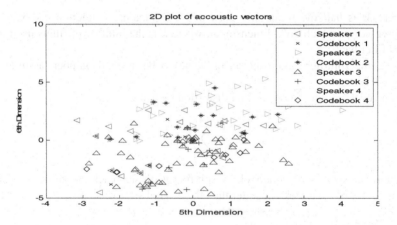

Fig. 3. Codebooks of 4 speakers

In the testing phase the system searches its database (which was created during Training) for the closest match. If the score of the closest match exceeds a preset threshold amount, then the input voice is not found in the training database and the system generates an alarm of unauthorised access. Figure 4 shows the flow diagram for Testing.

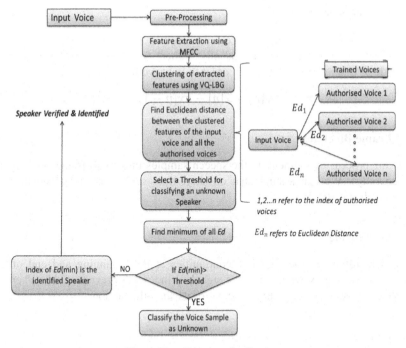

Fig. 4. Flow Diagram for Testing

4   Feature Extraction Using MFCC

Mel-frequency cepstral coefficients (MFCCs) are coefficients that collectively make up an MFC. It is an efficient feature extraction technique.

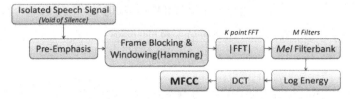

Fig. 5. Block diagram for MFCC Evaluation

It is expressed in the Mel-frequency scale, which is linearly spaced below 1000 Hz and logarithmically spaced above 1000 Hz. The shape of the vocal tract manifests itself in the envelope of the short time power spectrum, and the job of MFCCs is to accurately represent this envelope. The block diagram shown in Fig. 5 is explained as follows.

4.1   Pre-emphasis

Pre-emphasis refers to a system process designed to increase, within a band of frequencies, the magnitude of some (usually higher) frequencies with respect to the magnitude of other (usually lower), in order to improve the overall SNR.

$$y[n] = x[n] - ax[n - 1]$$

4.2   Frame Blocking

This is the process of segmenting the speech samples into small frames of length 30 ms (generally). We can assume that on short time scales the audio signal doesn't change much statistically.

4.3   Windowing

We need to minimize the signal discontinuities at the beginning and end of each frame. This is done by windowing the signal. Windowing prevents Gibbs Phenomenon. We have used the Hamming window for its smoother action.

4.4   FFT

The Fourier Transform converts each windowed frame from time to frequency domain. We have used Absolute FFT for better results.

4.5   *Mel* Filterbank

The formula for converting from frequency to *Mel* scale is:

$$B(f) = 1125 \ln(1 + f/700)$$

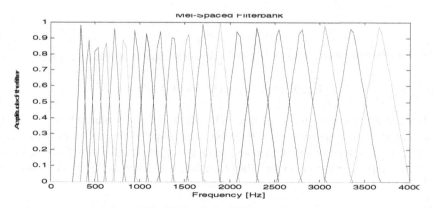

Fig. 6. Mel-Spaced Filterbank

The filter bank is designed with a starting frequency of 300Hz and stopping frequency of 4000 Hz.

4.6   Log Energy

After obtaining the filter bank it is multiplied to the energy of each filter to obtain filter bank energies. Once we have the Filter Bank energies, we take its logarithm. This is motivated by human hearing because we don't hear loudness on a linear scale.

4.7   DCT

The final step is to compute the DCT of the log Filter Bank energies. There are 2 main reasons this is performed. Because our Filter Banks are all overlapping, the Filter Bank energies are quite correlated with each other. The DCT de-correlates the energies.

$$c[n] = \sum_{i=1}^{M} log(y(i)).cos(\frac{\pi n}{M}(i - \frac{1}{2}))$$

5   Results and Discussion

Table 1 shows the Euclidean distance between the input voices (1st column) and all the voices trained. The 1st column accounts for the input voice of all the speakers and the 1st row accounts for the trained data. Clearly we get the least distance for the input voice corresponding to its original speaker.

Table 1. Euclidean Distance between the Clustered MFCC Vectors of Speakers

Trained Data → Input Data ↓	Ankit (Male)	Mihir (Male)	Deebya (Male)	Tintu (Male)	Rony (Male)	Puji (Female)
Ankit (M)	39.23	148.37	199.37	125.39	115.31	259.43
Mihir (M)	146.27	40.43	183.88	132.43	154.34	245.66
Deebya(M)	207.21	198.67	32.93	193.02	202.31	193.29
Tintu (M)	129.41	135.61	198.49	35.5	123.11	234.31
Rony (M)	110.19	157.96	189.41	133.32	40.01	252.2
Puji(F)	256.29	243.73	199.29	254.22	255.53	30.9

The ASR was tested for each speaker 40 times. The success rate of the ASR for each speaker is shown in table 2.

Table 2. Efficiency of the proposed method

Speaker→	Ankit (Male)	Mihir (Male)	Deebya (Male)	Tintu (Male)	Rony (Male)	Puji (Female)
Success Rate	92.5% (37/40)	92.5% (37/40)	97.5% (39/40)	95% (38/40)	92.5% (37/40)	97.5% (39/40)

6   Conclusion

MFCC represents the human voice better than any other method, thus finding a place in this system. The proposed method is extremely efficient as shown in both the tables. It could also successfully differentiate voices of twins with a marginal error. Currently there is a serious need to implement security systems which can identify wanted criminals just by their voice. Extracting effective features of speech and enhancing the emotion recognition accuracy for speaker recognition is our future work. This will be a huge advancement in Forensics.

References

1. Kekre, H.B., Bharadi, V.A., Sawant, A.R., Kadam, O., Lanke, P., Lodhiya, R.: Speaker Recognition using Vector Quantization by MFCC and KMCG Clustering Algorithm. In: 2012 International Conference on Communication, Information & Computing Technology (ICCICT), Mumbai, India, October 19-20 (2012)
2. Reynolds, D.A., Quatieri, T.F., Dunn, R.B.: Speaker Verification Using Adapted Gaussian Mixture Models. Digital Signal Processing 10, 19–41 (2000)
3. Miyajima, C., Hattori, Y., Tokuda, K., Masuko, T., Kobayashi, T., Kitamura, T.: Speaker Identification Using Gaussian Mixture Models Based On Multi-Space Probability Distribution. In: Proc. of ICASSP (2001)
4. Rabiner, L.R., Schafer, R.W.: Digital Processing of Speech Signals. Prentice-Hall, Englewood Cliffs (1978)
5. Reynolds, D.A.: An overview of automatic speaker recognition technology. In: Proc. IEEE Int. Conf. Acoust., Speech, Signal Process.
6. Mohanty, M.N., Routray, A., Kabisatpathy, P.: Voice Detection using Statistical Method. International Journal Engg. Techsci. 2(1), 120–124 (2011)
7. Linde, Y., Buzo, A., Gray, R.: An algorithm for vector quantizer design. IEEE Transactions on Communications 28, 84–95 (1980)
8. Huang, X., Acero, A., Hon, H.: Spoken Language Processing: A guide to theory, algorithm, and system development. Prentice Hall (2001)

A Survey on Power Aware Routing Protocols for Mobile Ad-Hoc Network

Samrat Sarkar and Koushik Majumder

Department of Computer Science & Engineering, West Bengal University of Technology,
BF 142, Sector I, Salt Lake, Kolkata 700064, India
koushik@ieee.org

Abstract. The Mobile Ad-Hoc network is a growing type of wireless network characterized by decentralized and dynamic topology. One of the main challenges in Mobile Ad-Hoc networks (MANETs) is that it has very limited power supply. To overcome the challenge, there are several power-aware routing protocols that have been developed in recent years. This paper describes a survey on some of those energy aware routing protocols for Mobile Ad-Hoc networks. The first category of power aware protocol schemes minimizes the total transmission power and the second category of schemes tries to increase the remaining battery level of every individual node to increase the lifetime of the entire network. The optimizations between these two objectives are important issues in power aware routing. This discussion focuses on different power saving algorithms and their development and modifications. After analyzing the existing works it has been seen that there are still several fields (using a dual threshold, passive energy saving etc) where we can give more focus in the future.

Keywords: Power Aware, Mobile Ad-Hoc Network, Routing Protocol.

1 Introduction

A network can be defined as a collection of different interconnected nodes that can be wired or wireless. A Mobile Ad-Hoc Network [1] is composed of a group of autonomous mobile wireless nodes without any centralized administration in which they forward the packets to each other in a multi-hop manner. Wireless multi hop Mobile Ad-Hoc networks offer communication facilities in areas where it is impossible for wired cable networks to reach. Mobile Ad-Hoc networks have proved to be useful in different areas like battlefield scenarios, in case of disaster recovery, cell phone, laptop, traffic control, space and astronomy.

MANET is infrastructure less, so the network topology may change very fast and unpredictably at any time. In MANETs there is limited bandwidth compared to the wired network. Another important issue in the Mobile Ad-Hoc network is power optimization as the network lifetime needs to be increased.

Most of the ad-hoc mobile devices today use lithium batteries. Usually the average lifetime of batteries in an idle phone is one day. A single node in the routing path going to 'dead' condition can cause the entire network to fail.

S.C. Satapathy, S.K. Udgata, and B.N. Biswal (eds.), *FICTA 2013*,
Advances in Intelligent Systems and Computing 247,
DOI: 10.1007/978-3-319-02931-3_35, © Springer International Publishing Switzerland 2014

A large number of researchers are trying to solve the problem of energy-efficient data transfer in the context mobile ad hoc networks [2]. The different existing protocols can be classified in the following two categories:

The first category of protocols deals with minimizing the total power requirements for the end to end transmission. The main advantage of this kind of protocol is that it always selects a minimum cost route very quickly. But the main disadvantage is that the nodes on that selected route are overused. So, critical nodes will 'die' soon by exhausting their battery life causing network failure.

The second category of protocols is based on battery aware routing algorithms that take care of battery life of individual nodes. In such manner, this category of protocols increases the battery lifetime of each node as well as the overall lifetime of the entire network.

2   Review of Existing Power Aware Protocol

2.1   Minimum Total Transmission Power Routing (MTTPR)

The Minimum Total Transmission Power Routing Protocol [3] is a basic power aware routing protocol that always tries to minimize the total transmission power of the entire network by selecting the minimum hop count route. This actually implements the metric 'minimize the energy consumed per packet' proposed in [3]. MTTPR, first calculates the total transmission power for all possible routes between source and destination. The total transmission power $P(n_i, n_j)$ between two hosts n_i and n_j for route l can be calculated from the following equation, $P_1 = \sum_{i=0}^{D-1} P(n_i, n_{(i+1)})$

Where 0 is source node and D is destination node.

Finally it selects the route with minimum total transmission power. Using this metric the minimum route k can be obtained from the following equation where A is the set containing all possible routes. $P_k = \min_{l \in A} p_l$

2.2   Minimum Battery Cost Routing (MBCR)

The MTTPR algorithm does not take care of battery life of every individual node, so MBCR algorithm is proposed by introducing an extra battery cost function [3] that is the inverse of remaining battery capacity. That means if the remaining battery power decreases the cost function will increase. This algorithm first finds the battery cost for each node of the network and finds the battery cost function. Let c_i^t is the remaining battery capacity of a host n_i at time t and let $f_i(c_i^t)$ is a battery cost function of node n_i. The cost function is the inverse of battery capacity that can be achieved from the following equation. $f_i(c_i^t) = \left(\frac{1}{c_i^t}\right)$

Now the total battery cost R_j for the route i with D nodes will be, $R_j = \sum_{i=0}^{D_{j-1}} f_i(c_i^t)$

From this equation we can find a route with the maximum remaining battery capacity by selecting a route i with having minimum battery cost.

$$R_i = \min \{R_j \mid j \in A\}$$

Here A is the set of all possible routes from source to destination.

The cost function is calculated based on the battery life remaining of individual nodes, so it takes care about individual battery life of nodes. If all nodes in the network have same battery charge remaining, it will choose the shortest hop.

2.3   Min-Max Battery Cost Routing (MMBCR)

MMBCR [3] is a modification of the MBCR algorithm in such a way that no critical node will be overused. Without summing the battery cost function of all nodes of every individual route, MMBCR finds the maximum battery cost among all nodes of different routes to find the critical nodes. The algorithm first finds the battery cost for each node and then finds the battery cost function. For every route, it selects maximum battery cost function among all nodes in that particular route. To find the battery cost R for route j the following equation is used, $R_j = \max_{i \in \text{route } j} f_i(c_i^t)$.

After finding the battery cost the desired route i can be obtained from the equation

$$R_i = \min\{ R_j \mid j \in A\}$$

The algorithm selects the route avoiding critical battery node, so the battery will be used more fairly to avoid network failure.

2.4   Conditional Max-Min Battery Capacity Routing (CMMBCR)

A hybrid approach CMMBCR was devised by C.K Toh [4] that gives a better routing algorithm based on the battery life. CMMBCR tries to minimize the total transmission power and also avoid the battery having low remaining capacity by adding an extra threshold to each battery node. This algorithm first finds the minimum battery capacity (R_i) for all nodes of each route. If $R_i \geq Y$ (chosen threshold value) is true for some or all routes between a source and destination, then the MTPR scheme is applied to select the route among all possible paths which satisfy the previous condition. If any path does not satisfy the condition then the route j is selected with the help of maximum battery remaining capacity by using the protocol MMBCR.

The battery capacity R_j^c for route j at time t can be defined by the following equation, $R_j^c = \min_{i \in \text{route } j} c_i^t$

The   threshold ranges between 0 and 100. Now let Q is the set containing all possible paths between source and destination at time t. Now, if A is the set of all routes between any two nodes at time t satisfying the following equation, $R_j^c \geq \gamma$.

$A \cap Q \neq \emptyset$ that means all nodes in some paths has remaining battery capacity higher than given threshold, and then it is needed to find a route by MTTPR scheme. Otherwise the route selection is done by MMBCR algorithm,

CMMBCR algorithm always depends upon the value of threshold γ chosen. If the value of the threshold is 0 it is identical to MTPR, because it will select shortest route. If the value of threshold is 100, then the condition $R_j^c \geq \gamma$ will never satisfy. So it is identical to MMBCR.

2.5   Power-Aware Multiple Access Protocol with Signaling (PAMAS)

Suresh Singh, Mike Woo and C. S. Raghavendra [3, 5] proposed a new algorithm for power saving in Mobile Ad-Hoc network using passive energy saving techniques. They presented five different metrics based on battery power that can reduce the routing cost from 5 to 30% over short-hop routing. PAMAS is a MAC-level protocol that is used to control the power off mechanism of that mobile node. The main idea behind PAMAS is that as maximum energy is wasted in overhearing, so PAMAS can save 40 to 70 % of battery power by turning off the node's radios, when they are not transmitting or receiving.

To save the power of a node the node can power off if any of the following conditions satisfies.

1. A node can power itself off if it is overhearing a transmission and does not have any packet to send.
2. A node can power off if at least one of its neighbor nodes is transmitting and at least one neighbor node is receiving a transmission.
3. The node can power off if all neighbors of a node are transmitting and the node is not a receiver.

In PAMAS the nodes attempt to capture the communication channel by exchanging RTS/CTS packets. This packet contains all the details about the transmission like duration, distance etc. A node can know this timing before going to the sleep mode. The RTS/CTS packet transmission is done on different signaling channel so it doesn't interfere with the ongoing transmission. The problem comes when a node is needed to transmit a data and it is in sleeping mode, so it cannot know about the time duration until it wakes up. To solve this problem, another protocol runs in the signaling channel that allows nodes to find the duration of remaining transmission.

2.6   Power Aware Ad-Hoc On-Demand Multipath Distance Vector (PAAOMDV)

Dr. A. Rajaram and J. Sugesh proposed a routing protocol named Power Aware Ad-hoc On-Demand Multipath Distance Vector (PAAOMDV) [6] that gives an optimum path between the power saving path and shortest path. The main Idea behind PAAOMDV is each node maintains an *Energy Reservation Table* (ERT) where each item is mapped with a route with all its details like destination id, amount of energy reserved, last operation time, route, and their functions. PAAOMDV applies a threshold value based on the remaining battery level at the time of route discovery to remove the very low remaining battery node. The operations of PAAOMDV include three steps.

The route discovery of PAAOMDV is based on the energy, E, that is defined as the minimum residual energy over its entire link.

$$E \equiv \min_{1 \leq h \leq H} \text{resudial energy}_h$$

Where h is the link number and H is the number of links or hops in the path. For calculating the residual energy all the information needed is available via the RREQ. The minimum residual energy in the entire path between two nodes i and d is used for the cost estimation. That is $E_{min}^{i,d} \equiv \min_{c \in path\_list_i^d} E_c$

All the paths between nodes n_i and n_d is denoted by $path\_list_i^d$. The route discovery of PAAOMDV is based on the RREQ or RREP packet. In case of a node n_d, if the neighbor node has a higher destination sequence number or shorter hop-count than the existing route for n_d, it updates the packet. Now if the destination sequence number and hop-count is equal with the existing route but with a greater E_i, E_{min}, the list of paths in node i and d's routing table is updated. So the path selection is dependent on both E_i, E_{min} as well as the destination sequence number and hop count. After Route establishment the data are sent from source to destination via that selected path. After each transmission the table is updated by subtracting the used energy for every used node. When an error occurs, the detected node sends a route error packet (RERR) to every node and they remove the corresponding item from their routing table and select an alternative path.

2.7   Energy Aware Ad-Hoc Routing (EAAR)

Another proposed energy aware routing algorithm is EAAR [7] that is based on naturally occurring ant's foraging behavior. This will not only optimize the effect of power consumption, but will also exploit the multi-path transmission properties of ant swarms. When a source node has some data to send, it broadcasts a ant (control packet) say F_s^d. Each intermediate nodes receive the replica of F_s^d that will be $F_s^d k$. Next will be $F_s^d KL$ (where k, L, ... are integers). All the journey information of the packet is stored in an array called j. Now when a node receives several ant of same generation it compares those ants. If the new arrived ant is a subset of previous one, then it is dropped. If the older received ant is a subset of younger received ant, then it is accepted. Else if, the new received ant is not any superset of any previous received ant then it is only accepted if the condition satisfies. $N \leftarrow \lambda * M$.

Where N is the hop count of previously received ant and M is the hop count of newly received ant. λ is a factor of taking decisions. After receiving the ant in a specific time, the ant travel backward to the source. At the time of moving backward each ant updates an entry in neighbors table $T_{n,d}^i$, where $T_{n,d}^i = \dfrac{MBR}{H}$

That is the inverse of the number of hops (H) multiplied by a minimum residual battery energy traveled by current backward node.

After completing path discovery the data session starts. When the first proactive backward ant is received by the host, it again sends another forward proactive ant to the first one. This always leads to a better path selection.

When a link failure occurs the node sends a control packet to all the neighbor nodes to remove the data from their entry.

3   Comparative Study of Power Aware Routing Protocols

These protocols have different goals, assumption and different mechanisms to achieve the goal. The basic protocol MTTPR always gives minimum total transmission power but it cannot satisfy the second goal of increasing network lifetime. The battery aware

	Advantages	Disadvantages
Minimum Total Transmission Power Routing (MTTPR)	It guarantees the minimum total transmission power of the network. This takes the least number of hops.	It does not deal with lifetime of each battery. The selected node on a specific route is exhausted quickly.
Minimum Battery Cost Routing (MBCR)	The routing is based on the remaining battery capacity of individual nodes. MBCR tries to minimize the total transmission power.	It does not assure that the selected route will be the shortest path. If a very low remaining battery lies on that specific route it will be exhausted quickly.
Min-Max Battery Cost Routing (MMBCR):	The battery of each node will be used in more efficient manner. Extends network lifetime by avoiding overuse of the most critical nodes.	There is no guarantee that minimum total energy path will be selected. Lifetime of all nodes decreases as it takes more power to deliver a packet.
Conditional Max-Min Battery Capacity Routing (CMMBCR)	CMMBCR can fulfill both goals of maximizing network and battery lifetime of each node. The threshold helps to avoid the critical node.	It takes more power to deliver a packet. Route selections take more time.
Power-Aware Multiple Access Protocol with Signaling (PAMAS)	Battery consumption minimizes by turning off the radio when nodes are not transmitting or receiving. This protocol saves 40-70% of battery life.	When a node is in sleep mode, it cannot hear about a new ongoing transmission. There is no proper wake up algorithm.

Power Aware Ad-hoc On-Demand Multipath Distance Vector (PAAOMDV):	Energy reservation table (ERT) is used instead of route cache. It not only deals with energy required in future.	Update in the table requires extra overhead. The average energy per packet is increased.
Energy Aware Ad-hoc Routing (EAAR)	EAAR is better compared to AODV and MMBCR in terms of energy conservation. This algorithm exploits the multi-path transmission properties of ant swarms.	The average energy per packet in high mobile conditions is not so good. In EAAR the packet delivery rate is high because time to judge the best route is high.

routing algorithms like MBCR, MMBCR, and CMMBCR take care of battery life of every individual node, but in some case the total transmission cost increases. The protocol PAAOMDV and CMMBCR use a special threshold function to filter the critical node for avoiding the network partition. A good approach is proposed by PAMAS, which use a passive power saving algorithm. Maximum energy is reduced in overhearing of others node's transmission, so idle node can put themselves off for a limited period of time. This protocol is very useful where the nodes are idle most of the time. PAAOMDV and EAAR use special packets like RERP, Ant for route discovery. The route is calculated based on different table and the cost is calculated by the Route discovery packets. PAAOMDV also deals with the energy required in the transmissions that give the cost function more accuracy.

4   Scopes for Improvement and Future Work

After surveying different power aware routing protocols it is seen that still there are future scopes in many directions that can solve the power aware problem more accurately.

The threshold can be used to remove critical nodes at the time of route discovery. Using dual threshold can help in routing when all nodes are below the first threshold. Secondly, the route discovery packet is transmitted to all the nodes in flooding manner that increases network traffic, cost, delay, and congregation. So, the routing algorithms can be designed in such a way that the routing decisions are taken locally with a limited number of nodes. Thirdly, link error is another problem in Mobile Ad-Hoc network. If any link error occurs when the data is transmitting; it needs to retransmit the entire data. So the link error should be taken into consideration to build the protocol with proper error control mechanism. Fourth, the transmission cost not only depends on the distance between two nodes but also on the load of the selected path. If the selected path is too congested, then the transmission cost increases. Hence, an alternate approach can be incorporated such that the route through lightly-loaded nodes is selected. This will result in less number of collisions thereby minimizing the energy required for the transmission. Fifthly, the optimal route from the battery life point of view will be such a route that ensures the even dissipation of energy by all the nodes in order to increase the overall network lifetime. Therefore, to build such a protocol for power aware routing, the formulation of the standard deviation can be taken into consideration by taking different node's battery cost as population set.

5   Conclusion

In this article we have made a survey on several power-aware routing protocols. As the nodes in Mobile Ad-Hoc network are free to move, so, a proper algorithm is needed for route selection. There should be a balance between the two goals of maintaining the minimum transmission power and increasing the network lifetime. We have carried out a detailed comparative analysis based on the advantages and disadvantages of those protocols. As a result we can conclude that the different protocols are suitable for different situations. MTTPR is useful when every node has enough battery power remaining, as it gives the shortest transmission cost. Among the basic battery aware protocols, CMMBCR and PAAOMDV give the best output as they always avoid the critical node. In case of multipath and congested scenario, EAAR give the better result because it has multicasting and congestion control algorithm. PAMAS can be used in that situation where maximum energy is wasted in overhearing.

As the mobile hosts have limited battery life so battery power should be used more efficiently to maximize the network life. More focus should be given to error control, power saving by switching of the node and congestion control. Although several researches have been done on power aware routing in these days, but this field is still in its infancy. More research on this work can lead to an appropriate route selection mechanism to extend the lifetime of both the nodes and the network.

References

1. Singh, A., Tiwari, H., Vajpayee, A., Prakash, S.: A Survey of Energy Efficient Routing Protocols for Mobile Ad-Hoc Networks. International Journal on Computer Science and Engineering 2(09), 3111–3119 (2010)
2. Goldsmith, A.J., Wicker, S.B.: Design Challenges for Energy-Constrained Ad Hoc Wireless Networks. IEEE Wireless Commun. 9(4), 8–27 (2002)
3. Singh, S., Woo, M., Raghavendra, C.S.: Power Aware Routing in Mobile Ad Hoc Networks. In: Proc. 4th Annual Int'l. Conf. Mobile Comp. and Net., pp. 181–190 (October 1998)
4. Toh, C.K.: Maximum Battery Life Routing to Support Ubiquitous Mobile Computing in Wireless Ad Hoc Networks. IEEE Commun. Mag. 39(6), 138–147 (2001)
5. Singh, S., Raghavendra, C.S.: PAMAS-Power Aware Multi-Access protocol with Signaling for Ad Hoc Networks. ACM Commun. Rev. 28(3), 5–26 (1998)
6. Rajaram, A., Sugesh, J.: Power Aware Routing for MANET Using On-demand Multipath Routing Protocol. IJCSI International Journal of Computer Science Issues 8(4(2)), 517–522 (2011)
7. Dhurandher, S.K., Misra, S., Obaidat, M.S., Gupta, P., Verma, K., Narula, P.: An Energy-Aware Routing Protocol for Ad-Hoc Networks Based on the Foraging Behavior in Ant Swarms. In: Proc. of the IEEE International Conference on Communications (ICC 2009), Dresden, Germany, June 14-18, pp. 1–5 (2009)

A New Trust Based Secure Routing Scheme in MANET

Mousumi Sardar and Koushik Majumder

Department of Computer Science & Engineering,
West Bengal University of Technology,
Kolkata, India
koushik@ieee.org

Abstract. A mobile ad hoc network (MANET) is a self-configuring network in which several wireless nodes temporary set up links between them. This dynamic nature of MANET makes it very challenging network as well as more vulnerable to several attacks. Thus establishing routing protocols to employ a secure route is difficult too. Trust mechanisms are incorporated in routing protocols to find secure route in an efficient way. In this paper, a new trust based routing approach is presented where packet forwarding ratio is taken into account to calculate direct trust and recommendation from neighbors to calculate indirect trust. The new approach relays the packets through the more trusted route as it considers both the direct and indirect trust.

Keywords: Trust, Dynamic Topology, Cryptographic mechanism.

1   Introduction

Mobile Ad-Hoc network (MANET) has no pre-defined infrastructure like the wired network. It is basically self-configuring network in which temporary paths are set up between the nodes during the packet relay. The nodes behave as host as well as router. The main features and characteristics of MANET [1] are: *Dynamic topology, Cooperation, Resource Constraints*.

Dynamic nature of MANET networks make it more vulnerable to attacks than wired networks. So to achieve secure communication between nodes, security becomes an important issue in MANET. Malicious nodes in the network perform different misbehavior like selective forwarding, packet dropping etc. This misbehavior causes performance degradation of the network. Hence, secure routing protocols are needed to design which is a more difficult and challenging too.

As Cryptographic mechanisms have high computational cost it can't be deployed in real MANET network. This mechanism can't identify the attacker nodes. It can only prevent the routing information from tampering. So a new approach is adopted in routing protocols to not only the data transmission but also the nodes. During routing, nodes and routes are selected on the basis of trust parameter which depends on the nodes behavior means how they perform. Trust on nodes may be determined by the direct or indirect communication with the nodes.

S.C. Satapathy, S.K. Udgata, and B.N. Biswal (eds.), *FICTA 2013*,
Advances in Intelligent Systems and Computing 247,
DOI: 10.1007/978-3-319-02931-3_36, © Springer International Publishing Switzerland 2014

In this paper, the remaining parts are organized as follows. We have discussed about trust in section 2 and related works in section 3. In section 4 our proposed work is elaborated and conclusion is done in section 5.

2   Trust Mechanism

Trust mechanism is introduced in the protocols to provide security in MANET. Trust is a value that is calculated on the basis of nodes action when needed. Trust is introduced to prevent from various attacks like wormhole, black-hole, Dos, selfish attack etc. Trust can be implemented in various ways such as by reputation, subjective logic, from opinion of nodes etc as there are no particular definitions of trust. According to (Marc Branchaud, Scott Flinn) trust has following properties:

- *Context Dependence*: In some specific context trust relationships are applicable.
- *Function of uncertainty*: Trust depends on the uncertainty of nodes action. It gives the probability of action performed by a node.
- *Quantitative value:* Trust can be assigned any type of numeric values discrete or continuous.
- *Asymmetric Relationship:* Trust relationship is asymmetric in nature. If node A trusts B and node B trust C that does not mean that A trusts C.

3   Related Work

In Trusted AODV [2] protocol node trust is calculated from the neighbor's opinion and route trust is calculated by generating R_ACK packets by the destination node in each entry in the routing table at some regular interval. Two additional control packets are used to calculate node trust: trust request packet (TREQ) and trust reply packet (TREP). This protocol gives better performance than AODV by eliminating malicious node but overhead increases due to use of additional control packets and indirect trust is only considered here.

The main idea of CONFIDANT [3] protocol is to identify non-cooperative nodes. A node selects a route based on trust relationships which are built up from experienced routing and packet forwarding behavior of other nodes. Each node monitors the behavior of all neighbor nodes. When any misbehaving node is found, alarm messages are sent to all other nodes in the network. As a result, all nodes in the network will be able to avoid that misbehaving node while selecting a route. But in this protocol the problem is that an attacker is able to send false alarm messages and can do false claim that a node is misbehaving.

Friendship Based AODV (FrAODV) [4] protocol is proposed based on AODV. There are two evaluation algorithms to evaluate forward and reverse path between source and destination: *RvEvaluate Algorithm, FwEvaluate Algorithm*. In this scheme, it is assumed that each node has identity can't be forged by any other malicious node and no of malicious node is less than the no of good nodes. In this proposed scheme every node has a list of friends with friendship values. This protocol gives better performance in terms of packet delivery ratio and routing overhead. But the end to end delay is not included in performance measurement metric. The delay is more here because two evaluation algorithms are used to establish path.

In Secure Routing Using Trust (SRT) [5] a secure routing using trust level is proposed. This scheme is based on node transition probability (NTP) and AODV. A trust rate is calculated as a parameter. When a node has data packet to send, it first floods control frame (beacon) in search of secure and reliable route. After broadcasting the first beacon trust rate is evaluated. This trust rate value divides the nodes of the network into 3 categories: ally list, associate list, acquaintance list. Where ally list implies level2, associate list implies level1 and acquaintance list implies level 0. If any node in the same level is not found trust is compromised by choosing a neighbor in the next lower level. The performance decreases in the presence of attacks except black hole. The trust is calculated on the basis of control packets only.

Friend Based Ad Hoc Routing Using Challenges To Establish Security (FACES) [6] achieves security in ad hoc network by sending challenges and sharing friend lists. In this scheme, the nodes are rated according to successful transmission of data and friendship with the other nodes. Depending on the rate of nodes, they are placed in the three different lists: *Question mark List, Unauthenticated List,* and *Friend List.* This algorithm has four steps: challenging neighbor, friends rating, sharing friends and route through friends. FACES is a hybrid protocol as the routing of data is on demand where as challenging and sharing occurs periodically. In this protocol control overhead is increased due to periodic flooding of challenge packet and periodic sharing of friend list.

In Trusted Based Security Protocol Routing [7] trust is calculated using a trust counter and success ratio. This protocol maintains confidentiality and authenticates the nodes based on digital signature. It detects the nodes which are misbehaving. This protocol can't detect authenticated malicious node. In this scheme, it is not explained that who increase the trust counter.

4   Proposed Work

In our scheme, a new trust based approach is proposed to find reliable, secure route for forwarding packets. In this scheme, route is discovered through RREQ and RREP message passing through the nodes in the network, like AODV. Some modification is done in AODV to make it more secure and reliable. In this new protocol, a simple modification is done in RREQ and RREP message by adding one additional field: *verification field* and the messages denote as mRREQ and mRREP. Every node maintains a trust-level table (Table1) to identify trusted nodes and a trust table corresponding to the node id. Every node monitors its neighbor node in this scheme. In this scheme, it is assumed that when any new node enters into the network it generates a key pair: public key and private key and then broadcasts its public key over the network. The new protocol has two phases which works as follows:

4.1   Route Discovery

When a node (S) wants to send packets, it first checks its routing table whether any route to the desired destination node (D) is available or not. If any route to the destination is found, it forwards the packet through that route. If the route is not found then S initiates route discovery by broadcasting mRREQ messages to its neighbor nodes. The verification field of RREQ message contains a secret value that source node generates, which is in turn encrypted by the destination nodes public key . After getting mRREQ message the neighbor nodes checks the message whether it is destined for it or not, if it is a destination node itself or it has a route to the destination, it generates mRREP message and sends back to the sender. If it has no route to the destination node then the node sets a reverse connection to the node from which it gets mRREQ message and rebroadcasts the message. Finally, when mRREQ message reaches to the destination, it first checks the verification field in mRREQ message using the private key of its own. If it decrypt successfully, it sends back mRREP message by again encrypting the same value that the destination gets after decryption. The destination node encrypts that value with the public key of sender which is already available to it. The mRREP message is sent back to the sender in the reverse path and forward path is set when mRREP message travels along the path. The destination node replies to all the mRREP messages it gets. When the source node gets mRREP message, the node first decrypts the value of verification field in mRREP message using its own private key. Then it verifies the value by comparing it with the value it sent to the destination. If both the values are the same then that mRREP is valid otherwise mRREP is rejected. There are two following cases during the route selection process:

Case1: when network is initialized the nodes don't have any trust value because the nodes are unfamiliar to each other. No node has any information about its neighbor nodes reliability. So for the initial case then it selects the node from which it gets the valid mRREP.

Case2: If the nodes in the network already have trust values, then the sender node selects the route that is verified successfully and having the highest trust value of the node from which it gets mRREP message.

4.2   Packet Forwarding and Trust Calculation

After the route is selected, the source node starts sending data packets through that route. After sending the packets, every time the sender node overhears the network. If the neighbor node of that route forwards the packet correctly, then the trust value of that node increases. If the node simply drops the packet then the trust value of that node decreases. And if the node doesn't send the data and crosses the predefined threshold of not forwarding data or generates packets abnormally for any malicious intention, the trust is decremented by setting trust value to 2. Then the source node selects the next trusted node. The trust value of a node is calculated on the basis of direct interaction of a node means direct trust and from the opinion of its neighbor nodes i.e. indirect trust. The nodes periodically exchange their opinions about the other neighbor nodes. Suppose a node let i send the packet to a node j then the direct trust (TD) on a node (j) is calculated as:

$$TD = \frac{\sum TPi - (\sum SPj - \sum DPj)}{\sum TPi}$$

where $\sum TP_i$ is the Total no. of packets sent by the node I, $\sum SP_j$ is the total no. of successful packet sent by the node j and $\sum DP_j$ is the total no. of dropped packets by the node j. For the indirect trust calculation, when the node i gets the opinion (i.e. the trust value) about the node j it calculates indirect trust (TID) as:

$$TID = \frac{\sum \text{trust values from neighbors about } j}{\text{no.of nodes which gives their opinion}}$$

Then the total trust ($Total_{trust}$) on a node j is evaluated as:

$$Total_{trust} = \frac{(TD + w.TID)}{2}$$

Where,

$$w = \begin{cases} 1, & Nunt = 0 \\ \dfrac{Nt}{Nunt}, & otherwise \end{cases}$$

Where *Nunt* is the no. of untrusted nodes and *Nt* is the no. of trusted nodes.

Table 1. Trust level

Trust_value	Meaning
0	Full Trust
1	Medium Trust
2	No Trust

5   Proposed Algorithm

Algorithm 1: Key Broadcasting

Step 1. Every node generates a key pair: public key and private key when the network is initialized or a node enters in a new network

Step 2. Broadcasts the public key over the network and keeps the private key secret

Step 3. End

Algorithm 2: Route discovery

Step 1. A node initiates route discovery when it has data packets to send to the another node

    1.1. Source node generates a secret value

    1.2. Encrypt the secret value by the public key of the destination node

       1.3. Add the encrypted value in the verification_field of mRREQ packet

       1.4. Broadcasts the mRREQ packet over the network

Step 2. After receiving mRREQ packet, every node

       2.1. Checks its routing table

       2.1.1. If the receiving node is destination node itself or it has route to the destination

          2.1.1.1. The node sends back mRREP packet to the source

       Else

          2.1.1.2. Rebroadcasts the packet

       End if

Step 3. The destination node receives the mRREQ packet

       3.1. Encrypt the secret value using the public key of the sender

       3.2. Add the encrypted value in the *verification_field*

       3.3. Gives replies to all mRREQ packets by sending the mRREP packets

Step 4. After receiving all mRREP packets, source node

       4.1. Decrypt the *verification_field* with its own private key

       4.2. Compares the decrypted value with the secret value which it sends to the destination

       4.2.1. If both the value matches

          4.2.1.1. Packet is valid and keep the mRREP packet

Else

          4.2.1.2. Drops the mRREP packet

            End if

       4.3. Case1: Initially source node selects that neighbor node from which it gets valid mRREP packet as no trust corresponding to the node is available

       4.4. Case2: Selects that neighbor node from which it gets valid mRREP packet with highest trust, as the trust is available

Step 5. End

Algorithm 2: Packet forwarding and Trust calculation:

Step 1. The source node sends the data packets through the selected node

Step 2. The sender node monitors the neighbor node after forwarding the packet

       2.1. If the node forwards or drops the data packets correctly

       2.2. Calculate the total trust on that node

       2.2.1. The sender node requests the neighbor node to give their opinion about that node

       2.2.2. After getting opinions from the neighbor nodes, indirect trust is calculated in the following way:

$$TID= \frac{\Sigma \text{trust values from neighbors about j}}{\text{no.of trusted nodes which gives their opinion}} \tag{1}$$

2.2.3. Calculate the weightage as:

$$w = \begin{cases} 1, & Nunt = 0 \\ \dfrac{Nt}{Nunt}, & otherwise \end{cases}$$

Where *Nunt* is the no. of untrusted nodes and *Nt* is the no. of trusted nodes.

2.2.4. The sender node calculates the direct trust as:

$$TD = \frac{\sum TPi - (\sum SPj - \sum DPj)}{\sum TPi} \tag{2}$$

Where, $\sum TP_i$ is the Total no. of packets sent by the node i, $\sum SP_j$ is the total no. of successful packet sent by the node j and $\sum DP_j$ is the total no. of dropped packets by the node j

2.2.5. Total trust on the node is calculated as:

$$Total_{trust} = \frac{(TD + w.TID)}{2} \tag{3}$$

Else
2.3. The node doesn't forward the data packets more than a predefined threshold value
2.3.1. Set the trust value of that node to 2
2.3.2. Selects the another trusted node to forward the packets
End if
2.4. Update the trust table
Step 3. End

6 Conclusion and Future Work

In this paper a new trust based scheme is proposed. Trust values are evaluated on each node based on packet forwarding ratio and opinions from others. If any untrusted node is found, then the route through that node is rejected. This scheme uses average function to calculate trust on a node which is very easy to calculate and has less computational time. As the packet forwarding ratio is taken into consideration so that the throughput is increased. Black hole attack is also prevented as the node selection depends upon the trust value of a node which in turn depends upon the packets forwarding by that node. This new scheme can also be implemented on other routing protocols and some techniques can be used for node authentication.

References

1. Aggelou, G.: Mobile Ad Hoc Networks. McGraw-Hill (2004)
2. Pushpa, A.M.: Trust based secure routing in AODV routing protocol. In: Proceedings of 2009 International Conference on Internet Multimedia Services Architecture and Applications (IMSAA), pp. 1–6. IEEE Press, USA (2009)

3. Buchegger, S., Boudec, J.L.: Performance Analysis of CONFIDANT Protocol. In: MOBIHOC 2002, June 9-11. EPFL Lausanne, Switzerland (2002)
4. Essia, T., Razak, A., Khokhar, R.S., Samian, N.: Trust-Based Routing Mechanism in MANET: Design and Implementation. Springer (June 18, 2011)
5. Edua Elizabeth, N., Radha, S., Priyadarshini, S., Jayasree, S., Naga Swathi, K.: SRT- Secure Routing using Trust Levels in MANETs. European Journal of Scientific Research 75(3), 409–422 (2012) ISSN 1450-216X
6. Dhurandher, S.K., Obidant, M.S., Verma, K., Gupta, P., Dhuradar, P.: FACES: Friendhip-Based Ad Hoc Routing Using Challenges to Establish Security in MANETs Systems. IEEE System Journal 5(2) (June 2011)
7. Sharma, S., Mishra, R., Kaur, I.: New trust based security approach for ad-hoc networks. IEEE (2010)
8. Bhalaji, N., Mukherjee, D., Banerjee, N., Shanmugam, A.: Direct Trust Estimated On Demand Protocol For Secured Routing In mobile Ad-Hoc Networks. International Journal of Computer Science & Security 1(5)
9. Sen, J.: A Distributed Trust Management Framework For Detecting Malicious Packet dropping Nodes In a Mobile Ad Hoc Network. International Journal of Network Security & Its Applications (IJNSA) 2(4) (October 2010)
10. Nagrath, P., Kumar, A., Bhardwaj, S.: Authenticated Routing Protocol Based On Reputation System For Ad-Hoc Network. International Journal on Computer Science and Engineering 2(9), 3095–3099 (2010)

IWDRA: An Intelligent Water Drop Based QoS-Aware Routing Algorithm for MANETs

Debajit Sensarma and Koushik Majumder

Department of Computer Science & Engineering, West Bengal University of Technology,
Kolkata, India
debajit.sensarma2008@gmail.com,
koushik@ieee.org

Abstract. Mobile ad-hoc network operates with no pre-setup infrastructure and is one of the most active research areas. In this kind of network mobile nodes are completely independent, self managed and they are highly dynamic in nature. Therefore, traditional routing cannot work properly in this environment. Besides this, Quality of Service (QoS) of the network is very important for real time and multimedia applications for providing better throughput. But providing QoS in routing is a challenging task. Thus in this paper we introduce a novel QoS aware multipath routing algorithm IWDRA, which is based on Intelligent Water Drop (IWD) algorithm and here packets follow the basic IWD properties among neighbor nodes. It provides better QoS of network which will increase network lifetime, network stability, packet delivery rate and it is also a highly adaptive routing which will support dynamic topology like MANET.

Keywords: MANET, Intelligent Water Drop (IWD), QoS Routing.

1 Introduction

MANET [1] is a collection of mobile nodes which communicate with each other by multi-hop links and the transmission medium is radio wave. In this network any mobile node communicates directly with the other mobile nodes which are in the transmission range using the radio wave and uses intermediate nodes to communicate with other nodes which are out of its range. Routing in MANET is a challenging task because here topology changes dynamically due to nodal mobility. Thus efficient routing protocols are needed for transmitting data from one node to another node. There are mainly three types of routing protocols. In proactive routing, a continuous traffic monitoring is being done to know the topological information among the network nodes. In reactive routing routes are established only when needed and hybrid routing which is a combination of the above two types of routing. Now a days, the demand of real time and quality of service (QoS) requirements increases drastically. QoS in the network aims to find routes that can provide the required quality needed for the applications. QoS routing is a technique that takes into account the appropriate link information and based on that statistics it selects the path that satisfies the QoS requirements. QoS routing is key part of a QoS mechanism as its main function is to

S.C. Satapathy, S.K. Udgata, and B.N. Biswal (eds.), *FICTA 2013*,
Advances in Intelligent Systems and Computing 247,
DOI: 10.1007/978-3-319-02931-3_37, © Springer International Publishing Switzerland 2014

find nodes that satisfies the application's requirements. Providing QoS in routing in MANET is a critical task because of its dynamic nature. Various routing protocols have been designed to tackle with the problem of QoS aware routing. In this paper, a new QoS aware routing protocol has been proposed which is based on the dynamic of river systems, actions and reactions which occurs among the water drops in river. The natural water drops are used to develop IWD and a better solution of the problem is reached with the cooperation of each IWDs. This algorithm is mainly a population-based constructive optimization algorithm. In our routing protocol each packet has the IWD properties and it takes the QoS metrics energy, buffer space to increase the network stability and packet delivery rate respectively. Besides this, delay, bandwidth is also considered to reduce the end to end delay and utilize the link capacity properly. The main feature of IWD is its "velocity" and this property of IWD makes this proposed algorithm better in comparison with other existing algorithms. The link which has better quality, the IWD packet of travelling through that link will gain more velocity than the other IWDs and will reach its goal faster than other IWDs. This helps in quick convergence to the better solution. Thus this protocol can find better route in less amount of time. The detailed description of IWD can be found in [2]. Also, in our routing an efficient route failure management technique is used to increase network throughput and lifetime of network which is very important for real time and multimedia applications.

The paper is organized as follows: In section 2 mathematical models is described. Section 3 illustrates the proposed routing algorithm and section 4 explains the performance analysis. Concluding remarks and future works are given in Section 5.

2   Mathematical Model

For mathematical analysis MANET is represented by a connected undirected graph. Let G (V, E) represents the mobile ad hoc network. Here V denotes the set of network nodes and E denotes the set of bidirectional links. QoS metrics with respect to each link e ∈E is delay (e) and bandwidth (e). With respected to node n ∈V , it is delay (n), energy (n) and buffer space (n) (i.e. the minimum remaining buffer space of node 'n'). Another QoS metric considered here is hop count. It is important because multiple hops are used for data transmission in MANET. So, it is necessary to find paths with minimum hops. The main motivation of this proposed algorithm is to find path from source to destination which will satisfy the QoS requirements such as delay, bandwidth, energy, buffer space and hop count.

Let, path (i, j) or R is entire path from node i to j where QoS constraints have to be satisfied.

From an arbitrary node i to an arbitrary node j, delay, bandwidth, energy, buffer space and hop count is calculated as-

$$\text{delay (path (i, j)) or D (R)} = \sum_{e \in P(i,j)} \text{delay}(e) + \sum_{n \in P(i,j)} \text{delay}(n)$$

where, delay (path (i,j)) is the transmission and propagation delay of the path(i,j) and delay (n) is the processing and queuing delay of node 'n' on path(i, j).

bandwidth(path(i,j)) or B(R)= $\min_{e \in P(i,j)}$ {bandwidth(e)}

where, bandwidth (e) is the available bandwidth of that link on path(i, j).

energy (path (i, j)) or E (R) = $\min_{n \in P(i,j)}$ { energy (n)}

where, energy (n) is the residual energy of node 'n' on path(i, j).

buffer size (path (i, j)) or BUFF (R) = $\min_{n \in P(i,j)}$ {buffer size (n)}

where, buffer size (n) is the available buffer space of node 'n' on path(i, j).

hop count (path (i, j)) or HC (R) = Number of nodes in the path.

3   Proposed Algorithm

This proposed routing algorithm is a QoS aware multipath adaptive routing algorithm. It has three phases namely Route finding phase, Route maintenance phase and Route failure management phase.

3.1   Route Finding Phase

Algorithm 1: Route Finding.

BEGIN

Step 1: Suppose Source S wants to send data to destination D with QoS constraints delay, bandwidth, buffer space, energy and it has no valid path exits in the cache.

Step 2: Source node creates an IWD_Req packet and sends the packet to its neighbor node by updating the soil and velocity of IWD_Req packet and soil of link between them. The soil and velocity is updated according to the given formula:

    i)   For each IWD_Req moving from node i to j the velocity vel $^{\text{IWD\_Req}}$ is updated by:

$$\text{(a)} \quad \text{vel}^{\text{IWD\_Req}}(t+1) = \text{vel}^{\text{IWD\_Req}}(t) + \frac{a_v}{b_v + c_v \text{soil}^2(i,j)} \quad (1)$$

        Where vel $^{\text{IWD\_Req}}(t+1)$ is the updated velocity of IWD_Req packet.

    ii) For each IWD_Req moving on path from node i to j, compute the ΔSoil (i, j) that the IWD_Req packet loads from the path by:

$$\text{(b)} \quad \Delta\text{soil}(i,j) = \frac{a_s}{b_s + c_s.\text{time}^2(i,j;\text{vel}^{\text{IWD\_Req}}(i,j))} \quad (2)$$

Such that,

$$\text{(c)} \quad \text{time}(i,j;\text{vel}^{\text{IWD\_Req}}(t+1)) = \frac{\text{HUD}(j)}{\text{vel}^{\text{IWD\_Req}}(t+1)} \quad (3)$$

Here,

$$\text{(d)} \quad \text{HUD}(j) = \frac{D(i,j)^{\lambda_D}}{B(i,j)^{\lambda_B} + E(i,j)^{\lambda_E} + \text{BUFF}(i,j)^{\lambda_{\text{BUFF}}}} \quad (4)$$

Here λ_B, λ_E, λ_D and λ_{BUFF} are the weight factors which indicate the relative significance of the QoS parameters during soil and velocity update on path (i, j).

iii) Update the Soil (i, j) of the path from node i to j traversed by that IWD_Req and also update the soil that the IWD_Req carries Soil^{IWD_Req} by:

(e) $\text{Soil}(i, j) = (1 - \rho_n) \cdot \text{Soil}(i, j) - \rho_n \cdot \Delta\text{Soil}(i, j)$

(f) $\text{Soil}^{IWD\_Req} = \text{Soil}^{IWD\_Req} + \Delta\text{Soil}(i, j)$

$$(5)$$

Step 3: When the IWD_Req packet reaches the destination, it will convert to the IWD_Rep packet and unicasted towards the source following the same path as in IWD_Req packet. In the returning phase IWD_Rep packet will collect the minimum buffer space and minimum energy of nodes and minimum bandwidth of the link in the path. Each node calculates the path preference probability by following formula:

(a)

$$P_{ijd} = \frac{[\text{Soil}(i,j)]^{\alpha 1} \cdot [D_{ijd}]^{\alpha 2} \cdot [\eta_{ijd}]^{\alpha 3} \cdot [B_{ijd}]^{\alpha 4} \cdot [E_{ijd}]^{\alpha 5} [\text{BUFF}_{ijd}]^{\alpha 6}}{\sum_{k \in N_i} [\text{Soil}(i,j)]^{\alpha 1} \cdot [D_{ikd}]^{\alpha 2} \cdot [\eta_{ikd}]^{\alpha 3} \cdot [B_{ikd}]^{\alpha 4} \cdot [E_{ikd}]^{\alpha 5} [\text{BUFF}_{ijd}]^{\alpha 6}} \quad (6)$$

Here,

(b) $\text{Soil}(i,j) = \dfrac{1}{\text{Soil}(\text{path}(i, j))}$

(c) $D_{ijd} = \dfrac{1}{\text{delay}(\text{path}(i, d))}$

(d) $\eta_{ijd} = \dfrac{1}{\text{hopcount}(\text{path}(i, d))}$

(e) $B_{ijd} = $ bandwidth $(\text{path}(i,d))$

(f) $E_{ijd} = $ energy $(\text{path } (i,d))$

(g) $\text{BUFF}_{ijd} = $ buffer size $(\text{path}(i,d))$.

(h) $\alpha 1$, $\alpha 2$, $\alpha 3$, $\alpha 4$, $\alpha 5$ and $\alpha 6$, are the tunable parameters which control the relative weights of link soil, delay, hop count, bandwidth, energy and buffer space respectively.

The node stores the path in the cache if it satisfies minimum Path Preference Probability and forwards the IWD_Rep packet otherwise it discards the IWD_Rep packet.

Step 4: When IWD_Rep packet reaches the source node it will calculate the Path preference probability and the path which satisfies minimum QoS constraints will store in the cache and also the Soil (i, j) value of IWD_Rep packet is stored. The path with better Path Preference Probability is selected for routing data packets.

Step 5: Whenever data packet reaches the destination node, the node will create an IWD_Update packet and send it towards the source node for updating the link quality. The soil updation is done according to the formula (5).

END

3.2   Route Maintenance Phase

Algorithm 2: Route Maintenance.

BEGIN
Step 1: When IWD_Update packet reaches intermediate node two cases can occur:
   i)   If the path preference probability of an intermediate node decreases, the
        ρ_n value become negative in eq. (5) applied as previous;

   ii) Otherwise ρ_n value remains same as before and soil updated according
        to eq. (5).
Step 2: When the IWD_Update reaches to the source the corresponding soil of the
path and the time is updated. Here two cases may arise:
   i)   If the updated soil value remain negative or zero (e.g. updated soil= Pre-
        vious soil-New soil ≤ 0) then no action taken and routing continued with
        existing path
   ii) Otherwise a new unexpired path with next best Path Preference Proba-
        bility is selected for routing.

END

3.3   Route Failure Management Phase

Here we assume route failure occurs for two reasons. Case 1: link break between
two nodes. Case 2: Node mobility.

Algorithm 3: Route Failure Management.

BEGIN
Case 1: If route failure occurs due to link break between two nodes, there can be
three possibilities:
   Case 1.1   If an intermediate node detects a route failure and it has a valid path
   to the destination, it will first search its cache for other valid routes, if it exists
   then the path with better path preference probability will be selected for
   routing. In parallel the node will send an Error packet towards the source
   node by setting more data bit to 0, which contains the information about
   invalid route and every node in the path along with the source node will delete the invalid route from the cache and routing will continued with the new
   path. When source node receives the Error packet if it has a valid path to the
   destination, that path is used for routing otherwise a new route finding phase
   started.
   Case 1.2   If Intermediate node has no valid path to the destination an
   IWD_upadate packet does not come for that route within the time and timer
   of that route will expired. Thus source will find that there is a problem with
   that route, i.e. there might be a link failure. At that time the source node will

stop sending the data with that path and the path which is unexpired and has better path preference probability will be selected for routing. In parallel the intermediate node where the link breakage occurs will send a Error packet towards the source node by setting more data bit to 1 if it has buffered data and whenever a node in the path has a valid, unexpired path to the destination, it will forward the Error packet by setting more data bit to 0 towards the source and also it will send an Route_ACK packet by setting Route exist flag to 1 to that effected node and the effected node will send the buffered data to the desired destination through the path in the stack. All invalid routes are deleted from the node cache in the path along with the source node.

Case 1.3   If network has no valid path to the destination, network partition occurs. At that time, the node where failure occurs will generate and send Error packet towards the source by setting more data bit to 1 if it has buffered data and when it reaches the source node, if more data bit remains 1then there is no valid path exists to the desired destination. All invalid paths will be deleted and the timer of source node for each path will expire.

Case 2: If route failure occurs due to nodal mobility then the node which is moved away from the transmission range will not give reply of Hello packet send by its neighbor. If the neighbor has a valid unexpired path in its cache then that path is selected for routing. Otherwise same procedure as Case 1.2 will be followed.

END

---

4   Performance Analysis

The main goal of this algorithm is to find a stable route from source to destination which will satisfy QoS constraints and will support multimedia, real time applications.

1. This is an adaptive routing algorithm and suitable for the network where mobility is high.
2. This algorithm finds multiple paths from source to destination which will satisfy QoS constraints delay, bandwidth, energy, buffer space and hop count.
3. Here, energy and buffer space of individual nodes are considered as QoS metrics which will increase the network lifetime and packet delivery rate respectively.
4. In this routing if a node does not satisfies the QoS constraints, then IWD_Rep packet is not forwarded further and thus it reduces the number of control packets that helps to utilize the bandwidth properly. Also power utilized properly.
5. The best feature of IWD is its velocity. The convergence of optimal path with satisfying QoS constraints can be found in finite and less amount of time as the velocity of IWD of the path with better path preference probability will be high and it will reach faster to the source node by making quick convergence to the better route.
6. The problem of stale cache does not exists because all routes status in the cache are updated time to time, so not invalid and unexpired route can exists in the node cache.

7. In this routing, IWD_Update packet is used to continuous monitoring the quality of the active route an whenever a change occur it can detect quickly. If the quality is degrading it can use another path with next better path preference probability for routing which will reduce the packet loss rate.
8. When route failure occurs the IWD_Update packet will not reach to the source node in time and the timer expires, so quickly the source node can select another better path for routing without waiting for the Error packet. It increases network lifetime and packet delivery rate as well as throughput of the network.
9. In this algorithm route failure is handled with respect to nodal mobility and link break between two nodes. Thus, it increases packet delivery rate and throughput of network.
10. It is a loop free routing. Here end to end delay, packet loss rate decreases and throughput of the network increases.

5   Conclusion and Future Works

In this paper a novel paradigm for QoS aware routing in MANETs based on the properties of intelligent water drop is proposed. IWDRA is an adaptive multipath routing algorithm and it is suitable in case of high degree of mobility. This algorithm is aware of node's remaining energy as well as buffer space which are very useful for increasing network lifetime and successful delivery of data packets. As it is a multipath routing bandwidth is utilized properly by enhancing efficient load balancing capability. It also reduces end to end delay at the time of routing. At first, in this routing, due to the velocity of the IWD packets the algorithm converges towards the proper path quickly which will satisfy the required QoS constraints and it reduces the route finding time. Secondly, in route failure management phase, a mechanism is used through which network converges towards the stable sate quickly and in a finite amount of time, which reduces the packet loss, increases network stability which is very important for real time and multimedia applications. Besides this, in route maintenance phase during data transmission always a route is monitored such that any degradation of quality of the route cannot effect the routing which also increases overall throughput of network.

In our future work we will simulate this proposed routing scheme and will compare it with the other QoS aware routing algorithms.

References

1. Macker, J., Corsen, S.: IETF Mobile Ad Hoc Networks (MANET) Working Group Charter, http://www.ietf.org/html.charters/manet-charter.html
2. Shah-Hosseini, H.: Optimization with the Nature-Inspired Intelligent Water Drops Algorithm. Evolutionary Computation (2009)
3. Rao, S.R.: An Intelligent Water Drop Algorithm for Solving Economic Load Dispatch Problem. International Journal of Electrical and Electronics Engineering 5(1), 43–49 (2011)

4. Kamkar, I., Akbarzadeh-T, M.-R., Yaghoobi, M.: Intelligent Water Drops a new optimization algorithm for solving the Vehicle Routing Problem. In: 2010 IEEE International Conference on Systems Man and Cybernetics (SMC), Istanbul, pp. 4142–4146 (2010)
5. Deepalakshmi, P., Radhakrishnan, S.: Ant Colony Based QoS Routing Algorithm for Mobile Ad Hoc Networks. International Journal of Recent Trends in Engineering 1(1), 459–462 (2009)
6. Deepalakshmi, P., Radhakrishnan, S.: An ant colony-based multi objective quality of service routing for mobile ad hoc networks. EURASIP Journal on Wireless Communications and Networking (2011)
7. Kannan, S., Kalaikumaran, T., Karthik, S., Arunachalam, V.P.: Ant Colony Optimization for Routing in Mobile Ad-Hoc Networks. International Journal of Soft Computing 5(6), 223–228 (2010)
8. Wankhade, S.B., Ali, M.S.: Route Failure Management Technique for Ant Based Routing in MANET. International Journal of Scientific & Engineering Research 2(9) (2011)
9. Sensarma, D., Majumder, K.: A Comparative Analysis of the Ant Based Systems for QoS Routing in MANET. In: Thampi, S.M., Zomaya, A.Y., Strufe, T., Alcaraz Calero, J.M., Thomas, T. (eds.) SNDS 2012. CCIS, vol. 335, pp. 485–496. Springer, Heidelberg (2012)
10. Wedde, H.F., et al.: BeeAdHoc: An energy efficient routing algorithm for mobile ad hoc networks inspired by bee behavior. In: Proceedings of the 2005 Conference on Genetic and Evolutionary Computation, Washington DC, USA (2005)
11. Bitam, S.: QoSBeeManet: A new QoS multipath routing protocol for mobile ad-hoc networks. In: GLOBECOM Workshops (GC Wkshps). IEEE (2010)
12. Sharvani, G.S., Cauvery, N.K., Rangaswamy, T.M.: Adaptive routing algorithm for MANETs: Termite. International Journal of Next-Generation Networks (IJNGN) 1(1), 38–43 (2009) (print) ISSN : 0975-7023, ISSN : 0975-7252
13. Shah-Hosseini, H.: Problem solving by intelligent water drops. In: Proceedings of IEEE Congress on Evolutionary Computation. IEEE, Swissotel The Stamford (2007)
14. Duan, H., Liu, S., Wu, J.: Novel intelligent water drops optimization approach to single UCAV smooth trajectory planning. Aerospace Science and Technology 13(8), 442–449 (2009)
15. Duan, H., Liu, S., Lei, X.: Air robot path planning based on Intelligent Water Drops optimization. IEEE International Joint Conference on Neural Networks, 1397–1401 (2008)

Design and Performance Analysis of D-STATCOM for Non-linear Load Composite Compensation

Gokulananda Sahu[1,*], Kamalakanta Mahapatra[1], and Subrat Kumar Sahu[2]

[1] Dept. of Electronics and Communication Engg. NIT, Rourkela, India
{gokulanandasahu,kmaha2}@gmail.com
[2] GHITM,Puri, India
Sahu487@gmail.com

Abstract. This paper investigates the design, analysis and simulation of a Distribution-STATic COMpensator(D-STATCOM) for non-linear load Composite(harmonic and reactive power) compensation on a three phase bus network. Composite compensation is achieved by implementation of a p-q controller, which monitors the load current and injects equal amplitude and opposite phase compensation currents to neutralize load reactive power and harmonics. This ensures the source current remains fundamental. This paper simulated results in MATLAB platform and showed that a D-STATCOM is suited for use in reactive power and harmonic compensation on any bus on a power system network.

Keywords: D-STATCOM, p-q, Harmonics, Unbalance.

1 Introduction

During the last decade, there has been sudden increase in the nonlinear load(Computers, Laser printers, SMPS, Rectifier etc.), which degrades the power quality causing a number of disturbances e.g. heating of home appliances , noise etc in power systems[1], [2] due to harmonics. These nonlinear load along with reactive power loads such as fan, pump, motors etc increase the burden on the power system. These loads draw lagging power factor currents and therefore give rise to reactive power burden in the distribution system. Excessive reactive power demand increases feeder losses and reduces active power flow capability in the power system. Sometimes their unbalance can worsen the system performance like affecting the active power flow capability of lines and operation of transformers. Therefore restoring the system for better functionality becomes a matter of concern for the utilities. To compensate the harmonics and reactive power due to non-linear load, a Distribution STATic COMpensator (D-STATCOM) is used [3]. The performance of DSTATCOM largely depends on the control algorithm used for reference current current extraction. The control algorithm used conventionally, were based on active and reactive power are found unsuitable for unbalance and harmonic conditions. Significant contribution for development of control algorithm was made by Budeanu and Fryze [2]. They provide power definition in frequency and time domain. They set the pathways for development universal set of power definitions which led to the

---

[*] Corresponding author.

S.C. Satapathy, S.K. Udgata, and B.N. Biswal (eds.), *FICTA 2013*,
Advances in Intelligent Systems and Computing 247,
DOI: 10.1007/978-3-319-02931-3_38, © Springer International Publishing Switzerland 2014

development of p-q theory by Akagi et al.[4]. In this paper performance of one algorithm such as instantaneous p-q theory is investigated in **three phase three wire system** for balanced source and nonlinear balanced and unbalanced Load. The measures of the performance is the source current total harmonic distortion and power factor.

Rest of the paper is organized as follows. In section 2 the system configuration is described. In section 3 brief discussion on p-q control theory is presented. In section 4 the performance indices used for evaluation are discussed. Simulation results are described in section 5. Finally in section 6, conclusion are drawn.

2   System Configuration

Fig.1 shows the basic circuit diagram of a D-STATCOM system with non -linear load connected three phase three wire distribution system. A nonlinear load is realized by using a three phase full bridge diode rectifier. A three phase voltage source converter (VSC) working as a D-STATCOM is realized using six insulated gate bipolar transistor (IGBTs) with anti-parallel diodes. At ac side, the interfacing inductors are used to filter high frequency components of compensating currents. The first harmonic load currents of positive sequence are transformed to DC quantities. The first harmonic load currents of negative sequence and all the harmonics are transformed to non-DC quantities and undergo a frequency shift in the spectrum. The voltage regulator in the converter DC side is performed by a proportional –integral(P-I) controller. Its input is the capacitor voltage error vdcref-vdc and through the regulation of the first harmonic active current of positive sequence . It is possible to control the active power flow in the VSI and thus the capacitor voltage vdc. The dynamics of each VSC are modeled by solving differential equations governing two modes of the inverter. The switching of the inverter is done by monitoring the reference and actual currents and comparison of error with the hysteresis band of hysteresis current controller(HCC)[6].

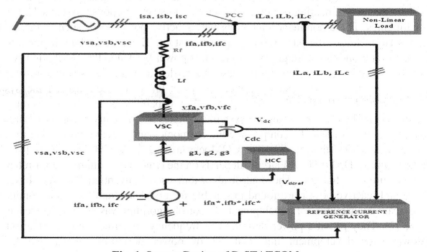

Fig. 1. System Design of D-STATCOM

The reference current generator(**p-q controller**) generates reference(compensating) currents which are fed to hysteresis current controller to generate the switching pulses for VSC.

3   Control Algorithm

3.1   Instantaneous p-q Theory

Instantaneous **p-q** Theory was initially proposed by Akagi[4]. This theory is based on the transformation of three phase quantities to two phase quantities in α-β frame and the Instantaneous active and reactive power is calculated in this frame [4],[5]. Sensed inputs v_{sa}, v_{sb} and v_{sc} & i_{La}, i_{Lb} and i_{Lc} are fed to the p-q controller shown in fig.2 and these quantities are processed to generate reference commands(i_{fa}^{*}, i_{fb}^{*}, i_{fc}^{*}) which are fed to a hysteresis based PWM current controller to generate switching pulses for D-STATCOM.

The system terminal voltages are given as

$$\left. \begin{array}{l} v_{sa} = V_m \sin(wt) \\ v_{sb} = V_m \sin(wt - 2\pi/3) \\ v_{sc} = V_m \sin(wt + 2\pi/3) \end{array} \right\} \tag{1}$$

and the respective load current are given as

$$\left. \begin{array}{l} i_{La} = \Sigma I_{Lan} \sin\{n(wt) - \theta_{an}\} \\ i_{Lb} = \Sigma I_{Lbn} \sin\{n(wt - 2\pi/3) - \theta_{bn}\} \\ i_{Lc} = \Sigma I_{Lcn} \sin\{n(wt + 2\pi/3) - \theta_{cn}\} \end{array} \right\} \tag{2}$$

In a.b and c coordinates a,b and c axes are fixed on the same plane apart from each other by $2\pi/3$. These phasors can be transformed into α-β coordinates using Clarke's transformation as follows.

$$\begin{bmatrix} v_\alpha \\ v_\beta \end{bmatrix} = \sqrt{\frac{2}{3}} \begin{bmatrix} 1 & -1/2 & -1/2 \\ 0 & \sqrt{3}/2 & -\sqrt{3}/2 \end{bmatrix} \begin{bmatrix} v_{sa} \\ v_{sb} \\ v_{sc} \end{bmatrix} \tag{3,}$$

$$\begin{bmatrix} i_\alpha \\ i_\beta \end{bmatrix} = \sqrt{\frac{2}{3}} \begin{bmatrix} 1 & -1/2 & -1/2 \\ 0 & \sqrt{3}/2 & -\sqrt{3}/2 \end{bmatrix} \begin{bmatrix} i_{La} \\ i_{Lb} \\ i_{Lc} \end{bmatrix} \tag{4}$$

Where α and β axes are the orthogonal coordinates . Conventional instantaneous power for three phase circuit can be defined as

$$p = v_\alpha i_\alpha + v_\beta i_\beta \tag{5}$$

Where p is equal to conventional equation

$$p = v_{sa}i_{sa} + v_{sb}i_{sb} + v_{sc}i_{sc} \tag{6}$$

Similarly , the instantaneous reactive power is defined as

$$q = v_{\beta}i_{\alpha} - v_{\alpha}i_{\beta} \tag{7}$$

Therefore in matrix form, instantaneous real and reactive power are given as

$$\begin{bmatrix} p \\ q \end{bmatrix} = \begin{bmatrix} v_{\alpha} & v_{\beta} \\ v_{\beta} & -v_{\alpha} \end{bmatrix} \begin{bmatrix} i_{\alpha} \\ i_{\beta} \end{bmatrix} \tag{8}$$

The α-β currents can be obtained as

$$\begin{bmatrix} i_{\alpha} \\ i_{\beta} \end{bmatrix} = \frac{1}{\Delta} \begin{bmatrix} v_{\alpha} & v_{\beta} \\ v_{\beta} & -v_{\alpha} \end{bmatrix} \begin{bmatrix} p \\ q \end{bmatrix} \tag{9}$$

Where $\Delta = v_{\alpha}^2 + v_{\beta}^2$

Instantaneous active and reactive powers p and q can be decomposed into an average(dc) and oscillatory

$$\left. \begin{array}{l} p = \bar{p} + \tilde{p} \\ q = \bar{q} + \tilde{q} \end{array} \right\} \tag{10}$$

Where $\bar{p}$ and $\bar{q}$ are the average dc part and $\tilde{p}$ and $\tilde{q}$ are the oscillatory (ac) part of these real and reactive instantaneous power. Reference currents are calculated to compensate the instantaneous reactive and the oscillatory component of the instantaneous active power. Therefore the reference compensating currents i_{fa}^* and $i_{f\beta}^*$ in α-β coordinate can be expressed as

$$\begin{bmatrix} i_{fa}^* \\ i_{f\beta}^* \end{bmatrix} = \frac{1}{\Delta} \begin{bmatrix} v_{\alpha} & v_{\beta} \\ v_{\beta} & -v_{\alpha} \end{bmatrix} \begin{bmatrix} -\tilde{p} \\ -q \end{bmatrix} \tag{11}$$

The oscillatory part of real power p is obtained by using 4th order low pass Butterworth filter of cut-off frequency 25 Hz. These currents can be transformed in abc quantities to find reference currents in a-b-c coordinates using reverse Clarke's transformation.

$$\begin{bmatrix} i_{fa}^* \\ i_{fb}^* \\ i_{fc}^* \end{bmatrix} = \sqrt{\frac{2}{3}} \begin{bmatrix} 1 & 0 \\ -1/2 & \sqrt{3}/2 \\ -1/2 & -\sqrt{3}/2 \end{bmatrix} \begin{bmatrix} i_{fa}^* \\ i_{f\beta}^* \end{bmatrix} \tag{12}$$

Fig. 2 shows the block diagram of p-q controller which is used to generate reference currents.

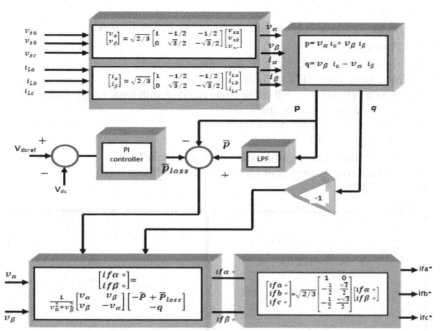

Fig. 2. p-q controller

4   Performance Indices

Total Harmonic Distortion

The total harmonic distortion (THD) [7]is used to define the effect of harmonics on the power system voltage. It is used in low-voltage, medium-voltage, and high-voltage systems. It is expressed as a percent of the fundamental and is defined as

$$THD(current) = \frac{\sqrt{\sum_{h=2}^{50} I_h^2}}{I_1}$$ (39)

According to IEEE-519 the permissible limit for distortion in the signal is 5%.

Power Factor

The ratio of average to apparent power is called as power factor.

Power Factor= P/S where P= Average power and S= Apparent power.

5   Results and Discussion

To investigate the performance of the D-STATCOM for p-q control algorithm, simulations are performed on matlab platform . A three phase three wire distribution system with parameters given below is considered for simulation. The performance of the control algorithm is evaluated based on two different cases.

System Parameters

Supply voltage : 50Vrms(L-N),50Hz, three phase balanced
Source impedance:Rs=0.1Ω,Ls=.5mH
Nonlinear load: Three phase full bidge diode rectifier with load(L=10mH,R_L=3.7Ω)
DC storage Capacitor Cdc=2000μF
Interface inductor Lf=2.2mH ,Rf=0.1Ω
DC Link voltage Vdc=100V

Case1- Balanced Source and balanced Non-Linear load
Case2- Balanced Source and Unbalanced Non-linear load.
SIMULATION
Case-1

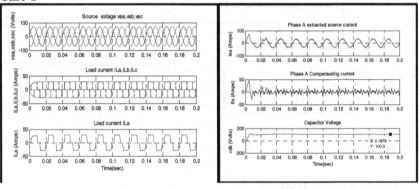

Fig. 3a Balanced source voltage, Load current ,Load currentPhase A

Fig. 3b Phase A extracted Soure current, Compensating current , DC link Capacitor Voltage

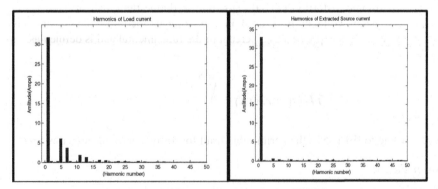

Fig. 3c Harmonics of Load current of Phase A

Fig. 3d Harmonics of extracted Source current of phase A

Case-2

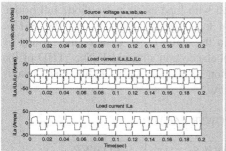
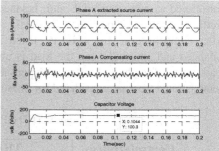

Fig. 4a Balanced source voltage, ,Unbalanced Load current,Load currentPhase A

Fig. 4b Phase A extracted Soure current, Compensating current and DC link Capacitor Voltage

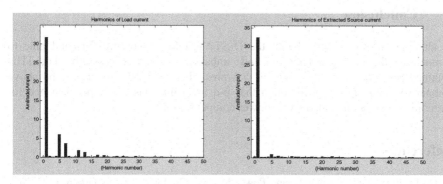

Fig.4c Harmonics of Load current of Phase A

Fig. 4d Harmonics of extracted Source current of phase A

Case-1   **Case-2**

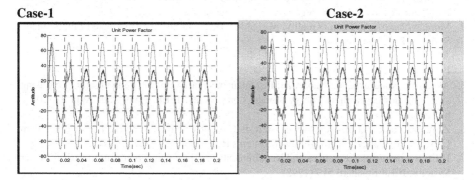

Fig. 5a Case 1 power factor

Fig. 5b Case 2 power factor

In **case 1** the source is assumed to be sinusoidal and balanced whereas the load is considered as non-sinusoidal and balanced with load as six pulse diode full bridge rectifier. Before compensation the THD of load current is found to be **23.2361%**. After compensation the THD is listed in the **Table1.** In **case 2** the source is balanced and sinusoidal but the load is unbalance non sinusoidal .The THD of the load current for phase A after compensation is summarized in **Table 1.**The results demonstrated here are considered for phase A.

Table 1. THD &POWER FACTOR

Control strategy	THD(%)		POWER FACTOR	
	CASE-1	CASE-2	CASE1	CASE2
After Compensation	3.0748	3.4863	0.9981	0.9971

6   Conclusion

In all cases it was observed that the D-STATCOM is working fine and able to compensate the nonlinear balanced and unbalanced load successfully. The THD obtained here are within the limit of 5% prescribed by IEEE 519. From the above comparison table it has been found that p-q algorithm has better performance for harmonic cancellation and reactive power compensation.

References

1. De La Rose, F.C.: Harmonics and Power Systems. Taylor & Francis Group, LLC, New York (2006)
2. Aredes, M., Akagi, H., Watanabe, E.H., Salgado, E.V., Encaracao, L.F.: Comparison between the p-r and p-q-r theories in three phase four wire systems. IEEE Trans. Power Electronics 24(4), 924–933 (2009)
3. Moreno-Munoz, A.: Power Quality: Mitigation Technologies in a Distributed Environment. Springer, London (2007)
4. Akagi, H., Watanabe, E.H., Aredes, M.: Instantaneous Power Theory and Applications to Power Conditioning. Wiley, Hoboken (2007)
5. Akagi, H., Kanazawa, Y., Nabae, A.: Instantaneous reactive power compensator comprising switching devices without energy storage components. IEEE Trans. Ind. Appl. IA-20(3), 625–630 (1984)
6. Zeng, J., Yu, C., et al.: A Novel Hysteresis Current Control for Active Power Filter with Constant Frequency. Electrical Power Systems Reasearch 68, 75–82 (2004)
7. IEEE Recommended Practices and Requirements /or Harmonic Control in Electrical Power Systems. IEEE Standard 519 (1992)

AODV Based Black-Hole Attack Mitigation in MANET

Subhashis Banerjee, Mousumi Sardar, and Koushik Majumder

Department of Computer Science and Engineering
West Bengal University of Technology
Kolkata, India
{mail.sb88,mousumi.sardar02}@gmail.com
koushik@ieee.org

Abstract. Packet dropping is a very dangerous attack in case of limited re-source networks like Mobile Ad-Hoc Network (MANET). During this attack the malicious node first claims that it has the freshest route to the destination, so the sender selects this as the coordinating node and starts sending data packets to the destination via this node. But afterwards it drops them rather forwarding to the destination. In this paper we give a very clever packet dropping or Black-hole attack detection and prevention technique. Here we use the notion of AODV's sequence number for identifying the Black-hole node in the network. Without using any extra packet or modifying any of the existing packet formats our method can efficiently detect and prevent the Black-hole or packet dropping attack in MANET. All the detection prevention are done by the originator node, so the originator need not relying on the other nodes in the network for this pur-pose. This method not only detects or prevents the Black-hole attack but is also capable to isolating the Black-hole node from the network.

Keywords: Black-hole attack, AODV, AODV sequence number, Pack dropping attack.

1   Introduction

The Mobile Ad-hoc network (MANET) is dynamically configured network of wire-less nodes without any pre-defined infrastructure or centralized authority like wired network. The nodes are very frequent in nature means they can join or leave the net-work at any time. In MANET, the nodes depend on each other for relaying packets. Due to this uncertain nature of nodes behaviour makes MANET more vulnerable to attacks (active and passive attack [1, 2, and 3]) than wired network. So security be-comes an important concern of the network for secure communication. Among the several types of network layer attack, one of the most frequently occurred attack is Black-hole attack. Black-hole attack is an active attack in which malicious node tries to form route towards the destination through itself and later drops packets that are forwarded through it. In this paper we present a mechanism to detect and prevent different types of Black-hole attack. Our mechanism is based on simple Ad-hoc On-demand Distance Vector routing (AODV) protocol.

S.C. Satapathy, S.K. Udgata, and B.N. Biswal (eds.), *FICTA 2013*,
Advances in Intelligent Systems and Computing 247,
DOI: 10.1007/978-3-319-02931-3_39, © Springer International Publishing Switzerland 2014

The remaining part is organized as follows: in section 2 we present an in depth discussion about the Black-hole attack with an example. Next in section 3 we give the literature review. In section 4, our proposed scheme for Black-hole attack detection and prevention is given. Section 5 contains the algorithm. The countermeasure of different types of Black-hole attack with example has been given in section 6 and we finally conclude the paper in section 7.

2   The Black-Hole Attack

During route discovery phase of the AODV [4] routing protocol the source node creates a RREQ packet and broadcasts it in the network. The RREQ packet contains the following information: *1. Destination IP, 2. Destination Sequence Number, 3. Originator IP, 4. Originator Sequence Number.* Sequence Number is a monotonically increasing integer value that is maintained by each originator node. The Sequence Number is used to represent the freshness of the information contained in the packet. During the route discovery phase, in the presence of Black-hole attack, when the malicious node receives RREQ packet, it sends back a RREP packet with a high Sequence Number to indicate that it has the fresher route towards the destination. The source node when receives that RREQ packet it selects that route for having high Sequence Number which is actually contained the malicious node in the path. Then the sender node starts to send packets through that path. On receipt the packets the malicious node start to drop the packets without forwarding it to the destination.

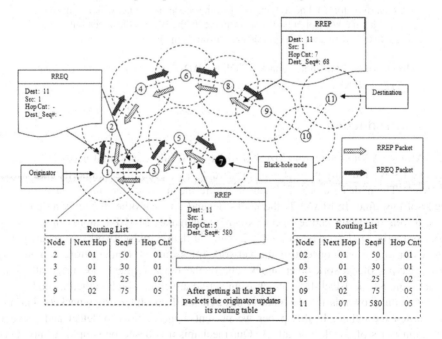

Fig. 1. Example of the Black-hole attack

Consider the MANET with 11 nodes shown in the fig.1. Node 1 is the sender and node 11 is the designated receiver. At first the sender starts the route discovery to discover the best route to the destination. It floods the route request (RREQ) packets in the network as shown in the picture. Now all other nodes including the Black-hole receive the RREQ packet. The node 9 and the Black-hole node (node 7) send the route reply (RREP) packets to the originator. But as displayed in the picture the black hole node does not have a valid path to the destination, and it sends a wrong RREP packet to the sender with a very high sequence number (here sq# is 580) which is much larger than the original one that the node 9 replied (which is 68). So according to the AODV protocol the sender should select the node for further communication that has the best route to the destination, and the freshness of rout is represented through the sequence number in AODV so the originator will select the Black-hole node for the future communication with the destination. And according to its nature the Black-hole node will start dropping the packets received from node 1 rather send it to the destination which is node 11.

3   Literature Survey

S. Marti et al. introduced Watchdog and Pathrater technique [5] to detect and mitigate misbehaviour at time of routing in the network. In Watchdog technique sender node goes into promiscuous mode and listens to its next neighbor node's transmission whether it has been forwarded on or not. If not the node is marked as misbehaving node when the no. of packets not forwarded by the node exceeds a certain threshold. Next based on this knowledge of misbehaving node Pathrater technique is used for choosing the most reliable path. But this technique might not detect a misbehaving node in the presence of a) ambiguous collisions, b) receiver collisions, c) limited transmission power, d) false misbehavior, e) collusion, and f) partial dropping.

S. S. Jain et al proposed a Neighbor monitoring and voting based mechanism [6] for detecting and removing the malicious nodes that launch Black-hole or Gray-hole attacks. In this mechanism the whole traffic is divided into a set of small data blocks. Before sending data packets sender node sends prelude message to inform destination about the incoming packet. After that it broadcasts monitor message to inform all the nodes in the path to start monitoring its next node. After that sender node starts to send data packets. On receiving the prelude message destination starts timer and counts the no. of data packets it received and sends back this information to source node by postlude message. In this way after getting postlude message within timeout period the sender node compares the no. of sent packets with no. of received packets. If both are same it sends the other data blocks to the destination. Otherwise it starts a malicious node detection and removal process by the help of voting mechanism. If the voteCount for a node exceeds predefined thresholdCount the node is marked as malicious node by the sender node. This method successfully detect and prevent the cooperative Black-hole and Gray-hole attacks in O(n) time but a hypothetical assumption that a neighbor node of any node is more trusted than malicious nodes make it quite impractical.

R. Shree et al. proposed Secure-ZRP [7] to detect and prevent Black-hole attack. A special packet 'bluff probe' is used to identify the Black-hole node. This packet contains a non-existent destination ID and broadcasts by the sender before forwarding the actual route request packet. The Black-hole node will sends back a reply when it receives that packet through intermediate node while the good nodes will forward the packet to its next neighbor as their routing table doesn't contain the fake destination ID. This is only detection technique that is applicable on both proactive and reactive protocol. But it cannot detect Gray-hole attack and every node has to maintain valid route table that impose too much overhead.

M. Al-Shurman et al. proposed Redundant Route Method and Unique Sequence Number Scheme [8] in which the safe route is selected on the basis of RREP packet observation that whether the two or more routes have same shared hop or not. If any route doesn't have any shared hops sender node waits for another RREP packet until a routes with shared node is identified or routing timer expires. From this shared nodes information sender prevents the Black-hole attack and selects the safe route.

A detailed literature survey on Wormhole attack and their existing countermeasures with a comparison can be found in our previous work [9].

After the survey work we find that many of the existing detection or prevention techniques modify the packet format by adding some extra fields in it or introduce some new packets. Many of them assume that the sender can control the intermediate nodes, and intermediate nodes will do many extra works like observing the behaviour of its neighbours in favor of the sender. But in the practical scenario this cannot be possible.

Here we propose a Black-hole detection and prevention technique that does not modify the packet format of existing routing protocol, and also does not introduce new packet. The sender can do all the detection and prevention process by itself. Our method only uses an extra route discovery phase for finding the Black-hole node. But this extra route discovery is an optimized route discover phase because during it the originator does not flood the complete network with the RREQ packets, it only multicast the RREQ packet along some routes from which it previously gets RREP packets.

4   Proposed Method for Black-Hole Attack Detection and Prevention

Our proposed Black-hole attack detection and prevention method consists of the following two phases: 1) Black-hole node identification phase. 2) Black-hole node removal phase.

4.1   Black-Hole Node Identification Phase

After the originator receives all the RREP packets it finds the packet which contains the largest sequence number from its cache. Now the originator creates new RREQ packets for the same destination node with a higher destination sequence number than the sequence number that the RREP packet contains which it receives previously from an intermediate node. Now the originator multicasts the packets through the route from which it gets the RREP packets.

According to the AODV protocol when a new RREQ sent by node S for a destination is assigned a higher destination sequence number. The intermediate nodes which know a route, but with a smaller sequence number, cannot send the RREP packet to the sender. Now all the intermediate nodes that receive the RREQ packet compare its destination sequence number for the same destination with the destination sequence number that the RREQ packet contains. As the sender used a false destination sequence number that is higher than all the destination sequence number that all the nodes have there should not be any RREP coming from any intermediate nodes. If any one of the intermediate node is malicious then according to its nature it will sends the RREP packets that have a higher destination sequence number than the RREQ packet contains for attracting the sender. When the sender receives that RREP packet, it confirms that this is a Black-hole node. The originator now selects the node that previously replied with the next height sequence number among the nodes that did not change its sequence number during the false RREQ propagation, for future communication to the receiver.

If there is no RREP during the false RREQ packet transmission the originator selects the node that has replied with the height sequence number during the first route discovery phase. When a Black-hole node is detected then the sender starts the Black-hole node removal process as follows:

4.2   Blackhole Node Removal Phase

Once the originator detects that there is a Black-hole node in the network, it adds it IP address in its malicious node table and avoid the node in future communication. And we assume that the nodes in the network periodically exchange the malicious node table, so other nodes in the network also aware of the Black-hole node.

5   Proposed Algorithm

Now we give the algorithm for detecting and preventing the Black-hole attack. Without any packet modification or imposing to much over head the sender can efficiently detect and prevent the Black-hole attack, as well as the Black-hole node is isolated from the network by our malicious node list exchange procedure. Our algorithm consists of the following two procedures:

1. Black-hole node identification during the route discovery phase of the AODV routing protocol.
2. Black-hole node removal from the network.

Procedure 1 : *Black-hole node identification during the route discovery phase of the AODV routing protocol*

Step 1: Originator initiates the route discovery by flooding the RREQ packets.

Step 2: Originator receives the RREP packets from other nodes in the network which have a valid path to the receiver.

Step 3: Originator stores all the RREP packets in its cache.

Step 4: Then the originator selects the RREP packet that has the maximum sequence number and extracts its sequence number in a variable called max.

Step 5: Now the originator creates new RREQ packets for the same destination node with a higher destination sequence number than the value of max.

Step 6: The originator also sets the hop count value of the RREQ packet to maximum of the distance of the replying nodes.

Step 7: Then it multicasts the new RREQ packet towards all paths from which it receives route replies during the first route discovery phase.

Step 8: Originator waits for a time span for RREP packets.

Step 9: If (*there is RREP packets with higher destination sequence number than that it sends*)

  9.1. The nodes that replied height sequence number is a Black-hole node and wants to carry out a packet dropping attack by giving the wrong information.

  9.2. Next the originator invokes the Black-hole node removal procedure (present in the next part of the algorithm).

  9.3. The originator selects the node that previously replied with the next height sequence number among the nodes that did not replied during the false RREP propagation, for future communication to the receiver.

  9.4. Sender starts to sends the data packets via the selected reliable route after the Black-hole removal process.

Step 10: Else

  // *There is no black hole node in the network*

  10.1. Sender selects the forward path that has been established during the first RREP propagation form the sender to the intermediate node that has replied with the freshest route to the destination.

  10.2. Sender starts to send the data packets via the selected route to the destination.

Step 11: End.

Procedure2 : *Black-hole node removal from the network*

Step 1: Add the IP address of the Black-hole node in the originators malicious node list.

Step 2: Every time after the malicious node list has been updated, the node shares its malicious node list with its neighbours.

Step 3: Nodes in the network avoid the nodes that are in the malicious node list for forwarding the data.

Step 4: End.

6   Black-Hole Attack Detection and Removal

Now we will explain how does our proposed method detect and remove the Black-hole node from the network with an example. Consider the MANET displayed in the fig. 1. After the route discovery phase, the originator receives all the RREP packets, and stores them in cache. Now it selects the packet with the highest sequence number and store the sequence number in a variable called max. In Our example the node 7 (Black-hole node) replied with the maximum Dest._Seq#, which is 580.

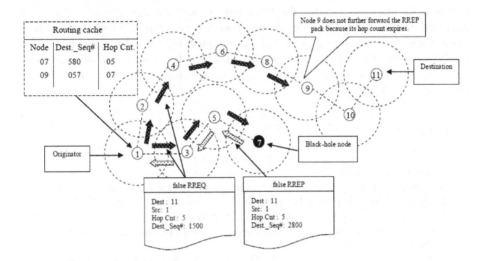

Fig. 2. Originator floods the false RREQ packs for finding the Black-hole node

Now according to our algorithm the sender starts the Black-hole node detection procedure. It first creates new RREQ packet with the Dest._Seq# set to a value greater than 580, which is in our example 1500 and sets the hop count to 5. Source selects the route through node 7 with hop count 5 because the node 9 has the maximum distance from the source. Now the originator sends the packet to node 11 via two different paths one is via 1→3→5→7, and another is via 1→2→4→6→8→9 (fig.2).

Consider the fig. 2 where the Black-hole node sends a false RREP packet corresponding to the false RREQ claiming that it has a better route towards the destination. In our example it sends a RREP packet that has a greater Dest._Seq# value (i.e., 2080) than the false one sent by the source. Now after receiving the RREP packet from the suspected node the originator gets confirm that this is a Black-hole node. So now the originator adds its id in the malicious node list and carries out the Black-hole node removal procedure. Next the originator selects node 9 for the future communication with the receiver, because it does not replied in the false route request.

Also note that our method is capable to detect the Black-hole nodes if there is more than one in the network. We think this is the main advantage of our method because as per our knowledge there is no detection or prevention scheme that can detect more than one Black-hole nodes if they present in the network.

7   Conclusion and Future Work

In this paper we present an initial work in detecting and mitigating the Black-hole attack in a theoretical and algorithmic point of view. Here we have proposed an efficient packet dropping attack prevention technique, which is the super-ordinate or generalized form of all types of Black-hole attacks. The main strength of the proposed method is that, it does not either modify the packet format of AODV or introduce any kind of new Black-hole detection packets like its predecessors. This method can efficiently detect all types of Black-hole attacks such as single and cooperative Black-hole attacks and also isolates the black-hole node from the network. The extra overhead that our method imposes on the network is very minimal with only an additional route discovery phase. And also we optimize the second route discovery by multicasting the RREQ packets without broadcasting that. As a result, the data packets reach the destination successfully, which is a sensitive issue in this type of limited resource network.

References

1. Nguyen, H.L., Nguyen, U.T.: A study of different types of attacks on multicast in mobile ad hoc networks. Ad Hoc Networks 6(1), 32–46 (2008)
2. Karmore, P., Bodkhe, S.: A Survey on Intrusion in Ad Hoc Networks and its Detection Measures. International Journal on Computer Science and Engineering, IJCSE (2011)
3. Rai, A.K., Tewari, R.R., Upadhyay, S.K.: Different Types of Attacks on Integrated MANET-Internet Communication. International Journal of Computer Science and Security, IJCSS 4(3) (2010)
4. Perkins, C.E., Belding-Royer, E.M., Das, S.R.: Ad hoc on-demand distance vector (AODV) routing. RFC 3561, The Internet Engineering Task Force, Network Working Group (2003), http://www.ietf.org/rfc/rfc3561.txt
5. Marti, S., Giuli, T.J., Lai, K., Baker, M.: Mitigating routing misbehavior in mobile ad hoc networks. In: Proceedings of the 6th Annual International Conference on MOBICOM, Boston, Massachusetts, United States, pp. 255–265 (2000)
6. Jain, S., Jain, M., Kandwal, H.: Advanced algorithm for detection and prevention of cooperative black and Grayhole attacks in mobile ad hoc networks. J. Computer Applications 1(7), 37–42 (2010)
7. Shree, R., Dwivedi, S.K., Pandey, R.P.: Design Enhancements in ZRP for Detecting Multiple Blackhole Nodes in Mobile Ad Hoc Networks. International Journal of Computer Applications 18(5), 6–10 (2011)
8. Al-Shurman, M., Yoo, S.M., Park, S.: Blackhole Attack in Mobile Ad Hoc Networks. In: Proceedings of the 42nd Annual ACM Southeast Regional Conference, ACM-SE'42, Huntsville, Alabama (April 2004)
9. Banerjee, S., Majumder, K.: A Comparative Study on Wormhole Attack Prevention Schemes in Mobile Ad-Hoc Network. In: Thampi, S.M., Zomaya, A.Y., Strufe, T., Alcaraz Calero, J.M., Thomas, T. (eds.) SNDS 2012. CCIS, vol. 335, pp. 372–384. Springer, Heidelberg (2012)

Path Planning Strategy for Mobile Robot Navigation Using MANFIS Controller

Prases Kumar Mohanty and Dayal R. Parhi

Robotics Laboratory,Department of Mechanical Engineering ,
National Institute of Technology, Rourkela
Odisha, India
pkmohanty30@gmail.com, dayalparhi@yahoo.com

Abstract. Nowadays intelligent techniques such as fuzzy inference system (FIS), artificial neural network (ANN) and adaptive neuro-fuzzy inference system (ANFIS) are mainly considered as effective and suitable methods for modeling an engineering system. The hallmark of this paper presents a new intelligent hybrid technique (Multiple Adaptive Neuro-Fuzzy Inference System) based on the combination of fuzzy inference system and artificial neural network for solving path planning problem of autonomous mobile robot. First we develop an adaptive fuzzy controller with four input parameters, two output parameters and five parameters each. Afterwards each adaptive fuzzy controller acts as a single takagi-sugeno type fuzzy inference system, where inputs are front obstacle distance (FOD), left obstacle distance (LOD), right obstacle distance (ROD) (from robot), Heading angle (HA) (angle to target) and output corresponds to the wheel velocities (Left wheel and right wheel) of the mobile robot. The effectiveness, feasibility and robustness of the proposed navigational controller have been tested by means of simulation results. It has been observed that the proposed path planning strategy is capable of avoiding obstacles and effectively guiding the mobile robot moving from the start point to the desired target point with shortest path length.

Keywords: ANFIS, obstacle avoidance, mobile robot, path planning.

1   Introduction

Path planning strategy and navigation of mobile robot has acquired considerable attention in recent years. The major issue in autonomous mobile robot is its navigational problem in uncertain and complex environment. If robot wants to travel among the unknown obstacles to reach a specified goal without collisions, then various sensors must be needed to identify and recognize the obstacles present in the real world environment. The sensor based motion planning approaches uses either global or local path planning depending upon the surrounding environment. Global path planning requires the environment to be completely known and the terrain should be static, on other side local path planning means the environment is completely or partially unknown for the mobile robot. Many exertions have been paid in the past to improve various robot navigation techniques.

S.C. Satapathy, S.K. Udgata, and B.N. Biswal (eds.), *FICTA 2013*,   353
Advances in Intelligent Systems and Computing 247,
DOI: 10.1007/978-3-319-02931-3_40, © Springer International Publishing Switzerland 2014

Recently there have been found many interesting research work have been developed by many researchers for path planning of mobile robots. Many authors have considered a controller with complete information of the environment [1-2].Due to the complexity and uncertainty of the path planning problem, classical path planning methods, such as road map approaches (Visibility Graph [3], Voronoi diagrams [4]), Grids [5], Cell decomposition [6] and artificial potential field [7] are not appropriate for path planning in dynamic environments. The artificial potential field method provides a simple yet effective technique to plan paths for robot. The major drawback is that robots are often trapped into a local minimum before reaching to the target. Among the soft intelligent techniques ANFIS is a hybrid model which combines the adaptability capability of artificial neural network and knowledge representation of fuzzy inference system [8]. There are many fuzzy logic methods using various implementation or in combination with other techniques [9-12]. Mobile robot path planning based on neural network approaches presented by many researchers [13-16]. Navigation of multiple mobile robots using Neuro-fuzzy technique addressed by Pradhan et al. [17]. In this navigational controller, output from the neural network given as input to the fuzzy controller to navigate the mobile robot successfully in the clutter environment. Experimental verifications also have been done with the simulation results to prove the validity of the developed technique. Navigation of mobile robots using adaptive neural-fuzzy system discussed by Nefti et al. [18]. In this paper different sensor based information given as input to the Sugeno–Takagi fuzzy controller and output from the controller is the robot orientation. Experimental results settle the importance of the methodology when dealing with navigation of a mobile robot in completely or partially unknown environment. To determine collision-free path of mobile robot navigating in a dynamic environment using Neuro-fuzzy technique presented by Hui et al. [19]. The performances of Neuro-fuzzy approaches are compared with other approaches (GA, Mamdani) and it was found that Neuro-fuzzy approaches are found to perform better than the other approaches. Control of mobile robot based on Neuro-fuzzy technique discussed by Godjevac and Steele [20]. In this paper they have shown how Neuro-fuzzy controllers can be achieved using a controller based on the Takagi-Sugeno design and a radial basis function neural network for its implementation.

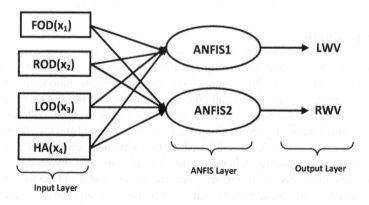

Fig. 1. Multiple ANFIS (MANFIS) Controller for Mobile Robot Navigation

In this paper, a new intelligent controller (MANFIS) has been designed for solving the path planning problem of mobile robot. Finally, simulation results are demonstrated to prove the authenticity of the proposed technique in various environments populated by variety of static obstacles.

2 Architecture of Multiple Adaptive Neuro-Fuzzy Inference System (MANFIS) for Mobile Robot Navigation

Adaptive network-based fuzzy inference system (ANFIS) is one of hybrid intelligent neuro-fuzzy system and it functioning under Takagi-Sugeno-type FIS, which was developed by Jang [8] in 1993. ANFIS has a similar configuration to a multilayer feed forward neural network but links in this hybrid structure only specify the flow direction of signals between nodes and no weights are connected with the links. There are two learning techniques are used in ANFIS to show the mapping between input and output data and to compute optimized of fuzzy membership functions. These learning methods are back propagation and hybrid. Parameters associated with fuzzy membership functions will modify through the learning process.

As for the prediction of left wheel velocity (LWV) and right wheel velocity (RWV) for mobile robot we assume that each adaptive neuro-fuzzy controller under consideration of four inputs i.e. Front obstacle distance (FOD) (x_1), Right obstacle distance (ROD) (x_2), Left obstacle distance (LOD) (x_3), Heading angle (HA) (x_4), and each input variable has five bell membership functions (MF) such as A_1(Very Near), A_2(Near), A_3(Medium), A_4(Far) and A_5(Very Far), B_1(Very Near), B_2(Near), B_3(Medium), B_4(Far) and B_5(Very Far), C_1(Very Near), C_2(Near), C_3(Medium), C_4(Far), and C_5(Very Far), D_1(Very Negative), D_2(Negative), D_3(Zero), D_4(Positive) and D_5(Very positive) respectively, then a Takagi-Sugeno-type fuzzy inference system if-then rules are set up as follows;

Rule: if x_1 is A_i and x_2 is B_i and x_3 is C_i and x_4 is D_i, then
f_n *(wheel velocity)* $= p_n x_1 + q_n x_2 + r_n x_3 + s_n x_4 + u_n$

A, B, C, and D are the fuzzy membership sets for the input variables x_1, x_2, x_3 and x_4 respectively.

where, i=1-5 and p_n, q_n, r_n, s_n and u_n are the linear parameters of function f_n and changing these parameters we can modify the output of ANFIS controller.

The function of each layer in ANFIS structure is discussed as follows:

Input Layer: In this layer nodes simply pass the incoming signal to layer-1. That is

$$\left.\begin{array}{l} O_{0,FOD} = X_1 \\ O_{0,ROD} = X_2 \\ O_{0,LOD} = X_3 \\ O_{0,TA} = X_4 \end{array}\right\} \tag{2.1}$$

First Layer: This layer is the fuzzification layer. Neurons in this layer complete fuzzification process. Every node in this stage is an adaptive node and calculating the membership function value in fuzzy set. The output of nodes in this layer are presented as

$$
\left.
\begin{array}{l}
O_{1,i} = \mu_{A_i}(X_1) \\
O_{1,i} = \mu_{B_i}(X_2) \\
O_{1,i} = \mu_{C_i}(X_3) \\
O_{1,i} = \mu_{D_i}(X_4)
\end{array}
\right\} \tag{2.2}
$$

Here $O_{1,i}$ is the bell shape membership grade of a fuzzy set S(A_i , B_i ,C_i and D_i) and it computing the degree to which the given inputs (X_1, X_2, X_3 and X_4) satisfies the quantifier S. Membership functions defined as follows;

$$
\mu_{A_i}(x) = \frac{1}{1 + \left[\left(\dfrac{x_1 - c_i}{a_i}\right)^2\right]^{b_i}} \tag{2.2(i)}
$$

$$
\mu_{B_i}(x) = \frac{1}{1 + \left[\left(\dfrac{x_2 - c_i}{a_i}\right)^2\right]^{b_i}} \tag{2.2(ii)}
$$

$$
\mu_{C_i}(x) = \frac{1}{1 + \left[\left(\dfrac{x_3 - c_i}{a_i}\right)^2\right]^{b_i}} \tag{2.2(iii)}
$$

$$
\mu_{D_i}(x) = \frac{1}{1 + \left[\left(\dfrac{x_4 - c_i}{a_i}\right)^2\right]^{b_i}} \tag{2.2(iv)}
$$

a_i, b_i and c_i are parameters that control the Centre, width and slope of the Bell-shaped function of node 'i' respectively. These are also known as premise parameters.

Second Layer: It is also known as rule layer. Every node in this layer is a fixed node and labeled as π_n. Every node in this stage corresponds to a single Sugeno-Takagi fuzzy rule. A rule node receives inputs from the respective nodes of layer-1 and determines the firing strength of the each rule. Output from each node is the product of all incoming signals.

$$
O_{2,n} = W_n = \mu_{A_i}(X_1).\mu_{B_i}(X_2).\mu_{C_i}(X_3).\mu_{D_i}(X_4) \tag{2.3}
$$

where W_n represents the firing strength or the truth value, of nth rule and n=1, 2, 3...636 is the number of Sugeno-Takagi fuzzy rules.

Third Layer: It is the normalization layer. Every node in this layer is a fixed node and labeled as N_n. Each node in this layer receives inputs from all nodes in the fuzzy rule layer and determines the normalized firing strength of a given rule. The normalized firing strength of the nth node of the nth rule's firing strength to sum of all rules's firing strength.

$$O_{3,n} = \overline{W}_n = \frac{W_n}{\sum_{n=1}^{625} W_n} \tag{2.4}$$

The number of nodes in this layer is the same the number of nodes in the previous layer that is 625 nodes. The output of this layer is called normalized firing strength.

Fourth Layer: Every node in this layer is an adaptive node. Each node in this layer is connected to the corresponding normalization node, and also receives initial inputs X_1, X_2, X_3 and X_4. A defuzzification node determines the weighted consequent value of a given rule define as,

$$O_{4,n} = \overline{W}_n f_n = \overline{W}_n \left[p_n(X_1) + q_n(X_2) + r_n(X_3) + s_n(X_4) + u_n \right] \tag{2.5}$$

Where $\overline{W}_n$ is a normalized firing strength from layer-3 and p_n, q_n, r_n, s_n, u_n are the parameters set of this node. These parameters are also called consequent parameters.

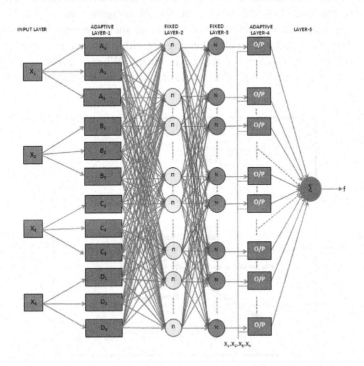

Fig. 2. The structure of ANFIS 1 network

Fifth Layer: It is represented by a single summation node. This single node is a fixed node and labeled as $\sum$. This node determines the sum of outputs of all defuzzification nodes and gives the overall system output that is wheel velocity.

$$O_{5,1} = \sum_{n=1}^{625} \overline{W} f_n = \frac{\sum_{n=1}^{625} W_n f_n}{\sum_{n=1}^{625} W_n} \tag{2.6}$$

3   Simulation Results and Discussion

We have been verified our proposed hybrid technique in two dimensional path planning through series of simulation experiments under completely or partially unknown environment. Our implementation was compiled using MATLABR2008a processing under Windows XP. All the simulation results were applied on PC with Intel core2 processor running at 3.0GHz, 4Gb of RAM and a hard drive of 160Gb.

The coordinates of the sides of the paths as well as coordinates of any static obstacles were known to the MANFIS controller. Knowing the coordinates of the robot, the current navigational controller can thus calculate the distances and heading angle of the robot, as if it was sensor. In current navigation model, we have been developed two main reactive behaviors: one to reach the target and the other avoiding obstacles. The simulated robot path planning algorithm has been developed and set obstacles at different position of the environment. When a robot is close to an obstacle, it must change its speed to avoid the obstacle. If a target is sensed by a

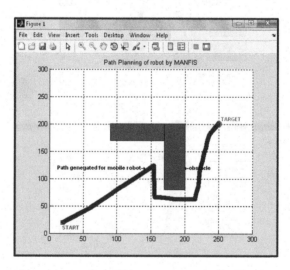

Fig. 3. Single robot escaping from corner end using MANFIS Controller

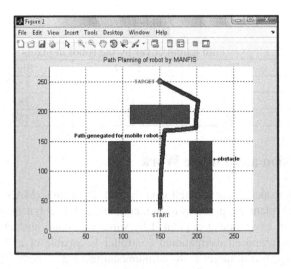

Fig. 4. Single robot escaping from corridor using MANFIS Controller

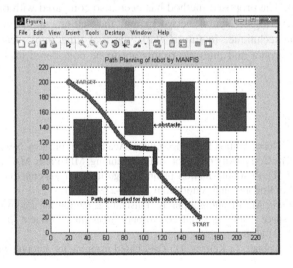

Fig. 5. Single robot escaping from maze environment using MANFIS Controller

mobile robot, it will decide whether it can reach that target, i.e. it will judge whether there are obstacles that will obstruct its path. If the path leading to the source is clear, the robot will turn and proceed towards the source. In Fig .3 to Fig. 5 shows the path created for mobile robot motion in various environments with considering different start and goal positions. It can be observed that, using sensory information, the mobile robot can reach successfully at the target object by efficiently using multiple types of reactive behaviors with proposed navigation algorithm. The simulation experiments were also compared with fuzzy logic and Neural Network and it is verified that using proposed technique the robot reached to the specified target in optimum path length (Table-1).

Table 1.

SL No.	Path length cover by the robot in pixels		
	Fig.3	Fig.4	Fig.5
MANFIS	178	136	150
FUZZY	201	158	188
NN	199	164	186

4   Conclusion and Future Work

In this research paper, a new intelligent hybrid algorithm (MANFIS) has been designed for path planning problem of autonomous mobile robot in an unknown or partially unknown environment populated by variety of static obstacles. It has been observed that the proposed navigational controller is capable of avoiding obstacles and effectively reached at target with optimum/shortest path length. The authenticity of the proposed technique has been verified and proven by simulation experiments using MATLAB. The proposed method has been also compared with other intelligent techniques (Fuzzy logic, Neural network) and the settlement in results show the efficiency of the technique. In future work, the real time implementation is to be carried out using robot and multiple robots are to be considered instead of a single mobile robot.

References

1. Latombe, J.C.: Robot Motion Planning. Kluwer Academic Publishers, New York (1990)
2. Canny, J.E.: The Complexity of Robot Motion Planning. MIT Press, MA (1988)
3. Lozano-Perez, T.: A simple motion planning algorithm for general robot manipulators. IEEE Journal of Robotics and Automation 3, 224–238 (1987)
4. Leven, D., Sharir, M.: Planning a purely translational motion for a convex object in two dimensional space using generalized voronoi diagrams. Discrete & Computational Geometry 2, 9–31 (1987)
5. Payton, D., Rosenblatt, J., Keirsey, D.: Grid-based mapping for autonomous mobile robot. Robotics and Autonomous Systems 11, 13–21 (1993)
6. Regli, L.: Robot Path Planning. Lectures Notes of Department of computer Science. Drexel University (2007)
7. Khatib, O.: Real time Obstacle Avoidance for manipulators and Mobile Robots. In: IEEE Conference on Robotics and Automation, vol. 2, pp. 505–505 (1985)
8. Jang, J.S.R.: ANFIS: Adaptive network-based fuzzy inference system. IEEE Transaction on System, Man and Cybernetics –Part b 23, 665–685 (1993)
9. Huq, R., Mann, G.K.I., Gosine, R.G.: Mobile robot navigation using motor schema and fuzzy content behavior modulation. Application of Soft Computing 8, 422–436 (2008)
10. Selekwa, M.F., Dunlap, D.D., Shi, D., CollinsJr, E.G.: Robot navigation in very cluttered environment by preference based fuzzy behaviors. Autonomous System 56, 231–246 (2007)

11. Abdessemed, F., Benmahammed, K., Monacelli, E.: A fuzzy based reactive controller for a non-holonomic mobile robot. Robotics Autonomous System 47, 31–46 (2004)
12. Pradhan, S.K., Parhi, D.R., Panda, A.K.: Fuzzy logic techniques for navigation of several mobile robots. Application of Soft Computing 9, 290–304 (2009)
13. Motlagh, O., Tang, S.H., Ismail, N.: Development of a new minimum avoidance system for behavior based mobile robot. Fuzzy Sets System 160, 1929–1946 (2009)
14. Velagic, J., Osmic, N., Lacevic, B.: Neural Network Controller for Mobile Robot Motion Control. World Academy of Science, Engineering and Technology 47, 193–198 (2008)
15. Singh, M.K., Parhi, D.R.: Intelligent Neuro-Controller for Navigation of Mobile Robot. In: Proceedings of the International Conference on Advances in Computing, Communication and Control, Mumbai, Maharashtra, India, pp. 123–128 (2009)
16. Castro, V., Neira, J.P., Rueda, C.L., Villamizar, J.C., Angel, L.: Autonomous Navigation Strategies for Mobile Robots using a Probabilistic Neural Network (PNN). In: 33rd Annual Conference of the IEEE Industrial Electronics Society, Taipei, Taiwan, pp. 2795–2800 (2007)
17. Pradhan, S.K., Parhi, D.R., Panda, A.K.: Neuro-fuzzy technique for navigation of multiple mobile robots. Fuzzy Optimum Decision Making 5, 255–288 (2006)
18. Nefti, S., Oussalah, M., Djouani, K., Pontnau, J.: Intelligent Adaptive Mobile Robot Navigation. Journal of Intelligent and Robotic Systems 30, 311–329 (2001)
19. Hui, N.B., Mahendar, V., Pratihar, D.K.: Time-optimal, collision-free navigation of a car-like mobile robot using neuro-fuzzy approaches. Fuzzy Sets and Systems 157, 2171–2204 (2008)
20. Godjevac, J., Steele, N.: Neuro-fuzzy control of a mobile robot. Neuro Computing 28, 127–142 (1999)
21. The Math Works Company, Natick, MA, ANFIS Toolbox User's Guide of MATLAB

Speech Emotion Recognition Using Regularized Discriminant Analysis

Swarna Kuchibhotla[1,*], B.S. Yalamanchili[2], H.D.Vankayalapati[3], and K.R. Anne[2]

[1] Research Scholar, Acharya Nagarjuna University, Guntur, A.P, India
[2] Department of Information Technology, VRSEC, Vijayawada, A.P, India
[3] Department of Computer Science and Engineering, VRSEC, Vijayawada, India

Abstract. Speech emotion recognition plays a vital role in the field of Human Computer Interaction. The aim of speech emotion recognition system is to extract the information from the speech signal and identify the emotional state of a human being. The information extracted from the speech signal is to be appropriate for the analysis of the emotions. This paper analyses the characteristics of prosodic and spectral features. In addition feature fusion technique is also used to improve the performance. We used Linear Discriminant Analysis (LDA), Regularized Discriminant Analysis (RDA), Support Vector Machines (SVM), K-Nearest Neighbor (KNN) as a Classifiers. Results suggest that spectral features outperform prosodic features. Results are validated over Berlin and Spanish emotional speech databases.

Keywords: Mel Frequency Cepstral Coefficients (MFCC), Pitch, Energy, Feature Fusion.

1 Introduction

The natural, fast and efficient way of communication between humans is Speech. Through speech one can recognize different characteristics of the speaker like gender, language employed, socio-economic back ground, speaker health state and emotional states etc. Speech signal mainly conveys two types of information [1]. First it contains the message that he or she wants to express, is included in linguistic concepts of speech and second is that we concentrate on the way how the message is delivered and these concepts are included in paralinguistic concepts of speech [2]. In speech recognition linguistic concepts are used in content recognition and paralinguistic concepts are used in speaker recognition and emotion recognition applications.

In Speech signal, the extracted features are mainly classified into two types, first one is prosodic features and second is spectral features. Energy, pitch, speech rate and voice quality are some examples of prosodic features. Mel frequency Cepstral Coefficients (MFCC), Linear Prediction Cepstral Coefficients (LPCC) are some examples of spectral features[4]. There is no solid theoretical basis relating to the characteristics of the speech to determine the best features for emotion identification

---

* Corresponding author.

S.C. Satapathy, S.K. Udgata, and B.N. Biswal (eds.), *FICTA 2013*,   363
Advances in Intelligent Systems and Computing 247,
DOI: 10.1007/978-3-319-02931-3_41, © Springer International Publishing Switzerland 2014

[3]. In addition to extracting prosody and spectral features we used feature fusion technique which is done by concatenating prosodic and spectral features.

Once the features are extracted these are given as input to the classifier. Here we used Linear Discriminant Analysis(LDA), Regularized Discriminant Analysis(RDA), Support Vector Machine(SVM) and K Nearest Neighbor(KNN) classifiers. A straight forward approach to classify the emotions is by using valence arousal space. In this paper we used LDA as a classifier for speech processing, but with this LDA we got some singularity problem so to eliminate this singularity we add regularization technique and named it as RDA and compare these results by implementing SVM and KNN also.

This paper is organized as follows Section 2 describes the emotional databases, section 3 describes the feature extraction procedures, Section 4 describes the classification algorithms, Section 5 describes the results obtained and its analysis finally Sections 6 describes the conclusion.

2   Speech Databases

In general Research deals with databases of acted, induced and completely spontaneous emotions [5] More number of speech databases are designed for speech emotion research such as Emo-DB (Berlin database of emotional Speech), SES(Spanish Emotional speech database), DES(Danish emotional speech database), English speech emotional databases etc..Among these databases, in this paper we used Berlin and Spanish emotional data bases which are acted emotional speech databases. From the total speech samples 2/3rd of the samples are used for training and 1/3 of the samples are used for testing.

2.1   Berlin Database of Emotional Speech

The Berlin emotional speech database is an open source speech database and contains seven basic emotions: anger, boredom, disgust, fear, happiness, sadness and neutral. All of the speech samples are simulated by ten professional native German actors (5 actors and 5 actresses). There are totally about 500 speech samples in this database. The length of the speech samples varies from 2 to 8 seconds.

2.2   Spanish Database of Emotional Speech

This database also contains basic emotions Anger, sadness, joy, fear, disgust, surprise and neutral. The corpus contains 184 different sentences for each emotion includes numbers, words, sentences in affirmative, exclamatory or interrogative forms and paragraphs. It comprises of recordings from two professional actors, one male and one female. In that we considered four emotions.

3   Feature Extraction

Initially speech signal is divided into several frames so that we are analyzing each frame in short time. A frame size of 256 samples and an overlapping of 100 samples is used. The features are usually estimated over simple statistics Mean, Variance, Minimum, Range, Skewness, and Kurtosis.

3.1   Energy Related Features

Speech is produced due to the excitation of vocal tract. The amplitude of the speech signal varies with time [1][2]. Representation of the energy of the speech signal reflects these amplitude variations. Energy values are calculated for each frame. Short time energy for a frame of n samples is given as

$$E = \sum_{i=1}^{n} x_i^2$$

where x_i is the speech signal for the ith frame.

3.2   Pitch Related Features

Pitch is an important property of a speech signal. Pitch represents vibrational frequency of sound source. Here we used Autocorrelation method to calculate pitch. Statistically, autocorrelation measures the degree of association between data points in a time series are separated by different time lags[1][2]. The motivation for the use of ACF is its capability in characterizing "self similarity" or periodicity. The short time auto-correlation function is given as

$$R_n(k) = \sum_{i=1}^{n} x(i)x(i+k) \qquad \text{where k is time lag}$$

where $x(i)$ is the speech signal for the ith frame and n is the total number of frames

Energy and pitch were estimated for each frame together with their first and second derivatives, providing 6 features per frame and the same six statistics are applied, totally we get 36 features

3.3   Mel Frequency Cepstral Coefficients (MFCC)

Mel frequency cepstral coefficients are most efficient in order to extract the correct emotional state of a human being which comes under spectral representation of speech. The cepstral representation of the speech spectrum provides a good representation of the local spectral properties of the speech signal [6]. The mel frequency mel(f) for a given frequency f is defined as

$$mel(f) = 2595 * \log 10(1 + \tfrac{f}{700})$$

Eighteen MFCC Coefficients were estimated for each and every frame and their first and second derivatives giving a total of 54 spectral features. The same six statistics are applied to these 54 coefficients. So totally 54*6=324 spectral features were extracted for each and every frame in a speech signal.

4   Classification of Speech Samples

The Extracted features are given as input to the classifier. The Purpose of the classifier is to assign speech sample to one of the four emotional classes based on set of measurements. Here we used LDA, RDA, SVM and KNN as classifiers. In our application training data is divided into four classes. A group of speech samples belonging to particular emotion are placed in each class. Here we considered four emotions namely Happy, Neutral, Anger and Sad.

4.1   Linear Discriminant Analysis (LDA)

LDA is a statistical method which minimizes the within class covariance and maximizes the between class covariance in the given data base of speech samples [7]. In other words it groups the speech signals belonging to one class and distinguishes the speech signals belonging to other classes. Classification is done based on spectral and prosodic features extracted from the speech signals.

4.2   Regularized Discriminant Analysis (RDA)

RDA is also a statistical classification method and an enhancement for LDA. Because of the high dimensionality speech samples and less number of classes LDA is suffered with Singularity problem. To overcome this problem we used RDA. In this method, the within covariance matrix S_w is regularized by biasing the diagonal components. The classification performance is improved by stabilizing the sample covariance matrix. It is somehow difficult in estimation of regularization parameter value as higher values of λ will disturb the information in the within class scatter matrix and lower values of λ will not solve the singularity problem in LDA [8].

4.3   Support Vector Machine (SVM)

SVM is used for speech emotion classification as it can handle linearly non separable classes. For multi class classification we used one against all approach. A set of prosody and spectral features are extracted from the speech signal and used as input to train the SVM. Among many possible hyper planes the SVM classifier selects a hyper plane which optimally separates the training speech samples with a maximum margin. To construct an optimal hyper plane one has to take into account a small amount of data, the so called support vectors[9], which determines this margin. As higher the margin the classification performance is more and vice versa.

4.4   K-Nearest Neighbour (KNN)

KNN is a non-parametric method for classifying speech samples based on closest training samples in the feature space. Similar to SVM and RDA the speech samples are given as input to the KNN classifier. Nearest Neighbor classification uses training samples directly rather than that of a model derived from those samples. It represents each speech sample in a d-dimensional space where d is the number of features. The similarity of the test sample with the training sample is compared using Euclidian distance. Once the nearest neighbor speech samples list is obtained the test speech sample is classified based on the majority class of nearest neighbors[10].

5   Performance Comparision of Different Classification Techniques

The four classification Algorithms LDA, RDA, SVM and KNN are implemented over Berlin and Spanish Databases using prosodic features, spectral features and their feature fusion, the results are shown in Fig. 1.This is a three dimensional represention of classifiers, feature and their efficiencies.

The Blue, Red and Green bars or front, middle and back rows comprises the efficiencies of prosodic features, spectral features and their feature fusion respectively. Each row represents efficiency with that particular feature. Each row consists of 8 bars or four pairs. Each pair corresponding to a classifier with two databases. For instance the first pair representing the performance of LDA classifier over Berlin (B) and Spanish (S) databases, second pair representing the performance of RDA classifier using these two databases, similarly the third and fourth pairs representing the efficiencies of SVM and KNN respectively.

By observing all the columns the column 3 has the highest recognition efficiency which is the representation of RDA using Berlin database. The performance of LDA is very low for all the features because of the singularity problem that occur due to high dimensional data. This is overcome using regularization technique which is represented by the columns 3 and 4 which gives the better classification performance. Similarly the classification performance of SV?M and KNN are comparable with RDA than that of LDA.

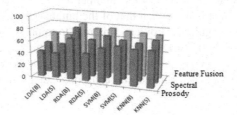

Fig. 1. Performance comparison of LDA, RDA, SVM and KNN

The percentage of efficiencies are represented clearly in Table 1.Basing on these results we observed that the overall efficiency is more for Berlin data base when compared with Spanish database.

Table 1. Algorithm wise efficiency representation. Here P-Prosody S-Spectral C-Feature Fusion

ALG	Berlin			Spanish		
	P(%)	S(%)	C(%)	P(%)	S(%)	C(%)
LDA	42.0	51.0	62.0	39.75	49.0	60.0
RDA	67.0	78.75	80.70	44.0	61.0	74.0
SVM	55.0	68.75	75.5	60.25	64.5	73.0
KNN	57.0	60.7	72.0	58.25	67.75	71.51

The emotion recognition rate using Berlin database for RDA is 67% wih prosodic features which suggest that prosodic features alone gives good recognition rate, to improve the performance further we used spectral features which gives a better recognition accuracy with a recognition rate of 78.75% further improvement is done by using feature fusion technique which gives a best recognition accuracy with a recognition rate of 80.7%. Similarly the recognition accuracy of SVM and KNN using feature fusion technique are 75.5% and 72% respectively. To Spanish database the efficiencies of RDA, SVM and KNN are 44%, 60.25% and 58.25% respectively. By comparing these we found that SVM has the good recognition accuracy with prosodic features. From Tabel1 it is observed that KNN has the good recognition accuracy using spectral features and finally using feature fusion the emotion recognition accuracy is nearly equal for RDA and SVM with an efficiency of 74% and 73% respectively and for KNN it is 71.51% as shown in Table 1.

6   Conclusion

In this paper we implement different classification Algorithms and compare the results. The results obtained by these algorithms are comparable with one another. We observed that RDA is best classifier; SVM and KNN are better classifiers when compared with LDA. Actually LDA is a very good classifier if the dimensionality of the data is less when compared with number of classes. In this paper our dimensionality of the data is more than that of the number of classes so we moved to RDA to overcome the singularity problem. In order to improve the efficiency further we have to use Voice Activity Detection (VAD) algorithm to discriminate voiced and unvoiced segments of speech. The performance of the classifier has to be improved in such a way that it reflects the emotions exactly in the valence arousal space.

References

1. Ververidis, D., Kotropoulos, C.: Emotional speech recognition: Resources, features, and methods. Speech Communication 48, 1162–1181 (2006)
2. Ayadi, M.E., Kamel, M.S., Karray, F.: Survey on speech emotion recognition: Features, classification schemes, and databases. Pattern Recognition 44, 572–587 (2011)
3. Luengo, I., Navas, E., Hernáez, I.: Feature Analysis and Evoluation for Automatic Emotion Identification in Speech. IEEE Transctions on Multimedia 12(6) (October 2010)
4. Cowie, R., Douglas-Cowie, E., Tsapatsoulis, N., Votsis, G., Kollias, S., Fellenz, W., Taylor, J.G.: Emotion Recognition in Human Computer Interaction. IEEE Signal Processing Magazine (January 2001)
5. Vogt, T., André, E., Wagner, J.: Automatic recognition of emotions from speech: A review of the literature and recommendations for practical realisation. In: Peter, C., Beale, R. (eds.) Affect and Emotion in Human-Computer Interaction. LNCS, vol. 4868, pp. 75–91. Springer, Heidelberg (2008)
6. Muda, L., Begam, M., Elamvazuthi, I.: Voice Recognition Algorithms using Mel Frequency Cepstral Coefficient (MFCC) and Dynamic Time Warping (DTW) Techniques. Journal of Computing 2(3) (March 2010) ISSN 2151-9617
7. Ji, S., Ye, J.: Generalized Linear Discriminant Analysis: A Unified Framework and Efficient Model Selection. IEEE Transactions on Neural Networks 19(10) (October 2008)
8. Ye, J., Xiong, T., Janardan, R., Bi, J., Cherkassky, V., Kambhamettu, C.: Efficient model selection for regularized linear discriminant analysis. In: Proc. CIKM Arlington, VA, pp. 532–539 (2006)
9. Cortes, C., Vapnik, V.: Support-vector networks. Machine Learning 20(3), 273–297 (1995)
10. Suresh, M., Ravikumar, M.: Dimensionality Reduction and Classification of Color Features data using SVM and KNN. International Journal of Image Processing and Visual Communication 1(4), 2319–1724 (February 2013)

An Extensive Selection of Features as Combinations for Automatic Text Categorization

Aamir Sohail[1], Chaitanya Kotha[1], Rishanth Kanakadri Chavali[1], Krishna Meghana[1], Suneetha Manne[1], and Sameen Fatima[2]

[1] Department of IT, VRSEC, Vijayawada, India
[2] Depratment of CSE, Osmania University, Hyderabad

Abstract. The merits of modern web search engines that intend to access such pages limit relatively to users' requirement relying highly on information retrieval techniques. For accessing most relevant user subject specific pages, building a categorization system that can analyse the content and present information precisely could be a good alternative. For Text Categorization, most of the researchers relied highly on trained dataset. Each trained dataset is usually large in size due to which most approximations, computations are time consuming. This makes the entire categorization system slow and inaccurate. The proposed method is novel and the number of features is used. This paper explores the effect of word and other values of word in the document, which express the features of a word in the document. The proposed features are exploited by tf-itf, position of the word and compactness. These features are combined and evaluated. The Experimental results showed a significant improvement in Text categorization process.

1   Introduction

In this modern world, Information retrieval and storage has been an essential requirement. This has led to collection of vast information through various media. The past decade has seen a tremendous increment in collection of data relevant to various fields such as Business transactions, Scientific data, Medical and personal data, Surveillance video and pictures, Satellite sensing games, digital data etc. Unfortunately, these massive collections of data stored on disparate structures very rapidly became more strenuous and pain staking. This has led to the discovery of "Text Mining".

Automatic extraction of new, previously unknown information usually from a large amount of different unstructured text resources by a computer is called [1] Text Mining. The field of text mining usually deals with texts whose function is the communication of factual information or opinions. The motivation for trying to extract information from such text automatically is compelling even if success is only partial. However, Information available on websites regarding particular context is too vast, so it is difficult to decide which category does it belongs to. In order to avoid such anomalies categorization of documents is done to identify their respective field. The best method for retrieval of accurate information is "Text Categorization".

S.C. Satapathy, S.K. Udgata, and B.N. Biswal (eds.), *FICTA 2013*,
Advances in Intelligent Systems and Computing 247,
DOI: 10.1007/978-3-319-02931-3_42, © Springer International Publishing Switzerland 2014

1.1   Text Categorization

Text categorization is the task of assigning predefined categories to natural language text [2]. Text Categorization may be formalized as the task of approximating the unknown target function: $\Phi: D \times C \rightarrow \{T, F\}$ that describes how documents ought to be classified, according to a supposedly authoritative expert) by means of a function, where $C = \{c1 \ldots c|C|\}$ is a pre-defined set of categories and D is a (possibly infinite) set of documents. The text categorization general view is shown in Fig. 1.

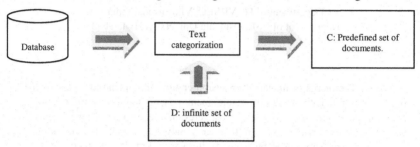

Fig. 1. General view of Text Categorization

The paper is organized as follows. In section 2, related research work on Text Categorization is discussed. The proposed approach is elaborated in section 3. The experimental results with observations are shown in section 4. Finally in section 5 conclusions are presented.

2   Related Research Work

Over the years, automated categorization technologies have used a succession of different algorithms. Some of the most longstanding technologies we could mention are semantics-based approaches, which had the disadvantage of high costs in human and financial terms if the classification system required updating. To counter this problem, several machine learning approaches are introduced. Several generations of statistics-based algorithms followed, which were able to produce ever more relevant results. Below is a presentation of some statistics based processing methods.

2.1   Machine Learning Approaches

This section describes three Machine Learning techniques that are common for Text Categorization: Naive Bayes categorizers, k- Nearest-Neighbour categorizers and Support Vector Machines. Machine Learning algorithms required to provide a set of examples from which the rules defining the machine behaviour are extracted. Automatic Text Categorization systems attempt to label documents according to ontology of classes defined by the user. The problem is a supervised task, that is the machine is tuned using a training set of labelled documents in order to minimize the error between the real target and the predict label.

Naive Bayes Classifier:A Naïve Bayes classifier is a probabilistic classifier based on applying Bayes' theorem with strong (naive) independence assumptions. A more descriptive term for the underlying probability model would be "independent feature model". In simple terms, a naive Bayes classifier assumes that the presence or absence of a particular feature is unrelated to the presence or absence of any other feature, given the class variable. Naive Bayes classifiers can be trained very efficiently in a supervised learning setting. However, there are few demerits which are to be stressed although. First, the naive Bayes classifier requires a large number of records to obtain good results. Second, when a predictor category is not present in the training data, naive Bayes assumes that a new record with that category of the predictor has zero probability. This can be a problem if this rare predictor value is important [5].

K-Nearest Neighbor: K-Nearest Neighbor classification is an instance-based learning algorithm that has shown to be very effective for a variety of problem domains including documents [6]. The key element of this scheme is the availability of a similarity measure that is capable of identifying neighbors of a particular document. There are various k-NN classification algorithms such as Weight Adjusted k-Nearest Neighbor where the weights of features are learned using an iterative algorithm. However there are few demerits of k-NN classification algorithm such as if the work is done at run-time, k-NN can have poor run-time performance if the training set is large. Another major drawback is similarity measure used in k-NN which uses all features in computing distances.

Support Vector Machines: SVMs are a new learning method introduced by V.Vapnik et al in the year 1979. They are well-founded in terms of computational learning theory and very open to theoretical understanding and analysis. One remarkable property of SVMs is that their ability to learn can be independent of the dimensionality of the feature space. SVMs measure the complexity of hypotheses based on the margin with which they separate the data, not the number of features. This means that we can generalize even in the presence of various features, if our data is separable with a wide margin using functions from the hypothesis space [7]. However, SVMs have their own limitations. Firstly, Support vector approach lies in the choice of kernel. Secondly, speed and size are highly limited both in training and testing. Thirdly, discrete data presents another problem. Lastly, it lacks transparency of results. [8].

2.2   N-Gram Based Text Categorization

In early 90s, Bag-Of-Bigrams (pairs of consequent words) was proposed by Lewis as a competitive representation [9]. While some of the researchers report significant improvement in text categorization results (Mladeni´c and Grobelnik), many of them show only marginal improvement or even a certain decrease [10] [13]. The reviewed literature resulted that, most of the researchers till now relied highly on training dataset or corpus to classify a test file to do Text Categorization. To overcome these drawbacks a novel method is proposed which does text categorization.

3   Proposed Method

The proposed method comes up with a new procedure of combining two or more features for text categorization. It does not require trained data unlike existing classifiers. Also has advantage of reduced time and space complexity.

3.1   Preprocessing

Case Folding: The whole document is converted to a unified font i.e. either capital case or lower case of alphabets. This is to get the uniqueness of all words in document.

Tokenization: The sentences in the document are segmented into tokens. These tokens are referred to be the simplest part of the document.

Stop Word Removal: some words in a document are extremely common and occur in a majority of the documents. In order to reduce search space and processing time, firstly, stop words are dealt separately by recognizing them in the stemmed file. Secondly, these stop words are removed from document.

Stemming: A number of stemming algorithms or stemmers have been developed to reduce a word to its stem or root form. In the proposed method used porter's algorithm for stemming the word to its root [11].

3.2   Selection of Features

There are many features which are used for text categorization like term frequency and inverse term frequency (tf-itf), compactness of the word, first appearance of word, based on title etc. Here text categorization is done using three main features which are term frequency and inverse term frequency (tf-itf), compactness of the word, first appearance of word and their combinations.

Term Frequency and Inverse Term Frequency (TF-ITF): The importance of a word can be measured by its term frequency which means it counts the number of occurrences of a particular term in the whole document by using following formula:

$$\text{TF}(t, d) = \frac{\text{COUNT}(t,d)}{\text{SIZE}(d)}$$

Considering only terms which have occurred many times will not be sufficient. There are also terms which occur less times in the document but have significant part in categorizing the document. In such case inverse term frequency is used which counts the significant words though they are less frequent. These terms are determined by using following formula:

$$\text{ITF}(t, d) = \frac{\log |d|}{|\{d \in D : t \in d\}|}$$

Now by combining both term and inverse term frequency to get the terms with combined values which gives the terms of most related to the document with high values.

$$\text{TF-ITF } (t, d) = \text{importance of } (t, d) * \text{ITF } (t, d)$$

Compactness of Appearance of the Word: A word is compact if it is concentrated in a specific part of a document and less compact if its appearances spread over the whole document. If the appearances of a word are less compact, the word is more likely to appear in different parts and more likely to be related to the theme of the document [12]. If the word is more frequent and important, the compactness of its appearances is low. Importance and the first appearance of a word show the earlier, the more important. These features are calculated with the following equations.

$$\text{Count } (t, d) = \sum_{i=0}^{n-1} c_i$$

$$\text{Centroid } (t, d) = \frac{\sum_{i=0}^{n-1} c_i \times i}{\text{count}(t,d)}$$

$$\text{Compactness } (t, d) = \frac{\sum_{i=0}^{n-1} c_i \times |i - \text{centroid}(t,d)|}{\text{count}(t,d)}$$

Position of First Appearance of the Word: Generally, authors start with abstract which gives the brief introduction about the document. This abstract contains the terms that are relevant to that domain. So position of first appearance of word is also given significant place in categorizing the document.

$$\text{First appearance } (t, d) = min_{i \in \{0 \dots n-1\}} \times c_i > 0? \, i : n$$

Combinations: As mentioned above each feature is capable of categorizing text documents. Through proposed method achieved better results when compared with the existing classifiers by using features combination. The combinations that are used here are a) tf-itf and compactness b) compactness and first appearance c) first appearance and tf-itf d) combination of three (tf-itf, compactness and first appearance).

4   Experimental Results and Observations

In this section, feature selection is explored on text documents. The features are combined as explained in the above section and their effects on documents are analysed. A sample documents are used to represent the effectiveness of each feature to categorize the document accurately. The ability of each feature and its combination to detect the document is presented in Table 2. The extent to which each document differed from the exact document is also analysed. This is done by calculating the number of words by which the document is matched to the relevant field or not.

The calculation is done with the following equation:

Difference {matched words of document with irrelevant field and matched words of document of relevant field}

The relevancy graph is shown in Fig.3. The document is termed as most relevant if it falls under "low "level of irrelevancy and vice versa.

Low [0-100] ----> represents the range of words by which a Particular document differed from original document.
Medium [100-500] ----> represents the range of words by which a particular document differed from original document.
High [500-1000] ----> represents the range of words by which a particular document differed from original document.

Table 2. Experimental results for sample random 10 documents

input	Tf-idf	Compactness	First appearance	First combination (tf-idf& compactness)	Second combination (compactness &first appearance)	Third combination (tf-idf&first appearance)	Three combination
Document -1 (Data Mining-dm)	✓	✓	✓	X[low-dm]	X[low-dm]	✓	X[low-dm]
Document -2 (Graphics -gr)	✓	X[med-gr]	X[high-gr]	✓	X[low-gr]	✓	✓
Document -3 (Artificial intelligence -ai)	✓	✓	X[med-ai]	✓	✓	✓	✓
Document -4 (software engineering -Se)	✓	✓	x[low-dm]	✓	✓	✓	✓
Document -5 (data mining -dm)	✓	X[low-dm]	X[low-se]	✓	✓	✓	✓
Document -6 (Bio informatics-Bi)	✓	✓	✓	X[med-bi]	✓	✓	✓
Document -7 (Data mining-dm)	✓	X[low-dm]	X[low-gr]	✓	✓	✓	✓
Document -8 (software engineering- Se)	✓	✓	X[med-dm]	✓	✓	✓	✓
Document -9 (Artificial intelligence-ai)	✓	✓	✓	✓	✓	✓	✓
Document -10 Graphics –gr)	✓	✓	X[med-gr]	✓	✓	✓	✓

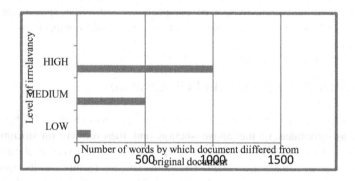

Fig. 2. Relevance graph for different selection of feature

Similarly, combinations are tried for documents of respective fields and the results attained are depicted to show the effectiveness of using combinations rather than using each feature alone to categorize documents. In Fig. 4, Fig 5 and Fig 6 Computer Graphics, Artificial Intelligence and Bioinformatics fields comprising of 30 sample documents each are shown.

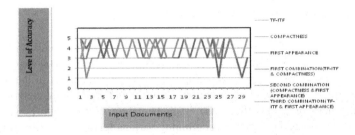

Fig. 3. Result graph for Computer Graphs field

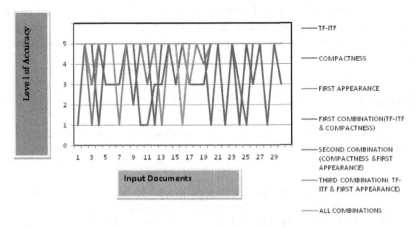

Fig. 4. Result graph for Artificial Intelligence field

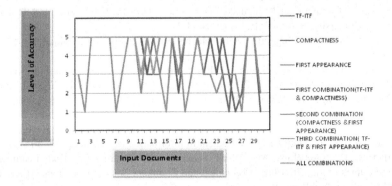

Fig. 5. Result graph for Bio- informatics field

The proposed method has achieved more accuracy when it comes to combinations especially when three features are combined and used for categorizing a document. The accuracy and efficiency had a par when it comes to combinations rather than using the features as individual ones.

5   Conclusions and Future Scope

Previous researches categorized data based on a single feature. The existing work on distribution features like compactness, First Appearance, Term Frequency are mainly relied on classifiers like navy Bayes, Support Vector machine etc. In this paper, the work was solely done with distribution features and their combinations without seeking the help of a classifier. The implementation of the features selection as combinations is done. The result of combining various features has led to an efficient way of identifying various documents besides decreasing the time complexity. Almost 100 documents of various fields were taken as input and tested upon all features and their combinations. The proposed method is unable to categorize the data based on phrases and paraphrases respectively. As a result, efficiency and   time -complexity remains a concern in a lighter vein. This supposition will be explored in the near future.

References

1. Ian, H.: Witten Computer Science, University of Waikato. Hamilton, New Zealand
2. Xue, X.-B., Zhou, Z.-H.: Distributional Features for Text Categorization. In: Fürnkranz, J., Scheffer, T., Spiliopoulou, M. (eds.) ECML 2006. LNCS (LNAI), vol. 4212, pp. 497–508. Springer, Heidelberg (2006)
3. Pattern Recognition and Machine Learning, Christopher Bishop. Springer (2006)
4. Duda, R.O., Hart, P.E., Stork, D.: Pattern Classification. Wiley and Sons
5. Ng, A.Y., Jordan, M.I.: On Discriminative vs. Generative Classifiers: A comparison of Logistic Regression and Naive Bayes. Neural Information Processing Systems (2002)
6. Li, B., Yu, S., Lu, Q.: An Improved k-Nearest Neighbor Algorithm for Text Categorization Institute of Computational Linguistics Department of Computer Science and Technology Peking University, Beijing, P.R. China, 100871
7. Auria, L., Rouslan: Support Vector Machines (SVM) as a Technique for Solvency Analysis
8. Joachims, T.: Text categorization with support vector machines: Learning with many relevant features. In: Nédellec, C., Rouveirol, C. (eds.) ECML 1998. LNCS, vol. 1398, pp. 137–142. Springer, Heidelberg (1998)
9. Lewis, D.D.: An Evaluation of Phrasal and Clustered Representations on a Text Categorization Task. In: Proceedings of the 15th Annual International ACM SIGIR Conference on Research and Development in Information Retrieval, pp. 37–50 (1992)
10. Mladeni, D., Grobelink, M.: Word Sequences as Features in Text Learning. In: Proceedings of the 17th Electro Technical and Computer Science Conference (ERK 1998). IEEE section, Ljubljana (1998)
11. Xue, X.-B., Zhou, Z.-H.: Distributional features for text categorization. In: Fürnkranz, J., Scheffer, T., Spiliopoulou, M. (eds.) ECML 2006. LNCS (LNAI), vol. 4212, pp. 497–508. Springer, Heidelberg (2006)
12. Yang, Y., Pedersen, J.O.: A Comparative Study on Feature Selection in Text Categorization. In: Proc. of Int'l Conf. on Machine Learning, pp. 412–420 (1997)
13. Zečević, A.: On feature distributional clustering for text categorization. In: Proceedings of the Student Research Workshop Associated with RANLP, pp. 145–149. Hissar, Bulgaria (September 13, 2011)

Performance Investigation of DMV (Detecting Malicious Vehicle) and D&PMV (Detection and Prevention of Misbehave/Malicious Vehicles): Future Road Map

Megha Kadam[1] and Suresh Limkar[2]

[1] Department of Computer Engineering, GHRCEM, Pune, India
[2] Department of Computer Engineering, AISSMS IOIT, Pune, India
{megha.desai1,sureshlimkar}@gmail.com

Abstract. Recently the method called DMV (detecting malicious vehicle) for VANET is presented by many researchers in order to efficiently detect and isolate the malicious vehicle from the VANET in order to reduce the information leakage. This method is purely based on the concept of certificate authority, clustering, while lists monitoring etc. However as per the performances observed during the practical studies, these methods efficiently reduces the packet drops, however consumes more time for its processing and affecting other performance parameters like end to end delay, throughput and jitter which is the most vital for any kinds of wireless communications. Therefore to overcome these issues we have proposed method called Detection and Prevention of Misbehave/Malicious Vehicles D&PMV [1]. In our previous work we have observed that this method not affecting the performances of throughput, end to end delay and jitter. Therefore this method becomes efficient and more secured as compared to DMV. The architecture of both methods presented along with their algorithms for performance investigation. This paper emphasis on the discussion and analysis over both the methods in order to show that our proposed method gives much better performance as compared to existing methods.

Keywords: Abnormal behavior, Vehicular Ad Hoc Networks, Honest vehicle, Secure communication, malicious vehicle, Detection, Prevention, DSR, MANET.

1   Introduction

More commonly VANETs are used for the intelligent transportation systems worldwide. These kinds of networks are built using the short range radio communication among the available vehicles in the VANET as well as infrastructure used at road. For security and authentication, there are many numbers of certificate authorities present in VANET. Every CA is handling the task of vehicle authentications those are already registered to particular VANET network region. Hence in short, vehicle authentication is done by validating the CA [2]. The communication between vehicles is mainly done for traffic information, current road

S.C. Satapathy, S.K. Udgata, and B.N. Biswal (eds.), *FICTA 2013*,
Advances in Intelligent Systems and Computing 247,
DOI: 10.1007/978-3-319-02931-3_43, © Springer International Publishing Switzerland 2014

conditions, notifications for emergency incident etc. This all information is vital and needs to be exchanged between vehicles. This exchange mechanism is based on the use of routing protocols such as AODV, DSR, and DSDV etc. This protocols works according to routing table, distances, forwarding mechanisms. As like MANET, VANET networks are also vulnerable to attacks like misbehaving nodes, selfish nodes etc. which are ready to damage your networks personal information [5, 6]. Attackers or malicious vehicles perform in several ways and have different objectives such as attackers eavesdrop the communication between vehicles, drop, and change or inject packets into the network [3].

To provide the security for such problems in VANET, recently many methods presented, however each of them associated with their limitations. In the recent time studied the Detecting Malicious Vehicle (DMV) which is effective and efficiently reduces the packet drops ratio in the network [4]. In [4] the design of novel misbehavior detection scheme at the application layer, called DMV, and tag vehicles using their distrust values.

In DMV, verifiers operate independently from each other. In addition, DMV can improve the performance of verifier selection at high speeds. In DMV, a number of vehicles are located in a cluster. Each cluster has one main cluster-head and one spare cluster-head, where the spare cluster-head is the trustiest vehicle after the main cluster-head. However this method is associating with limitations. The throughput is becomes very less due to this high end to end delay and jitter occurs. To overcome this limitation we proposed new approach called D&PMV [1] in which DMV is extended in order to detect and correctly mitigate presence of malicious vehicles from the network by keeping the up to date performances of throughput, delay and jitter. This approach improves these performances as compared to DMV.

This paper discusses both the papers with their algorithms, flowcharts &working. This paper is organized as section 2 discusses DMV algorithm in detail, section 3 discusses the method called D&PMV along with its architecture, section 4, the performance metrics, network scenarios used and comparative studies presented followed by comparative result.

2   Detection of Malicious Vehicles (DMV)

DMV algorithm standing based on below three approaches:

1. Abnormal Behavior: This means that vehicle node acted for dropping or duplicating packets in VANET to mislead other vehicles or abolish important messages based on their personal aims.

2. Honest (normal) Vehicles: A vehicle that has a normal behavior, where a vehicle with normal behavior forwards messages correctly or generates right messages.

3. Malicious Vehicles: If the abnormal behavior of vehicle V is repeated in such a way that Td of V becomes greater than threshold value σ, vehicle V will be called a malicious vehicle. The parameter Td represents the distrust value of behavior of vehicle V when it forwards messages.

In DMV as per [4], verifiers operate independently from each other. In addition, DMV can improve the performance of verifier selection at high speeds. In DMV, a number of vehicles are located in a cluster. Each cluster has one main cluster-head and one spare cluster-head, where the spare cluster-head is the trustiest vehicle after the main cluster-head [4]. Each vehicle is monitored by some of its trustier neighbors which are defined as verifier nodes. If verifier nodes observe an abnormal behavior from vehicle V (a node that drops or duplicates packets), they increase the distrust value of vehicle V. The identification code of vehicle V is reported to its Certificate Authority (CA) as a malicious node when its distrust value becomes higher than a threshold value.

Following algorithm shows the monitoring process for vehicle V when it joins to a cluster which is given in [4].

Algorithm 1: Malicious vehicle detection

Step 1 : Vehicle V joins the network
Step 2 : Get the cluster keys
Step 3 : Assigning the verifiers to newly joined vehicle V.
Step 4 : Start monitoring behavior of vehicle V.
Step 5 : If (verifier detecting the abnormal behavior of vehicle V)
              Report to cluster_head (CH)
                  goto step 6;
                  else
                  goto step 4;
Step 6 : CH modifies the distrust value (d)

Step 7 : If distrust value less than or equal to threshold (t) value (which is set for each new vehicle once that join the network) then update the white list of network else generates warning message to cluster agents of CH
              if (d <= t)
              goto step 4
              else
              goto step 8
Step 8 : Alarm generation in order to provide the warning message among all the other vehicles under the same cluster head
Step 9 : Finally Isolation of Detected malicious vehicle.

3   Detection and Prevention of Misbehave/Malicious Vehicles (D&PMV)

This approach presents improved performance of VANET network under the presence of malicious vehicles in the VANET. In this method we have extended above said method by adding the concepts of caching during the routing paths build up process of routing protocol. Following algorithm defines the steps which are used.

This method first scan every routing path during its creation for the existence of black hole node, if its available then we have to simply reconstruct the new path by ignoring the black hole node. This helps in reduction of packet drops. After that, remove the cache contents those having footprints of black hole nodes. This saves more time and hence resulted into less end to end delay and jitter time.

Algorithm 2: Malicious Vehicle Detection

Step 1	:	Vehicle V joins the network
Step 2	:	Get the cluster keys
Step 3	:	Assigning the verifiers to newly joined vehicle V.
Step 4	:	Start monitoring behavior of vehicle V.
Step 5	:	If (verifier detecting the abnormal behavior of vehicle V)

Report to cluster_head (CH)

goto step 6;
 else
goto step 4;

Step 6	:	CH modifies the distrust value (d)
Step 7	:	If distrust value less than or equal to threshold (t) value (which is set for each new vehicle once that join the network) then update the white list of network else generates warning message to cluster agents of CH

if (d <= t)
goto step 4
else
goto step 8

Step 8	:	Alarm generation in order to provide the warning message among all the other vehicles under the same cluster head
Step 9	:	If alarm is generated

Retrieve information of malicious vehicles
else
goto step10

Step 10	:	Pares all communication paths in network
Step 11	:	Select path from parsed paths

If malicious vehicle exists then
 Dump that path and continue
Else
Add current route in secondary cache and continue

Step 12	:	Select path from secondary cache for communication

Figure 1 shows the architecture of proposed approach.

4   Performance Investigation

Concerning to the aim of the paper, we modified the existing DSR protocol with the functionality of black hole attack detection along with its prevention for both DMV and D&PMV algorithms. Mobile nodes in the MANET acts as host vehicles and router node means nodes in the MANET are responsible for both data forwarding and routing mechanisms. But because of few malicious nodes which acts as misbehaving & selfish nodes, data packets not delivered to the destination and dropped by such nodes. These nodes are called as black hole attacker nodes.

We analyzed the performance of existing DSR protocol with this new modified security enabled DSR protocol using the performance metrics like throughput, delay and jitter. We used NS2 simulator for this work.

As far as performance investigation concern, we used following network conditions for VANET network. Table 2 showing the same:

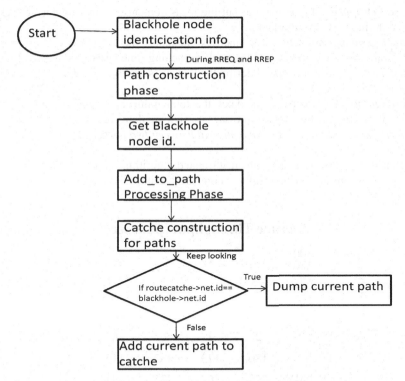

Fig. 1. Proposed process of D&PMV algorithm

Table 1. Comparison of proposed systems simulation parameters with [4]

Number of Nodes	30
Traffic Patterns	CBR (Constant Bit Rate)
Network Size	500 x 500
Max Speed	10 m/s
Simulation Time	100s
Transmission Packet Rate Time	10 m/s
Pause Time	2.0s
Routing Protocol	DSR(DMV)/DSR (P&DMV)
Number of Misbehaving Nodes	2/4/6/8/10

Performance Parameters

1) *Packet Dropped*: This is nothing but the measurement of total number of packets dropped during network simulation.

2) *End to end packet delay*: This metrics calculates the time between the packet origination time at the source and the packet reaching time at the destination. Here if any data packet is lost or dropped during the transmission, then it will not consider for the same.

3) *Throughput*: This metrics calculates the total number of packets delivered per second, means the total number of messages which are delivered per second.

4) *Jitter: This is nothing but the time interval between two packets in queue.*

Based on above network conditions, investigated algorithms and performance metrics of the same, we got following results of comparative analysis:

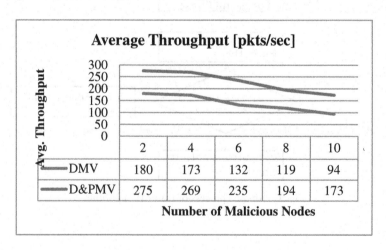

Fig. 2. Throughput Performance Analysis

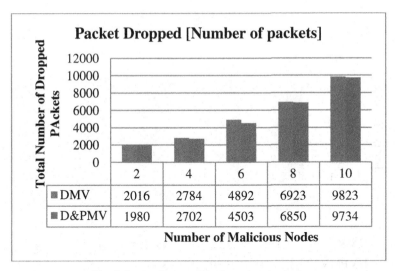

Fig. 3. Packet Dropped Performance Analysis

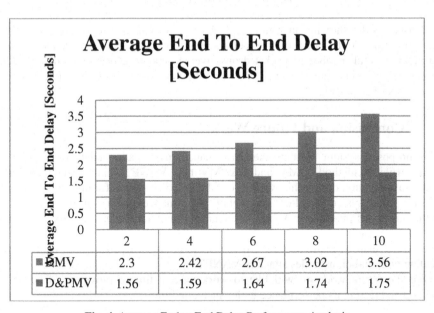

Fig. 4. Average End to End Delay Performance Analysis

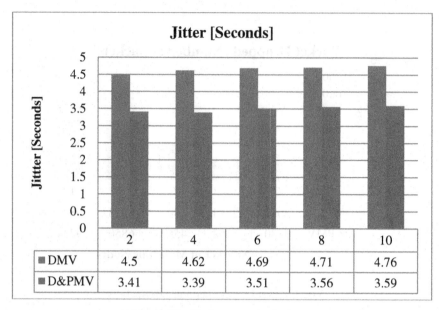

Fig. 5. Jitter Performance Analysis

Above results showing in figure 2, 3, 4 & 5, we can now conclude that DMV algorithm for VANET security is less efficient in terms of parameters such as average throughput, total number of packet drops, average end to end delay and jitter. Our proposed method is outperforming DMV and showing how it is efficient to use in real time VANET applications.

5   Conclusion and Future Work

This proposed system presents the performance investigation analysis in order to claim efficient method. DMV and D&PMV two VANET security methods were under investigation during this work. From the practical results for both this methods, our proposed system resulted into better performance under varying number of malicious nodes presence in VANET. Increasing malicious nodes resulted into more packet drops and less throughput. The main aim of this research paper is to present future road map by claiming best efficient method out of two available methods for dealing with malicious nodes attack in VANET.

For the future work we will like to work on real time deployment of proposed method and testing its performance under such conditions.

References

1. Kadam, M., Limkar, S.: D and PMV: New approach for detection and prevention of misbehave/Malicious vehicles from VANET. In: Satapathy, S.C., Udgata, S.K., Biswal, B.N. (eds.) Proceedings of the International Conference on Frontiers of Intelligent Computing: Theory and Applications (FICTA) 2013. AISC, vol. 247, pp. 287–295. Springer, Heidelberg (2014)

2. Abdulhamid, H., Tepe, K.E., Abdel-Raheem, E.: Performance of DSRC systems using conventional channel estimation at high velocities. Int. J. Electron. Commun, 556–561 (2007)
3. Artimy, M.: Local density estimation and dynamic transmission-range assignment in Vehicular Ad Hoc Networks. IEEE Trans. Intell. Transp. Syst. 8(3), 400–412 (2007)
4. Daeinabi, A., Rahbar, A.G.: Detection of malicious vehicles (DMV) throughmonitoring in Vehicular Ad-Hoc Networks. Springer (April 5, 2011)
5. Ghosh, M., Varghese, A., Gupta, A., Kherani, A.A., Muthaiah, S.N.: Detecting misbehaviors in vanet with integrated root-cause analysis. Ad Hoc Networks 8(7), 778–790 (2010)
6. Papadimitratos, P., Buttyan, L., Holczer, T., Schoch, E., Freudiger, J., Raya, M., Ma, Z., Kargl, F., Kung, A., Hubaux, J.P.: Secure vehicular communicationsystems: Design and architecture. IEEE Wireless Communication Magazine, 100–109 (2008)
7. Park, S., Aslam, B., Turgut, D., Zou, C.C.: Defenseagainst sybil attack in vehicular ad hoc network based on roadside unitsupport. In: MILCOM, pp. 1–7 (2009)
8. Parno, B., Perrig, A.: Challenges in security vehicular networks. In: HotNets-IV (2005)
9. Klaus, P., Nowey, T., Mletzko, C.: Towards a security architecturevfor vehicular ad hoc networks. In: First International Conference on Availability, Reliability and Security (2006)
10. Biswas, S., Haque, M.M., Misic, J.V.: Privacy andanonymity in vanets: A contemporary study. Ad Hoc & Sensor Wireless Networks 10(2-3), 177–192 (2010)
11. Boneh, D., Boyen, X., Shacham, H.: Short group signatures. In: Franklin, M. (ed.) CRYPTO 2004. LNCS, vol. 3152, pp. 41–55. Springer, Heidelberg (2004)

Performance Analysis of OFDM Based DAB Systems Using BCH-FEC Coding Technique

Arun Agarwal[1] and Kabita Agarwal[2]

[1] Department of Electronics and Communication Engineering
ITER College, Siksha 'O' Anusandhan University
Bhubaneswar-751030, Odisha, India
arun.agarwal23@gmail.com
[2] Department of Electronics and Telecommunication Engineering
CV Raman College of Engineering
Bhubaneswar-751030, Odisha, India
akkavita22@gmail.com

Abstract. Radio broadcasting technology in this era of compact disc is expected to deliver high quality audio programmes in mobile environment. The Eureka-147 Digital Audio Broadcasting (DAB) system with coded OFDM technology accomplish this demand by making receivers highly robust against effects of multipath fading environment. In this paper, we have analyzed the performance of DAB system conforming to the parameters established by the ETSI (EN 300 401) using time and frequency interleaving, concatenated Bose-Chaudhuri-Hocquenghem coding and convolutional coding method in different transmission channels. The results show that concatenated channel coding improves the system performance compared to convolutional coding.

Keywords: DAB, OFDM, Multipath effect, concatenated coding.

1 Introduction

Digital Audio Broadcasting (DAB) was developed within the European Eureka-147 standard [1] to provide CD quality audio programmes (mono, two-channel or multichannel stereophonic) along with ancillary data transmission (e.g. travel and traffic information, still and moving pictures, etc.) to fixed, portable and mobile receivers using simple whip antennas [2]. The reception quality of analog AM/FM systems on portable radio is badly affected by Multipath fading (reflections from aircraft, vehicles, buildings, etc.) and shadowing [3]. These systems also suffer from interference from equipment, vehicles and other radio stations. DAB uses coded orthogonal frequency division multiplexing (COFDM) technology to combat the effects of Multipath fading and inter symbol interference (ISI). Additionally the VHF frequency band available for the sound broadcasting throughout the world has either saturated or fast approaching saturation. There is a need for more spectrally efficient broadcasting technology apart from conventional analog systems. Since DAB uses OFDM technology therefore the system can operate in single frequency networks (SFNs) providing the efficient usage of available radio frequency spectrum.

S.C. Satapathy, S.K. Udgata, and B.N. Biswal (eds.), *FICTA 2013*,
Advances in Intelligent Systems and Computing 247,
DOI: 10.1007/978-3-319-02931-3_44, © Springer International Publishing Switzerland 2014

Earlier work focused more on the effect of protection levels in diverse transmission channels, design and implementation of DAB channel decoder (physical layer) on FPGA hardware [5, 13, 14]. There are many forms of concatenated coding techniques. In this paper we propose a BCH coding based concatenated channel coding technique for improved performance of DAB system in different transmission channels.

In this paper we developed a DAB base-band transmission system based on Eureka-147 standard [1]. The design consists of energy dispersal scrambler, QPSK symbol mapping, convolutional encoder (FEC), time interleaver, D-QPSK modulator with frequency interleaving and OFDM signal generator (IFFT) in the transmitter side and in the receiver corresponding inverse operations is carried out along with fine time synchronization [4] using phase reference symbol. DAB transmission mode-II is implemented. A frame based processing is used in this work. Bit error rate (BER) has been considered as the performance index in all analysis. The analysis has been carried out with simulation studies under MATLAB environment.

Following this introduction the remaining part of the paper is organized as follows. Section 2 introduces the DAB system standard. Section 3, provides brief overview of the DAB system. In Section 4, the details of the modeling and simulation of the system using MATLAB is presented. Then, simulation results have been discussed in Section 5. Finally, Section 6 provides the conclusions.

2   Introduction to DAB System

The working principle of the DAB system is illustrated in conceptual block diagram shown in Fig. 1. At the input of the system the analog signals such as audio and data services are MPEG layer-II encoded and then scrambled. In order to ensure proper energy dispersal in the transmitted signal, individual inputs of the energy dispersal scramblers is scrambled by modulo-2 addition with a pseudo-random binary sequence (PRBS), prior to convolutional coding [1]. The scrambled bit stream is then subjected to forward error correction (FEC) employing punctured convolutional codes with code rates in the range 0.25-0.88. The coded bit-stream is then time interleaved and multiplexed with other programs to form Main Service Channel (MSC) in the main service multiplexer.

The output of the multiplexer is then combined with service information in the Fast Information Channel (FIC) to form the DAB frame. Then after QPSK mapping with frequency interleaving of each subcarriers in the frame, $\pi/4$ shifted differential QPSK modulation is performed. Then the output of FIC and MSC symbol generator along with the Phase Reference Symbol (PRS) which is a dedicated pilot symbol generated by block named synchronization symbol generator is passed to OFDM signal generator. This block is the heart of the DAB system. Finally, the addition of Null symbol to the OFDM signal completes the final DAB Frame structure for transmission.

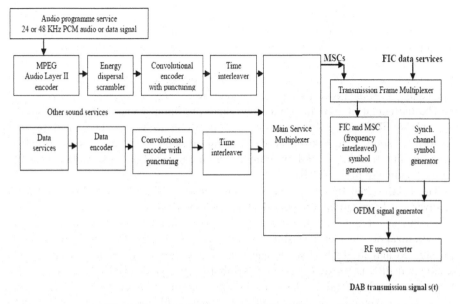

Fig. 1. Complete DAB transmitter block diagram [1]

3   Overview of DAB System

The Eureka-147 DAB system consists of three main elements. These are Source coding, Channel coding, multiplexing and transmission frame and COFDM. These technical aspects make the system very reliable, multi-programme, providing robust reception on mobile receivers using simple antennas.

3.1   Source Coding

Source coding employs MUSICAM (Masking Pattern Universal Sub-band Integrated Coding And Multiplexing) audio coding that uses the principle of Psycho acoustical masking as specified for MPEG-2 Audio Layer-II encoding. This exploits the knowledge of the properties of human sound perception, particularly, the spectral and temporal masking effects of the ear. Principle of MUSICAM audio coding system is that it codes only audio signal components that the ear will hear, and discards any audio component that, according to the Psycho acoustical model, the ear will not perceive [7]. This technique allows a bit rate reduction 768 Kbit/s down to about 100 Kbit/s per mono channel, while preserving the subjective quality of the digital audio signal. This allows DAB to use spectrum more efficiently and delivering high quality sound to the listeners.

3.2   Channel Coding, Multiplexing and Transmission Frame

Channel coding is based on punctured convolutional forward-error-correction (FEC) which allows both equal and Unequal Error Protection (UEP), matched to bit error sensitivity characteristics [1]. Using rate compatible punctured convolutional (RCPC) codes, it is possible to use codes of different redundancy in the transmitted data stream in order to provide ruggedness against transmission distortions, without the need for different decoders [2]. Basic idea of RCPC channel coding is to generate first the mother code. The daughter codes will be generated by omitting certain redundancy bits, the process known as puncturing.

The individual programme (audio and data) are initially encoded, error protected by applying FEC and then time interleaved. These outputs are then combined together to form a single data stream ready for transmission. This process is called as Multiplexing. In DAB several programmes are multiplexed into a so-called ensemble with a bandwidth of 1.536 MHz

The DAB signal frame has the structure shown in Fig. 2 that helps in efficient receiver synchronization. The period TF of each DAB transmission frame is of 24 ms or an integer multiple of it. According to system standard the first symbol is the Synchronization channel consisting of Null symbol during which no information is transmitted and the PRS symbol. The null symbol is used to estimate rough frame timing and PRS is the dedicated pilot symbol having predetermined modulation for fine time synchronization. The next symbol is the FIC channel which carries Multiplex Configuration Information (MCI). It has fixed symbols which are known to the receivers to decode any of the sub-channels instantly. The FIC is made up of FIBs (Fast Information Block). The FIBs contains 256 bits. The FIC data is a non-time-interleaved channel with fixed equal error protection [1] code rate (1/3).

The last symbol is the MSC channel that carry audio and data service component. It forms the main payload of the DAB frame. The MSC is a time interleaved data channel divided into a number of sub-channels which are individually convolutionally coded, with equal or unequal error protection (EEP or UEP). Each sub-channel may carry one or more service components. The MSC is made up of CIFs. The CIF contains 55296 bits. The smallest addressable unit of the CIF is the Capacity Unit (CU), comprising 64 bits. Therefore, the CIF contains 864 CUs.

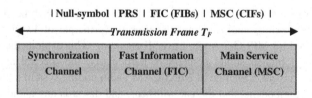

Fig. 2. DAB transmission signal frame structure

3.3   COFDM

DAB uses COFDM technology that makes it resistant to Multipath fading effects and inters symbol interference (ISI). OFDM is derived from the fact that the high serial bit stream data is transmitted over large (parallel) number sub-carriers (obtained by dividing the available bandwidth), each of a different frequency and these carriers are orthogonal to each other. OFDM converts frequency selective fading channel into N flat fading channels, where N is the number of sub-carriers. Orthogonality is maintained by keeping the carrier spacing multiple of 1/Ts by using Fourier transform methods, where Ts is the symbol duration. Since channel coding is applied prior to OFDM symbol generation which accounts for the term 'coded' in COFDM.

3.4   DAB Transmission Modes and System Parameters

The Eureka 147 DAB [1] system has four transmission modes of operation named as mode-I, mode-II, mode-III, and mode-IV, each having its particular set of parameters. The use of these transmission modes depends on the network configuration and operating frequencies. This makes the DAB system operate over a wide range of frequencies from 30 MHz to 3 GHz.

The details of DAB system parameters for all the four transmission modes are shown in Table-1. All the four DAB modes have same signal bandwidth of 1.536 MHz, 2 bits per carrier per symbol (D-QPSK modulation) and sampling frequency of 2.048 MHz. It may be seen from Table I that Transmission mode-II has 384 sub-carriers at 4 KHz spacing. OFDM symbol length (Ts) is 312 μs. If channel impulse response is < 62 μs then there will be no ISI. Orthogonality between sub-carriers is maintained if sinusoids have integer number of cycles in Ts given by equation below

Table 1. System parameters for Four DAB transmission modes

System Parameter	Mode -I	Mode -II	Mode -III	Mode -IV
No. of sub-carriers	1536	384	192	768
OFDM symbols/frame	76	76	153	76
Transmission frame duration	196608 T	49152T 24 ms	49152 T 24 ms	98304 T 48 ms
Null-symbol duration	2656 T 1297 ms	664 T 324 μs	345 T 168 μs	1328 T 648 μs
OFDM symbol duration	2552 T 1246 ms	638 T 312 μs	319 T 156 μs	1276 T 623 μs
Inverse of carrier spacing	2048 T 1 ms	512 T 250 μs	256 T 125 μs	1024 T 500 μs
Guard interval	504 T 246 μs	126 T 62 μs	63 T 31 μs	252 T 123 μs
Max. RF	375 MHz	1.5GHz	3 GHz	750MHz
Sub-carrier spacing	1 kHz	4 kHz	8 kHz	2 kHz
FFT length	2048	512	256	1024

$$\int_0^{T_s} \sin(2\pi f_c t)\,\sin(2\pi 2 f_c t)\,dt = 0 \tag{1}$$

Where is the sub-carrier frequency. Orthogonality in frequency domain is kept by taking FFT of 4 ms segment (mode-II) which is equivalent to convolving spectrum with sinc and sampling at multiples of 4 KHz. Mode-II can be designed to operate as a SFN where all the transmitter radiate over same frequency identical OFDM symbols at the same time. Optimal spacing between transmitters can be evaluated from the guard time interval Tg and velocity of light c as given by following equation

$$s = T_g \times c \qquad (2)$$

Equation (2) gives transmitter spacing of 18.6 Km for mode-I.

4   The Simulation Model

Fig. 3 presents the complete block diagram of the DAB system which was modeled and simulated by us in MATLAB environment. The main objective of this simulation study is to evaluate the BER performance of the DAB system using concatenated coding technique. The simulation parameters are obtained from Table I for transmission mode-II. A frame based processing is used in this simulation model. The system model was exposed to AWGN channel, Rayleigh fading channel and Rician channel for performance analysis. The important blocks of the simulation model is discussed in detail as follows:

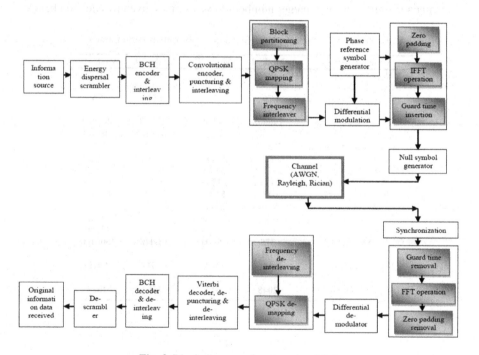

Fig. 3. Block diagram of system simulated

4.1   Concatenated Channel Coding

The virtually error free channel can be achieved by concatenated coding using convolutional code as 'inner code' together with a BCH code as 'outer code' given in Fig. 4. This technique improves the BER performance of the DAB system in different transmission channels as will be shown in Section V. The outer coder is accomplished with BCH code (511,439, t = 8). The principal advantage of BCH code is the ease with which they can be decoded using an algebraic method called as Syndrome decoding. This allows very simple electronic hardware to perform the task and the decoder can be made small and low powered. BCH codes are very flexible allowing control over the selection of the block length and belong to a class of multiple random error correcting codes. It is a polynomial code over a finite field called Galois array with a particularly chosen generator polynomial. These are cyclic codes.

The inner coding is based on punctured convolutional forward-error-correction (FEC) which allows both Equal and Unequal Error Protection. We have used convolutional encoder with constraint length 7 with octal forms of generator polynomials as 133, 171, 145 and 133, respectively [1].

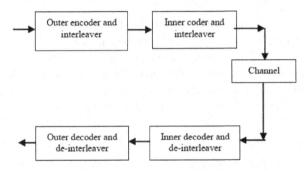

Fig. 4. Block diagram of concatenated coding

4.2   Synchronization

Fine time synchronization or symbol timing synchronization [4] is performed by calculating the Channel Impulse Response (CIR) based on the actually received time frequency PRS and the specified PRS stored in the receiver. Multiplication of received PRS with complex conjugate of PRS at the receiver results in cancellation of the phase modulation of each carrier. The phase reference symbol can be converted to impulse signal or CIR can be obtained by an IFFT operation of the resultant product as illustrated in following formula:

$$\text{CIR} = \text{IFFT}\{\text{Received PRS} \cdot \text{PRS*}\} \tag{3}$$

Where PRS* is the complex conjugate of the phase reference symbol. The peak of the impulse signal obtained from (4) will give position of the start of the PRS compared to a set threshold (T) providing correct symbol timing as well as frame timing.

4.3   Viterbi Decoding

To minimize the transmission errors due to channel impairments the DAB system in the transmitter employed powerful rate compatible punctured convolutional code (RCPC) with constraint length 7 and mother code rate of 1/4 for channel coding. For decoding these codes the Viterbi algorithm [11] will be used, which offers best performance according to the maximum likelihood criteria. The input to the Viterbi decoder will be hard-decided bits that are '0' or '1', which is referred to as a hard decision. Computational requirements or complexity of Viterbi decoder grow exponentially as a function of the constraint length (L), so it is usually limited in practice to constraint length of L = 9 or less.

5   Simulation Results and Discussion

In this section we have presented the simulation results along with the bit error rate (BER) analysis for AWGN channel, Rayleigh fading channel and Rice channel. The results are shown for transmission mode-II and the simulation parameters are taken as per the DAB standard [1].

First of all the correctness of the DAB simulation model given in Fig. 3 will be tested. Fig. 5 presents the system BER performance in AWGN channel. It can be seen from Fig. 5 that both experimental and theoretical BER plots are same and almost overlapping each other. This justifies that the DAB system model simulated is perfectly implemented. The result also indicates that to achieve a BER of 10^{-4} theoretical $\pi/4$ D-QPSK needs an additional SNR of 4.3 dB compared to theoretical BPSK.

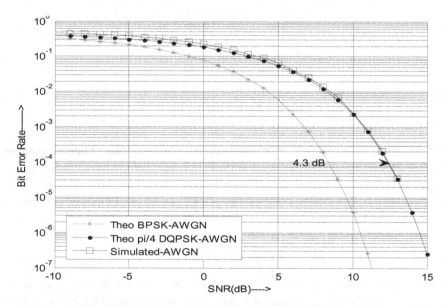

Fig. 5. BER performance of DAB mode-II in AWGN channel

The performance of DAB system with FEC coding is analyzed next. No puncturing was applied. Decoding was done with Viterbi algorithm. Fig. 6 presents the result for the DAB system with FEC coding. From the Fig. 6 it can be seen that the use of the channel coding improves the BER performance of the DAB system. It can be evaluated from above figure that to achieve a BER of 10^{-4} coded DAB system without puncturing gives a coding gain of approximately 8 dB compared with the uncoded system.

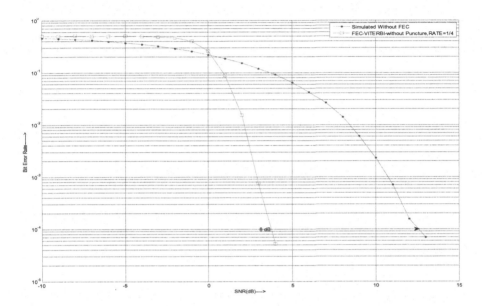

Fig. 6. BER performance with and without FEC coding in AWGN channel

The performance analysis using concatenated channel coding technique is investigated next. Here we use BCH coding as the outer coding and convolutional coding as the inner coding. Codeword length was taken as 511 and Message length to be 439 for error correcting capability of eight. Fig. 7 presents the results for BCH-FEC coding in AWGN channel. Fig. 7 presents that concatenated coding (employing outer BCH coding and inner convolutional coding) improves the BER performance marginally compared with only convolutional coding. A coding gain of about 0.5 dB is observed in AWGN channel.

After analyzing the BER performance in AWGN channel, the performance analysis in Rayleigh fading channel was investigated next. Fig. 8 presents the results for BCH-FEC coding in fading channel with Doppler frequency 40 Hz (i.e., v= 48 km/hr). Fig. 8 reveals that concatenated coding (employing outer BCH coding and inner convolutional coding) improves the BER performance compared with only convolutional coding. It provides a coding gain of about 1 dB.

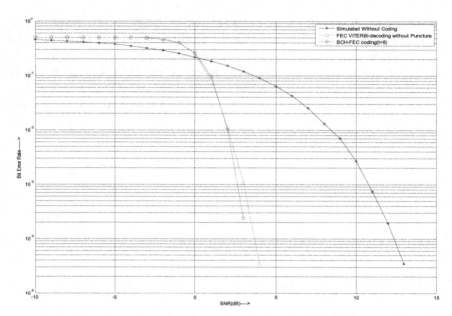

Fig. 7. BER performance with concatenated coding in AWGN channel

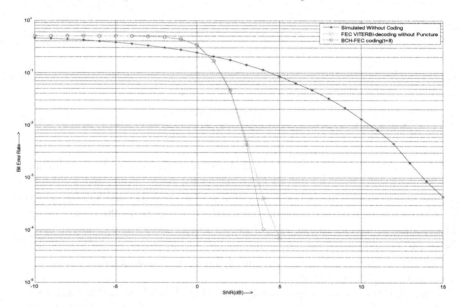

Fig. 8. BER performance with concatenated coding in Rayleigh fading channel

Similarly the performance of the system with concatenated coding in Rician channel will be investigated next. Fig. 9 presents the results for BCH-FEC coding in a Rice channel.Fig. 9 indicates that concatenated coding (employing outer BCH coding and inner convolutional coding) improves the BER performance marginally compared with only convolutional coding. It provides a coding gain of about 0.5 dB in Rice channel.

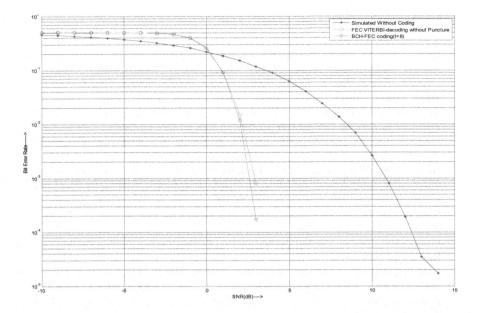

Fig. 9. BER performance with concatenated coding in Rice fading channel

6   Performance Analysis

The proposed concatenated channel coding provides an improved BER performance in different channels for OFDM-based DAB system. From the simulation results it is observed that FEC is practically well suited for channel coding giving a coding gain of about 8 dB in AWGN channel, 12 dB in Rayleigh fading channel and 8.2 dB in Rician channel. Results shows that system employing BCH combined with FEC coding gave a coding gain of 0.5 dB in AWGN channel & Rice channel and coding gain of 1 dB in fading channel for the to achieve a BER of 10^{-4}.

7   Conclusion

A simulation based performance analysis of DAB system using concatenated coding technique is described in this paper to evaluate the effectiveness of DAB system as a radio broadcasting technology in different transmission channels. Comparison was made between the performances of the concatenated coding with only convolutional coding. The parameter of functional blocks can be changed to visualize the change in system performance. According to simulations, DAB appears to be suitable radio broadcasting technology for high performance in diverse transmission channels.

References

1. ETSI, Radio Broadcasting Systems; Digital Audio Broadcasting (DAB) to mobile, portable and fixed receivers, EN 300 401, V1.3.3 (2001-05) (April 2001)
2. Hoeg, W., Lauterbach, T.: Digital Audio Broadcasting-Principles and Applications. John Wiley & Sons, Ltd. (2001)
3. Kozamernik, F.: Digital Audio Broadcasting – radio now and for the future. EBU Technical Review, no.265 Autumn (1995)
4. Ernest, P.P.: DAB implementation in SDR, University of Stellenbosch. Master's thesis (December 2005)
5. Gaetzi, L.M., Hawksford, M.O.J.: Performance prediction of DAB modulation and transmission using Matlab modeling. In: IEEE International Symposium on Consumer Electronics – Proceedings, pp. 272–277 (2004)
6. Bilbao, H.U.: Dab Transmission System Simulation, Linkoping Institute of Technology. Master's thesis (August 2004)
7. Bower, A.J.: DIGITAL RADIO–The Eureka 147 DAB System. Electronic Engineering BBC (April 1998)
8. ETSI TR 101 496-3, Digital Audio Broadcasting (DAB); Guidelines and rules for implementation and operation; Part 3: Broadcast network, V1.1.2 (2001-2005) (2001)
9. Harada, H., Prasad, R.: Simulation and Software Radio for mobile communications. Artech House (2003)
10. Singh, R.P., Sapre, S.D.: Communication Systems, 2nd edn. Tata McGraw-Hill Education Pvt. Ltd. (2007)
11. Proakisp, J.G.: Digital Communications, 3rd edn. McGraw-Hill (1995)
12. MATHWORKS, http://www.mathworks.com/CommunicationsToolbox/ ErrorDetectionandCorrection/Channels/ AWGNchannel&Fadingchannels
13. Yang, K.-S., Yu, C.-T., Chang, Y.-P.: Improved Channel Codec Implementation and Performance Analysis of OFDM based DAB Systems. In: Proceedings of the 2006 International Wireless Communications and Mobile Computing Conference, IWCMC 2006, pp. 997–100 (2006)
14. Holisaz, H., Mehdi Fakhraie, S., Moezzi-Madani, N.: Digital Audio Broadcasting System Modeling and Hardware Implementation. In: Proceedings of the IEEE Asia-Pacific Conference on Circuits and Systems, APCCAS, pp. 1814–1817 (2006)
15. Agarwal, A., Patra, S.K.: Performance prediction of Eureka-147 DAB system Using Interleaving and different Coding rates. In: Proceedings of the IEEE International Conference on MEMS, Nano & Smart Systems, ICMENS, vol. 403-408, pp. 4119–4125 (2011) ISBN-13: 978-3-03785-312-2
16. Agarwal, A., Agarwal, K.: Design and Simulation of COFDM for high speed wireless communication and Performance analysis. IJCA International Journal of Computer Applications 2, 22–28 (2011) ISBN: 978-93-80865-49-3

A Study of Incomplete Data – A Review

S.S. Gantayat[1], Ashok Misra[2], and B.S. Panda[3,*]

[1] GMRIT, Rajam, Andhra Pradesh
[2] CUTM, Parlakhemundi, Odisha
[3] MITS Engineering College, Rayagada, Odisha
`sasankosekhar.g@gmrit.org`, `amisra1972@gmail.com`,
`bspanda@sify.com`

Abstract. Incomplete data are questions without answers or variables without observations. Even a small percentage of missing data can cause serious problems with the analysis leading to draw wrong conclusions and imperfect knowledge. There are many techniques to overcome the imperfect knowledge and manage data with incomplete items, but no one is absolutely better than the others.

To handle such problems, researchers are trying to solve it in different directions and then proposed to handle the information system. The attribute values are important for information processing. In the field of databases, various efforts have been made for the improvement and enhance of database query process to handle the data. The different researchers have tried and are trying to handle the imprecise and/or uncertainty in databases. The methodology followed by different approaches like: Fuzzy sets, Rough sets, Boolean Logic, Possibility Theory, Statistically Similarity etc.

Keywords: Data, Uncertainty, Incomplete Information, Missing Data, Expert Systems.

1 Introduction

Data is a rough idea through which an object cannot be identified. Organized collection of related data is served as information. Today information acquisition and processing is a very difficult part for the users due to the unavailability of complete information from the source or the data may be corrupted at the time of transmission.

The growth of the size of data and number of existing databases far exceeds the ability of humans to analyze this data, which creates both a need and an opportunity to extract knowledge from databases [*Han and Kamber, 2008 & Kantadzic, 2005*]. Medical databases have accumulated large quantities of information about patients and their medical conditions. Relationships and patterns within this data could provide new medical knowledge. Analysis of medical data is often concerned with treatment of incomplete knowledge, with management of inconsistent pieces of information and with manipulation of various levels of representation of data.

---

* Corresponding author.

S.C. Satapathy, S.K. Udgata, and B.N. Biswal (eds.), *FICTA 2013*,
Advances in Intelligent Systems and Computing 247,
DOI: 10.1007/978-3-319-02931-3_45, © Springer International Publishing Switzerland 2014

Now-a-days the uncertainty in the datasets is the major problem to get complete information of a particular object or to develop an Expert system to retrieve accurate information from the existing one.

There are several works carried out to handle incomplete information to retrieve information from an incomplete dataset. Some of them are based on Rough sets, Fuzzy sets, Statistical methods, Bayesian Classifiers, Similarity etc. But till now no result came to implement these methods due to the drawbacks in their methodologies or to handle large datasets.

It is known that Rough sets and Fuzzy sets handle the uncertainty in the datasets. Rough set concerns on the classification of data from the datasets giving the core and reducts to identify the importance of a particular or group of data, while Fuzzy sets concerns on the degree of membership of a particular data based on the partial existence in a dataset. Bayesian classifier uses the Bayesian network to show the flow of the information in a dataset and their relationships. The similarity relation measures the similarity between the data in a particular dataset either using the tolerance relation or Fuzzy relation, which is a new direction to find out the exact information of a data.

Leading to draw wrong conclusions and imperfect knowledge, there are many techniques to manipulate the imperfect knowledge and manage data with missing items, but no one is absolutely better than the others. Different situations require different solutions. So it is required to find a suitable method do handle the missing values from the databases.

2   Problems with Incomplete Information

Most information systems usually have some missing values due to unavailability of data or after processing the data. Missing values minimizes the quality and quantity of classification rules generated by a data mining system. These values could influence the coverage percentage and number of rules generated and lead to the difficulty of extracting useful information from the data sets.

It is reported that firewalls, antiviral and other security software control the process of downloading from network and writing data to disks. These programs interfere in opening network connections and in file creation process. Several firewalls and some antiviral software mix the data of different connections, and thus they mix file parts. Also many other cases are known when security programs caused a data corruption. Some customers report on high CPU load and timeouts etc., and some other customers report on data corruption.

It is the common problem existing from the beginning of the data handling in different applications.

It is observed that assigning a "*null*" value to all missing attribute values may cause a serious effect in data analysis and decision analysis because the missing or incomplete values are just "*missed*" but they do exist and have an influence on the decision.

3   Review of Research Directions on Incomplete Information

In this section some of the concepts of the earlier research works on missing or incomplete data is discussed.

It is known that the intelligent techniques of data analysis are mainly based on quite strong assumptions (some knowledge about dependencies, probability distributions. Large number of experiments), are unable to derive conclusions from incomplete knowledge, or cannot manage inconsistent pieces of information [*Lavrajc, Keravnou and Zupan, 1997*].

3.1   Imputation

Imputation is a class of methods by which estimation of the missing value or its distribution is used to generate predictions from a given model. In particular, either a missing value is replaced with an estimation of the value or alternatively the distribution of possible missing values is estimated and corresponding model predictions are combined probabilistically, when various imputation treatments for missing values in historical or training data are available that may also be deployed at prediction time. However, some treatments such as multiple imputations [*Rubin, 1987*] are particularly suitable to induction. In particular, multiple imputation (or repeated imputation) is a Monte Carlo approach that generates multiple simulated versions of a data set in which each are analyzed and the results are combined to generate inference.

3.2   Predictive Value Imputation (PVI)

With value imputation, missing values are replaced with estimated values before applying a model. Imputation methods vary in complexity. For example, a common approach in practice is to replace a missing value with the attribute's mean or mode value (for real-valued or discrete-valued attributes, respectively) as estimated from the training data. An alternative is to impute with the average of the values of the other attributes of the test case. More rigorous estimations use predictive models that induce a relationship between the available attribute values and the missing feature. Most commercial modeling packages offer procedures for predictive value imputation. The method of surrogate splits for classification trees [*Breiman et al., 1984*] imputes based on the value of another feature, assigning the instance to a subtree based on the imputed value. As noted by Quinlan (1993), this approach is a special case of predictive value imputation.

3.3   Distribution-Based Imputation (DBI)

For an estimated distribution over the values of an attribute, one may estimate the expected distribution of the target variable, that is, the weighting the possible assignments of the missing values. This strategy is common for applying classification trees in AI research and practice, because it is the basis for the missing value treatment implemented in the commonly used tree induction program, C4.5 [*Quinlan, 1993*]. Specifically, when the C4.5 algorithm is classifying an instance, and a test regarding a missing value is encountered, the example is split into multiple pseudo-instances each with a different value for the missing feature and a weight corresponding to the estimated probability for the particular missing value, based on the frequency of values at this split in the training data. Each pseudo-instance is routed down the appropriate tree branch according to its assigned value. Upon reaching a leaf node, the class-membership probability of the pseudo-instance is assigned as the frequency of the class in the training instances associated with this leaf. The overall estimated probability of class membership is calculated as a weighted average of class membership probabilities over all pseudo-instances. If there is more than one missing value, the process recourses with the weights combining multiplicatively. This treatment is fundamentally different from value imputation because it combines the classifications across the distribution of an attribute's possible values, rather than merely making the classification based on its most likely value.

3.4   Unique-Value Imputation (UVI)

There are some methods which uses the simple substitution of the values from historic information and in some cases, rather than estimating an unknown feature value it is possible to replace each missing value with an arbitrary unique value. Unique-value imputation is preferable when the following two conditions hold [*Ding and Simonoff, 2006*],

(i) the fact that a value is missing depends on the value of the class variable, and

(ii) this dependence is present both in the training and in the application or test data.

3.5   Reduced-Feature Models

Imputation is required when the model being applied to an attribute whose value is missing in the test instance. For imputing with the average of other features may difficult, but in certain cases it is a reasonable choice. For example, for surveys and subjective product evaluations, there may be very little variance among a given subject's responses, and a much larger variance between subjects for any given question. a new classification tree could be induced after removing from the training data the features corresponding to the missing test feature. This reduced-model approach may potentially employ a different model for each test instance. This can be accomplished by delaying model induction until a prediction is required, a strategy presented as "*lazy*" classification tree [*Friedman et al., 1996*]. Alternatively, for reduced-feature modeling one may store many models corresponding to various patterns of known and unknown test features.

3.6   Replacing Missing Data

There are many approaches to handle missing attribute values in data mining [*Grzymala-Busse 1991; Grzymala- Busse et al. 1999b; Michalski et al. 1986; Quinlan 1993*]. So far we experimented with the closest fit algorithm for missing attribute values, based on replacing a missing attribute value by an existing value of the same attribute in another case that resembles as much as possible the case with the missing attribute values [*Grzymala-Busse et al. 1999*].

3.7   Rough Sets

The main advantage of rough set theory, introduced by Z. Pawlak in 1982 [*Pawlak 1982, 1991; Pawlak et al. 1995*], is that it does not need any preliminary or additional information about data (like probability in probability theory, grade of membership in fuzzy set theory, etc.). In rough set theory approach inconsistencies are not removed from consideration. Instead, lower and upper approximations of the concept are computed. On the basis of these approximations, LERS computes two corresponding sets of rules: certain and possible, using algorithm LEM2 [*Grzymala-Busse, 1992*].

3.8   Tolerance Relation

In the literature there are several extensions of relations such as tolerance relation, non-symmetric relation, and valued tolerance relation. The problem of missing values handling within the rough set framework has been already discussed in the literature [*Grzymala-Busse, 1991*]. These approaches consider alternative definitions of the discernibility which reflect various semantics of missing attribute values. In [*Grzymala-Busse and Hu, 2001, Grzymala-Busse and Goodwin, 1999*], the authors performed computational studies on the medical data, where unknown values of the attributes were replaced using probabilistic techniques. Recently, Greco et al. used a specific definition of the discernibility relation to analyze unknown attribute values for multicretria decision problems [*Greco, Matarazzo and Slowinski, 1999*]. In [*Stefanowski and Tsouki`as, 1999*] two different semantics for incomplete information "missing values" and "absent values" were discussed also; they introduced two generalizations of the rough set theory to handle these situations. In [*Nakata and Sakai, 2005*] the authors examined methods of valued tolerance relations. They proposed a correctness criterion to the extension of the conventional methods which is based on rough sets for handling missing values.

3.9   LERS

In the research, the main data mining tool was LERS (Learning from Examples based on Rough Sets), developed at the University of Kansas [*Grzymala-Busse, 1992*] LERS has proven its applicability having been used for years by NASA Johnson Space Center (Automation and Robotics Division), as a tool to develop expert systems of the type most likely to be used in medical decision-making on board the International Space Station. LERS handles inconsistencies using rough set theory. LERS is used to find the missing values in a dataset.

3.10   Similarity Relation

[*E. A. Rady, et al, 2007*] introduced a new definition of similarity relation MSIM which depends on some important conditions concerning the number of missing values. They enhanced and developed Krysckiewiczs' work *[Krysckiewiczs, 1998, 1999 and Krysckiewiczs and Rybinski, 2000]* by introducing some essential restrictions and conditions on the similarity between objects and introducing the modified similarity relation (MSIM). The essential point of MSIM is making the generalized decisions which have taken over the whole set of attributes a reasonable combination of decisions, so that the reducts can be computed easily and the generalized decision has a valuable meaning. Also, a discernibility matrix is defined, reducts for IIS and for DT are derived, and the decision rules are introduced with the set approximations are defined.

3.11   Other Methodologies

There are so other techniques are used to overcome the problem of the incomplete or missing data. The intelligent techniques used in medical data analysis are neural network [*Wolf, Oliver, Herbert and Michael, 2000*], Bayesian classifier [*Cheesemen and Stutz, 1996*], classification model [*Saar-Tsechansky and Foster, 2007 & Setiono, 2000*], decision/classification trees [*Hassanien, 2003, Ding and Simonoff, 2006*], Cluster-Based Algorithms [*Fujikawa and Ho, 2002*], Statistical Analysis [*Little and Rubin, 2002*], Rough set Model [*Grzymala-Busse, 1997, 1999, 2001, 2004, 2005 etc.*], some of these techniques already discussed above.

4   Significance of the Study

In current era, all the information is stored in digital media. The information may be corrupted; when there will be a noise introduced in the digital media. Sometimes it may happen that the data may be partially lost during the transmission, failure of storage media etc. In the case of complete loss of information we cannot retrieve all the data.

In real life application everyone is facing so many problems to handle missing or incomplete data. For example, consider the bio-data of employees, market research etc., where in most cases it is found that, in the bio-data of an employee some information provided are missing. In market research, it is also found that some questions have no answer or may be by mistake eliminated. In such situations it is required to retrieve actual information to help the industries, the organizations and the decision makers, for decision making for their strategic actions.

So it is required to develop a new kind of algorithm or methodology to handle the missing attribute values or ill known data under imprecise and/or uncertainty for retrieving information using some real and online dataset so that for any decision making process it may be used. It can further be applied in the field of Expert system development to make decisions in different areas either in the education or in the industry.

5   Conclusion

In this paper we have studied some proposed algorithms given in the different research outcomes and to find why these algorithms are not efficient to handle the uncertainty in the incomplete datasets or information tables. It is required to try to eliminate any drawbacks available in these methodologies either giving a modified algorithms or propose a new algorithm based on the feasibility of the datasets and to handle the missing attribute values or ill known data or missing data under imprecise and/ or uncertainty for retrieving information.

These algorithm(s) will be helpful to find the approximated missing values of the attributes of the materials, to check the consistencies of the chemical properties, identifying the unknown behaviour of an object from a group of data, identifying the part of the missing signal from a digital signal, etc.

References

[Han J. and M. Kamber, 2001] Han, J., Kamber, M.: Data Mining: Concepts and Techniques. Morgan Kaufmann, San Francisco (2001)

[Grzymala-Busse, 2004] Grzymala-Busse, J.W.: Three Approaches to Missing Attribute Values- A Rough Set Approach. In: Workshop on Foundations of Data Mining, Associated with 4th IEEE International Conference on Data Mining, Brighton, UK (2004)

[Grzymala-Busse and Hu, 2001] Grzymala-Busse, J.W., Hu, M.: A Comparison of Several Approaches to Missing Attribute Values in Data Mining. In: Ziarko, W.P., Yao, Y. (eds.) RSCTC 2000. LNCS (LNAI), vol. 2005, pp. 378–385. Springer, Heidelberg (2001)

[Grzymala-Busse, 2005] Grzymala-Busse, J.W.: Incomplete Data and Generalization of Indiscernibility Relation, Definability, and Approximations. In: Ślęzak, D., Wang, G., Szczuka, M.S., Düntsch, I., Yao, Y. (eds.) RSFDGrC 2005. LNCS (LNAI), vol. 3641, pp. 244–253. Springer, Heidelberg (2005)

[Grzymala-Busse and Goodwin, 2001] Grzymala-Busse, J.W., Goodwin, L.K.: Coping with Missing Attribute Values Based on Closest Fit in Preterm Birth Data: A Rough Set Approach. Computation Intelligence 17(3), 425–434 (2001)

[Grzymala-Busse and Wang, 1997] Grzymala-Busse, J.W., Wang, A.Y.: Modified algorithms LEM1 and LEM2 for rule induction from data with missing attribute values. In: Proc. of the Fifth International Workshop on Rough Sets and Soft Computing (RSSC 1997) at the Third Joint Conference on Information Sciences (JCIS 1997), Research Triangle Park, NC, March 2-5, pp. 69–72 (1997)

[Kerdprasop, Saiveaw and Pumrungreong, 2003] Kerdprasop, N., Saiveaw, K.Y., Pumrungreong, P.: A comparative study of techniques to handle missing values in the classification task of data mining. In: 29th Congress on Science and Technology of Thailand, Khon Kaen University, Thailand (2003)

[Kryszkiewicz, 1995] Kryszkiewicz, M.: Rough set approach to incomplete information systems. In: Proceedings of the Second Annual Joint Conference on Information Sciences, Wrightsville Beach, NC, September 28-October 1, pp. 194–197 (1995)

[Little and Rubin, 2002] Little, R.J., Rubin, D.B.: Statistical Analysis with Missing Data (2002)

[Kantadzic, 2003] Kantadzic, M.: Data Mining: Concepts, Models, Methods & Algorithms. John Wiley & Sons, NY (2003)

[Nakata and Sakai, 2005] Nakata, M., Sakai, H.: Rough sets handling missing values probabilistically interpreted. In: Ślęzak, D., Wang, G., Szczuka, M.S., Düntsch, I., Yao, Y. (eds.) RSFDGrC 2005. LNCS (LNAI), vol. 3641, pp. 325–334. Springer, Heidelberg (2005)

[Quinlan , 1989] Quinlan, J.R.: Unknown attribute values in induction. In: Proc. Sixth Intl. Workshop on Machine Learning, pp. 164–168 (1989)

[Slowinski and Stefanowski, 1989] Slowinski, R., Stefanowski, J.: Rough classification in incomplete information systems. Mathematical and Computer Modelling 12(10-11), 1347–1357 (1989)

[Stefanowski and Tsouki'as, 1999] Stefanowski, J., Tsoukiàs, A.: On the extension of rough sets under incomplete information. In: Zhong, N., Skowron, A., Ohsuga, S. (eds.) RSFDGrC 1999. LNCS (LNAI), vol. 1711, pp. 73–82. Springer, Heidelberg (1999)

[Stefanowski and Tsouki'as, 2001] Stefanowski, J., Tsoukiàs, A.: Incomplete information tables and rough classification. Computational Intelligence 17(3), 545–566 (2001)

[Wu, Wun and Chou, 1997] Wu, C.-H., Wun, C.-H., Chou, H.-J.: Using association rules for completing missing data. In: HIS, pp. 236–241. IEEE Computer Society (2004)

[Lavrajc, Keravnou and Zupan, 1997] Lavrajc, N., Keravnou, E., Zupan, B.: Intelligent Data Analysis in Medicine and Pharmacology. Kluwer Academic Publishers (1997)

[Cheesemen and Stutz, 1996] Cheesemen, P., Stutz, J.: Bayesian classification (AutoClass): theory and results. In: Fayyad, U.M., Piatetsky-Shapiro, G., Smyth, P., Uthunsamy, R. (eds.) Advances in Knowledge Discovery and Data Mining. AAAI Press/MIT Press (1996)

[Saar-Tsechansky and Foster, 2007] Maytal, S.-T., Provost, F.: Handling Missing Values when Applying Classification Models. Journal of Machine Learning Research 8, 1625–1657 (2007)

[Ding and Simonoff, 2006] Ding, Y., Simonoff, J.: An investigation of missing data methods for classification trees. Working paper 2006-SOR-3, Stern School of Business, New York University (2006)

[E. A. Rady et al., 2007] Rady, E.A., Abd El-Monsef, M.M.E., Abd El-Latif, W.A.: A Modified Rough Set Approach to Incomplete Information Systems. Research Article Received (October 30, 2006) (revised January 27, 2007) (accepted March 27, 2007)

[Fujikawa and Ho, 2002] Fujikawa, Y., Ho, T.-B.: Cluster-Based Algorithms for Dealing with Missing Values. In: Chen, M.-S., Yu, P.S., Liu, B. (eds.) PAKDD 2002. LNCS (LNAI), vol. 2336, pp. 549–554. Springer, Heidelberg (2002)

[Greco et al., 2000] Greco, S., Matarazzo, B., Slowinski, R.: Rough set processing of vague information using fuzzy similarity relations. In: Calude, C.S., Paun, G. (eds.) Finite Versus Infinite: Contributions to an Eternal Dilemma. Discrete Mathematics and Theoretical Computer Science (London), pp. 149–173. Springer, London (2000)

[Kryszkiewicz, 1998] Kryszkiewicz, M.: Rough set approach to incomplete information systems. Information Sciences 112, 39–49 (1998)

[Kryszkiewicz, 1998a] Kryszkiewicz, M.: Rules in Incomplete Information Systems. Information Sciences 113(3-4), 271–292 (1999)

Application of Sensor in Shoe

Pritee Parwekar, Akansha Gupta, and Shrey Arora

Department of Computer Science
Jaypee Institute of Information Technology
pritee.parwekar@jiit.ac.in,
{akanshadav,smokinshrey}@gmail.com

Abstract. This paper presents the development of portable wearable system for sports related data capture and analysis in order to enhance the athletic performance. The sensors that are being used include 4-FSRs (Force Sensitive Resistors), accelerometer, GPS module (GTPA010) and pulse sensor. Arduino duemilanove 328 has been used as a microcontroller. The wearable system that was developed can be put into a mount which can be subsequently attached to a shoe. This wearable system provides an inexpensive alternative to the rather bulky and expensive motion sensing labs which are used for real time data analysis. Sensors produce qualitative results, without the accuracy needed for true biomechanical analysis and thereby facilitate to provide broad estimates of quantities attractive to the hobbyist.

Keywords: sensors, body sensor networks.

1   Introduction

Olympic Games at London have showcased some of the finest sporting achievements. One of the most admirable which has achieved legendary status is that of Usain Bolt. He has created history by successfully defending his 100m and 200m for a brisk 19.42 seconds to win the 200 metres while he clocked 9.63 to easily dismiss the field in the 100m sprint event. This has been the nucleus of our inspiration to consider this area of sports for research.

Numerous systems have been developed for the enhancement of athletic performance and the most noticeable among them is the motion sensing labs. These labs are equipped with expensive state of the art facilities, and with elaborate gadgets connected to the subjects for monitoring parameters. Further, the visual gait analysis, used for evaluation of the athletes' style of walking or running largely depends on the skill of the observer assessing the athlete. This method of assessing the athlete is time consuming and due to the subjectivity due to the human observer, the system is qualitative and inconsistent with users. Direct motion measurement systems or motion capture technology is another prominent method for data capture and analysis for performance evaluation of the athlete, but this system also has its own set of limitations. Both these systems have an elaborate lab setup for facilitating the camera field view or camera distortion and connectivity with other sensors. Thus, there is a limitation for replication of the natural habitat in which the sportsman is expected to perform.

S.C. Satapathy, S.K. Udgata, and B.N. Biswal (eds.), *FICTA 2013*,   409
Advances in Intelligent Systems and Computing 247,
DOI: 10.1007/978-3-319-02931-3_46, © Springer International Publishing Switzerland 2014

Bringing in an inexpensive portable system, which could provide reliable and repeatable data remotely, is the way ahead, for capturing more realistic data and free from the geographical limitations of a lab setup. There is considerable research in progress for making such systems portable. The advantage of such system is that it would be inexpensive, portable and reduce the subjectivity associated with the observer's and also alleviate the performance anxiety of the athlete associated with a lab performance.

2   Literature Survey

In 2007 S. Tatsuta *et al* [1] developed a human behaviour recognition system using wearable sensors which can distinguish between human gestures such as walking, talking, running, standing etc. This system consist of Foot pressure sensing shoes, sound sensing glasses, motion sensing watch, , pen-shaped ceiling sensor and location positioning mobile phones which were all transmitting data by wireless communication.

In 2008 Scarborough *et al* [2] developed an integrated wireless system for gait analysis outside the confines the traditional motion laboratory. The system consists of orthogonal accelerometers, orthogonal gyroscopes, force sensors, bidirectional bend sensors, dynamic pressure sensors and electric field height sensors. The system detects the orientation of the feet and the heel strike caused by the interaction with surface.

In a more recent research Zakaria *et al* [3] developed a wearable instrument where the dynamics of human lower limb have been measured. The developed system in this case is relatively cheap as it involves two axis accelerometers, gyroscopes and force sensitive resistors. De Rossi *et al* [4] developed an in shoe device which constituted of 64 sensitive elements with battery that provides high frequency data acquisition, wireless transmission and 7 hours autonomy.

Many commercial systems have also been developed which provide athlete with information with results generated from various sensors. The companies like Reebok, Nike, Adidas etc. Are involved in developing these[5]. For medical purposes companies like Tekscan have developed instrument which helps to ensure effective offloading to reduce plantar ulceration risk and also identify and correct patho-mechanical dysfunctions but was only focusing on pressure aspects of gait analysis [6].

Our research seeks to develop a system that is inexpensive which not only helps to enhance the athletic performance but also provides alternative to various commercially available systems. The system comprises of accelerometer [7], 4 FSRs[8], GPS module[9] (GTPA010) and pulse sensor[10] which are cheaper than other available systems. Moreover this system does not require any motion sensing laboratory and can be used in regular condition. It not only gives statistics of performance but also helps in sending alert messages. It will also help trainers to monitor athlete's performance without their presence at actual training location.

Table 1. System Components

SENSOR TYPE	PARAMETER	MODEL NO.	PLACEMENT
Accelerometer	Velocity	ADXL345	Backside of the feet, above the toe.
4-FSRs (Force Sensitive Resistors)	Force distribution under foot	FSR – 0.5"	4 metatarsal points on feet
GPS module	Locate the position of the subject	GTPA010	Should be attached to the mount, but should be in open air for 10 minutes.
Pulse rate sensor	To measure heartbeat and human pulse	Grove ear clip heart rate sensor	The clip should be attached to the body part where there is dense capillary content.

3   Setup/Model

The proposed solution has four sensors each focusing on different aspects of athletic performance. FSR is used to measure the force distribution under the foot while the subject is running or walking. Pulse sensor acquires readings of heart rate of athlete while wearing the shoe that has been attached with the developed circuit at the time of training. Accelerometer acquires the data on required speed and actual speed obtained. These three sensors have been used with the perspective of enhancement of speed in mind which is very important in athletic events such as 100m sprint. GPS helps locating the position of the athlete in an outdoor cross country situation, so that in case of emergency the athlete can be easily tracked. This would also free the athlete from the intruding monitoring devices and setup and help the analysts sit a lab environment while they impervasively monitor the athlete.

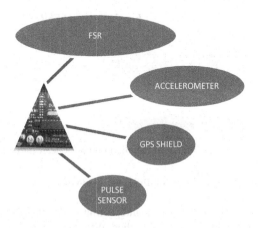

Fig. 1. Sensors on the circuit

4   Implementation

- A single FSR is connected with arduino as per circuit diagram and the results are recorded. Arduino[11] is microcontroller based on atmega. It is an easy to use and easy to code controller and also reduces cost of the proposed model. No issues have been reported while implementing like faulty readings from FSR or in series if one stops working then readings of other FSR also stopped etc.
- Next step is to connect 4 FSR in series as per requirement and obtain output. Use FSR in series so as to obtain cumulative reading of whole foot.
- Further a pulse sensor is attached to arduino according to the circuit diagram to obtain heartbeat of person in different conditions like running, walking and still. Plot the output of heart rate sensor using software serial chart.

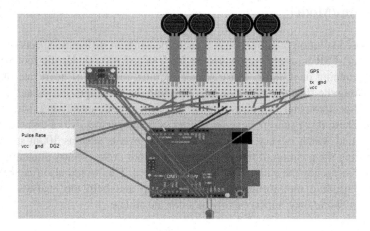

Fig. 2. Circuit of FSR

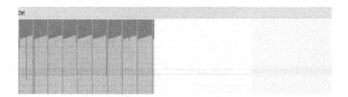

Fig. 3. Serial chart output of heart rate sensor

Average Heart rate output = 98 beats/sec

- The next step involves integrating code of FSR and pulse rate sensor i.e. running both sensors together. While programming the integrated codes care has to be taken to check the flow of program for correct readings. The algorithm has o take care that in every loop flow it calculates both FSR readings and Pulse sensor reading once only. After 1 minute only pulse sensor readings average is to be calculated.

- Next step is to use serial chart to plot the output of pulse sensor to obtain the real time plot of pulses of athlete where FSR and grove heart rate sensor were integrated in single circuit.
- In the next action the GPS is connected on arduino as shown in circuit (fig 4). This should be undertaken outdoors where open skies are available and as far as possible in clear weather to ensure proper line of sight with the satellites to obtain good GPS readings.

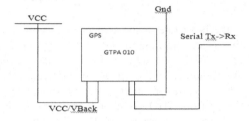

Fig. 4. GPS – arduino schematic circuit

- Accelerometer is connected according to spi connection scheme so that it can work in coordination with other sensors and to have more reliable readings and obtained desired results.
- An integrated code is made for connecting accelerometer, FSR, pulse sensor. GPS has not been considered for connection in this setup to remove influence of environmental factors.
- The processing tool is used to create text file of the readings collected from serial communication. Then java code is used to mail this file using javamail1.4 [12] to the required mail id and code for SMS has been taken from [13].
- Base station is to be embedded with java code to send message and email. Base station in this experimental setup i.e. laptop can be mobile. Only requirement is to have an internet connection. The arduino board is hard connected to the base station through cable.

5   Results

A wearable system was developed comprising of many sensors attached as a mount to the shoe measuring various statistics in integrated manner for athlete was built using the sensor architecture described above. The four sensors are stacked together and attached to a universal shoe attachment, it has been found to work better outside a room as GPS readings improve because of better satellite communication. It can take load of about 100 kg. The FSRs at the heel are compressed signifying weight transfer. The java program sends mail and message which were readily reaching their target phone numbers and emails. This wearable system will also help in analysing result as we can get actual readings of athlete.

Table 2. Heart rate of the subject

123466	grid_h_origin_color = #CCC	20	[_default_]
679		125449	min=-1
18	grid_v_origin = 0	662	max=1
124125	grid_v_step = 10	**Heart_rate_is: 91**	
659	grid_v_color = #KKK	0	[Field1]
19	grid_v_origin_color =	126103	color=red
124787	transparent	654	min=0
662		1	max=255
		126749	
		646	[Field2]
			color=blue

The table 2 depicts the heart rate of the subject. It is coming out to be 91 bpm . Now the normal resting heart rate range is 60 – 100 bpm. This can help to detect athletic heart syndrome (AHD) or athletic bradycardia. Athletic bradycardia is common in sports in which human heart is enlarged and human heart rate is said to be lower than normal. This is common among endurance athletes and heavy weight trainers. A person suffering from AHD would be having heart beat around 40-60 bpm. The readings depicted in the table 1 are around 91 of the subject. So if they were less than 60 there might be a chance that the subject might be suffering from athletic heart syndrome (AHD).

Table 3. Readings of force sensitive resistors (FSR)

Table 4. Readings of the accelerometer ADXL 345

Force in Newtons = 0

Analog reading = 492
Voltage reading in mV = 2404
FSR resistance in ohms = 10798
Conductance in micrMhos: 92
Force in Newtons: 1

Analog reading = 776
Voltage reading in mV = 3792
FSR resistance in ohms = 3185
Conductance in micrMhos: 313
Force in Newtons: 3

Analog reading = 814
Voltage reading in mV = 3978
FSR resistance in ohms = 2569
Conductance in micrMhos: 389
Force in Newtons: 4

87	92	93
90	95	95
92	97	97
94	99	99
96	102	102
98	105	103
101	108	106
103	110	108
106	113	110
107	114	111

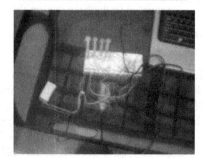

Fig. 5. circuit of FSR and grove heart rate sensor

Table 3 depicts the readings of force sensitive resistors (FSR) that are placed at various metatarsal points on the feet that help to calculate foot plantar pressure. Constant updates of the pressure exerted by the feet can help detect the lower extremity musculoskeletal disorders like tendonitis.

Table 4 shows the readings of the accelerometer ADXL 345 of the subject. They depict the acceleration caused by the motion or shock.

Fig 6 depicts the readings of GPS. It will detect the location of the subject by sending the coordinates during outdoor regimen. The GPS has GTPA010 chip on it. The GPS receiver talks to the computer using serial port through NMEA protocol. These four sensors are worn by the subject in the form of mount attached to the shoe. These sensors send the data continuously over certain period of time to the base station which are analysed by the experts to diagnose any medical issue that can arise due to irregularity in the data and also enhance the performance of the athlete.

 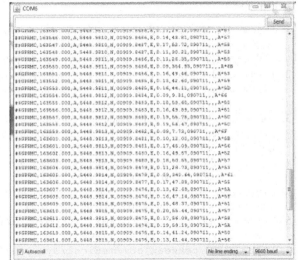

Fig. 6. Outputs of ADXL 345 and FSR   **Fig. 7.** Outputs of GPS

6   Conclusion

In this paper we have discussed the use of four sensors that can be put in a mount which is further attached to the shoe for real time data analysis. Each of these sensors are performing certain task that helps to detect issues that might exist with the body of the athlete which can cause decrease in the performance. If these issues are diagnosed at right time can be cured with proper rehabilitation process. Moreover these can also be used as parameters to enhance the athletic performance.

The wearable system will also help us to send various alert messages and email of daily reporting to the athlete trainer which can monitor athlete performance in case not present at training sessions. Current work is focused on collection of the data, sending the obtained data to various concerned people through messages to any authorized person or email. Visual demonstration of sensor based shoes that it gives training details is up to satisfactory precision.

7   Future Work

Improvement in location information using assisted GPS is being pondered upon which would enable reading inside enclosed spaces. We could also look up for better FSR with increased load tolerance to more than 20% at least. In the next phase we would be applying a wireless module between the designed system and base station so as to communicate wirelessly.

References

1. Mizuno, H., Nagai, H., Sasaki, K., Hosaka, H., Sugimoto, C., Khalil, K., Tatsuta, S.: Wearable sensor system for human behavior recognition. Department of Human and Engineered Environmental Studies. The University of Tokyo, Kashiwa, Chiba, Japan
2. Bamberg, S.J.M., Benbasat, A.Y., Scarborough, D.M., Krebs, D.E., Paradiso, J.A.: Gait analysis using a shoe-integrated wireless sensor system. IEEE Transactions on Information Technology in Biomedicine 12(4) (July 2008)
3. Parikesit, E., Mengko, T.L.R., Zakaria, H.: Wearable Gait Measurement System Based on Accelerometer and Pressure Sensor. In: Biomedical Engineering, School of Electrical Engineering & Informatics ITB, Bandung, Indonesia
4. De Rossi, S.M.M., Lenzi, T., Vitiello, N., Donati, M., Persichetti, A., Giovacchini, F., Vecchi, F., Carrozza, M.C.: Development of an in-shoe pressure-sensitive device for gait analysis. In: 33rd Annual International Conference of the IEEE EMBS Boston, Massachusetts USA, August 30 - September 3 (2011)
5. http://support-en-us.nikeplus.com/app/answers/detail/a_id/30875/p/3169,3700,last (accessed on February 23, 2013)
6. http://www.tekscan.com/medical/system-fscan1.html (last accessed on February 23, 2013)
7. Accelerometer, http://www.dimensionengineering.com/info/accelerometers (last accessed August 28, 2012)
8. Force sensitive resistors, http://www.tekscan.com/flexible-force-sensors (last accessed August 28, 2012)
9. GPS, http://www.rhydolabz.com/index.php?main_page=product_info&cPath=122&products_id=424 (last accessed on February 23, 2013)
10. Pulsesensor, http://www.makershed.com/Pulse_Sensor_for_Arduino_5_Volt_p/mkre1.htm (last accessed on February 23, 2013)
11. Arduino homepage, http://arduino.cc/ (last accessed on February 23, 2013)
12. Javamail 1.4, http://www.oracle.com/technetwork/java/javamail-1-4-140512.html (last accessed on February 23, 2013)
13. http://jtechbits.blogspot.in/2011/06/sending-sms-through-way2sms-in-java.html, http://stackoverflow.com/questions/3177616/how-to-attach-multiple-files-to-an-email-using-javamail (last accessed on February 23, 2013)

Robot Assisted Emergency Intrusion Detection and Avoidance with a Wireless Sensor Network

Pritee Parwekar and Rishabh Singhal

Department of Computer Science
Jaypee Institute of Information Technology
NOIDA, India
{pritee2000,singhalrishabh1991}@gmail.com

Abstract. In the century of wars with the unseen and many a times undefined enemy, military challenge ranges from active battles to covert operations. The loss of intelligent and trained soldier in military actions is becoming increasingly strategically costlier. Replacement of the human soldier with a mechanical one especially in hostile environments is the failure of military. The paper has contributed towards implementation of robotics in military applications for intrusion detection and avoidance. The sensors will detect any human intrusion in restricted areas and will inform the base station consisting of a microcontroller the zone in which the intruder is present. This microcontroller will communicate with our remote bot consisting of a microcontroller and will give the zone number in which the intrusion took place thus instructing the bot to go and check the intrusion in specified area.

Keywords: Robot assisted military devices.

1   Introduction

Military robots are autonomous robots or remote controlled devices used for military applications. We all recognize that being a soldier is a dangerous job, but some of the tasks that soldiers are required to perform are much more dangerous than others. Walking through minefields, deactivating unexploded bombs or clearing out hostile buildings, for example, are some of the most dangerous tasks a soldier is required to perform in the line of duty. What if one could send unmanned objects like robots to do these jobs instead of humans? The collateral damage of a mishap could result in loss of equipment, rather than loss of life. And we could always build more robots cheaper and faster with growing technology.

This quest being the stimulus, the paper focuses on military applications and research, as the title suggests, applying the concept of robotics and sensors for the military applications such as patrolling and searching. The proposed bot will proactively move to the area of intrusion in a fixed path and simultaneously track it's position and beam this location information to the base station. The base station will soon inform the remote station through wireless communication if an intrusion occurs in the specific area. The bot will start moving towards the intruded area and when the

S.C. Satapathy, S.K. Udgata, and B.N. Biswal (eds.), *FICTA 2013*,   417
Advances in Intelligent Systems and Computing 247,
DOI: 10.1007/978-3-319-02931-3_47, © Springer International Publishing Switzerland 2014

bot reaches to it's specified zone it will be detected by the IR/LDR sensor in that area and hence it will stop. Now we can easily drive our bot from base station using a simple touch screen in the intruded zone to know what's going wrong in that area and capture images. The bot can be driven through touch screen from our base station since we are using CC2500 Module for wireless communication, and now from the captured images we can detect any threat using image processing techniques in Open cv.

2   Related Work

Hybrid Sensitive Robot Algorithm for Intrusion Detection [1] based on bio-inspired robots uses a qualitative stigmergic mechanism in which each robot is endowed with a stigmergic sensitivity level facilitating the exploration and exploitation of the search space. The Adaptive Neuro-Fuzzy Inference System (ANFIS) [2] can classify the alerts with high accuracy and reduce number of false positive alerts considerably. Further this algorithm learns from output values and tunes IF-THEN rules. DogoIDS: A Mobile and Active Intrusion Detection System for IEEE 802.11s [3] is an intrusion detection in WMNs that harness an active probing detection technique in combination with a dedicated mobile node platform. It has applicability as active and mobile IDS for wireless multihop networks. Though it has been developed for IEEE 802.11s networks, it can find applications in wireless sensor networks.

3   Setup of Model

The system consists of Atmega-16 microcontroller and development board fig.1, Infrared sensors fig.2, cc2500 wireless modules fig.3, adaptors for power supply, touch screen, 12 volt DC motors to drive the bot, these are shown in the block diagram fig. 4.

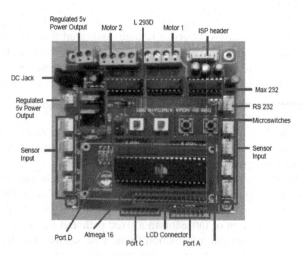

Fig. 1. Photograph of development board

Fig. 2. Infrared sensor

All of controlling commands are sent from the base station computer to robot via cc2500 wireless modules. The High Speed CC2500 Based Wireless module is a plug and play replacement for the wired Serial Port (UART) supporting baud rates upto 38400. This CC2500 based Wireless module allow engineers of all skill levels to quickly and cost-effectively add wireless capabilities to virtually any product.

Fig. 3. CC2500 wireless module

The controlling commands compose of the locomotion control and robot's navigation path. In order to monitor video in real time, a wireless surveillance camera is mounted on the remote bot. The controlling commands are generated from the base station. The codes are transmitted with the help of cc2500. On the receiver side cc2500 is configured as receiving purpose. A separate power station is setup for the independent power delivery to the motors and to the on board. The ir sensors are used for intrusion detection in different zones.

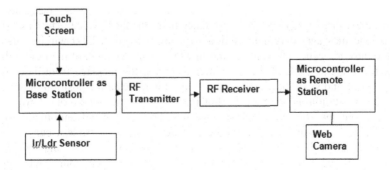

Fig. 4. Block diagram of setup

4   Implementation

The restricted area is divided into 2 zones let them be named as zone1 and zone2.The ir sensors are placed in different zones after calibrating their values, these calibrated values are used in our code to detect whether the intrusion took place or not and now the microcontroller at the base station will automatically calculate the mean value or threshold value for intrusion detection. When the intruder enters the restricted zone and passes by the ir sensor the value of the sensor greater than the threshold value would be transferred to the base station and the base station will be informed about the zone in which the intrusion took place. The fig. 5 shows the BS as base station RS as remote station Z1 as zone 1 and Z2 as zone 2, Ir 1 and Ir 2 to detect intrusion in zone 1 and 2 respectively, s1 and s2 are ir sensors to check the location of the remote station while in motion. These sensors are connected to the base station with the help of underground wires.

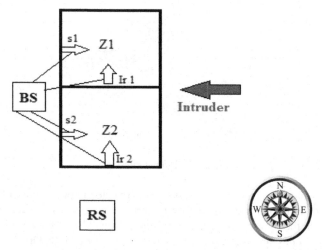

Fig. 5. (Ir1, Ir2, s1, s2 are all Infrared Sensors ; BS & RS are Base & Remote Station)

If the intruder enters in the field from the east it will surely pass the sensors Ir 1 or Ir 2 situated in the south of zone 1 and 2 respectively. The base station will be informed about the intrusion due to the change in values of ir sensor, since due to the intrusion the value of sensors will be greater than threshold value. The base station will pass the message to remote station with the help of cc2500 wireless modules about the intruded zone. The remote station which is in the south of the field will start moving towards the north of the field after receiving the message from base station. The sensors s1 and s2 in the west of zone 1 and 2 will locate the position of patrolling bot or remote station as soon as the bot passes by the sensor s1 or s2 we will be informed about the zone in which our bot has reached and will automatically stop when passes the sensor s1 or s2 of the intruded zone. The camera can be placed on the bot which will give live streaming video of the surroundings and now we can drive our wireless bot to any direction using touch screen connected with the microcontroller at base station.

5   Limitations

If somehow the intruder after passing Ir 1 or Ir 2 the sensors which are being used to detect intrusion, comes in front of the sensors s1 or s2 which are being used to locate the position of the bot and our bot while in motion has not reached the intruded zone than it may stop before it's destination, since the values more than the threshold values of sensor s1 and s2 will be passed to the base station thus instructing the remote station that you have reached your destination so stop (though our remote station is far away from the intruded zone). Thus to overcome this limitation we have to work manually, since we have a camera on our bot we will be watching the live streaming videos of the surroundings and on the basis of this we have to judge the location of the bot and drive our bot using touch screen connected to base station.

6   Result

The proposed robot was successfully designed and implemented, cc2500 works effectively within range of 40 meters (line of sight). With the help of Bascom [4][5] as the coding platform and Atmega-16 [6] as the microcontroller the intrusion detection and avoidance was achieved. Thereby providing safety to restricted areas and reducing the task for human soldiers in risky areas such as nuclear plants and national borders. Instead of IR we can also use ultrasonic sensors.

Fig. 6. Snapshot of the entire setup

7   Conclusion and Future Work

The patrol robotic system with intrusion detection and avoidance using wireless sensor network has been briefly described. Its performances were observed to be excellent in unstructured environments. The development of this mini-bot robot can be adapted to fit many other applications easily by changing the top part of the robot. Single camera can be mounted on the bot and various tasks can be done using image processing in open-cv. It can capture images and send to the laptop situated at the base station and we can use these images to detect severity of threat. If the threat detected is severe than these images can be forwarded to other systems in the base station by creating server and client structure in java using a wi-fi network, thus informing other high officials regarding the intrusion.

References

1. Pintea, C.-M., Pop, P.C.: Sensor networks security based on sensitive robots agents: A conceptual model. In: Herrero, Á., Snášel, V., Abraham, A., Zelinka, I., Baruque, B., Quintián, H., Calvo, J.L., Sedano, J., Corchado, E. (eds.) Int. Joint Conf. CISIS'12-ICEUTE'12-SOCO'12. AISC, vol. 189, pp. 47–56. Springer, Heidelberg (2013)
2. Orang, Z.A., et al.: Using Adaptive Neuro-Fuzzy Inference System in Alert Management of Intrusion Detection Systems. I. J. Computer Network and Information Security 11, 32–38 (2012)
3. do Carmo, R., Hollick, M.: DogoIDS: A Mobile and Active Intrusion Detection System for IEEE 802.11s Wireless Mesh Networks. Technical Report TR-SEEMOO-2012-04
4. http://www.mcselec.com/index.php?option=com_content&task=view&id=14&Itemid=41
5. http://www.mcselec.com/?option=com_content&task=view&id=14&Itemid=41
6. http://www.atmel.com/Images/doc2466.pdf

Interaction between Internet Based TCP Variants and Routing Protocols in MANET

Sukant Kishoro Bisoy[1] and Prasant Kumar Patnaik[2]

[1] SOA University, Bhubaneswar and Faculty Member, C.V. Raman College of Engg.,
Bhubaneswar, India
sukantabisoyi@yahoo.com
[2] School of Computer Engineering, KIIT University, Bhubaneswar, India
patnaikprasantfcs@kiit.ac.in

Abstract. Due to the presence of mobility in the mobile ad hoc network (MANET), the interconnections between nodes are likely to change. Dynamic nature of MANET makes TCP more aggressive in case of packet loss and retransmits a lost packet and unnecessary causes an energy loss. TCP carries 95% of the internet traffic to transport data over the internet. Hence, it is of utmost importance to identify the most suitable and efficient TCP variants that can perform well in MANET. Main objective of this paper is to find suitable routing protocols for TCP variants and analyze the performance differential variation in terms of throughput, packet loss rate and energy consumption. Simulations result using NS2 shows that, OLSR is best routing protocol with respect to throughput and packet loss ratio irrespective of TCP variants and it provides a lower packet loss rate (15% to 20%) than others in most situations. DSDV protocol consumes 15% to 25 % less energy than AODV and 5 to 10% than OLSR.

Keywords: AODV, DSDV, OLSR, TCP-Newreno, TCP-Sack1, TCP-Vegas.

1 Introduction

MANET forms a random network by consisting of mobile nodes which communicates over wireless path. This kind of network is more appropriate where networking infrastructure is not available and set up time is very less and temporary network connectivity is required. In this every device acts as a router or gateway in wireless ad hoc network. Since each device in MANET is battery operated it becomes utmost important to save energy consumption. Many routing protocols are available in MANET. Among them AODV [1] and OLSR [2] is standardized by IETF MANET working groups [3]. AODV is a reactive routing protocol. OLSR and destination sequenced distance vector (DSDV) [4] are proactive routing protocols. All of these routing protocols work at network layer.

TCP [5] is reliable protocol of transport layer which provides in-order delivery of data to the TCP receiver. It can work well when we combine wired with wireless. In fact TCP has its variants of protocols like TCP-Newreno, TCP-Sack1 and TCP-Vegas. All these were proposed to improve the performance of TCP protocols. So far

S.C. Satapathy, S.K. Udgata, and B.N. Biswal (eds.), *FICTA 2013*,
Advances in Intelligent Systems and Computing 247,
DOI: 10.1007/978-3-319-02931-3_48, © Springer International Publishing Switzerland 2014

lot of research work has done on routing protocols and TCP variants. Some researcher say TCP-Vegas is better than TCP- Newreno in some scenario. Author [6], compared the performance of TCP on reactive routing protocol (AODV) and proactive routing protocol (DSDV). In this paper our aim was to find suitable routing protocols for TCP variants and analyze the TCP variants performance in terms of throughput, packet loss and energy consumption in order to find the appropriate one of them for the mobile ad hoc network environments.

The rest of the paper is structured as follows. Section 2 explains the Routing Protocol in MANET. Section 3 will present TCP and its variants. Section 4 will present simulation set up and section 5 will explain result and analysis. Finally, we conclude our work in section 6.

2   Related Work

Author [7] proposed a protocol called I-AODV to show that the minimum hops always may not be the criteria to choose the best path. They considered node status is the best criteria to select next hop in the routing protocol.

Author in [8] studied the inter layer interaction between MAC and physical layer and demonstrated the performance differences between dynamic source routing (DSR) and AODV. The performance varies because they adopt different mechanism for routing.

Proactive routing protocol performs better than reactive at the cost of higher routing load [9]. In order to show that AODV, DSR and OLSR routing protocols in MANETs were considered and evaluated with various network conditions.

In a multi hop network the quantity of message length and node mobility affects the performance of ad hoc network. Author studied the performance of position based routing over a framework called communication-theoretic [10]

3   Routing Protocols in MANET

Dynamic nature of MANET makes the routing complicated and route failure occurs frequently. So node mobility is main source of route failures in wireless network. In addition to node mobility, channel contention may be the other reason of route failure. Routing protocols of MANET broadly classified in two different categories based on how they discover the route. One is proactive protocols and another is reactive protocols.

3.1   Destination-Sequenced Distance-Vector

It is a proactive routing protocol whose routing method based on the Bellman-Ford algorithm [11]. Each node maintains a routing table to keep distance information to other node. The routing table is updated when there is a change of network topology and informed other nodes periodically about the change. Each entry has sequence number to indicate its freshness and loop free. Also it helps to mark stale route. The sequence number is incremented by a node after sending each message to other.

3.2   Optimized Link State Routing

It is a proactive routing protocol [2], which periodically updates its routing table and exchange information with other. In this every node do not broadcast route message in the network. Only multipoint relay(MPR) node is supposed to send the message. The MPR nodes are selected by neighbor nodes and informed others in their control message. Main purpose of MPR node is to reduce the overhead of network and to minimize the re-transmissions required. Among the two control messages used by OLSR namely Hello and topology control(TC) message, hello message is used to find link information and neighbors node information within 1-hop distance and TC message used to inform own advertise neighbors.

3.3   Ad Hoc On- Demand Distance Vector

It is a reactive routing protocol which discovers route on demand when a packet needs to be send by a source. Route discovery process starts by sending route request(RREQ) packet to their neighbors. Then neighbor forward the RREQ to their neighbor and so on. This sending process is continued by every neighbor node until the destination gets the message or they have a route to destination. On either case nodes reply back with a route reply(RREP) message. In case of route breakage the intermediate node discover another new route or send a route error (RERR) message to the source. Upon receiving RERR the source node tries to get new route by invoking again route discovery process.

4   Internet Based TCP and Its Variants

TCP carries 95% of the internet traffic to transport data over the internet. Basic objective of TCP is to provide reliable communication over an unreliable network layer and to deliver the data from sender to receiver correctly and in order. So far lot of research work has done on TCP and several version of TCP have been implemented [12][13]. Different routing mechanism have great impact on the performance of TCP[9]. Each TCP variants have different mechanism for congestion control and loss recovery.

4.1   Newreno

TCP-Reno has been modified in two versions. The first version is Newreno and second version is SACK. By making small changes at the sender side of Reno get Newreno [14]. Through its new improved fast recovery algorithm, it recovers multiple packet losses per window and minimizes as many as transmission timeout events [15]. Therefore Newreno version is more suitable than Reno in the mobile wireless environment where packet losses may occur in bursts.

4.2   Sack1

It is an extension of TCP Reno which notifies which segment is received correctly
and corrects the multiple dropped segments problem. It is suitable for a network
where multiple segments lost in a single window of TCP. It also retains the properties
of   TCP-Tahoe and TCP-Reno. In SACK acknowledgment is sent selectively rather
sending cumulative acknowledge [16]. Therefore it is possible to send more than one
lost segment per RTT.

4.3   Vegas

Another modification of Reno is TCP Vegas. It has different mechanism to control
the congestion which helps to detect the congestion before any segment loss occurs
[17]. It relies on measured RTT values of sent packets. It adopts different bandwidth
estimation schemes. TCP Vegas accurately estimates the available bandwidth in the
network by finding the difference between expected flow rates and actual flow rates.
Then accordingly it estimates the level of congestion and updates the window size.

5   Simulation Set Up

We study the performance of ad hoc routing protocol that may affect TCP
performance with three parameters: the choice of TCP variants, node mobility and
pause time. We evaluate the performance using NS2 simulator [18].

5.1   Simulation Tool and Parameter

In order to analyze the interaction, NS-2 is used over random way point mobility
model. We considered three ad hoc routing protocols such as OLSR, DSDV, AODV
and three TCP variants such as Newreno, Sack1 and Vegas. We create a random
network of size 30 nodes and each node moves from one random location to another.
Among 30 nodes, five TCP-connections were established containing fixed pairs of
sender and receiver. Each connection stays for 300 sec long. The pause time, speed
and parameters used for or simulations are shown in table 1.

Table 1. Parameter values

PARAMETER	VALUE	PARAMETER	VALUE
Channel type	Wireless channel	Receiving Power	1.4 w
MAC	802.11	Idle Power	1.1 w
Routing Protocol	OLSR,DSDV, AODV	Sleep Power	0.002 w
Pause Time	0, 10, 20 Sec	Transmission Range	250 m
Speed	10, 20, 30, 40,50 m/s	TC-INTERVAL	5 sec
Packet type and size	FTP and 512 bytes	Hello-INTERVAL	3 sec
Initial Energy	100 Joule	Topography	1000 m X 1000m
Transmission Power	1.65 w	Time of simulation	300 Sec.

5.2   Energy Consumption Model

Energy consumption of a network can be measured by a linear model [19]. Overall energy cost of network includes sending and receiving traffic and idle and sleep mode. But idle power consumption is bit important to total energy cost. So minimizing the energy consumption of network interface becomes utmost important. In an ad-hoc configuration, an 802.11 network interface card may be in any of three energy states: transmit, idle, and receive [20]. Our energy measurement is based on linear model proposed by Feeney. Energy consumption for transmit, receive, idle and sleep mode are given in table 1.

6   Result and Analysis

In this section, we present the results gathered from the simulation experiments from various scenarios. In order to study the performance of above protocols 5 pairs of nodes are selected as 5 flows. Each pair runs for 300 seconds. The performance metrics used to analyze the performances are aggregated throughput (kbps), packet loss and consumption of energy.

6.1   Throughput Measurement

First we study the performance of the ad hoc routing protocols with TCP variants to analyze the interaction between the network and transport layers within ad hoc network environment. Initially throughput of Vegas is measured over OLSR, DSDV and AODV with different node mobility and pause time. We found Vegas protocol over OLSR gives better performance than DSDV and AODV with respect to throughput (see figure 1). But irrespective of ad hoc routing protocol performance, throughput decreases with increase of node mobility. Because when node mobility increases the link failure occurs more frequently this causes more packet loss and decreases throughput. OLSR finds next hop immediately towards destination and start

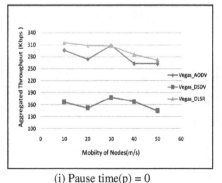
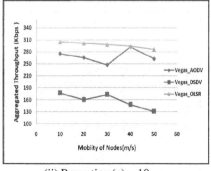

(i) Pause time(p) = 0   (ii) Pause time(p) = 10

Fig. 1. Vegas over OLSR, DSDV and AODV at (i) p=0 (ii) p=10

transmitting packet before timeout occurs. So OLSR protocol recovers from link losses more quickly than reactive one. When link broken occurs, AODV takes more time to recover a route as a result more segment loss occurs.

Next, we calculated the throughput of TCP-Sack1 and TCP-Newreno over OLSR, DSDV and AODV. Still OLSR routing protocol provides good performance than DSDV and AODV for different node mobility (see figure 2 and 3). It can be concluded that OLSR is better routing protocol than DSDV and AODV with respect to throughput without regards to TCP variants which shows same result as in [21]. Then we measured the throughput of Vegas, Sack1 and Newreno over selected ad hoc routing protocols. As figure 4 and 5 suggests when OLSR and DSDV are used as routing protocols, TCP-Newreno shows better performance than TCP-Vegas and TCP-Sack1. Performance of TCP-Vegas over OLSR and DSDV is not good as Newreno because it rely on accuracy of measurement RTT value. In case of link change, OLSR forces the every node to update their routing tables, which helps the protocol to select a new route. DSDV protocol updates their routing table when the topology changes in the network.

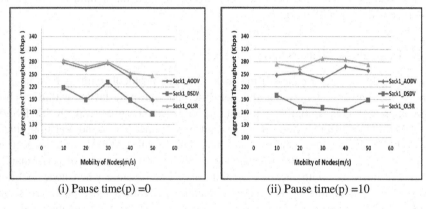

(i) Pause time(p) =0   (ii) Pause time(p) =10

Fig. 2. TCP-Sack1 over OLSR, DSDV and AODV at (i) p=0 (ii) p10

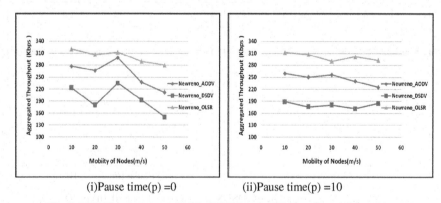

(i)Pause time(p) =0   (ii)Pause time(p) =10

Fig. 3. TCP-Newreno over OLSR, DSDV and AODV at (i) p=0 (ii) p=10

But AODV protocol follows the same path until the path is broken. If topology change occurs frequently due to node movement, inaccuracy estimation of base RTT in Vegas results performance degradation. So it can't perform well in proactive routing protocol. On the other hand, when AODV is used as routing protocol, TCP-Vegas gives better performance than TCP-Newreno and TCP-Vegas (see Figure 6). Because TCP-Vegas maintain its windows size in an appropriate range which helps to reduce the unnecessary route discovery. Newreno decreases the throughput due to increases of window size aggressively and hidden node problem [20]. As our result says there is no much throughput difference between TCP variants with any ad hoc routing protocols.

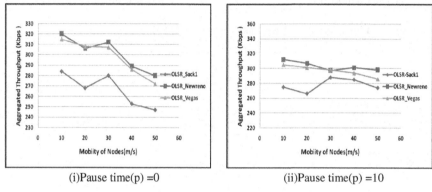

(i)Pause time(p) =0   (ii)Pause time(p) =10

Fig. 4. Newreno, Sack1 and Vegas over OLSR at (i) p=0 (ii) p=10

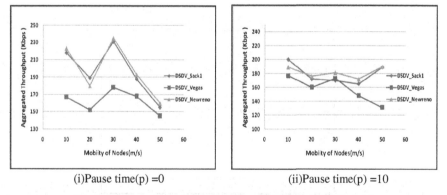

(i)Pause time(p) =0   (ii)Pause time(p) =10

Fig. 5. Newreno, Sack1 and Vegas over DSDV at P=0 (ii) p=10

6.2   Packet Loss Measurement

Next we measure the packet loss rate of TCP-variants over OLSR, DSDV and AODV. As figure 7 shows packet loss range varies from 2.0 to 8.0 for all TCP variants we have considered. In case of lossy link OLSR shows better performance than DSDV and AODV which happens due to random node mobility. In most

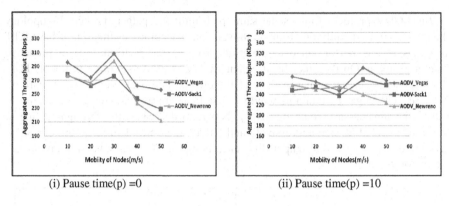

(i) Pause time(p) =0   (ii) Pause time(p) =10

Fig. 6. Newreno, Sack1 and Vegas over AODV at p=0 (ii) p=10

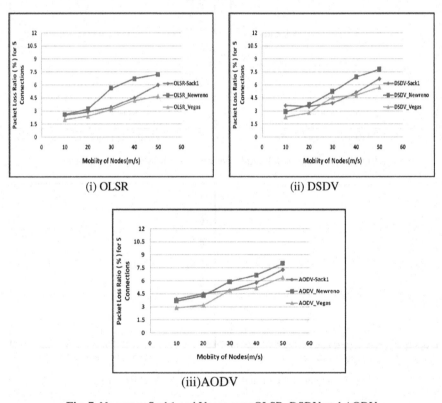

(i) OLSR   (ii) DSDV

(iii)AODV

Fig. 7. Newreno, Sack1 and Vegas over OLSR, DSDV and AODV

situation Vegas shows better performance among considered TCP variants and provides lower packet loss(see figure 7). Because Vegas accurately estimate the available bandwidth in the network by finding the difference between expected flow rates and actual flow rates. In the other hand, TCP variants like Sack1 and Newreno are not good as Vegas for growing mobility.

6.3   Energy Consumption Measurement

Initially we measure the energy consumption of OLSR, DSDV and AODV routing protocols for different node mobility. Our result shows that DSDV consumes 15 to 25 % less energy than AODV and 5 to 10% than OLSR (see figure 8). Our result shows that reactive protocols consume more energy per received bit when AODV is used as a routing protocol. Among TCP variants TCP-Vegas consumes more energy per received bit because it relies on the measurement of propagation received bit than the proactive ones (OLSR and DSDV) (see figure 9 and 10) because of their routing overhead caused by route discovery, route maintenance on-demand. TCP-Newreno and TCP-SACK achieve the best performance in terms of energy consumed per delay in the network.

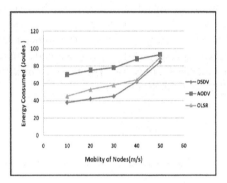

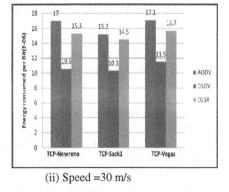

Fig. 8. Energy consumed by routing protocols at different node mobility

Fig. 9. Energy consumed per bit received at speed 10 m/s

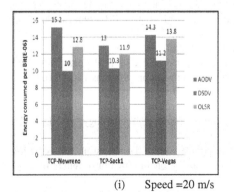

(i)   Speed =20 m/s   (ii) Speed =30 m/s

Fig. 10. Energy consumed per received bit at (i) Speed 20 m/s (ii) Speed 30 m/s

7   Conclusions

Aim of this work was to find suitable routing protocol for particular TCP variants with respect to throughput and energy consumption. We found that TCP performance more sensitive by random nature of ad hoc network of high mobility. So selecting appropriate routing protocol is utmost important. Our result shows that irrespective of TCP variants, OLSR is better routing protocol than AODV and DSDV. Among TCP variants Newreno performs better for OLSR and Vegas perform better for AODV protocol. We conclude that Vegas is the best protocol among TCP variants due to its lower packet loss rate (15% to 20%) in most situations and OLSR is best routing protocol with respect to packet loss irrespective of TCP variants.

In the other hand, AODV consume more energy per received bit than the proactive ones (OLSR and DSDV) because of their routing overhead caused by route discovery, route maintenance on-demand. TCP-Vegas have the worst performance in terms of energy consumption per received bit because it relies on the measurement of propagation delay in the network. For all TCP variants, DSDV performs better than others with respect to energy consumption. Our result shows that DSDV consumes 15 to 25 % less energy than AODV and 5 to 10% than OLSR.

References

1. Perkins, C.E., Belding-Royer, E., Das, S.R.: Ad-hoc on-demand distance vector (AODV) routing. IETF RFC 3561 (2003)
2. Clausen, T., Jacquet, P., Laouiti, A., Minet, P., Muhlethaler, P., Qayyum, A., Viennot, L.: Optimized Link State Routing Protocol(OLSR). IETF RFC 3626
3. Internet Engineering Task Force.: Manet working group charter,
   http://www.ietf.org/html.charters/manet-charter.html
4. Parkins, C.E., Bhagwat, P.: Highly Dynamic Destination Sequence Distance Vector Routing (DSDV) for mobile computers. In: Proc. of ACM SIGCOMM 1994, London, UK (1994)
5. Postel, J.: Transmission Control Protocol. RFC 793 (1980)
6. Papanastasiou, S., Ould-Khaoua, M.: Exploring the performance of TCP Vegas in Mobile Ad hoc Networks. International Journal of Communication Systems 17(2), 163–177 (2004)
7. Feng, D., Zhu, Y.: An Improved AODV Routing Protocol Based on Remaining Power and Fame. In: International conference on Electronic Computer Technology, pp. 117–121 (2009)
8. Lin-zhu, W., Ya-qin, F., Min, S.: Performance comparison of Two Routing Protocols for Ad Hoc Networks. In: WASE International Conference on Information Engineering, pp. 260–262 (2009)
9. Mbarushimana, S., Shahrabi, A.: Comparative study of reactive and proactive routing protocols performance in mobile ad hoc networks. In: 21st International Conference on Advanced Information Networking and Applications Workshops (AINAW 2007), pp. 679–684 (2007)

10. Vijaya, I., Mishra, P.B., Dash, A.R., Rath, A.K.: Influence of Routing Protocols in Performance of Wireless Mobile Adhoc Network. In: Second International Conferences on Emerging Applications of Information Technology (EAIT), pp. 340–344 (2011)
11. Ford Jr, L.R., Fulkerson, D.R.: Flows in Networks. Princeton Univ. Press (1962)
12. Jacobson, V.: Congestion avoidance and control. Computer Communication Review 18(4), 314–329 (1988)
13. Jacobson, V.: Modifed TCP Congestion Avoidance Algorithm. Technical report (1990)
14. Hoe, J.: Start-up Dynamics of TCP's Congestion Control and Avoidance Scheme. Master's thesis. MIT (1995)
15. Floyd, S., Henderson, T., Gurtov, A.: The NewReno Modification to TCP's Fast Recovery Algorithm. RFC 3782 (2004)
16. Fall, K., Floyd, S.: Simulation-based comparison of tahoe, reno, and sack tcp. Computer Communication Review 26, 5–21 (1996)
17. Brakmo, L., O'Malley, S., Peterson, L.: TCP Vegas: New Techniques for Congestion Detection and Avoidance. In: Proc. of ACM SIGCOMM, New York, USA, pp. 24–35 (1994)
18. Information Sciences Institute, The Network Simulator Ns-2, University of Southern California, http://www.isi.edu/nanam/ns/
19. Freeney, L.M.: An Energy Consumption Model for Performance Analysis of Routing Protocols for Mobile Ad hoc Networks. Mobile Networks and Applications 6(3), 239–249 (2001)
20. Feeney, L.M., Nilsson, M.: Investigating the energy consumption of a wireless network interface in an ad hoc networking environment. In: Proc. IEEE INFOCOMM, Anchorage AK (2001)
21. Kim, D., Bae, H., Song, J.: Analysis of the Interaction between TCP Variants and Routing Protocols in MANETs. In: Proc. of the 2005 International Conference on Parallel Processing Workshops (ICPPW 2005), pp. 380–386 (2005)

Symbolic Data Analysis for the Development of Object Oriented Data Model for Sensor Data Repository

Doreswamy and Srinivas Narasegouda

Department Of P.G Studies and Research in Computer Science, Mangalore University, Mangalagangothri, Mangalore, India
doreswamyh@yahoo.com,srinivasnpatil@gmail.com

Abstract. Data generated by sensors, need to be stored in a repository which is of large in size. However, data stored in sensor data repository consists of inconsistent, inaccurate, redundant and noisy data. Deployment of data mining algorithm on such sensor datasets declines the performance of the mining algorithm. Therefore, data preprocessing techniques are proposed to eliminate inconsistent, inaccurate, redundant, noisy data from the datasets and symbolic data analysis approach is proposed to reduce the size of the data repository by creating a symbolic data table. The data stored in a more comprehensible manner through symbolic data table is modeled as object oriented data model for mining knowledge from the sensor data sets.

Keywords: symbolic data analysis, data mining, object oriented data model, sensor data repository.

1 Introduction

The development in technologies is generating a gigantic amount of data every day. Such data can be converted into useful information and knowledge using data mining tools. Most of the data mining techniques look for patterns in tables to discover the knowledge. However, as the technology developed, the data is not restricted to mere numeric form. It has become more complex in nature. And such complex data may not fit in traditional database management systems. Hence, it is necessary to develop a data repository system which can easily manage simple and complex data. One way of managing such data would be to develop an object oriented data repositories.

The development of an object oriented sensor data repository is needed in order to provide an accurate and precise data for the both researchers and developers to test their algorithms on realistic sensor data. It facilitates different researchers to evaluate the performance of their algorithms by testing them on the standard datasets [12]. An object oriented approach for sensor data repository also provides a platform for fusion of heterogeneous sensor data from different platforms. Such approach acts as source of information in visualizing real world applications [9].

S.C. Satapathy, S.K. Udgata, and B.N. Biswal (eds.), *FICTA 2013*,
Advances in Intelligent Systems and Computing 247,
DOI: 10.1007/978-3-319-02931-3_49, © Springer International Publishing Switzerland 2014

The paper is organized as follows. Section 2 describes the need for an object oriented data model and their existing implementations in the real world. Section 3 describes the related work. The proposed object oriented data model for sensor data repository is described in section 4. Section 5 presents the experimental results and the conclusion is given in section 6.

2 Object Oriented Approach

The traditional database management system may not fit well in all applications especially when the applications represent the real world entities along with their behavior and involving complex data types. The best way of dealing with such problem is to develop an object oriented database management system (OODBMS) which is a fusion of object oriented principles and database management system principles. In OODBMS, database is accessed by objects in a transparent approach. The OODBMS has several advantages over traditional database management systems [11].

Many applications have been developed using object oriented concepts. An object oriented methodology was used to analyze, design and develop a prototype of office procedures [10]. An object oriented approach was proposed to develop conceptual modeling of data warehouses, multidimensional databases and online analytical processing applications [13]. It provided a theoretical basis for the use of OO databases and object-relational databases in data warehouses, multidimensional databases and online analytical processing applications. Object oriented data models such as OMT-G [6] and IFO [14] was suggested to represent the geographic spatial data. OMT-G was able to overcome the limitations regarding primitives to represent spatial data, and to provide more sufficient tools for modeling geographic applications. From literature review we can clearly say that object oriented approaches offer many advantages over traditional methods for database designers to handle complex data.

3 Related Work

The vast applications of sensors generate massive amount of data. Such data are in need of standard repositories for storing and managing an accurate and precise data for the development of research in sensor network applications. XML based Environmental Markup Language was proposed by German Federal Environmental Agency for representing the environmental data [3]. A sensor data repository was developed for autonomous driving research with an objective to enhance the research in sensor processing for autonomous vehicle by providing an accurate and precise data for the researchers to test their algorithms. The data was collected using a NIST High Mobility Multi-purpose Wheeled Vehicle and stored in a relational database [12]. In 2006 SensorBase.org was proposed similar to web blogs for storing and sharing sensor network data [7]. Similar to SensorBase.org, CRAWAD is another group which stores the sensor data [1]. An object oriented approach was used to develop Object Oriented World Model to store different sensor data and fuse them into one platform [4]. In maritime surveillance, data are collected from different sensors and platforms. An object oriented approach was used to develop sensor data fusion architecture namely Object Oriented World Model for providing an information source for the maritime surveillance application through visualization module and to track analysis or behavior recognition [9].

4   Proposed Model

In existing sensor data repositories, data is stored either in a traditional database system or object oriented database system. Data analysis on a very large database is difficult and moreover data in real world is inconsistent, redundant and noisy. Data accuracy is not guaranteed and it is found to be less reliable. Data mining results on such data are imprecise and inaccurate. Hence, we propose symbolic data mining approach for object oriented sensor data repository by combining symbolic data analysis and object oriented concepts. Using symbolic data analysis we created a symbolic data table and data preprocessing techniques are applied to increase the accuracy and reliability of the data. And using object oriented concepts an object oriented data model (OODM) for sensor data repository is proposed. In any sensor network each sensor represents a real world entity which has its own behavior in the real world. Sensors are time and location related. Hence, while developing a repository model we have to consider time and location at which the data was generated. We have conducted our experiments on Intel Berkeley Research lab dataset [2]. We defined 4 classes namely $Class\_Sensor\_Location, Class\_Aggregate\_Connectivity, Class\_Lab\_Datset$ and $Class\_OODM$. In each class we defined attributes and methods. OODM class contains methods for processing the data. Then we created objects of each class and called the method to process the data. Object oriented programming concepts such as class, object, encapsulation, polymorphism and inheritance have been applied, which provides the facilities for processing the user queries. Fig 1 shows the class diagram of our proposed model. The proposed model is processed in two stages namely creating symbolic data table and data preprocessing. Each stage is explained in detail in the following sections.

4.1   Symbolic Data Table

In our approach we have combined symbolic data analysis and data mining techniques. Symbolic data analysis provides a symbolic data table in which data is stored in a more comprehensible manner compared to traditional data tables i.e., classical data on a random variable p is represented in a p-dimensional space by a single point where as symbolic data represents a random variable p with measurements are p-dimensional hypercube in space [5].

Converting standard table to symbolic table: In classical data if Y_j is the jth random variable, $j = 1, ...p$ measured on an individual $i \in \Omega$ where $\Omega = 1, 2, ...n$

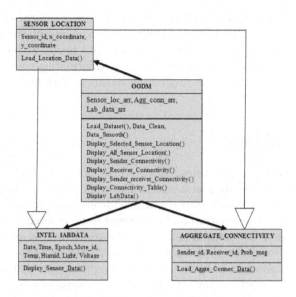

Fig. 1. Class diagram of the proposed model

then $Y_j = x_{ij}$ represents a particular value for i on j^{th} variable Y_j. And n x p matrix $X = (x_{ij})$ represents the entire dataset. In symbolic data table, if Y_j measured on a category $w_u \in E$ where $E = w1, w2, ...w_n$ then $Y_j(w_u) = \xi_{uj}$ represents a particular value and $\xi = \xi_{uj}$ represents the complete set. In classical data x_{ij} in X takes one value where as in symbolic data table ξ_{uj} takes several values. Figure 2 shows the transformation of classical data to random variable in symbolic data table. In any dataset two random multi-valued variables can be mapped together to form a category which contains all the data belonging to the mapped category. In our experiment we have considered multi-valued random variables *date* and *mote_id* to form a category. U_i is the set containing all distinct values of random variable *mote_id* where $i = 1, 2...n$. V_j is the set containing all distinct values of random variable *date* where $i = 1, 2...m$. In our dataset n and m represents the random variable *mote_id* and *date* respectively. In each set, each distinct value is assigned a serial number starting from 1. i.e., *mote_id* is assigned from 1 to 54 and *date* is assigned from 1 to 36. To implement symbolic table we have used Dictionary class in $C\#$ which takes two parameters *key* and *data*. A unique key is generated for storing and retrieving the data for each category by using the equation (1).

$$Key_{ij} = (U_i - 1)(m) + V_j \tag{1}$$

4.2 Data Preprocessing

Data preprocessing techniques improves the quality of the data and mining results. There are numerous techniques available in data preprocessing to increase

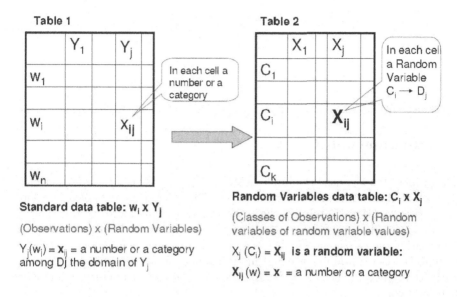

Fig. 2. Standard data to random variable. (Image taken from [8])

accuracy and reliability of data and to reduce the data size without jeopardizing the data mining results. Following sections describes the data preprocessing techniques applied in our experiment.

Data Cleaning: In data cleaning process, inconsistent data, missing data are handled using data mining techniques. If the number of attribute values missing is more in a row, then the entire row is discarded. Data which are consistent and accurate are stored in the symbolic data table by using the *key* value generated by the Equation (1).

Data Smoothing: In most of the real-time applications data are noisy. We applied data mining's bin by means method for smoothing the data. We considered consecutive readings taken in a time period by sensor and each attribute is distributed into buckets or *bins* of size 3. Mean value of each *bin* is calculated and each original value is replaced by mean value of the *bin*. Since, all three values in a *bin* contains the same value i.e., mean value of *bin*, two out of three readings are considered as redundant and hence discarded.

Data Reduction: In environmental monitoring, features such as temperature, humidity are likely to be constant for a period of time. Hence, before storing any value we checked whether the previous reading is identical or not. This is done using the *Minkowski* distance formula shown in equation (2). If the readings are identical then the data is discarded otherwise data is stored into the symbolic data table.

$$d(x,y) = \left(\sum_{i=1}^{n} |x_i - y_i|^p \right)^{1/p} \tag{2}$$

Where $p = 1, X = (x_1, x_2, ...x_n)$ and $Y = (y_1, y_2, ...y_n)$. X and Y represent the set of attributes.

5   Experimental Result

The proposed model has been implemented using $C\#$. We conducted experiment on Intel Berkeley Research lab dataset [2]. This dataset consists of 3 text files. (1) Sensor location containing the information such as mote id and their x-coordinates and y-coordinates. (2) Aggregate Connectivity Data containing the information such as the probability of a message reaching from sender to receiver. (3) Lab Dataset containing the information about the sensor data readings taken with attributes such as date, time, epoch count, mote id, temperature, humidity, light and voltage.

The proposed model stores and manages the complex sensor data using an object oriented concept. First, symbolic data table is created then inconsistent and inaccurate data are handled using data mining techniques. Noisy data are removed by smoothing the data using binning method. And finally, redundant data are removed. In our experiment we found that approximately 71.61% of inconsistent, noisy and redundant data were removed. The details of the data reduction at different stages are shown statistically in Table 1. The reduced dataset from OODM, stores very less amount of data compared to the original dataset without losing the information.

Table 1. Dataset details at different stages

Different stages of data preprocessing	Data reduced in %
Data reduced in Data Cleaning	4.0466%
Data reduced in Data Smoothing	63.9251%
Data reduced in Data Reduction	3.6445%
Total reduction	**71.6162%**

The graphical user interface (GUI) of the model provides the visualization of the data and provides a form through which user can enter the queries. As this is an object oriented model, there is no need for complex structured query language. We have tested the model to answer queries such as location of sensors, probability of a message reaching from sender to receiver, and details of sensor readings of the user specified sensor. Fig 3 shows the details of sensor location, aggregate connectivity of sensors and Fig 4 shows the sensor data of a specified sensor.

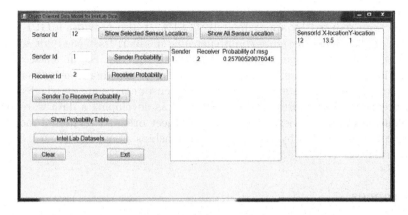

Fig. 3. Graphical representation of dataset

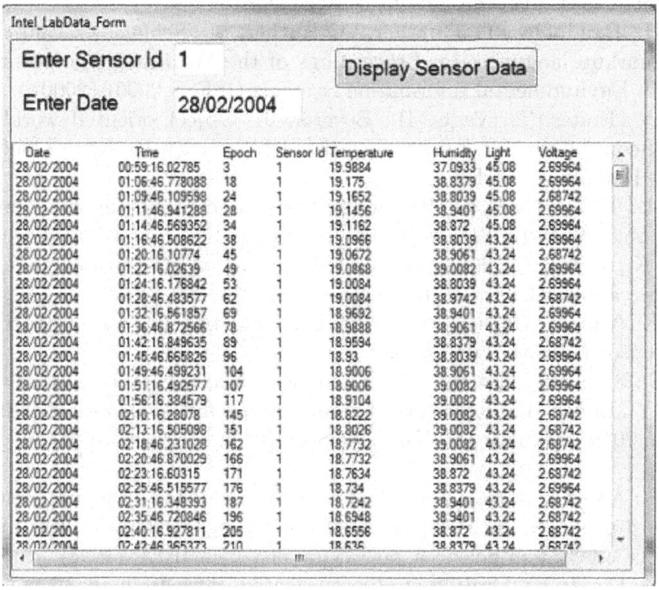

Fig. 4. Graphical representation of dataset

6   Conclusion

The proposed object oriented data model using symbolic data analysis approach provides a sensor data repository for storing and managing sensor data. It provides data preprocessing techniques to ensure the accuracy and consistency of the data by handling inconsistent, redundant and noisy data. OODM reduces the original database size without losing the information. As the data is stored in symbolic data table, in future it is easy to aggregate data for summarizing and creating a data warehouse. Our experimental result shows that database size has been reduced by approximately 71.61%. It also provides a GUI to display the sensor data in a more comprehensible manner. User need not to know structured query language in order to process the queries. So far, basic queries such as location of sensors, probability of a message reaching from sender to receiver, and details of sensor data of a specified sensor can be answered effectively. In future we plan to address other issues such as developing a data warehouse, processing complex queries, developing an object oriented relational model to reduce query processing by identify the relationships among the classes, objects and attributes.

References

1. CRAWDAD, http://crawdad.cs.dartmouth.edu/
2. Intel Berkeley Research lab dataset,
   http://db.csail.mit.edu/labdata/labdata.html
3. Arndt, H., Bandholtz, T., Gunther, O., Ruther, M., Schutz, T.: Eml-the environmental markup language. In: Proceedings of the Workshop Symposium on Integration in Environmental Information Systems (ISESS 2000) (2000)
4. Bauer, A., Emter, T., Vagts, H., Beyerer, J.: Object oriented world model for surveillance systems. In: Future Security: 4th Security Research Conference, pp. 339–345. Fraunhofer Verlag (2009)
5. Billard, L., Diday, E.: Symbolic data analysis: conceptual statistics and data mining, vol. 654. Wiley (2012)
6. Borges, K., Davis, C., Laender, A.: Omt-g: An object-oriented data model for geographic applications. GeoInformatica 5(3), 221–260 (2001)
7. Chang, K., Yau, N., Hansen, M., Estrin, D.: Sensorbase.org-a centralized repository to slog sensor network data (2006)
8. Diday, E.: Symbolic data analysis of complex data: Several directions of research
9. Fischer, Y., Bauer, A.: Object-oriented sensor data fusion for wide maritime surveillance. In: 2010 International Waterside Security Conference (WSS), pp. 1–6. IEEE (2010)
10. Frank, U.: An object-oriented methodology for analyzing, designing, and prototyping office procedures. In: Proceedings of the Twenty-Seventh Hawaii International Conference on System Sciences, vol. 4, pp. 663–672. IEEE (1994)
11. Obasanjo, D.: An exploration of object oriented database management systems, http://www.25hoursaday.com/WhyArentYouUsingAnOODBMS.html
12. Shneier, M., Chang, T., Hong, T., Cheok, G., Scott, H., Legowik, S., Lytle, A.: Repository of sensor data for autonomous driving research. In: Proceedings of SPIE, vol. 5083, pp. 390–395. Citeseer (2003)
13. Trujillo, J., Palomar, M., Gomez, J., Song, I.: Designing data warehouses with oo conceptual models. Computer 34(12), 66–75 (2001)
14. Worboys, M., Hearnshaw, H., Maguire, D.: Object-oriented data modelling for spatial databases. Classics from IJGIS: Twenty Years of the International Journal of Geographical Information Science and Systems 4(4), 119 (2006)

Fault Detection in Sensor Network Using DBSCAN and Statistical Models

Doreswamy and Srinivas Narasegouda

Department Of P.G Studies and Research in Computer Science, Mangalore University, Mangalagangothri, Mangalore, INDIA.
doreswamyh@yahoo.com,srinivasnpatil@gmail.com

Abstract. In any sensor network one of the major challenges is to distinguish between the expected data and unexpected or faulty data. In this paper we have proposed a fault detection technique using DBSCAN and statistical model. DBSCAN is used to cluster the similar data and detect the outliers whereas statistical model is used to build a model to represent the expected behaviour of the sensor nodes. Using the expected behaviour model we have detected the faults in the data. Our experimental results on Intel Berkeley research lab dataset shows that faults have been successfully detected.

Keywords: Fault detection, DBSCAN, Statistical model, Sensor Network.

1 Introduction

Wireless sensor networks consist of hundreds of sensor devices. Each sensor device has capability to sense, compute and transmit the data. These sensors are used for monitoring an environment by deploying them in the space where the monitoring it to be done. Generally sensors are used to monitor temperature, humidity, light, voltage etc [10] in various types of applications such as environmental monitoring, security, surveillance, disaster management etc.

However, sensor devices have some limitations such as limited storage capacity, bandwidth, power supply and computational capabilities. Due to these restrictions sensor networks are open to the elements such as faults, leading to the untrustworthy data which can affect the quality of the data [4].

Faults in sensor data can occur due to various reasons such as low battery, hardware malfunctioning, exceeding the range of the device etc. And any data which is faulty can affect the results of decision making process. Hence, it is desirable to detect the fault in sensor network.

In sensor network fault can be defined as the data generated by a sensor device which is either below or beyond the expected values. This brings a great challenge in machine learning process for identifying and distinguishing between the expected and unexpected behaviour. Data mining techniques such as clustering, Bayesian algorithms, and statistical methodologies are very helpful in identifying the faults in the sensor network. Literature review reveals that numerous

S.C. Satapathy, S.K. Udgata, and B.N. Biswal (eds.), *FICTA 2013*, 443
Advances in Intelligent Systems and Computing 247,
DOI: 10.1007/978-3-319-02931-3_50, © Springer International Publishing Switzerland 2014

methodologies have been proposed by many researchers for fault detection in sensor networks.

The rest of the paper is organized as follows. Section 2 describes the literature review. The proposed model is presented in Section 3. Experimental results are explained in Section 4 and conclusion is given in Section 5.

2   Related Work

Fault detection in sensor network has been an interesting topic of research for many researchers. Many researchers have proposed various fault detection methodologies. In [5] statistical techniques and cross-validation techniques were combined to find online sensor faults in which statistical approach were used for identifying the probability of a faulty sensor. A Bayesian algorithm for fault detection by exploiting the spatiotemporal correlation among the measurements was proposed by [6]. The concept of external manager was introduced in [12] for fault detection. Even though the external manger works adequately, but the communication cost between the nodes and external manager may decline the sensor energy. [3]Developed a framework for the identification of faults in which each sensor node confirms their status by comparing its value with the median of its neighbors. A distributed fault detection technique was developed by [2] in which each sensor identifies the fault by comparing its sensed value with its nearest neighbors. However, this method is quite complex as sensor nodes has to compare their reading with neighbors. [9] Used Bayesian and Neyman-Pearson methods to develop energy efficient fault tolerant mechanism. In their approach they used minimum number of neighbors to reduce the communication cost. In [7] communication graph was used to define the neighborhoods and threshold test was used for identifying the faulty sensors. [13] Suggested fuzzy data fusion technique for identifying the faults in the sensor networks. [8] Proposed an evolving fuzzy classifier which is set of fuzzy rules defined and updated using incremental clustering procedure. [11] Suggested that different types of faults can be identified by studying environmental features, statistical features and data centric view.

3   Proposed Model

Faults in sensor network can be defined as the unexpected data generated by sensor nodes. There are various reasons due to which a sensor node may generate faulty data. And these faults can be identified by studying the *environmental features, statistical features*, and *data-centric view*.

In *environmental features* we considered physical certainties. The physical certainties define the basic laws of science. For example relative humidity is always calculated in terms of percentage. Hence, we have considered the readings between 0 and 100 as expected results and readings which are less than 0 or greater than 100 are considered as faulty.

Data centric view can be defined as the faults identified based upon the characteristics of the data such as outliers. In our proposed model we have used DBSCAN (Density Based Spatial Clustering of Applications with Noise) to identify the outliers. DBSCAN forms the clusters based on the density among the data points. In DBSCAN, to measure the distance between the data points Euclidean distance formula (1) has been used.

$$d(x,y) = \sqrt{\sum_{i=1}^{n}(x_i - y_i)^2} \tag{1}$$

Where $x_i = (x_1, x_2, x_3 \ldots x_n)$ and $y_i = (y_1, y_2, y_3 \ldots y_n)$ represents the temperature and humidity values respectively.

The main advantage of DBSCAN is it requires only two parameters namely minimum number of points to form a cluster and the ε (epsilon) value to identify the ε-neighborhood which is within the radius of ε. Unlike K-means clustering user need not mention the number of cluster to be formed in advance. This feature of DBSCAN ensures that dense objects will be clustered. The DBSCAN algorithm used for $C\#$ is given below.

```
DBSCAN_Algorithm(points, eps_val, minpts)
   Clusters=Get_clusters(points, eps_val, minpts)
End //End of DBSCAN Algorithm

Get_cluster(points, eps_val, minpts)
   If points==null
     Return null
   Assign cluster_id=1
   For each point in points
     If p.cluster_id is unclassified
       Extend_cluster(points,p,cluster_id,eps_val,minpts)
       Cluster_id=cluster_id+1
     End If
   End For
   For each point p in points
     If p.cluster_id > 0
       Add p to clusters
     End If
   End For
   Return clusters
End //End of Get_cluster

GetRegion(points, point p, eps_val)
   For each p in points
     Distance=Euclidean_distance(p,points)
     If(Distance <= eps_val)
       Add points to region
```

```
    End If
  End For
  Return region
End //End of Get_Region

Extend_cluster(points, p, cluster_id, eps_val, min_pts)
  Seeds=GetRegion(points,p,eps_val)
  If Seeds count < min_pts
    Mark p.cluster_id as noise
    Return false
  Else
    For each point in Seeds
      Seeds.cluster_id=cluster_id
    End For
    Remove p from Seeds
    While seed count is > 0
      Current_p=Seeds[0]
      Result=GetRegion(points,current_p,eps_val)
      If(result count >= minpts)
        For each point in result
          result_p=result[i]
            If(result_p.cluster_id is unclassified OR
                   result.cluster_id is noise)
              If (result_p.cluster_id is unclassified)
                add result_p to Seeds
              End If
              result_p.cluster_id=cluster_id
            End If
         End For
        End If
        Remove Current_p from Seeds
      End While
      Return true
    End Else
  End// End of Extent_cluster
```

Even faults have some similar patterns. Hence, after detecting the outliers using DBSCAN there is a probability that one or more clusters may contain faulty data. In order to detect faulty data in clusters we have used statistical model. In statistical model we have used statistical measures such as mean and standard deviation to model the expected behavior. We considered the first day readings of [1] to model the expected behavior. Mean and standard deviation of temperature and humidity are calculated for individual sensor nodes. The expected minimum and maximum values are modeled using the formulae 2 , 3 , 4 , 5.

$$min_{temp} = \mu_{temp} - k\sigma_{temp} \qquad (2)$$

$$max_{temp} = \mu_{temp} + k\sigma_{temp} \qquad (3)$$

$$min_{humid} = \mu_{humid} - k\sigma_{humid} \qquad (4)$$

$$max_{humid} = \mu_{humid} + k\sigma_{humid} \qquad (5)$$

Where μ and σ represent the mean and standard deviation respectively. In the monitored environment temperature may rise up to 35^0 Celsius and humidity may rise up to 65%. By experiment we found that when k value is 4.5, any temperature up to 35^0 Celsius and humidity up to 65% is considered as normal or expected.

4   Experimental Result

The proposed fault detection technique has been implemented using visual studio 2010 with $C\#$. We have selected [1] dataset for experimental purpose. It contains seven features namely time, epoch count, mote id, temperature, humidity, light, and voltage. It has 2.3 million observations taken in 36 days. The dataset has been reduced to 629274 observations by applying data mining preprocessing techniques. Firstly, missing data are handled using data cleaning process where an entire observation is discarded if more number of attributes are missing. Secondly, data smoothing's binning method is used for smoothing the data. In binning method observations are divided into bins of size three. Each value is replaced by its mean value and then two out of three readings are discarded as they are identical. Finally we used Minkowski distance formula to check whether consecutive readings are identical or not. If they are identical then the redundant data generated is discarded.   The fault detection has been applied on these

Table 1. Results obtained from the proposed model

Dataset Details	Readings for Days 1-12	Readings for Days 13-24	Readings for Days 25-36
No of observations	319758	239322	70194
Physical certainties	191	2755	29967
No of clusters	4	4	12
No of Outliers	109	764	639
Data removed using statistical feature	842	1090	6564
No of faults removed	**1142**	**4609**	**37170**
Total no of faults removed in 36 days	**42921**		
No of observations in the final Dataset	**586353**		

subsets in three stages. In the first stage using the physical certainties, expected range of the feature are defined to identify the fault. In particular we have defined the humidity readings which are less than 0 or exceeding 100 are considered as faulty. In second stage outliers are detected by studying the characteristics of the data. For outlier detection DBSCAN is used with input parameters, minimum number of points to form a cluster 25 and ε value 1.75. In the final stage, expected behaviour of the features are modelled by applying statistical measures on the historical data. The statistical model is developed using the formulae (2),(3),(4) and (5) and any data which is above the range of expected maximum or below the expected minimum value are considered as faulty data. Table 1 shows the details of the fault detected and removed in different stages of the proposed model. Expected behaviour of the features such as temperature and humidity are shown graphically where feature is plotted against sensor id i.e., Fig 1 shows the expected minimum and maximum temperature and Fig 2 shows the expected minimum and maximum humidity.

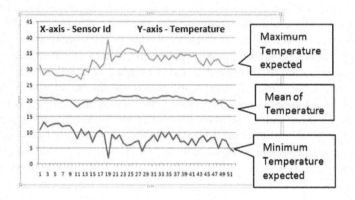

Fig. 1. Graph showing expected temperature

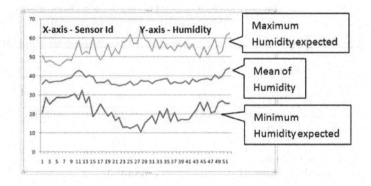

Fig. 2. Graph showing expected Humidity

5   Conclusion

In this paper we have proposed fault detection in sensor network by studying the environmental features, statistical features, and data-centric view.

An environmental feature defines the physical certainties and expected behavior of the phenomenon. Physical certainties are used to define the expected range of the features such as any humidity value below 0 or exceeding 100 is considered as fault. Expected behavior of the phenomenon is modeled by applying the statistical measures on the historical data. Any data which do not comply with the expected behavior is considered as fault. In data centric view, characteristics of data such as outliers are detected using the DBSCAN (Density Based Spatial Clustering of Applications with Noise).

The faulty data may contain information and it can be interpreted by analysing the data. Hence, in future we wish to analyse the faulty data to extract the useful information if there exist any.

References

1. Intel Berkeley Research lab dataset,
   http://db.csail.mit.edu/labdata/labdata.html
2. Chen, J., Kher, S., Somani, A.: Distributed fault detection of wireless sensor networks. In: Proceedings of the 2006 Workshop on Dependability Issues in Wireless Ad Hoc Networks and Sensor Networks, pp. 65–72. ACM (2006)
3. Ding, M., Chen, D., Xing, K., Cheng, X.: Localized fault-tolerant event boundary detection in sensor networks. In: Proceedings of the 24th Annual Joint Conference of the IEEE Computer and Communications Societies, INFOCOM 2005, vol. 2, pp. 902–913. IEEE (2005)
4. Gaber, M.: Data stream processing in sensor networks. Learning from Data Streams, p. 41 (2007)
5. Koushanfar, F., Potkonjak, M., Sangiovanni-Vincentelli, A.: On-line fault detection of sensor measurements. In: Proceedings of IEEE Sensors, vol. 2, pp. 974–979. IEEE (2003)
6. Krishnamachari, B., Iyengar, S.: Distributed bayesian algorithms for fault-tolerant event region detection in wireless sensor networks. IEEE Transactions on Computers 53(3), 241–250 (2004)
7. Lee, M., Choi, Y.: Fault detection of wireless sensor networks. Computer Communications 31(14), 3469–3475 (2008)
8. Lemos, A., Caminhas, W., Gomide, F.: Adaptive fault detection and diagnosis using an evolving fuzzy classifier. Information Sciences (2011)
9. Luo, X., Dong, M., Huang, Y.: On distributed fault-tolerant detection in wireless sensor networks. IEEE Transactions on Computers 55(1), 58–70 (2006)
10. Ma, X., Yang, D., Tang, S., Luo, Q., Zhang, D., Li, S.: Online mining in sensor networks. In: Jin, H., Gao, G.R., Xu, Z., Chen, H. (eds.) NPC 2004. LNCS, vol. 3222, pp. 544–550. Springer, Heidelberg (2004),
    http://dx.doi.org/10.1007/978-3-540-30141-7_81

11. Ni, K., Ramanathan, N., Chehade, M., Balzano, L., Nair, S., Zahedi, S., Kohler, E., Pottie, G., Hansen, M., Srivastava, M.: Sensor network data fault types. ACM Transactions on Sensor Networks (TOSN) 5(3), 25 (2009)
12. Ruiz, L., Siqueira, I., Wong, H., Nogueira, J., Loureiro, A., et al.: Fault management in event-driven wireless sensor networks. In: Proceedings of the 7th ACM International Symposium on Modeling, Analysis and Simulation of Wireless and Mobile Systems, pp. 149–156. ACM (2004)
13. Shell, J., Coupland, S., Goodyer, E.: Fuzzy data fusion for fault detection in wireless sensor networks. In: 2010 UK Workshop on Computational Intelligence (UKCI), pp. 1–6. IEEE (2010)

Fast SSIM Index for Color Images Employing Reduced-Reference Evaluation

Vikrant Bhateja, Aastha Srivastava, and Aseem Kalsi

Department of Electronics and Communication Engineering,
Shri Ramswaroop Memorial Group of Professional Colleges (SRMGPC),
Lucknow-227105(U.P.), India
{bhateja.vikrant,aasthasrivastava49,aseemkalsi}@gmail.com

Abstract. Image quality assessment employing reduced-reference estimation approaches evaluate the perceived quality with only partially extracted features of the reference image. The primary aim of these approaches is to make objective evaluation flexible enough; accommodating the effect of any new distortion introduced in the image. Based on this concept, the paper proposes a fast approach for quality assessment of color images by modifying the SSIM index. The methodology involves sub-band decomposition of color images in wavelet domain for extracting statistical features. The computational complexity during estimation of features is reduced in this work by using the gradient magnitude approach. Noteworthy reduction in computational time is observed with the proposed index and the evaluation is also found coherent with full-reference and reduced-reference SSIM.

Keywords: computational time, CSIQ database, Discrete Wavelet Transform (DWT), Human Visual System (HVS), structural distortion.

1 Introduction

With the recent advancement and growth of internet around the globe the demand for online digital images is increasing rapidly. This trend requires development of fast measures for quantitative evaluation of image quality. These evaluation approaches help in parameter optimization of various image processing algorithms for performance enhancement. For those image processing applications which require the end results to be viewed by human observers; it is expected that the subjective analysis (for estimating the visual quality of images) should be performed effectively with due precision. The use of colored images simplifies the task of subjective analysis as the Human visual system (HVS) is highly adaptable to the changing shades in color than in the gray scale [1]-[3]. This further leads to accuracy in determination of subjective score for such images but in turn makes the objective analysis highly complex. This is so because objective analysis requires the assessment to be a function of multiple types of distortions [4] and it is also expected to be in due coherence with HVS. Objective evaluation of images can be broadly categorized into three categories: Full Reference (FR), No Reference (NR) and Reduced Reference (RR) approaches to IQA.

S.C. Satapathy, S.K. Udgata, and B.N. Biswal (eds.), *FICTA 2013*,
Advances in Intelligent Systems and Computing 247,
DOI: 10.1007/978-3-319-02931-3_51, © Springer International Publishing Switzerland 2014

FR approaches require access to the complete reference image for its evaluation. Some common FR approaches proposed in the past are Mean Squared Error (MSE), Just Noticeable Difference (JND), Peak Signal-to-Noise Ratio (PSNR) and Structural Similarity (SSIM). MSE, PSNR, and JND have been used for their simplicity but provided quality assessment of images only as a function of error measurement between the reference and the transformed pixels. IQA using these fundamental measures were generally incoherent with the HVS and hence their different variants are proposed in literature [5]-[8]. Moreover, SSIM corresponds well with the HVS because it emphasizes more on structural distortions [9]-[11]. However, it is not possible to access the entire reference image in case of online applications. Also, IQA employing FR approaches requires greater computational times as the entire image is referred. This motivates the development of NR IQA approaches that could predict the quality without accessing the original image. Although, these approaches are limited to assessment for only a particular distortion; and introduction of new distortion adversely affects the perceptual quality evaluation [12]-[17]. Last category of IQA using RR approaches is a compromise between FR-IQA and NR-IQA. RR evaluation methods are effective for online applications as they do not require the entire reference image; but only a few features which could be easily transmitted along with the distorted image [17]-[19]. But, with the growing advancements in the field of communication involving huge amount of data transfer; calls for improvement in RR methods to suit the application requirements. Tao et al. [20] proposed a RR method that measures the quantity difference of the visual sensitivity coefficients in contour domain by employing city-block distance. Lin et al. [21] proposed RR technique using Generalized Gaussian Density (GGD) which performs reorganisation of Discrete Cosine Transform; which is a complex attribute when applied for gray scale images. Hence, in the proposed paper a RR evaluation method for color images is presented which modifies the SSIM index to reduce its computational complexity. The process involves sub-band decomposition of images employing Discrete Wavelet Transform (DWT) along with the application of gradient magnitude approach for estimation of features in each sub-band. The remaining paper is structured as follows: Section 2 describes the proposed IQA method. Section 3 contains the obtained results and its discussion. Section 4 draws the conclusions.

2   Proposed Methodology for RR-IQA

The motivation of this work is to develop a computationally efficient RR evaluation framework for color images that is coherent with HVS and generalized for a multiple distortion types, i.e. compression and blurring. The first step in this process is to decompose the reference as well as the distorted images into four sub-bands employing DWT using an appropriate wavelet family. This is so because the RR-IQA approach is based on the extraction of certain features from the reference as well as the distorted images. The second step involves extraction of reduced reference features from the sub-images of two images. The two features extracted from the decomposed sub-bands are divergence and standard deviation. Hence, a total of eight reduced reference features are extracted, i.e. two from each sub-band. The procedure for the extraction of these features is explained in the paragraphs to follow. The value of divergence parameter is calculated between the reference (X) and the distorted (Y) images as:

$$\hat{d}(x^n \| y^n) = p(x^n) - p(y^n) \tag{1}$$

where: $p(x^n)$ and $p(y^n)$ are the probability distribution functions (PDF) of the n-th sub-band of the reference and distorted images respectively and $\hat{d}(x^n \| y^n)$ denotes the divergence between the sub-bands. The second feature to be extracted is the standard deviation which is computed using gradient magnitude [22] which is approximated as follows:

$$G_{xn}(i,j) = \max\left\{\left|f_{xn}(i+1,j+1) - f_{xn}(i,j)\right|, \left|f_{xn}(i+1,j) - f_{xn}(i,j+1)\right|\right\} \tag{2}$$
$$+ \frac{1}{4}\min\left\{\left|f_{xn}(i+1,j+1) - f_{xn}(i,j)\right|, \left|f_{xn}(i+1,j) - f_{xn}(i,j+1)\right|\right\}$$

where: G_{xn} is the gradient magnitude of the n-th sub-band of the reference image. Similarly, the gradient magnitude of all the sub-bands is calculated for the distorted image using (2). In (2) $f_{xn}(i,j)$ is the pixel location in the n-th sub-band of the reference image. The computation of standard deviation is time consuming as it requires a large number of addition and multiplication operations. The gradient magnitude approach reduces the complexity as it requires only 4 subtractions and 1 addition for a 2 x 2 window. This is followed by computation of standard deviation using the gradient magnitude. Once, the values of divergence and standard deviation parameters are estimated for the reference and distorted images; next step in the procedure is the computation of distortion measure (ϕ) which is stated as follows:

$$\phi = \eta\left(\sigma_{Gx}, \sigma_{Gy}\right)\log\left(1 + \sum_{n=1}^{N}\left|\hat{d}\left(x^n \| y^n\right)\right|\right) \tag{3}$$

It can be observed in (3) that ϕ is a function of $\hat{d}$ and an additional factor $\eta\left(\sigma_{Gx}, \sigma_{Gy}\right)$. This factor is a function of standard deviation of the two images and is introduced for enhancing the effect of structural distortions in the measurement of ϕ as these distortions affect the quality of the image more than the non-structural distortion types. As the structural dissimilarity between the images increases, η also increases. Therefore, this leads to an increase in the value of ϕ. This is followed by the calculation of slope factor (α) which is computed for the images. The value of α is based on the linear relationship between distortion measure and FR SSIM. In this work, it is required to study ϕ for blurring and compression. The proposed measure (Q) for these distortions is obtained by making use of different values of α. Thus, the final expression for Q can be stated as:

$$\begin{pmatrix} Q_b \\ Q_j \end{pmatrix} = 1 - \begin{pmatrix} \alpha_b \\ \alpha_j \end{pmatrix}\phi \tag{4}$$

where: $.002 \le \alpha_b \le .03$ and $.01 \le \alpha_j \le .04$ are the slope factors for blurring and compression respectively. Q_b and Q_j are the quality measure for blurring and compression respectively.

3 Results and Discussions

In this work, the validation of the proposed index has been performed by evaluation of the distorted images of the CSIQ database [23]. Figure 1 and 2 contains the reference and distorted images respectively used in this work for simulation. The color images are divided into four sub-bands each using DWT employing db1 wavelet family. This is followed by the extraction of the two statistical features divergence and standard deviation as described in (1)-(2). The distortion measure ϕ was obtained using (3). The final quality index (Q) for different distortions is obtained as in (4).

Img1   Img2

Fig. 1. Reference Images from CSIQ Database [23]

In this paper the two state-of-art indices i.e. SSIM and RR SSIM along with the proposed measure Q are tested against the Mean Opinion Score obtained by subjective analysis of images [23]. Table 1 enlists the estimation of proposed quality measure (Q) along with the SSIM and RR SSIM for various distortions. These measures are verified according to their Pearson Correlation Coefficient (PLCC) calculated between Q and the subjective score and are shown in table 2. From this table it is observed that for compression Q has a higher correlation as compared to the normal SSIM and RR SSIM measures except for img 5 which shows better correlation for SSIM in case of compression. However, for blurring Q shows lower correlation than normal SSIM but are better than RR SSIM measure.

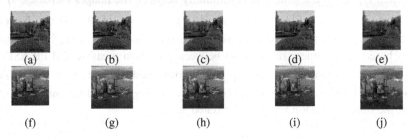

| (a) | (b) | (c) | (d) | (e) |
| (f) | (g) | (h) | (i) | (j) |

Fig. 2. Distorted Images from CSIQ Database. 3(a)-(e) Blurring and 3(f)-(j) Compression obtained at various levels of distortions.

Table 1. Estimation of Quality Measures (SSIM, RR SSIM and Proposed Q) for Various Distortions

Distortion Type	Quality Measures	Img 1	Img 2	Img 3	Img 4	Img 5
Blurring	SSIM	0.9998	0.9998	0.9996	0.9999	0.9996
	RR SSIM	0.9978	0.9974	0.9975	0.9977	0.9950
	Q	0.9948	0.9910	0.9927	0.9935	0.9876
Compression	SSIM	0.9996	0.9997	0.9994	0.9958	0.9880
	RR SSIM	0.9987	0.9986	0.9983	0.9973	0.9921
	Q	0.9951	0.9898	0.9929	0.9952	0.9929

Table 2. PLCC Comparison Between SSIM, RR SSIM and Proposed Measure Q

Distortion type	Quality Measures	Img 1	Img 2	Img 3	Img 4	Img 5
Blurring	SSIM	0.9091	0.9500	0.9374	0.8602	0.8757
	RR SSIM	0.8619	0.9217	0.8873	0.7974	0.9320
	Q	0.8653	0.9242	0.8831	0.7980	0.9220
Compression	SSIM	0.9442	0.8983	0.9571	0.9707	0.9535
	RR SSIM	0.9551	0.9012	0.9641	0.9543	0.8456
	Q	0.9656	0.9089	0.9569	0.9642	0.8805

Table 3. Computational Time (milli-secs) for Various Distortions

Distortion Type	Measure	Img 1	Img 2	Img 3	Img 4	Img 5
Blurring	SSIM	507.4	389.8	393.3	389.7	409.6
	RR SSIM	399.6	292.2	442.7	493.8	404.8
	Q	325.2	147.5	322.7	321.1	315.8
Compression	SSIM	419.4	398.3	390.7	389.9	399.9
	RR SSIM	372.0	379.7	351.9	356.04	334.2
	Q	325.3	322.1	323.0	328.4	320.5

Finally, to show that the proposed quality measure is faster in operation than the other measures the computational time for the measures was calculated. The simulations are performed on a laptop with Microsoft Windows 7 Home Basic and Intel® Core™ i5-3210M /2.50GHz. It can be observed from table 3 that the computation time of Q is much less as compared to the other measures. The computational time for img 2 distorted by blurring is reduced to more than half of the time required by FR SSIM whereas in case of compression marginal reduction in time is observed. Overall an approximate reduction of 25% - 30% in computational time for FR SSIM and 15%

- 20% reduction for RR SSIM measure is observed. A higher reduction in computational time is observed for FR SSIM because it is a FR measure and Q makes use of both RR and gradient magnitude approach. Thus, the proposed measure can find applications in evaluating the performance of various image denoising algorithms [24]-[25] contaminated with noises like Gaussian [26]-[27], speckle [28]-[31], impulse [32] noises or their mixtures [33].

4   Conclusion

In this paper, a quality measure (Q) is proposed which is based on the design principle of the SSIM approach and proves to be an efficient HVS based quality measure. The performance of Q has been experimentally verified for various distortion types like blurring and compression. The experiments have been performed for color images by extracting features in wavelet domain. The effectiveness of the proposed measure is justified from good correlation obtained between the evaluation process using Q and the human perception results. An added advantage of Q is that it makes use of a fast approach for evaluation of the images which reduces its computational complexity thereby achieving real-time performance and therefore has a potential to be employed in visual communication applications.

References

1. Lukac, R.: Adaptive Color Image Filtering based on Center-Weighted Vector Directional Filters. Springer Transaction on Multidimensional Systems and Signal Processing 15(2), 169–196 (2004)
2. Gupta, P., Srivastava, P., Bharadwaj, S., Bhateja, V.: A HVS based Perceptual Quality Estimation Measure for Color Images. ACEEE International Journal on Signal & Image Processing (IJSIP) 3(1), 63–68 (2012)
3. Gupta, P., Srivastava, P., Bhardwaj, S., Bhateja, V.: A Novel Full Reference Image Quality Index for Color Images. In: Satapathy, S.C., Avadhani, P.S., Abraham, A. (eds.) Proceedings of the InConINDIA 2012. AISC, vol. 132, pp. 245–253. Springer, Heidelberg (2012)
4. Gupta, P., Tripathi, N., Bhateja, V.: Multiple Distortion Pooling Image Quality Assessment. International Journal on Convergence Computing 1(1), 60–72 (2013)
5. Gupta, P., Srivastava, P., Bharadwaj, S., Bhateja, V.: A Modified PSNR Metric based on HVS for Quality Assessment of Color Images. In: Proc. of IEEE International Conference on Communication and Industrial Application (ICCIA), Kolkata (W.B.), India, pp. 96–99 (2011)
6. Wang, Z., Bovik, A.C.: Mean Squared Error: Love It or Leave It? IEEE Signal Processing Magazine, 98–117 (2009)
7. Sheikh, H.R., Sabir, M.F., Bovik, A.C.: A Statistical Evaluation of Recent Full Reference Image Quality Assessment Algorithms. IEEE Transactions on Image Processing 15(11), 3440–3451 (2006)
8. Jain, A., Bhateja, V.: A Full-Reference Image Quality Metric for Objective Evaluation in Spatial Domain. In: Proc. of IEEE International Conference on Communication and Industrial Application (ICCIA), Kolkata (W. B.), India, pp. 91–95 (2011)

9. Wang, Z., Lu, L., Bovik, A.C.: Video Quality Assessment based on Structural Distortion Measurement. Signal Processing Image Communication 19(2), 121–132 (2004)
10. Wang, Z., Bovik, A.C., Sheikh, H.R., Simoncelli, E.P.: Image Quality Assessment: From Error Visibility to Structural Similarity. IEEE Transactions on Image Processing 13(4), 600–612 (2004)
11. Wang, Z., Bovik, A.C.: A Universal Image Quality Index. IEEE Signal Processing Letters 9(3), 81–84 (2002)
12. Sheikh, H.R., Bovik, A.C., Cormack, L.: No-Reference Quality Assessment using Natural Scene Statistics: JPEG2000. IEEE Transactions on Image Processing 14(11) (1918-1927)
13. Ferzli, R., Karam, L.J.: A No-Reference Objective Image Sharpness Metric based on the Notion of Just Noticeable Blur (JNB). IEEE Transaction Image Processing 18(4), 445–448 (2009)
14. Jaiswal, A., Trivedi, M., Bhateja, V.: A No-Reference Contrast Measurement Index based on Foreground and Background. In: Proc. of IEEE Second Students Conference on Engineering and Systems (SCES), Allahabad, India, pp. 460–464 (2013)
15. Trivedi, M., Jaiswal, A., Bhateja, V.: A No-Reference Image Quality Index for Contrast and Sharpness Measurement. In: Proc. of IEEE Third International Advance Computing Conference (IACC), Ghaziabad (U.P.), India, pp. 1234–1239 (2013)
16. Trivedi, M., Jaiswal, A., Bhateja, V.: A Novel HVS Based Image Contrast Measurement Index. In: Mohan, S., Suresh Kumar, S. (eds.) ICSIP 2012. LNEE, vol. 222, pp. 545–555. Springer, Heidelberg (2012)
17. Wang, Z., Bovik, A.C.: Reduced and No-Reference Image Quality Assessment: The Natural Scene Statistic Model Approach. IEEE Signal Processing Magazine 28(6), 29–40 (2011)
18. Sheikh, H.R., Bovik, A.C.: Image Information and Visual Quality. IEEE Transactions on Image Processing 15(2), 430–444 (2006)
19. Rehman, A., Wang, Z.: Reduced-Reference Image Quality Assessment by Structural Similarity Estimation. IEEE Transactions on Image Processing 21(8), 3378–3389 (2012)
20. Tao, D., Li, X., Lu, W., Gao, X.: Reduced-Reference IQA in Contourlet Domain. IEEE Transactions on Systems, Man, and Cybernetics, Part B: Cybernetics 39(6), 1623–1627 (2009)
21. Ma, L., Li, S., Zhang, F., Ngan, K.N.: Reduced-reference Image Quality Assessment using Reorganized DCT-based Image Representation. IEEE Transactions on Multimedia 13(4), 824–829 (2011)
22. Zhen, L., Yong, S., Jingjing, Z., Xiangming, W., Tao, S.: Video Quality Assessment based on Fast Structural Similarity Index Algorithm. In: Proc. of IEEE Fourth International Conference on Ubiquitous and Future Networks (ICUFN), pp. 336–339 (2012)
23. Larson, E.C., Chandler, D.M.: Categorical Image Quality (CSIQ) Database (2010), http://vision.okstate.edu/csiq
24. Singh, S., Jain, A., Bhateja, V.: A Comparative Evaluation of Various Despeckling Algorithms for Medical Images. In: Proc. of (ACM ICPS) CUBE International Information Technology Conference & Exhibition, Pune, India, pp. 32–37 (2012)
25. Gupta, P., Srivastava, P., Bharadwaj, S., Bhateja, V.: A New Model for Performance Evaluation of Denoising Algorithms based on Image Quality Assessment. In: Proc. of (ACM ICPS) CUBE International Information Technology Conference & Exhibition, Pune, India, pp. 5–10 (2012)
26. Jain, A., Bhateja, V.: A Novel Detection and Removal Scheme for Denoising Images Corrupted with Gaussian Outliers. In: Proc. of IEEE Students Conference on Engineering and Systems (SCES 2012), Allahabad (U.P.), India, pp. 434–438 (2012)

27. Jain, A., Bhateja, V.: A Versatile Denoising Method for Images Contaminated with Gaussian Noise. In: Proc. of (ACM ICPS) CUBE International Information Technology Conference & Exhibition, Pune, India, pp. 65–68 (2012)
28. Gupta, A., Tripathi, A., Bhateja, V.: Despeckling of SAR Images via an Improved Anisotropic Diffusion Algorithm. In: Satapathy, S.C., Udgata, S.K., Biswal, B.N. (eds.) Proceedings of Int. Conf. on Front. of Intell. Comput. AISC, vol. 199, pp. 747–754. Springer, Heidelberg (2013)
29. Gupta, A., Tripathi, A., Bhateja, V.: Despeckling of SAR Images in Contourlet Domain using a New Adaptive Thresholding. In: Proc. of (IEEE) 3rd International Advance Computing Conference (IACC 2013), Ghaziabad (U.P.), India, pp. 1257–1261 (2013)
30. Bhateja, V., Tripathi, A., Gupta, A.: An Improved Local Statistics Filter for Denoising of SAR Images. In: Thampi, S.M., Abraham, A., Pal, S.K., Rodriguez, J.M.C. (eds.) Recent Advances in Intelligent Informatics. AISC, vol. 235, pp. 23–29. Springer, Heidelberg (2014)
31. Bhateja, V., Singh, G., Srivastava, A.: A Novel Weighted Diffusion Filtering Approach for Speckle Suppression in Ultrasound Images. In: Satapathy, S.C., Udgata, S.K., Biswal, B.N. (eds.) Proceedings of the International Conference on Frontiers of Intelligent Computing: Theory and Applications (FICTA) 2013. AISC, vol. 247, pp. 455–462. Springer, Heidelberg (2014)
32. Jain, A., Singh, S., Bhateja, V.: A Robust Approach for Denoising and Enhancement of Mammographic Breast Masses. International Journal on Convergence Computing 1(1), 38–49 (2013)
33. Jain, A., Bhateja, V.: A Novel Image Denoising Algorithm for Suppressing Mixture of Speckle and Impulse Noise in Spatial Domain. In: Proc. of (IEEE) 3rd International Conference on Electronics & Computer Technology (ICECT 2011), Kanyakumari (India), vol. 3, pp. 207–211 (2011)
34. Gupta, A., Ganguly, A., Bhateja, V.: An Edge Detection Approach for Images Contaminated with Gaussian and Impulse Noises. In: Mohan, S., Suresh Kumar, S. (eds.) ICSIP 2012. LNEE, vol. 222, pp. 523–533. Springer, Heidelberg (2012)

A Novel Weighted Diffusion Filtering Approach for Speckle Suppression in Ultrasound Images

Vikrant Bhateja, Gopal Singh, and Atul Srivastava

Dept. of Electronics & Communication Engg., SRMGPC, Lucknow-227105 (U.P.), India
{bhateja.vikrant,gopal.singh13492,atul.srivastava216}@gmail.com

Abstract. Ultrasound images mainly suffer from speckle noise which makes it difficult to differentiate between small details and noise. Conventional anisotropic diffusion approaches tend to provide edge sensitive diffusion for speckle suppression. This paper proposes a novel approach for removal of speckle along with due smoothening of irregularities present in the ultrasound images by modifying the diffusion coefficient in anisotropic diffusion approach. The present work proposes a diffusion coefficient which is a function of difference of instantaneous coefficient (of variation) and the coefficient of variation for homogeneous region. The finally reconstructed image is obtained by weighted addition of the response of proposed anisotropic diffusion filter and the Laplacian filtered image. Simulation results show that performance of the proposed approach is significantly improved in comparison to recently developed anisotropic diffusion filters for speckle suppression.

Keywords: Anisotropic diffusion, Speckle suppression, Laplacian, Instantaneous coefficient of variation.

1 Introduction

In the near past, ultrasound imaging has emerged as the gold standard for doctors and radiologists for the detection of cysts (both benign and malignant) and cancerous tumors. This is because of its several advantages over other imaging modalities. It is non-invasive, harmless, and efficient in terms of cost and accuracy [1]. However, ultrasound images are contaminated with an inherent noise called 'speckle' which tends to have a granular effect on the image, thereby degrading its visual quality [2]. For simplifying the therapeutic decision making and diagnosis, the ultrasound images should have minimum amount of noise. This calls for the development of speckle filtering techniques over past decades. The conventional methods for speckle reduction include the Lee filtering [3] and Kuan filtering [4]. In Lee filtering, the multiplicative speckle noise is converted into additive noise before filtering. In Kuan filter, the filtering action varies according to the image statistics based on non stationary mean and variance image model. In progression, techniques involving anisotropic diffusion [5] were proposed employing variable diffusion coefficient. The first work in this domain was Perona-Malik anisotropic diffusion (PMAD) filter [6] in which a variable coefficient of diffusion was used in the standard scale-space paradigm so that it has a larger value in the homogeneous regions. Detail-Preserving Anisotropic Diffusion (DPAD) [7] is based on the extension of Frost's and Kuan's linear minimum mean square error filters used for a multiplicative noise. The Speckle-Reducing Anisotropic Diffusion (SRAD) [8] is based on the partial differential equations (PDE). Oriented Speckle-Reducing Anisotropic Diffusion (OSRAD) [9] is a technique in which a vector is associated with the SRAD filter to achieve directional filtering. Ramp-Preserving Perona-Malik model (RPPM) [10] makes the use of a diffusion coefficient chosen to avoid the staircasing effect and preserve edges. You-Kaveh's models (YKM) [11] approximates the noisy image with a piecewise planar image. The Adaptive Window Anisotropic Diffusion [12] exploits a variable size and orientation window. The work of A. Gupta *et al.* [13-15] was based on the despeckling of SAR images and the work of A. Jain *et al.* [16-18] dealt with denoising of the mixture of different noises in medical images. All the above mentioned approaches perform well. However, limitations like edge blurring, over-smoothing and greater number of iterations are present. Hence, the proposed work aims to reduce the complexity, provide better filtering and develop a computationally efficient approach by using the diffusion process without performing multiple iterations. The performance is shown under section of results and discussions. The paper is structured as follows: in Section 2, the novel diffusion filtering approach is proposed. Section 3 presents the results and discussions and in Section 4, the paper has been concluded.

2   Proposed Diffusion Filtering Approach

The diffusion filtering techniques have proved to be superior over conventional techniques in terms of their speckle suppression ability, feature preservation and edge enhancement. These techniques make the use of fundamental anisotropic diffusion equation given by Perona and Malik [5] stated as:

$$I_t = c\Delta I + \nabla c . \nabla I \tag{1}$$

where: I is the input ultrasound image, c is the conduction or diffusion coefficient, ∇ is the gradient operator and Δ is the Laplacian operator. By the judicious choice of parameter c, the diffusion process can be controlled. The aim is to make the diffusion coefficient approach unity in the interior of homogeneous region (to enhance diffusion in this region) and zero at the boundaries (to stop diffusion at boundaries avoiding blurring). However, problems like staircasing effect and blurring have encouraged the researchers to develop better approaches. This has led to the evolution of better performing diffusion coefficients. In this context, the SRAD filtering approach [8] uses the diffusion coefficient as a function of local gradient magnitude and Laplacian operators for edge preservation. In this proposed filtering approach, a modified diffusion coefficient has been presented to improve the performance of SRAD filters. The new diffusion coefficient is a non linear function of coefficient of variations. If $p(x,y)$ denotes the instantaneous coefficient of variation and $p_o(x,y)$ denotes the instantaneous coefficient of variation in the homogeneous region, then the diffusion coefficient proposed in this work can be stated as:

$$c(p) = \frac{1}{\sqrt{1+(p-p_o)^n}} \tag{2}$$

where: n denotes the power index raised to the difference of coefficients of variation. Significant improvement in speckle suppression can be attained by approximating the value of n to be greater than 3. The coefficient p can be mathematically represented as:

$$p = \sqrt{\frac{Var(I_{i,j})}{\overline{I}_{i,j}^2}} = \sqrt{\frac{\frac{1}{4}\sum_{q=1}^{4}(I_q - \overline{I}_{i,j})^2}{\overline{I}_{i,j}^2}} \tag{3}$$

where: $\overline{I}_{i,j} = \frac{1}{4}\sum_{q=1}^{4} I_q$ is the average intensity of a pixel considering the four nearest neighborhood pixels. Coefficient p_0 is given by:

$$p_o = \left(\frac{R}{\sqrt{2}}\right) MAD\left(\| \nabla \log I_{i,j^t} \|\right) \tag{4}$$

$MAD(.)$ is called the median absolute deviation, $\| \quad \|$ and $| \quad |$ are the magnitude of gradient and the absolute value respectively. R is a constant whose value is 1.4826 [19]. The modified diffusion coefficient leads to improved isotropic diffusion in homogeneous regions of the ultrasound images (speckled). The instantaneous coefficient of variation therefore evaluates to larger values on high contrast regions and lower values on homogeneous regions. Hence, in homogeneous regions p is taken close to p_0 to make $c(p)$ approach unity and for the edges, the value of p is large so that $c(p)$ is made as low as possible. Diffusion filtering approaches are implemented using multiple iterations of diffusion equation which at times to the computational load and also degrades the quality of reconstructed image. Further, the present work addresses this issue by weighted addition of filtered response in the manner suggested. The first step involves speckle filtering using (1)-(4). The reconstructed image obtained is then denoted as I_1. Secondly, another image is generated which is composed of the noisy image added to its weighted Laplacian given by (5):

$$L(I) = a.c\left(\| \nabla I \|\right)\Delta I \tag{5}$$

where: a is a constant whose value is less than unity. Its lower value ensures that the edges are preserved. The inhomogeneous weight $c\left(\| \nabla I \|\right)$ is used to reduce diffusion near edges. When this weighted Laplacian is added to the image, smoothed output image is obtained. This image I_2 is given by:

$$I_2 = I + L(I) \tag{6}$$

The finally reconstructed image I_{final} is obtained by weighted addition of two images I_1 and I_2 generated in the first and second step. This is shown mathematically as:

$$I_{final} = K_1.I_1 + K_2.I_2 \qquad (7)$$

where: K_1 and K_2 are the weights for the images whose values are to be determined experimentally based on the values of evaluation parameters.

3   Results and Discussions

3.1   Evaluation Parameters

Two state-of-art evaluation parameters Peak Signal-to-Noise Ratio (*PSNR*) [21] and the Structural Similarity Index (*SSIM*) [20] are used for performance evaluation. The higher value of *PSNR* denotes the better quality of reconstructed ultrasound image. The luminance, contrast and structural similarity functions are combined to generate Structural Similarity (*SSIM*) Index. Its value ranges from zero to unity where the value zero corresponds to zero structural similarity and unity represents exact similarity. Mathematically, it is given as:

$$SSIM = \left(\frac{2\mu_x\mu_y + C_1}{\mu_x^2 + \mu_y^2} \right) \cdot \left(\frac{2\sigma_x\sigma_y + C_2}{\sigma_x^2 + \sigma_y^2 + C_2} \right) \left(\frac{2\sigma_{xy} + C_3}{\sigma_x\sigma_y + C_3} \right) \qquad (8)$$

where: μ_x and μ_y are the mean values of images x and y, σ_x and σ_y are the local standard deviations while σ_{xy} is the cross correlation of x and y after removing the mean. The values of parameters C_1, C_2 and C_3 in this equation are taken in accordance with [20]. The performance of the proposed approach has also been evaluated based on Coefficient of Correlation (*CoC*) which can show the similarity between actual and expected results. *CoC* can be given by:

$$CoC = \frac{1}{N-1} \sum_{i=1}^{N} (x_i - \mu_x)(y_i - \mu_y) \qquad (9)$$

where: x_i and y_i are the ith samples in the images x and y and N is the total number of pixels. The value of *CoC* approaching unity denotes better preservation of features between the input and output ultrasound images. Image quality assessment measures those used above and some proposed recently [22]-[27] can be used for evaluation of speckle suppression algorithms.

3.2   Simulated Results

The input ultrasound images for this experiment are taken from [28]. In the simulation process, (1)-(4) and (5)-(6) are used to generate the first and second image respectively, as described in the previous section. Then, (7) is used to generate the final reconstructed image. Finally, (8)-(9) are used for performance evaluation of the proposed model. The value of n in (2) is taken 4 and a in (5) is taken 0.3. In (4), it is clear that p_o uses the logarithm of the input ultrasound image. So, in order to avoid a mathematical error which may occur for the pixel with a value 0, a negligibly small value is added to every pixel (which has almost no effect on the visibility). To avoid over-enhancement and over-smoothing, in (7) $K_1+K_2=1$ is satisfied. Also to ensure that speckle is removed as much as possible $K_1>K_2$ is also ensured. In this experiment, values taken are $K_1=0.8$ and $K_2=0.2$ as these values produced the better edge preservation and smoothing. Figure 1 shows the images which have been corrupted by the speckle noise of different variances and the denoised images by the proposed approach.

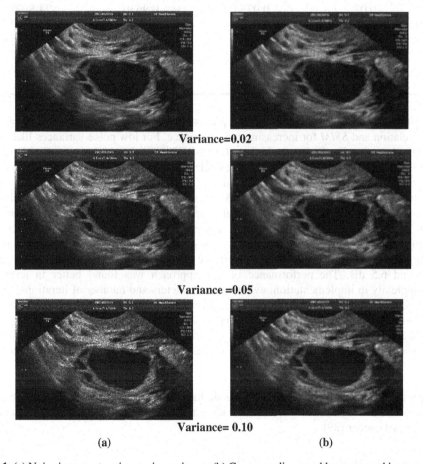

Variance=0.02

Variance =0.05

Variance= 0.10

(a)   (b)

Fig. 1. (a) Noisy images at various noise variances (b) Corresponding speckle suppressed images

It can be seen that the level of speckle is considerably reduced and the visualization of the ultrasound images is also improved to a great extent in the reconstructed images. It also shows that while the speckle noise is increasing, the performance of the approach is still able to preserve the details of the image and is not found to be ineffective at larger amount of noise. Table 1 shows the performance of proposed approach in terms of various performance evaluation metrics described earlier. The values of performance parameters are obtained at various noise variances in increasing order from 0.00 to 0.10.

Table 1. Performance Evaluation of Proposed Speckle Filtering Approach at Different Noise Variances for Images in Fig.1

Noise Variance	SSIM	CoC	PSNR (in dB)
0.00	0.9764	0.9680	27.1052
0.01	0.9750	0.9656	25.7898
0.02	0.9734	0.9630	24.8098
0.03	0.9718	0.9607	23.9323
0.04	0.9708	0.9588	23.2691
0.05	0.9688	0.9557	22.6148
0.08	0.9645	0.9496	21.3690
0.10	0.9619	0.9453	20.6450

It is clear from the table that the proposed method maintains a high degree of correlation and *SSIM* for increasing noise variance. For low noise variances like 0.01-0.05, the value for *CoC* is as high as 0.968-0.958 showing high correlation between the output ultrasound image and the speckle-free image. At high noise variance, the performance is even more commendable. *CoC* of around 0.95 is considered to be appreciable at such high noise. The *SSIM* index shows that at low noise variance, more than 97% of the structural features are preserved and it drops very slowly to 96.2% for 0.10 noise variance, still a very good value of *SSIM*. The *PSNR* does not drop drastically with increasing noise densities. It can be seen that as the noise variance is increased from 0.00 to 0.10, the *PSNR* only undergoes a total change of around 6.5 dB. The performance of this approach was found better in terms of complexity in implementation, evaluation parameters and number of iterations. In the work of G. Liu *et at.* [12], the *SSIM* index for the model was around 0.65 after 60 iterations. The proposed approach produces high *SSIM* equivalent to 0.97 without using further iterations. In the other models shown, SRAD had a *SSIM* of 0.35, Anisotropic Wiener filter had *SSIM* 0.5 and DPAD had *SSIM* of 0.35 after 60 iterations. The OSRAD method produced its best *SSIM* of 0.47 at 2-3 iterations but it was still less from the proposed model. All these results show that the proposed approach can be very helpful in denoising the ultrasound images and simplifying the work of radiologists and doctors in reading the images for computer-aided detection of breast cancer [29].

4   Conclusion

Speckle is the major undesirable artifact (noise) present inherently in the ultrasound images. Its removal from these images is an important and complicated process needed for further processing. In this paper, a novel approach is presented which makes the use of a novel diffusion coefficient and in-homogeneously weighted

Laplacian to generate the reconstructed image. The parameters used are determined experimentally and those which provide better results are chosen. The visibility of features of the ultrasound image is highly improved and over-enhancement of intensities and over-smoothing has also been taken care of. The method is efficient and produces fruitful results without performing large number of iterations. Future possibilities in this method are the improvements in the approach based on the textural features of the ultrasound image.

References

1. Kremkau, F.W.: Diagnostic ultrasound: principles and instruments. Saunders, New York (2003)
2. Gobbi, D.G., Comeau, R.M., Peters, T.M.: Ultrasound probe tracking for real-time ultrasound/MRI overlay and visualization of brain shift. In: Taylor, C., Colchester, A. (eds.) MICCAI 1999. LNCS, vol. 1679, pp. 920–927. Springer, Heidelberg (1999)
3. Lee, J.S.: Digital enhancement and noise filtering by use of local statistics. IEEE Transactions on Pattern Analysis and Machine Intelligence 2(2), 165–168 (1980)
4. Kuan, D.T., Sawchuck, A.A., Strand, T.C., et al.: Adaptive noise smoothing filter for images with signal dependent noise. IEEE Transactions on Pattern Analysis and Machine Intelligence 7(2), 165–177 (1985)
5. Fabbrini, L., Greco, M., Messina, M., Pinelli, G.: Improved anisotropic diffusion filtering for SAR image despeckling. Electronics Letters 49(10), 672–674 (2013)
6. Perona, P., Malik, J.: Scale-space and edge detection using anisotropic diffusion. IEEE Transactions on Pattern Analysis and Machine Intelligence 12(7), 629–639 (1990)
7. Aja-Fernández, S., Alberola-López, C.: On the estimation of the coefficient of variation for anisotropic diffusion speckle filtering. IEEE Transactions on Image Processing 15(9), 2694–2701 (2006)
8. Yu, Y., Acton, S.: Speckle reduction anisotropic diffusion. IEEE Transactions on Image Processing 11(11), 1260–1270 (2002)
9. Krissian, K., Westin, C.F., Kikinis, R., Vosburgh, K.: Oriented speckle reducing anisotropic diffusion. IEEE Transactions on Image Processing 16(5), 1412–1424 (2007)
10. Chen, Q., Montesinos, P., Sun, Q.S.: Ramp preserving Perona-Malik model. Signal Processing 90, 1963–1975 (2010)
11. You, Y., Kaveh, M.: Fourth order partial differential equations for noise removal. IEEE Transactions on Image Processing 9(10), 1723–1730 (2000)
12. Liu, G., Zeng, X., Tian, F., Li, Z., Chaibou, K.: Speckle reduction by adaptive window anisotropic diffusion. Signal Processing 89, 2233–2243 (2009)
13. Gupta, A., Tripathi, A., Bhateja, V.: Despeckling of SAR Images via an Improved Anisotropic Diffusion Algorithm. In: Satapathy, S.C., Udgata, S.K., Biswal, B.N. (eds.) Proceedings of Int. Conf. on Front. of Intell. Comput. AISC, vol. 199, pp. 747–754. Springer, Heidelberg (2013)
14. Gupta, A., Tripathi, A., Bhateja, V.: De-Speckling of SAR Images in Contourlet Domain Using a New Adaptive Thresholding. In: Proc. of (IEEE) 3rd International Advance Computing Conference (IACC), Ghaziabad (U.P.), India, pp. 1257–1261 (2013)
15. Bhateja, V., Tripathi, A., Gupta, A.: An Improved Local Statistics Filter for Denoising of SAR Images. In: Thampi, S.M., Abraham, A., Pal, S.K., Rodriguez, J.M.C. (eds.) Recent Advances in Intelligent Informatics. AISC, vol. 235, pp. 23–29. Springer, Heidelberg (2014)

16. Jain, A., Singh, S., Bhateja, V.: A Robust Approach for Denoising and Enhancement of Mammographic Breast Masses. International Journal on Convergence Computing 1(1), 38–49 (2013)

17. Jain, A., Bhateja, V.: A Novel Image Denoising Algorithm for Suppressing Mixture of Speckle and Impulse Noise in Spatial Domain. In: Proc. of (IEEE) 3rd International Conference on Electronics and Computer Technology (ICECT), Kanyakumari, India, vol. 3, pp. 207–211 (2013)

18. Singh, S., Jain, A., Bhateja, V.: A Comparative Evaluation of Various Despeckling Algorithms for Medical Images. In: Proc. of (ACMICPS) CUBE International Information Technology Conference & Exhibition, Pune, India, pp. 32–37 (2012)

19. Yu, Y., Acton, S.: Edge detection in ultrasound imagery using the instantaneous coefficient of variation. IEEE Transactions on Image Processing 13(12), 1640–1655 (2004)

20. Wang, Z., Bovik, A.C., Sheikh, H.R., Simoncelli, E.P.: Image quality assessment: from error visibility to structural similarity. IEEE Transactions on Image Processing 13(4), 600–612 (2004)

21. Alain, H., Djemel, Z.: Image quality metrics: PSNR vs. SSIM. In: International Conference on Pattern Recognition, pp. 2366–2369 (2010)

22. Gupta, P., Srivastava, P., Bharadwaj, S., Bhateja, V.: A HVS based Perceptual Quality Estimation Measure for Color Images. ACEEE International Journal on Signal & Image Processing (IJSIP) 3(1), 63–68 (2012)

23. Gupta, P., Srivastava, P., Bharadwaj, S., Bhateja, V.: A Novel Full-Reference Image Quality Index for Color Images. In: Satapathy, S.C., Avadhani, P.S., Abraham, A. (eds.) Proceedings of the InConINDIA 2012. AISC, vol. 132, pp. 245–253. Springer, Heidelberg (2012)

24. Gupta, P., Tripathi, N., Bhateja, V.: Multiple Distortion Pooling Image Quality Assessment. International Journal on Convergence Computing 1(1), 60–72 (2013)

25. Gupta, P., Srivastava, P., Bharadwaj, S., Bhateja, V.: A Modified PSNR Metric based on HVS for Quality Assessment of Color Images. In: Proc. of IEEE International Conference on Communication and Industrial Application (ICCIA), Kolkata (W.B.), India, pp. 96–99 (2011)

26. Jain, A., Bhateja, V.: A Full-Reference Image Quality Metric for Objective Evaluation in Spatial Domain. In: Proc. of IEEE International Conference on Communication and Industrial Application (ICCIA), Kolkata (W. B.), India, pp. 91–95 (2011)

27. Bhateja, V., Srivastava, A., Kalsi, A.: Fast SSIM Index for Color Images Employing Reduced-Reference Evaluation. In: Satapathy, S.C., Udgata, S.K., Biswal, B.N. (eds.) Proceedings of the International Conference on Frontiers of Intelligent Computing: Theory and Applications (FICTA) 2013. AISC, vol. 247, pp. 447–454. Springer, Heidelberg (2014)

28. http://www.gehealthcare.com/usen/ultrasound/voluson/ international/_files/img/p8-imagequality/main/Image-4VP8_Ovarian-Cysts.jpg

29. Bhateja, V., Urooj, S., Pandey, A., Misra, M., Lay-Ekuakille, A.: A Polynomial Filtering Model for Enhancement of Mammogram Lesions. In: Proc. of IEEE International Symposium on Medical Measurements and Applications (MeMeA 2013), Gatineau (Quebec), Canada, pp. 97–100 (2013)

Detection of Progression of Lesions in MRI Using Change Detection

Ankita Mitra[1], Arunava De[2], and Anup Kumar Bhattacharjee[3]

[1] Department of Electronics and Communication, Dr. B.C. Roy Engineering College,
Durgapur, India
ankimitra.2009@gmail.com
[2] Department of Information Technology, Dr. B.C. Roy Engineering College,
Durgapur, India
arunavade@yahoo.com
[3] Department of Electronics and Communication, National Institute of Technology,
Durgapur, India
akbece12@yahoo.com

Abstract. Change detection is a process of identifying the changes in a state of an object over time. We use the phenomena of change detection to detect the changes occurring in MRI of brain having cancerous and non cancerous lesions. A Hybrid Particle Swarm Optimization algorithm that incorporates a Wavelet theory based mutation operation is used for segmentation of lesions in Magnetic Resonance Images. The segmented lesions are the Region of Interest. This method of using change detection algorithm would be helpful in detecting changes in Region of Interests of MRI with lesions and also to view the progress of treatment for cancerous lesions.

Keywords: Region of Interest, Particle Swarm Optimization, Magnetic Resonance Imaging, Entropy, Multi-resolution Wavelet Analysis, Hybrid Particle Swarm Optimization, Wavelet Mutation.

1 Introduction

This paper proposes a scheme to understand the progress of cancerous and non cancerous lesions in MRI of brain. We deal with multi-modal MRI images. The histogram of multi-modal images has multiple peaks as opposed to two peaks for a bi-modal image. We separate the lesions from the healthy part of an MRI image. The lesions form the ROI. Using change detection we identify the change occurring in the lesion of brain.

In our technique we segregate the image based on levels of intensity, because lesion portion of the MRI image will have a different intensity value with that of a multimodal MRI image without lesions. Particle Swarm optimization (PSO)[1] is a population based stochastic optimization algorithm. Reference [2] found that using Entropy maximization using PSO algorithm gives better results in comparison with other methods

S.C. Satapathy, S.K. Udgata, and B.N. Biswal (eds.), *FICTA 2013*,
Advances in Intelligent Systems and Computing 247,
DOI: 10.1007/978-3-319-02931-3_53, © Springer International Publishing Switzerland 2014

[3]. PSO converges sharply in the earlier stages of the search process and to overcome this problem of convergence we perform optimization using Hybrid Particle Swarm Optimization with Wavelet Mutation Operation (HPSOWM) algorithm to get a threshold to segregate the lesions from the healthy tissues.

Change detection is used to compare the progression of the lesions in the MR image of brain. The change masks are the pixels sets that significantly different in the previous images as compared with the present image. Estimating the change mask is a step towards change understanding. The change mask is the ROI of the MR image with lesions. We discard the unimportant changes such as those induced by cameras, sensor noise, illumination variation etc.

2   Related Work

Reference [6] introduced an energy criterion formulated by intensity-based class uncertainty and region homogeneity. The threshold is selected by minimizing the energy. Reference [7] applied the idea of maximizing the between class variance in histogram-based thresholding. The method shows satisfactory results in various applications. However, it tends to split the larger part when the sizes of object and background are unequal [8]. Reference [9] took advantage of the knowledge about the range of background proportion to the ROI to confine the range of threshold selection and achieved reliable results in segmenting magnetic resonance (MR) and computed tomography (CT) images of the human brain. Reference [10] demonstrated that threshold can be obtained by optimizing the weighted sum of the within-class variance and the intensity contrast.

Change detection is the process of identifying differences in different images over time [11]. During the 1970's advancements in image processing techniques resulted in the development of image differencing and image rationg[12]. Image ratio-ing [11] is also done between two images which is similar to image differencing.

3   Proposed Algorithm

3.1   Preprocessing : Algorithm to Get the ROI

Entropy Maximization Using Hybrid PSO. First we compute the normalized histogram h(n) for a gray image f(x,y).

$$P_n = h(n) = f_n / N , n = 0,1,2,...255.$$ (1)

Where f_n is the observed frequency of gray level n (or f_n is the number of pixel that having gray level n) and N is the total number of pixels in the picture. For multimodal image we want to divide the total image into (k+1) number of homogeneous zones and for that we consider the threshold gray levels at t_1, t_2, t_3,.......t_k.

Shannon Entropy is defined as

$$H = -\sum_{n=0}^{t_1} P_{1n} \ln P_{1n} - \sum_{n=t_1+1}^{t_2} P_{2n} \ln P_{2n} .... - \sum_{n=t_n+1}^{255} P_{kn} \ln P_{kn}. \qquad (2)$$

Where

$$P_{1n} = P_n / \sum_{n=0}^{t_1} P_n \ \text{ for } 0 \le n \le t_1, \ P_{2n} = P_n / \sum_{n=t_1+1}^{t_2} P_n \ \text{ for } t_1 < n \le t_2,$$

$$P_{kn} = P_n / \sum_{n=t_k+1}^{255} P_n \ \text{ for } t_k < n \le 255$$

We have obtained the function H for the threshold gray level t_1 to t_k using HPSOWM algorithm [4],[5],[13]. Now we apply this expert knowledge that the gray level of diseased lesion zone of a MRI image vary from range say T_x to T_y. We optimize the threshold range using PSO and Wavelet Mutation which is explained in subsequent section.

Optimization Using HPSOWM Algorithm

We optimize the basis functions obtained using Entropy maximization technique using the concept of HPSOWM algorithm. PSO algorithm in the Pseudo code of HPSOWM given in Fig.1 is the standard PSO algorithm [1]

The Hybrid PSO algorithm with Wavelet Mutation as discussed in [4-5], [13] is a variation of PSO and the Hybrid PSO is used for optimizing the initial value of threshold to be used for segmenting the MRI image.

Hybrid Particle Swarm Optimization with Wavelet Mutation Operation (HPSOWM)

In the early stages PSO works fine but it creates problems when it nears the optimal stage. If the current position of the particle coincides with the global best position and if its inertia weight and velocity are different from zero then the particle will move away. When velocities are very close to zero all the particles will stop moving once they catch up with the global best particle which results in premature convergence. This is stagnation of PSO. Reference [15] proposed a hybrid PSO with the integration of Genetic Algorithm's mutation with a constant mutating space. However in that approach, the mutating space is kept unchanged, as a result the space of permutation of particles in PSO remains unchanged.

HPSOWM overcomes the deficiencies of PSO algorithm. In this algorithm larger mutating space is set in the early stages of the search to find a solution from the solution space whereas it fine tunes to a better solution in the later stage of the search by setting a smaller mutating space based on the properties of Multi-resolution Wavelet Analysis [14]. The Pseudo code is given in Fig. 1, in which the mutation on particles is performed after updating their velocities and positions.

The HPSOWM algorithm is used for optimizing the initial value of threshold. Using the threshold value we get the ROI which contains the lesions , the results of the process is shown in Fig.5(a) and Fig.5(b). The original un-segmented brain MRI images of the patients is displayed in Fig .3(a) , 3(b),4(a),4(b).

```
begin
        t→0                        // iteration number
        Initialize X(t)       // X(t): Swarm for iteration t
        Evaluate f(X(t)) // f(·): fitness function
while (not termination condition) do
        begin
        t→t+1
        Perform the process of PSO
        Perform mutation operation
        Reproduce a new X(t)
        Evaluate f(X(t))
        end
end
```

Fig. 1. Pseudo code for HPSO with Wavelet operation

3.2 Change Detection Using Image Differencing

Previously change detection were based on the signed difference image $E(x) = Y_2(x) - Y_1(x)$. This type of change detection methods are still in use. The thresholding is done on the difference image. The change mask is generated according to the following equation [16]:

$$C(x) = \begin{cases} 1, if \ |C(x)| > \tau \\ 0, otherwise \end{cases}$$

The threshold τ is chosen empirically. This algorithm is the simplest differencing algorithm that is employed by us to view the progress of lesions in MRI of brain.

The segmentation of the lesions are carried out using HPSOWM algorithm. The segmentation of the MR image results in the ROI. We then filter the ROI to get the actual lesion devoid of the background information. The lesions of the two MR images are compared using change detection. We apply the differencing method to get the results as depicted in fig 3(c), 4(c), 5(c). Fig.2 depicts the flow diagram of the system.

4 Results and Discussions

A perfect balance between the exploration of new regions and the exploitation of the already sampled regions in the search space is expected in HPSOWM. This balance, which critically affects the performance of the HPSOWM, is governed by the right choices of the control parameters, e.g., swarm size (n_p), the probability of mutation (p_m), and the shape parameter of wavelet mutation (WM) ($\xi_{\omega m}$) . Changing the parameter $\xi_{\omega m}$ will change the characteristics of the monotonic increasing function of WM. The dilation parameter a will take a value to perform fine tuning faster as $\xi_{\omega m}$ increases. In general, if the optimization problem is smooth and symmetric, it is easier to find the solution, and the fine tuning can be done in early iteration. Thus, a larger value of $\xi_{\omega m}$ can be used to increase the step size (σ) for the early mutation [4,5,13].

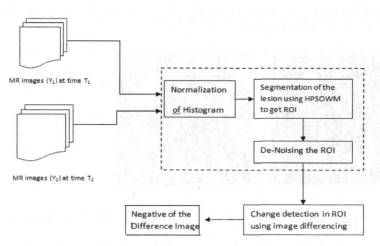

Fig. 2. Flow Diagram of the proposed Change Detection technique

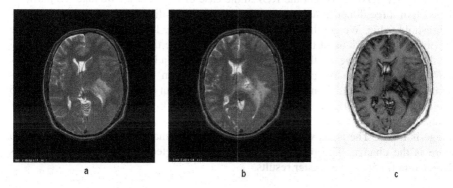

Fig. 3. a) Un- segmented brain MRI-1, b) Un- segmented brain MRI-2 , c) Change detection output using Differencing

We have considered data from two patients for experimentation purpose. Fig.3(a) belong to a person undergoing chemotherapy whereas fig 4(a) dataset belongs to a patient with bone tumor.

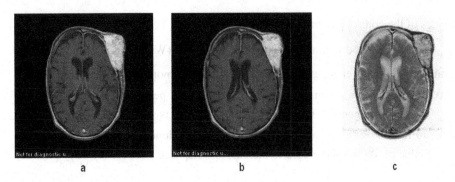

Fig. 4. a) Original un- segmented brain MRI-1 b) Original un- segmented brain MRI-2 c) Change detection output using differencing

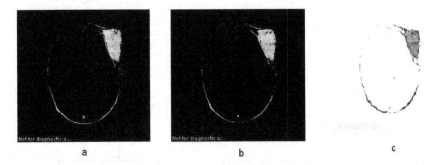

Fig. 5. a) ROI-1 b) ROI-2 c) Change detection output using differencing

To make the change easier to detect we have found ROI using Entropy Maximization using HPSOWM. To get the ROI in the case of Figure 4(a), we take ten particles for each of three dimensional spaces taken, where each dimension of space represents a threshold value. We get three threshold values 25,115, 160. Using expert knowledge we observe that lesions of the MRI image lie above the threshold value of 160 Using the threshold value we get the ROI as depicted in Fig.5(a,b).

We have applied the changed detection algorithm with and without the pre-processing step. Figure 3 and 4 shows the results without preprocessing whereas Fig.5 depicts the results with the pre-processing step. Figure 3(c),4(c),5(c) shows the results of change detection using image differencing. The negative of the resulting difference image is found. The areas with dark shades show the changed areas. Darker the area more is the change. The result shows that change detection with preprocessing as depicted in fig 5(c) gives better results.

5   Conclusion

The problem of premature convergence in PSO is taken care using the concept of HPSOWM. The proposed method is very useful in diagnosis and treatment of MR images with lesions. The proposed method of preprocessing and subsequent method of change detection can be very helpful in detecting changes in MR images during the course of treatment. Any improvement in the patient condition can be identified easily using this method of change detection. Different optimization algorithms may be used and a comparative study of the results can be made in the diagnosis and treatment of lesions in a broad range of human organs

References

1. Rao, S.S.: Engineering optimization:Theory and practice, 4th edn., pp. 709–711. John Wiley and Sons (2009)
2. De, A., Das, R.L., Bhattacharjee, A.K., Sharma, D.: Masking based segmentation of diseased MRI images. In: Proceedings of the IEEE International Conference on Information Science and Applications, ICISA 2010, Seoul chapter, Seoul, Korea, pp. 230–236 (2010)
3. Kabir, Y., Dojat, M., Scherrer, B., Forbes, F., Garbay, C.: Multimodal MRI Segmentation of Ischemic Stroke lesions. In: Proceedings of the 29th Annual International Conference of the IEEE EMBS, Cite Internationale, Lyon France (2007)
4. De, A., Bhattacharjee, A.K., Chanda, C.K., Maji, B.: MRI Segmentation using Entropy Maximization and HybridParticle Swarm Optimization with Wavelet Mutation. In: Proceedings of World Congress on Information and Communication Technologies (WICT 2011), Mumbai, pp. 362–367 (2011)
5. De, A., Bhattacharjee, A.K., Chanda, C.K., Maji, B.: Hybrid Particle Swarm Optimization with Wavelet Mutation based Segmentation and Progressive Transmission Technique for MRI Images. International Journal of Innovative Computing, Information and Control 8(7(B)), 5179–5197 (2012)
6. Saha, P.K., Udupa, J.K.: Optimum image thresholding via class uncertainty and region homogeneity. IEEE Trans. Pattern Anal. Mach. Intell. 23(7), 689–706 (2001)
7. Otsu, N.: A thresholding selection method from gray- level histograms. IEEE Trans. Syst. Man Cybern. 9(1), 62–66 (1979)
8. Kittler, J., Illingworth, J.: On threshold selection using clustering criteria. IEEE Trans. Syst. Man Cybern. 15(5), 652–655 (1985)
9. Hu, Q., Hou, Z., Nowinski, W.L.: Supervised range- constrained thresholding. IEEE Trans. Image Process 15(1), 228–240 (2006)
10. Qiao, Y., Hu, Q., Qian, G., Luo, S., Nowinski, W.L.: Thresholding based on variance and intensity contrast. Pattern Recognition 40, 596–608 (2007)
11. Singh, A.: Digital change detection techniques using remotely-sensed data. Int. J. Remote Sens. 10, 989–1003 (1989)
12. Lunetta, R.S., Elvidge, C.D.: Remote Sensing change Detection: Environmental Monitoring Methods and Applications. Ann Arbor Press, Chelsea (1998)
13. Ling, S.H., Iu, H.H.C., Leung, F.H.F., Chan, K.Y.: Improved Hybrid Particle Swarm Optimized Wavelet Neural Network for Modeling the Development of Fluid Dispensing for Electronic Packaging. IEEE Transactions on Industrial Electronics 55(9), 3447–3460 (2008)
14. Duabechies, I.: Ten lectures on Wavelets. Society for Industrial and Applied Mathematics, Philadelphia (1992)
15. Ahmed, A.A.E., Germano, L.T., Antonio, Z.C.: A hybrid particle swarm optimization applied to loss power minimization. IEEE Transactions on Power Systems 20(2), 859–866 (2005)
16. Radke, R.J., Andra, S., Al-Kofahi, O., Roysam, B.: Image Change Detection Algorithms: A Systematic Survey. IEEE Trans. Image. Process 14, 294–307 (2005)

Optimal Covering Problem for a MANET through Geometric Theory

Jagadish Gurrala and Jayaprakash Singampalli

Department of Computer Science and Engineering,
Anil Neerukonda Institute of Technology and Sciences,
Visakhapatnam, Andhra Pradesh, India
{jagadish1215,jayaprakash.singampalli}@gmail.com

Abstract. A reliable routing protocol with the purpose of attaining high percentage of data delivery in the mobile ad hoc networks is presented. The routing protocol presented for MANETs employs multi-grid routing scheme adaptively uses varying cell sizes, unlike single-grid based protocols. In a dense network, a small-cell node is employed to serve more alternative cells for a path. Meanwhile, a large-cell can be used to allow the probability of seamless data forwarding when the network is sparse. The paper simplifies complex coverage problem step by step. By means of math modeling, theoretical analysis and formula deducting, classical geometric theories and the method of mathematics induction are adopted.

Keywords: Mobile Ad-hoc Networks(MANET), data forwarding, sensing.

1 Introduction

A MANET can be a standalone network or it can be connected to external networks (Internet).The main two characteristics of MANET are mobility and multi hop and hence multi hop operation requires a routing mechanism designed for mobile nodes. In mobile ad-hoc networks where there is no infrastructure support as is the case with wireless networks, and since a destination node might be out of range of a source node transmitting packets; a routing procedure is always needed to find a path so as to forward the packets appropriately between the source and the destination [1]. Within a cell, a base station can reach all mobile nodes without routing via broadcast in common Wireless networks. In the case of ad-hoc networks, each node must be able to forward data for other nodes.Therefore the requirements of the protocol for MANETare loop free paths, optimal path, dynamic topology maintenance etc.

2 Related Work

Reactive Routing Protocol: Reactive routing protocol is an on-demand routing protocol for mobile ad-hoc networks. The protocol comprises of two main functions of route discovery and route maintenance. Route discovery function is responsible for

S.C. Satapathy, S.K. Udgata, and B.N. Biswal (eds.), *FICTA 2013*,
Advances in Intelligent Systems and Computing 247,
DOI: 10.1007/978-3-319-02931-3_54, © Springer International Publishing Switzerland 2014

the discovery of new route, when one is needed and route maintenance function is responsible for the detection of link breaks and repair of an existing route. Reactive routing protocols, such as the AODV [2], the DSR [3], do not need to send hello packet to its neighbor nodes describes the various notations used in this paper to represent the documents and related concepts. frequently to maintain the coherent between nodes. Another important feature of reactive routing protocol is that it does not need to distribute routing information and to maintain the routing information which indicates about broken links [4]. Both the neighbor table and routing information would be created when a message needed to be forwarded and nodes maintain this information just for certain lifetime. When communication between two nodes completes, nodes discard all these routing and neighbor information.

If another message needs to be forwarded, same procedure continues.

3   Our Approach to Find the Optimal Path

The illustration in Figure-1, theoretical hypotheses are shown below:

Hyp.1 a node's detecting ability is omnidirectional, that is , its coverage range is a disk whose radius is r and whose area is D ($D=\pi r^2$).

Hyp.2 in a MANET covering field, all nodes radio power is uniform, that is, the radio radius r of all MANET's is equal.

Hyp.3 in a MANET covering field, all nodes are in the same plane.

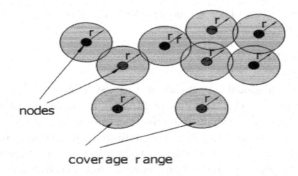

nodes

coverage range

Fig. 1. MANET's *sensing range* and MANET's covering field *sensor field*

According to above hypotheses, the minimum number of nodes, that is demanded in order to entirely and seamlessly cover the MANET's covering field field to illustrate in figure 1 is the number if coverage area of every node in it is maximal.

Expression in math is:

For $\forall x \in F, \exists i$, so as to $x \in D_i$, and $\bigcup_{j \in N} D_j$ is maximal, or max $\bigcup_{j \in N} D_j$

Where x is any point in the sensor field, F is the sensor field, D is disk of every node, N is number of nodes and U is union.

3.1   Theortical Study for MANET' Sensing Range

Theorem 1. Area of inscribed equilateral triangle is maximal in all inscribed triangle of a circle.

Proof: To illustrate in figure 2, in circle A, $\triangle C1C2C3$ and $\triangle C1'C2C3$ are inscribed triangle of it. $C1P1 \perp C2C3$ and $C1P1$ is through the center of A. $C1'P2 \perp C2C3$ and $C1'$ is any point except for point C1.

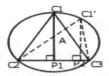

Fig. 2. Graphic illustration of Theorem 1

Obviously, for $\forall C1' \in$ A, if $C1' \neq C1$, then $|C1P1| > |C1'P1|$ ($|C1P1|$ denotes the length of segment C1P1).

Moreover $\because |C2C3| = |C2C3|$

$\therefore$ area of $\triangle C1C2C3 > \triangle C1'C2C3$, $\because C1'$ is discretional

$\therefore$ area S of $\triangle C1C2C3$ is maximal .Here, $|C1C2| = |C1C3|$, Moreover,

$\because$ symmetry

$\therefore$ in a similar way, $|C2C1| = |C2C3|$ and $|C3C1| = |C3C2|$

$\therefore$ $|C1C2| = |C2C3| = |C3C1|$

$\therefore$ $\triangle C1C2C3$ is equilateral triangle and its area is maximal.

Theorem 2. To illustrate in figure 3, if seamless topology area of 3 seamless topology disks: D1, D2 and D3 get maximum, then 3 circles: C1, C2 and C3 correspondingly encircling 3 disks must intersect at only point A. That is, $D1 \cap D2 \cap D3 \neq \Phi$, if max(D1UD2UD3), then $C1 \cap C2 \cap C3 = \{A\}$.

Proof: Max(D1UD2UD3)$\rightarrow$min(D1$\cap$D2$\cap$D3)

$\because$ D1$\cap$D2$\cap$D3$\neq\Phi$

According to the definition of intersection

$\therefore$ C1$\cap$C2$\cap$C3 ={A}

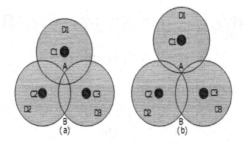

Fig. 3. Graphic illustration of Theorem-2

Theorem 3. Seamless topology area of 3 seamless topology disks: D1, D2 and D3 is maximal and its value is $\frac{4x + 3\sqrt{3}}{2}r^2$ if 3 circles: C1, C2 and C3 correspondingly encircling D1, D2 and D3 intersect at point A and $\triangle$C1C2C3 is equilateral triangle.

Proof: Let max(D1UD2UD3) be true, according to theorem 2, then circle C1, circle C2 and circle C3 must intersect at point A.
  Furthermore, $\because$ |AC1| = |AC2| = |AC3| = r

  $\therefore$ point C1, C2 and C3 are concyclic

  $\therefore$ $\triangle$C1C2C3 is inscribed triangle of circle A let area of $\triangle$C1C2C3 be maximal, according to theorem 1, then $\triangle$C1C2C3 must be equilateral triangle.

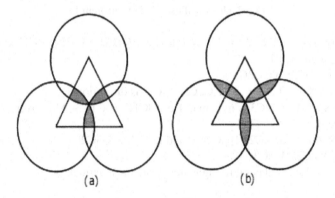

Fig. 4. Graphics illustration of Theroem-3

  Since area of $\triangle$C1C2C3 get maximum, here, area of gray field in figure 4(a) is minimal to a certainty. That is, area of gray field in figure 4(b) is minimal, i.e. max(D1UD2UD3).
  Area S1 of gray field in figure 5(b):

$$S1 = \text{area of sector AC2A' - area of } \triangle AC2P3$$
$$= \frac{30^0 \prod r^2}{360^0} - \frac{1}{2} |C2P3| * |AP3|$$
$$= \frac{\prod r^2}{12} - \frac{1}{2} * \sqrt{r^2 - (r/2)^2} * (r/2) = \frac{2\prod - 3\sqrt{3}}{6} r^2 \quad (1)$$

$\because$ Symmetry

  $\therefore$ Area S2 of gray field in figure 5(c)

$$S2 = 4*S1 = \frac{2\prod - 3\sqrt{3}}{6} r^2 \quad (2)$$

$\because$ Symmetry

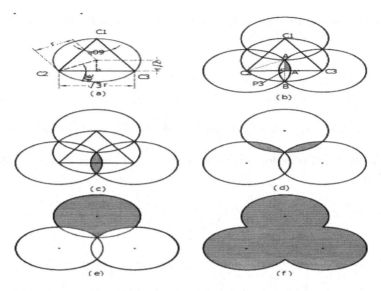

Fig. 5. Graphic illustration of Computing Area

∴ Area S3 of gray field in figure 5(d)

$$S3=2*S2=\frac{2\prod-3\sqrt{3}}{3}r^2 \tag{3}$$

∴ Area S4 of gray field in figure 5(e)

$$S4=D-S3 = \prod r^2-\frac{2\prod-3\sqrt{3}}{3}r^2 = \frac{\prod+3\sqrt{3}}{3}r^2 \tag{4}$$

∴ Area S5 of gray field in figure 5(f)

$$S5=3*S4+3*S2=3*\frac{\prod+3\sqrt{3}}{3}r^2+3*\frac{2\prod-3\sqrt{3}}{6}r^2=\frac{4\prod+3\sqrt{3}}{2}r^2 \tag{5}$$

∴ The problem proves to be true

From the process of proving, we notice that $\angle AC2B=60°$, ∴ $360°/\angle AC2B = 360°/60°= 6$, i.e. A circle   C1 is exactly covered by 6 circles   C2, C3, C4, C5, C6, and   C7. The case is illustration in figure 6.

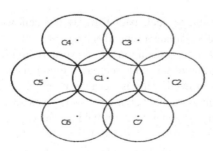

Fig. 6. Graphic illustration that one circle is seamlessly covered by 6 circles

3.2   Minimum Number of Nodes in Manet's Covering Field That Is Covered Entirely And Seamlessly

According to the above theorem 3, in a given sensor field F, the illustration in Figure-7 shows the topology graph. From the figure , it is known that a node is added every time, then the increment δ of coverage area is

$$\delta = D - 3 * S_2 = \prod r^2 - 3 * \frac{2\prod - 3\sqrt{3}}{6} r^2 = \frac{3\sqrt{3}}{2} r^2 \qquad (6)$$

∴ the number N of nodes in the transmission field (Some of boundary nodes are ignored) is

$$N = \frac{F}{\delta} = \frac{F}{\frac{3\sqrt{3}}{2} r^2} = \frac{2F}{3\sqrt{3} \, r^2} \qquad (7)$$

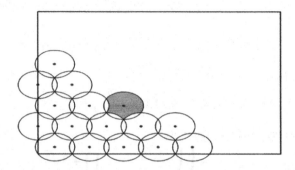

Fig. 7. Graphic illustration of counting the number of nodes

4   Multi-grid Reliable Routing

The edges of the local graphs for area constructed in our sensin based routing are weighted with our proposed reliability metrics. These areas at each node are merged to form a global graph. When a source node has a data to send, the routing protocol as described in [5] attempts to discover a connected path of reliable cells in the global graph of the small grid. The source node runs Dijkstra's algorithm on the snapshot of the global graph to compute such path towards the cell where the destination is located. The motivation to initiate a route discovery around a node of smallest region is because it provides a representation of network region at a higher resolution. However, if the local graph of this node regions is found disconnected, the nodes with the larger area size is utilized. This is because it might be difficult to discover a connected set of areas in a graph when the network is sparse and the node mobility is increases the probability of generating a reliable connected path of cells. A path, if discovered in the source node is cached and used to forward data until it observes the changes in the grid. The corresponding node identifications are inserted in the header of the data packet, along with the transmission area length of the selected node and forwarded towards the next hop cell using a neighbor selection protocol. Whenever the source route is modified due to change in the node, forwarding nodes use the l length embedded in the header to identify the nodes of a particular region. Neighbor selection uses shortest routing algorithms as notified from the source to select a node from the neighboring cell that can forward the packet toward the destination. A neighbor lookup table is maintained based on the periodic HELLO information to update the list of neighbors belonging to the particular node. The potential relay nodes of the neighbor node transmission region for each region are ordered according to their maximum lifetime in the designated region and the validity of the node. A forwarding node first checks the node identity and the node corresponding to the transmission region area information provided in the packet header. Then the look-up table is fetched to find a best relay node in the next reliable tramission area. If the selected neighbor is non-existent due to mobility or other reasons, the region of larger node that maps to the current node identity is selected. The probability of finding potential forwarding neighbor increases with the larger area of the nodes. The selection of a node in the larger tramission area is done by finding the location of a smaller circle using Eq. (6) and (7). Another relay node in the larger neighborhood is thus selected to forward data towards the destination. Further, even if the larger transmission is empty or forwarding data fail, sender node refreshes the global graph and re-runs Dijkstra's algorithm to find a new path towards the destination.

4.1   Path Re-construction and Routing Loop

In our proposed scheme, we compute a path as a sequence of nodes and thus routing loop does not exist. However, when the nodes in the path do not have neighbor for forwarding data, path reconstruction might induce loop formation. New path selection from the failure point onward might create a path with the nodes that already exist in the previous path. Two measures are taken to avoid routing loop: First, the neighbor information about the failure is immediately refreshed through the HELLO packets. However, this information is not immediately carried throughout the network. The newly re-constructed path is compared with the existing in the data packet. If the path length in the data packet is smaller than the new path, we wait for w seconds before creating the new path for the data. The w h seconds is equal to the product of HELLO packet interval and the number of hops the data packet has traversed. The disadvantage of this method is that it adds the delay at the cost of successfully delivering the data.

5   Conclusion

In this paper, we demonstrated the mathematically the MANET's optimal covering of the nodes and the fastest way of data transfer and the simulations are done in NS 2.33. network simulator.

References

1. Gorantala, K.: Routing protocols in Mobile ad- hoc networks, master's thesis report, Ume° a University, Sweden (June 15, 2006)
2. Perkins, C., Royer, E.: Ad-hoc on-demand distance vector routing. In: Proc. 2nd IEEE Workshop on Mobile Computing Systems and Applications, pp. 90–100 (1999)
3. Johnson, D., Maltz, D.: Dynamic source routing in ad hoc wireless networks. In: Imielinski, Korth (eds.) Mobile Computing, vol. 353, Kluwer Academic, Boston (1996)
4. Ding, R., Yang, L.: A reactive geographic routing protocol for wireless sensor networks. Beihang University, China (1963)
5. Shrestha, D.M., Kim, C., Ko, Y.-B.: A Reliable Multi-Grid Routing Protocol for Tactical MANETs. In: RACS 2011. ACM (November 2011) ISBN: 978-1-4503-1087

Hierarchical Divisive Clustering with Multi View-Point Based Similarity Measure

S. Jayaprada, Amarapini Aswani, and G. Gayathri

Department of Computer Science and Engineering,
Anil Neerukonda Institute of Technology and Sciences,
Visakhapatnam, Andhra Pradesh, India
{jayaprada.suri,gayathri.ganivada}@gmail.com,
aswani.amarapini@yahoo.com

Abstract. Clustering is task of grouping a set of objects in such a way that objects in the same group are more similar to each other than to those in other groups. In this paper, we introduce hierarchical divisive clustering with multi view point based similarity measure. The hierarchical clustering is produced by the sequence of repeated bisections. The bisecting incremental k-means with multi view point based similarity measure is used in the clustering. We compare our approach with the existing algorithms on various document collections to verify the advantage of our proposed method.

Keywords: Hierarchical Clustering, Document Clustering, Text Mining, Similarity Measure.

1 Introduction

Clustering is the process of grouping a set of objects into clusters. Each cluster is a collection of objects which are similar. Document clustering is very important in text mining and it is used in many applications in information retrieval and knowledge management. There have been many clustering algorithms proposed in every year. According to the recent study [2], k-means is the simple and effective algorithm still remains one of the top 10 data mining algorithms. In [3], the bisecting k-means is said to be better than simple k-means for the sparse and high dimensional document datasets. The bisecting k-means is used in the hierarchical clustering to build the hierarchy of clusters.

In cluster methods we have two main approaches. One approach in document clustering is partitional clustering algorithms which are k-means, bisecting k-means, incremental k-means. Another approach is hierarchical clustering which proceed by stages producing a sequence of partitions, each corresponding to a different number of clusters which can be either 'agglomerative', or 'divisive'. The agglomerative clustering is the bottom-up approach. In the divisive hierarchical clustering all the documents are taken initially as a single cluster then cluster is divided into sub clusters based on the similarity. The divisive hierarchical clustering is top-down approach. In this clustering we use a bisecting k-means algorithm to split the clusters.

S.C. Satapathy, S.K. Udgata, and B.N. Biswal (eds.), *FICTA 2013*,
Advances in Intelligent Systems and Computing 247,
DOI: 10.1007/978-3-319-02931-3_55, © Springer International Publishing Switzerland 2014

2   Related Work

The Table-1 describes the various notations used in this paper to represent the documents and related concepts. Each document in a d-corpus corresponds to an m-dimensional vector d, where m is the total no of terms that the document corpus has. All the documents in the datasets are pre-processed and it converts into Term Frequency-Inverse Document Frequency (TF-IDF) and normalized to have a unit length.

Table 1. Notations

Notation	Description
n	number of documents
m	number of terms
c	number of classes
k	number of clusters
d	document vector, $\|d\|=1$
$s = \{d_1, \ldots d_n\}$	set of all documents
S_r	set of all documents in the cluster r
$D = \sum_{d \in s} d_i$	composite vector of all documents
$D_r = \sum_{d \in s_r} d_i$	composite vector of the cluster r
$C = D/n$	centroid vector of all the documents
$C_r = D_r/n_r$	centroid vector of cluster r, $n_r = \mid S_r \mid$

2.1   Single View Point Similarity

A few most popular traditional similarity measures using single view point are reviewed firstly. For the document clustering the similarity measures are given in [11]. The most popular distance measure is the Euclidean distance measure. The Euclidean distance between the two objects is defined as

$$dis(d_i, d_j) = \| d_i - d_j \| . \tag{1}$$

Traditional k-means algorithm uses this measure [2]. The objective of k-means is to minimise the Euclidean distance between objects of a cluster and its cluster's centroid.

$$\min \sum_{r=1}^{k} \sum_{d_i \in s_r} \| d_i - d_j \| . \tag{2}$$

However, for data in sparse and high dimensional space such as that in document clustering, cosine similarity is more widely used. Similarity of two document vectors d_i and d_j, $sim(d_i, d_j)$ is defined as cosine of angle between them. For unit vectors, this is equal to their inner product.

$$sim(d_i, d_j) = \cos(d_i, d_j) = d_i{}^t d_j \qquad (3)$$

Cosine similarity is used in spherical k-means [5] and the objective function of the spherical k-means is maximizing the cosine similarity between the documents in a cluster and it's Centroid.

$$\max \sum_{r=1}^{k} \sum_{d_i \in s_r} \frac{d_i{}^t C_r}{\| C_r \|} \qquad (4)$$

In [6], the empirical study was conducted to compare a variety of criterion functions for document clustering.

2.2   Multi View Point-Based Similarity Measure

The cosine similarity (3) is can be expressed in the following without changing its meaning:

$$sim(d_i, d_j) = \cos(d_i - 0, d_j - 0) = (d_i - 0)^t (d_j - 0) \qquad (5)$$

Where '0' is the 'vector 0' that represents the origin point. According to this formula, the measure takes 0 as one and only reference point. The similarity between two documents d_i and d_j is determined w.r.t the angle between the two points when looking from origin.

If we are looking from many different view points to the two documents then we can measure how close or distinct pair of points. From a third point d_h, the directions and the distances d_i and d_j are indicated, by the difference vectors $(d_i - d_h)$ and $(d_j - d_h)$.looking from various reference points d_h to the view d_i, d_j and working on their difference vectors, we define similarity between the two documents as

$$sim(d_i, d_j) = \frac{1}{n - n_r} \sum_{d_h \in s \setminus s_r} sim(d_i - d_h, d_j - d_h) \qquad (6)$$

As described by the above equation, similarity of two documents d_i and d_j given that they are in the same cluster-is defined as the average similarities measured relatively from all the views of other documents outside the cluster. We call this proposal as MultiViewpoint-based Similarity or MVS [1]. The relative similarity is defined by the dot product of the difference vectors, we have

$$MVS(d_i, d_j \mid d_i, d_j \in s_r)$$
$$= \frac{1}{n - n_r} \sum_{d_h \in s \setminus s_r} (d_i - d_h)^t (d_j - d_h)$$
$$= \frac{1}{n - n_r} \sum_{d_h} \cos(d_i - d_h, d_j - d_h) \| d_i - d_h \| \| d_j - d_h \| \qquad (7)$$

The similarity between two points d_i and d_j inside the cluster S_r , viewed from a point d_h outside this cluster, is equal to the product of the cosine of the angle between d_i and d_j looking from d_h and the Euclidean distances from d_h to these points. The overall similarity between d_i and d_j is determined by taking average over all the viewpoints not belonging to cluster S_r . The MVS is compared with the cosine similarity measure for the document collections in [1]. In that paper, the comparison had done with two document data sets reuters7 and k1b and also describes the advantages of MVS.

3   Hierarchical Divisive Clustering Using MVS(HD-MVS)

Different criterion functions are used in clustering algorithms are clearly described in [9], for single view point based similarity. But in this paper we are using multi view point based similarity measure. The two criterion functions are described in [1]. They are given as follows

$$I_R = \sum_{r=1}^{k} \frac{1}{n_r^{1-\alpha}} \left[\frac{n + n_r}{n - n_r} \| D_r \|^2 - \left(\frac{n + n_r}{n - n_r} - 1 \right) D_r{}^t D \right] \cdot \tag{8}$$

$$I_v = \sum_{r=1}^{k} \left[\frac{n + \| D_r \|}{n - n_r} \| D_r \| - \left(\frac{n + \| D_r \|}{n - n_r} - 1 \right) \frac{D_r{}^t D}{\| D_r \|} \right] \cdot \tag{9}$$

Where α called as regulating factor, which has some constant value ($\alpha \in [0,1]$) .In the formulation of I_R , a cluster quality is measured by the average pairwise similarity between the documents within the cluster. However the I_R can be lead to sensitiveness to the size and tightness of the clusters. To prevent this, an alternative approach is to consider similarity between each document vector and its cluster's Centroid instead. I_V calculates the weighted difference between the two terms: ‖Dr‖and D^{t}Dr/‖Dr‖, which again represent an intracluster similarity measure and an intracluster similarity measure, respectively.

The incremental k-way function with multi view point based similarity measure for document clustering is described in[1]. In that the accuracy of the clustering is also mentioned. The criterion function I_v gives the best result than I_R , So in the proposed method we only consider I_v as the objective function for incremental algorithm. Consider the expression of I_v in (9) depends only on n_r and can be written in a general form

$$I_v = \sum_{r=1}^{k} I_r(n_r, D_r) \tag{10}$$

Where $I_r(n_r, D_r)$ corresponds to the objective value of cluster r. We can construct a hierarchical clustering solution with partitional algorithms. Incremental K-Means is the better partitional clustering algorithm compared with the k-means. At initialization, k arbitrary documents are selected to be seeds from which initial partitions are formed. Refinement is a procedure that consists of a number of iterations. During each iteration, the n documents are visited one by one in a totally random order. Each document is checked if its move to another cluster results in improvement of the objective function. If yes, the document is moved to the cluster that leads to the highest improvement. If no clusters are better than the current cluster, the document is not moved. The clustering process terminates when an iteration completes without any documents being moved to new clusters.

Unlike traditional k-means, this algorithm is a stepwise optimal procedure. While k-means only updates after all n-documents have been reassigned, the incremental clustering algorithm updates immediately whenever each document is moved to new cluster. Since every move when happens increases the objective function value, convergence to a local optimum is guaranteed.

The incremental k-way is given below:

1: Procedure Initialization
2: Select k seeds $s_1, s_2 \ldots s_n$ randomly
3: $cluster[d_i] <- p = \arg\max_r \{ s_r^t, d_i \}, \forall\ i = 1, \ldots n$
4: $D_r <- \sum\limits_{d_i \in s_r} d_i, n_r < - \mid s_r \mid, \forall\ r = 1, \ldots, k$
5: End procedure
6: Procedure Refinement
7: Repeat
8: $\{ v[1:n] \} < -$ random permutation of $\{1, \ldots n\}$
9: for $j < -1 : n$ do
10: $i < -v[j]$
11: $p < -cluster[d_i]$
12: $\Delta I_p < - I(n_p - 1, D_p - d_i) - I(n_p, D_p)$
13: $q < - \arg\max_{r, r \neq p} \{ I(n_r + 1, D_r + d_i) - I(n_r, D_r) \}$
14: $\Delta I_q < - I(n_q + 1, D_q + d_i) - I(n_q, D_q)$
15: If $\Delta I_p + \Delta I_q > 0$ then
16: Move d_i to the cluster $q : cluster[d_i] < -q$
17: Update D_p, n_p, D_q, n_q
18: end if
19: end for
20: until no move for all n documents
21: end procedure

The incremental k-means clustering is used in the bisecting k-means to bisect the clusters into two clusters. This algorithm is used in bisecting algorithm at each bisecting step to form hierarchy of clusters. This will give the dendrogram of divisive clustering algorithm.

Algorithm. Bisecting Incremental K-Means

Input: K=2 in Incremental K-Means, S: $(d_1, d_2, ... d_n)$ document collection

Output: A hierarchy of clusters (leaf clusters contain a single documents)

Step 1. Treat all the documents as one initial cluster.

Step 2. Pick a leaf cluster C (or initial) to split. Choose the cluster with the least overall similarity.

Step 3. Bisecting Step: Use Incremental K-Means to split cluster C into two sub-clusters, C_1 and C_2

Step 4. Add the two clusters that are produced from the partition to the list of leaf clusters (candidate clusters to split).

Step 5. Repeat steps 2, 3 and 4 until each cluster at the bottom of the hierarchy contains a single document.

4   Document Collection

In our experiment we have used the large and high dimensional datasets for document clustering which are already available in [7]. The data corpus that we used for experiments consists of 20 bench mark document data sets of which 10 were tested. These data sets from CLUTO [7] had been used for experimental testing in previous papers, and their source and the origin had been described in detail [8], [9]. Table 2 summarizes their characteristics. The corpora present a diversity of size, number of classes and class balance. They were all preprocessed by standard procedures, including stop-word removal, stemming, and removal of too rare as well as too frequent words, TF-IDF weighting and normalization. Document datasets are large and high dimensional.

Table 2. Document Data Sets

Data	Source	c	n	m	Balance
k1a	WebACE	20	2,340	13,859	0.018
la1	TREC	6	3,204	17,273	0.290
re0	Reuters	13	1,504	2,886	0.018
reviews	TREC	5	4,069	23,220	0.099
wap	WebACE	20	1,560	8,440	0.015
la12	TREC	6	6,279	21,604	0.282
new3	TREC	44	9,558	36,306	0.149
sports	TREC	7	8,580	18,324	0.036
tr11	TREC	9	414	6,424	0.045
reuters7	Reuters	7	2,500	4,977	0.082

c:# of classes, n:# of documents, m: # of words

Balance= (smallest class size)/(largest class size)

5   Cluster Quality

The quality of various clustering algorithms can be evaluated with regards to both internal and external measures. Internal measures compare different sets of clusters without reference to external knowledge. The cohesiveness of a cluster, which is called "overall similarity" and is based on the pairwise similarity of the documents in a cluster, is an internal measure. Contrary to internal measures, external measures evaluate the clustering quality by comparing the clusters produced from clustering algorithms against already defined classes. The most common external measure is F-Measure.

5.1   F-Measure

F-Measure is more suitable for measuring the effectiveness of not only partitional but also of hierarchical clustering. F-Measure combines the precision and recall ideas from information retrieval area. For each manually labeled category (topic) T, we assume that a cluster C corresponding to the topic T will be formed somewhere in the hierarchy. To find the cluster C corresponding to category T, traverse the hierarchy computing precision, recall and F-Measure.

For any category T and cluster C, we define:

$$P(C,T) = \frac{N}{|C|} \quad . \tag{12}$$

$$R(C,T) = \frac{N}{|T|} \quad . \tag{13}$$

Where N is the number of members of category T in cluster C, $|C|$ is the number of documents in cluster C, $|T|$ is the number of documents in category T. For hierarchical clustering, we consider the cluster with the highest F-Measure to be the cluster corresponding to the category T. The overall F-Measure for the hierarchy is computed by taking the weighted average of the F-Measure for each topic T and is defined as

$$overall\_F-Measure = \frac{\sum_{T \in S} |T| * F(T)}{\sum_{T \in S} |T|} \quad . \tag{14}$$

5.2   Experimental Results

The hierarchical divisive clustering with multi view point based similarity measure is applied on many document datasets which are already mentioned in section5. The F-measure is calculated for all the datasets and compared with the already existing algorithms. Most of the datasets gives the best results with hierarchical clustering with MVS.

Table 3. Clustering results in F-Measure

Data	HD-MVS	MVSC- I_v	k-means	Spk-means
k1a	0.550	0.592	0.502	0.545
la1	0.694	0.723	0.565	0.719
re0	0.609	0.458	0.421	0.421
reviews	0.658	0.748	0.644	0.730
wap	0.564	0.571	0.516	0.545
la12	0.803	0.735	0.559	0.722
new3	0.658	0.547	0.500	0.558
sports	0.743	0.804	0.499	0.702
tr11	0.695	0.728	0.705	0.719
reuters7	0.786	0.775	0.658	0.718

6   Conclusion

In this paper, the hierarchical divisive clustering uses a patitional approach to bisect the cluster is described. And also uses the multi view point based similarity measure in this clustering algorithm. The proposed algorithm is suitable for the large text documents. The results proved that the multi view point similarity measure in hierarchal clustering algorithms gives the best cluster quality. We also compared the proposed algorithm with the existing algorithms to prove that this clustering algorithm is more efficient than other.

References

1. Nguyen, D.T., Chen, L., Chan, C.K.: Clustering with Multi-Viewpoint Based Similarity Measure. IEEE Transactions on Knowledge and Data Engineering PP (2011)
2. Wu, X., Kumar, V., Quinlan, J.R., Ghosh, J., Yang, Q., Motoda, H., McLachlan, G.J., Ng, A., Liu, B., Yu, P.S., Zhou, Z.-H., Steinbach, M., Hand, D.J., Steinberg, D.: Top 10 Algorithms in Data Mining. Knowledge Information Systems 14(1), 1–37 (2007)
3. Steinbach, M., Karypis, G., Kumar, V.: A Comparison of Document Clustering Techniques. In: KDD Workshop on Text Mining (2000)
4. Manning, C.D., Raghavan, P., Schutze, H.: An Introduction to Information Retrieval. Cambridge Univ. Press (2009)
5. Dhillon, Modh, D.: Concept Decompositions for Large Sparse Text Data Using Clustering. Machine Learning 42(1/2), 143–175 (2001)
6. Zhao, Y., Karypis, G.: Empirical and Theoretical Comparisons of Selected Criterion Functions for Document Clustering. Machine Learning 55(3), 311–331 (2004)
7. Karypis, G.: CLUTO a Clustering Toolkit. technical report, Dept. of Computer Science, Univ. of Minnesota (2003),
   http://glaros.dtc.umn.edu/~gkhome/views/cluto

8. Zhong, S., Ghosh, J.: A Comparative Study of Generative Models for Document Clustering. In: Proc. SIAM Int'l Conf. Data Mining Workshop Clustering High Dimensional Data and its Applications (2003)
9. Zhao, Y., Karypis, G.: Criterion Functions for Document Clustering: Experiments and Analysis. Technical Report, Dept. of Computer Science, Univ. of Minnesota (2002)
10. Zhao, Y., Karypis, G.: Evaluation of hierarchical clustering algorithms for document datasets. In: Proc. of Int. Conf. on Inf. & Knowledge Management, pp. 515–524 (2002)
11. Huang, A.: Similarity Measures for Text Document Clustering. In: NZCSRSC 2008, Christchurch, New Zealand (April 2008)

Discovery of Knowledge from Query Groups

Sunita A. Yadwad and M. Pavani

Department of Computer Science,
Anil Neerukonda Institute of Technology and Sciences,
Visakhapatnam,Andhra pradesh, India
sunitaay@sify.com, pavani.csebtech@gmail.com

Abstract. Now a day's web usage is increasing and most of the queries submitted to server are informational rather than navigational. Users access complicated informational and task-oriented goals like arranging for travelling, managing banking transactions, or planning their decisions on buying new products. However, the primary option for accessing data on-line continues to be through keyword search. A complex task such as managing bank transactions should be broken down into a number of subtasks (queries) over a period of time. As an example, a user may first search on target accounts branch names, etc. So there is need for maintaining user search history which incorporates a sequence of four queries displayed in reverse timely order together with their corresponding U.R.L clicks. This paper explains the how to maintain user search history and missing knowledge from query group. This paper uses one pass algorithm in order to generate knowledge from user search groups.

Keywords: Query groups, Sliding window model, knowledge discovery.

1 Introduction

The web access in world drastically increased for searching and managing the bank transactions, travel arrangements product purchases. The above complex task can be decomposed into queries over period of time. The search history is a feature which allows to getting their searches by storing information of query and click. So there is need for organizing search histories that can be helpful to user and identifying group of similar queries. The advantage of dynamic query grouping allows the system to be better understanding of a session and search experience.

The issues related in structuring of user search history into set of query groups as an automation and dynamic activity. Query group is a set of queries and clicks of one user, these queries of and that are similar attempt get common information. These groups are dynamically updated whenever new query stream comes. So there is a need for maintaining streaming model. There are three models for processing stream data (1) Landmark model (2) time Fading model (3) Sliding window model. In this paper we use sliding window model in which we store all <session time, query> transaction up to limited storage. When storage is full it removes least recent

S.C. Satapathy, S.K. Udgata, and B.N. Biswal (eds.), *FICTA 2013*,
Advances in Intelligent Systems and Computing 247,
DOI: 10.1007/978-3-319-02931-3_56, © Springer International Publishing Switzerland 2014

transaction and store new transaction. Discovering knowledge from query group is an extension to work done by Hwang etal, in which it generates query group by using dynamic query grouping algorithm and extracts association among query groups. In this, we propose a framework for extracting knowledge from user search history and develop a model for maintaining query stream dynamic updating.

The remaining section of this paper is organized as follows. Section 2 reviews the related work. Section 3 describes Preliminaries. Section 4 gives frame work for discovering knowledge group from the dataset. Section 5 presents evaluation results. Finally, we conclude paper in Section 6.

2   Related Work

The work of Hwong [1] et al described about structuring of user search history into query groups. The main objective of their work is to determine whether two keyword questions belong to same task. Jones et al [2] and Boldi et al [3] proposed sections to problems to the same search task . They [2] use search session data and use the classification approach based on time, text and user search logs. The work of Hwong et al differs from previous works in following aspects-first the query-log characterizing in [2] [3] are extracted from co-relation statistics of two Queries. The work of [1] additionally considers pair of Queries common URL clicks and used Query session graphs. Some of the existing works considered problems of how to segment a Query stream into sessions. The most of the segmentations were time based.

The work of Beeferman et al focused on online Query grouping, he found Query classes to be useful for a Repository. The problem of Query clustering can be accomplished offline. The proposed work used single window process model [5] for extracting associations among Query groups. Whereas existing Landmark [6] and time fading [7]. The proposed algorithm will extend works of [8] [6] and it will require one scan Query dataset, it shared flexible update new stream.

In their work A. Spink, et al described a   user's single session with a Web based search engine [6] or information retrieval (IR) system that consist of accessing information topics. They have presented findings from two studies. First, the study of two-query search sessions on the search engine. Second a study of keyword search sessions. They have examined the degree of search.. They have proposes an approach to interactive information retrieval (IR) contextually within a multitasking framework. The implications of our findings for Web design and further research are discussed in further sections.

3   Preliminaries

Group Query: A Query group is an ordered list of queries Q_i together with the corresponding set of clicked url's C_i of Q_i. A Query group is denoted as $Q = (\{Q_i, C_i\}, \{Q_k, C_k\})$.

The following is the Algorithm for select Best Query Group method:

Input: The current singleton query group T_c containing the current query Q_c and set of clicks C_c
A set of existing query group $T= \{T_1,.....,T_m\}$.
A similarity threshold $R_s, 0 \leq R_s \leq 1$
Output:
The query group Q that best matches Q_c, or a new one if necessary.
$Q = \phi$
$R_{max} = Rs$
for i= 1 to m
if $S(T_c, T_i) > R_{max}$
   $Q = T_i$
   $R_{max} = S (T_c, T_i)$
   If $Q = \phi$
   $T = T \cup Tc$
   $Q = T_c$
   Return S.

Time: S_{time} (t_c, t_i) is defined as the inverse of the time interval (e.g., in seconds) between the   times that Q_c and Q_i are issued, as follows:

$$S_{time} (t_c, t_i) = \frac{1}{|time(Qc)-time(Qi)|}$$

The queries Q_c and Q_i are the most recent queries in t_c and t_i respectively. Higher S_{time} values imply that the queries are temporally closer.

Jaccard: $S_{jaccard} (t_c, t_i)$ is defined as the fraction of common words between Q_c and Q_i as follows:

$$S_{jaccard} (t_c, t_i) = \frac{|words(Qc) \cap words(Qi)|}{|words(Qc) \cup words(Qi)|}.$$

Levenshtein: S_{edit} (t_c, t_i) is defined as $1- D_{edit}(t_c, t_i)$. The edit distance D_{edit} is the number of character insertions, deletions, or substitutions required to transform one sequence of characters into another, normalized by the length of the longer character sequence

CoR: $S_{cor} (t_c, t_i)$ is the Jaccard coefficient of Q_c's set of retrieved pages retrieved(Q_c) and Q_i's set of retrieved pages retrieved(Q_i) and is defined as:

$$S_{cor} (t_c, t_i) = \frac{|retrived(Qc) \cap retrived(Qi)|}{|retrived(Qc) \cup retrived(Qi)|}$$

4   Knowledge Discovery from Keyword Query Groups

4.1   Architecture

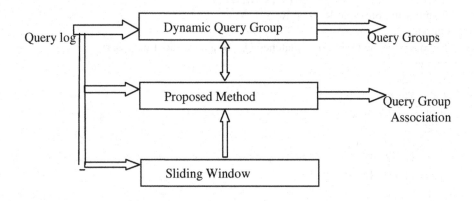

Fig. 1. Proposed Model

The above figure represents architecture proposed framework in which it takes Query search log as input and produce Query groups by using algorithm called Dynamic Query groups. From Query groups and Query data set our proposed model generates Query group association rules which can be helpful to user for future decision making for storing stream new session, Query transaction and update Query groups.

4.2   Stream Processing

A data stream is an ordered collection of session dada that arrives in particular order. An effective data stream mining process must be one scan processing due to the unbounded characteristics of session query streams. Some transactions are moved through the sliding window, it cannot access them again after process completion. It is necessary to retain the accuracy of the algorithm when the window processing streams.   The sliding window is divided into a collection of disjoint equal-sized blocks.

The creation of the sliding window is as follows: The size of the window is X/E, where X is a constant and E = E - X/N. The System should adjust X in order to change the size of the block and assume that there are B transactions being processed in main memory. B will change in proportion to the available memory. Large blocks with large labels contain recent transactions. FS is defined as a collection of synopses over the window in which each of the windows maintains significant transactions of one block. Stream processing model here is to find a way to extract <Session, Query> set for finding association among the query groups. The Sliding Windows model store and update query groups in sliding window. Data streams within the session query sliding window are   processed at the time.

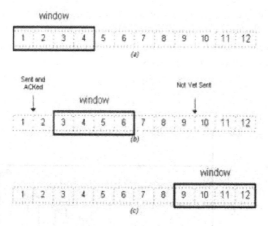

Fig. 2. sliding window processing

4.3   One Pass Algorithm

Objective: It finds query group association sets from data streams based on sliding window

  Inputs   : Data Stream DS, minimum support s, sliding window size |w|
  Output   : Query group relationships

1.   Initially <Session, Query> repository and sliding windows are empty
2.   While Ti comes
      2.1 If (i==1) then
              2.1.1. Create Pattern Tree with root labeled "ROOT"
              2.1.2. Insert_ptree (ROOT, T1)
              2.1.3. Update Item set counts in Query group Repository.
      2.2 Else if (i<=|w|) then
              /*sliding window is not full*/
              2.2.1. Insert_ptree (ROOT, T1)
              2.2.2. Update Item set counts in Query group Repository.
      2.3 Else
              2.3.1. remove (ptree, Ti-w)
              2.3.2. insert (ptree, Ti)
              2.3.3. Find frequent item sets (Ti-w)
              2.3.4. Find frequent item sets (Ti)
              2.3.5. Update counts.
      2.4. Query group association Output

5   Experimental Results

The experiments were conducting using algorithm was written in java, Input is generated using random data generator. The random data streams are T5I4D100K, where T denotes the average transaction size, I denote the average item size and D is the number of transactions, respectively. We generated query group tree from incoming data streams and this tree was converted to Query group association and applied to one scan algorithm for transactions which are in side sliding window. Experiments performed on various parameters as shown in table below.

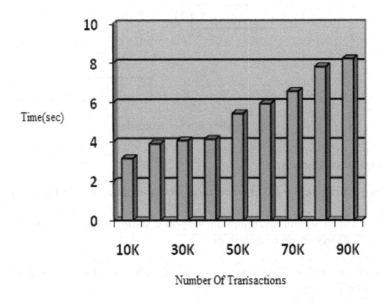

Number Of Transactions

Parameter	Range
No of Distinct Items	5–100
Sliding Window Size	10-500
Average Number of items	4-10
Threshold	0.025-0.0025

Fig. 3. Parameters

6   Conclusion

The query grouping and click graphs contain useful information on user behavior. In this paper we investigated the usefulness of the discovered knowledge extracted from these query groups and provide query suggestions and biasing the ranking of search results. This paper explained how to maintain user search history and missing knowledge from query group. This paper uses one pass algorithm in order to generate knowledge from user search groups. The experimental results support objective of the work.

Acknowledgement. We are very much thankful to Dr.Suresh Chandra Satapathy for his most valuable suggestion and encouragement

References

1. Spink, A., Park, M., Jansen, B.J., Pedersen, J.: Multitasking during Web Search Sessions. Information Processing and Management 42(1), 264–275 (2006)
2. Boldi, P., Bonchi, F., Castillo, C., Donato, D., Gionis, A., Vigna, S.: The Query-Flow Graph: Model and Applications. In: Proc. 17th ACM Conf. Information and Knowledge Management, CIKM (2008)
3. Baeza-Yates, R., Tiberi, A.: Extracting Semantic Relations from Query Logs. In: Proc. 13th ACM SIGKDD Int'l Conf. Knowledge Discovery and Data Mining, KDD (2007)
4. Barbakh, W., Fyfe, C.: Online Clustering Algorithms. Int'l J. Neural Systems 18(3), 185–194 (2008); Levenshtein, V.I.: Binary Codes Capable of Correcting Deletions, Insertions and Reversals. Soviet Physics Doklady 10, 707–710 (1966)
5. Sahami, M., Heilman, T.D.: A Web-based Kernel Function for Measuring the Similarity of Short Text Snippets. In: Proc. the 15th Int'l Conf. World Wide Web (WWW 2006), pp. 377–386 (2006)
6. Chang, J.H., Lee, W.S., Zhou, A.: Finding Recent Frequent Itemsets Adaptively over Online Data Streams. In: ACM SIGKDD Int'l Conf. on Knowledge Discovery and Data Mining (August 2003)
7. Charikar, M., Chen, K., Farach-Colton, M.: Finding Frequent Items in Data Streams. Theoretical Computer Science (January 2004)
8. Dora Cai, Y., Pape, G., Han, J., Welge, M., Auvil, L.: MAIDS: Mining Alarming Incidents from Data Streams. In: Int'l Conf. on Management of Data (June 2004)

D-Pattern Evolving and Inner Pattern Evolving for High Performance Text Mining

B. Vignani and Suresh Chandra Satapathy

Dept. of CSE,
Anil Neerukonda Institute of Technology and Sciences, India
Vignani.boppana@gmail.com
sureshsatapathy@ieee.org

Abstract. Many data mining techniques have been introduced to perform different information tasks to mine useful patterns in text documents. However, the way to use effectively and update discovered patterns is still a research issue, particularly within the domain of text mining . Text mining methods adopt term based approach and phrase based approach. Phrase based approach performs better than the term based as phrases carry more information. In this paper we have tendency to propose a new methodology to enhance the utilization of the effectively discovered patterns by including the process of D-pattern evolving and inner pattern evolving.

Keywords: text mining, pattern mining, pattern deploying, pattern evolving.

1 Introduction

In recent years the growth of digital data is improved due to which knowledge discovery and data mining has attracted a great deal of attention. It turns the data into useful information and knowledge, which is extracted from the massive quantity of data. Knowledge discovery is an interdisciplinary area focusing upon methodologies for extracting useful knowledge from the data[4]. So, it is said that data mining is an essential step in the process of knowledge discovery. Many data mining techniques have been proposed for the purpose of developing efficient mining algorithms to find the suitable patterns. How to effectively use this pattern is still a research issue.

In this paper we effectively use and update the patterns and apply it to the field of text mining. Text mining is also referred as text data mining. It is the discovery of interesting knowledge in text documents and also finding the unknown information, by automatically extracting the knowledge from different written resources. It is used to denote all the tasks that try to extract useful information by finding patterns from large texts. Text mining share many characteristics with data mining but differ in many ways.

- Many knowledge discovery algorithm defined in context of data mining are inapplicable for textual application.
- Special mining task such as concept relationship analysis, are unit distinctive to text mining.

- The unstructured form of the full text necessitates special linguistic preprocessing for extracting main features of text.

It is challenging issue to find the accurate text documents of what users want. Information retrieval provides many methods to solve this challenging issue.

Information retrieval deals with the representation, storage, organization of and access to information items. The representation of the information should be provided to the user with easy access by which he can understand[8].IR retrieves as many relevant documents as possible and by filtering irrelevant documents at same time. IR systems doesn't provide adequately of what users need. Several text mining ways are developed so as to retrieve the relevant document.

Term based approach consist of keywords ie single terms, these keywords consist of same meaning for different terms. It is categorized as Synonym and Homonym.

Due to this disadvantage it is said that phrased based approach performs better than term based as phrases carry more information. By using phrases as index terms, a document that contains a phrase that occurs in the request would be ranked higher than a document that just contains its constituent words in unrelated contexts[3].

Although phrases are less ambiguous the disadvantages are

- Phrases have inferior statistical properties to terms,
- They have low frequency of occurrence, and
- There are large numbers of redundant and noisy phrases among them,
- The theory of computing probabilities based on term dependencies is not practical.

In order to solve the above problem one solution is proposed ie pattern mining based approach which adopts the concept of closed sequential patterns and pruned non closed patterns from the representation with an attempt to reduce the size of the feature set by removing noisy patterns. Another challenge for the data mining based methods is that more time is spent on uncovering data from the information, consequently less significant enhancements are created compared with information retrieval strategies.

In this paper we propose a d-pattern and inner pattern evolving for text mining to effectively use the discovered patterns. The d-pattern algorithm is used to discover the patterns from the positive documents, then it calculates the support and evaluates the weight. In the inner pattern evolving the noisy documents are identified ,these error patterns will be removed and shuffling is done.

2   Related Work

In their work Many types of text representations have been proposed in the past. A well known one is the bag of words that uses terms as elements within the vector of the feature space. Bag-of-words is a difficult keyword representation in the area of information retrieval. In Rocchio [2]classifiers the tf*idf weighting scheme is used for text representation. In addition to TFIDF, the global IDF and entropy weighting scheme is proposed in the information retrieval, it increases the performance by a mean of 30 percent. Different weighting schemes are given in text categorization for the bag of words representation.

Text classification[1] is a task of classifying a collection of documents into completely different classes from the predefined set.The problem of the bag of words approach is a way to choose a limited number of features among an enormous set of terms so as to extend the system's efficiency and avoid overfitting. In order to reduce the number of features, several spatiality reduction approaches have been conducted by the utilization of feature selection techniques, such as Information Gain, Mutual Information, Chi-Square, Odds ratio, and so on. Details of these selection functions were explicited in feature selection and feature extraction.

Various data mining techniques have been used for text analysis by extracting the identical terms as phrases form document collections. Text mining from ontology[5] has become essential means for discovering the knowledge and also for building the information intensive systems. In their work they have described for mining ontology's from text resources. The difficult problem associated with this analysis is to identify what kind of ontology is often mechanically discovered from information sets to illustrate significant descriptions for user profile. Due to this problem the pattern evolving technique was introduced to develop the performance of term based ontology methods.

Pattern mining is a data mining method that involves finding the existing patterns in data. In their work they have focused on discovering the patterns from the huge collection of data. How to effectively use the patterns is still a research issue. In the field of text mining, pattern mining techniques can be used to find various text patterns, such as sequential patterns, frequent itemsets, cooccurring terms and multiple grams, noun phrase, key phrase for increasing a illustration with these new types of features. Nevertheless, the challenging issue is the way to effectively cope up with the massive quantity of discovered patterns.

Support Vector Machine

SVM is a supervised learning model with associated learning algorithms that analyze data and recognize patterns used for classification analysis[9]. SVM considers a collection of input file and predicts for each given input which of two possible classes form the output.

A Support Vector Machine (SVM) is a discriminative classifier formally defined by a separating hyperplane. In other words, given labelled training information, the algorithm outputs an associate degree optimum hyperplane that categorizes new examples.

Pattern Taxonomy Model

In this model we consider a document. Each document splits into paragraphs. Here we consider each paragraph as a single term. We apply the data mining method to the frequent patterns from these transactions and generate pattern taxonomies[7]. The unnecessary patterns are removed during the pruning phrase.

We calculate the absolute and relative support to estimate the significance of patterns.

$$\text{Supp}_a(p) = |\{S | S \in d, P \text{subset} S\}|$$
$$\text{Supp}_r(P) = \text{Supp}_a(P)/|d|$$

In pattern pruning schema the redundant patterns are removed ie the patterns that have same frequency are eliminated as they always occur together in a document.

In this model it states that if <t3,t4>,<t3,t6> is a sub sequence of <t3,t4,t6> then the shorter element such as <t3,t4> and <t3,t6> should be removed as it has same frequency

Parapgraph	Terms
dp_1	$t_1\ t_2$
dp_2	$t_3\ t_4\ t_6$
dp_3	$t_3\ t_4\ t_5\ t_6$
dp_4	$t_3\ t_4\ t_5\ t_6$
dp_5	$t_1\ t_2\ t_6\ t_7$
dp_6	$t_1\ t_2\ t_6\ t_7$

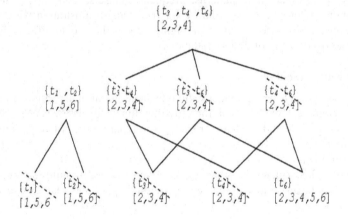

Pattern Deploying Model

We have pattern deploying method[6] and d-pattern.

The patterns can be discovered with the help of weight function to assign the value for each pattern according to its frequency.

Weight(d)=$\sum_{t\epsilon T} support(t)\ T(t, d)$

Where T(t,d)=1;

PDM is proposed to solve the problem of inappropriate evolution of patterns, discovered using data processing ways. This approach is proposed to handle some problems like reducing feature space and its complexity etc and may additionally notice the specificity of patterns by combining two or more words. D-pattern takes the pattern support obtained within the section of pattern discovery which is considered when we deploy patterns into common hypothesis space. A probability function is introduced for the feature significance. The detailed process is presented in the SPmining algorithm[7]

D-pattern mining describes the training process of finding the set of d-patterns and it is also used to discover all the patterns from the positive documents. For each positive documents the SPmining algorithm rule is termed to provide rise the set of closed sequential patterns.

Closed sequential pattern: it develops several efficient search space pruning methods[10].

It can be based on two steps:

a. It generates the complete closed sequence, a superset of frequent sequence and stores it in an exceedingly prefix sequence lattice.
b. It does pruning before to eliminate non closed sequence.

In closed sequential patten it first removes the unnecessary terms and then calls clospan by doing depth first search and builds the prefix lattice.

Later it calculates the support and the weight is evaluated.

$d_1 = $ <(carbon,2), (emiss,1),(air,1),(pollut,1)>
$d_2 = $ <(greenhous,1),(global,2),(emiss,1)>
$d 3 = $ <(greenhous,1),(global,1),(emiss,1)>
$d_4 = $ <(carbon,1), (air,2), (antarct,1)>
$d_5 = $ <(emiss,1),(global,1),(pollut,1)>

Doc.	Pattern taxonomies	Sequential patterns
d_1	$PT_{(1,1)}$	$\{\langle carbon \rangle_4 , \langle carbon, emiss \rangle_3\}$
	$PT_{(1,2)}$	$\{\langle air, pollut \rangle_2\}$
d_2	$PT_{(2,1)}$	$\{\langle greenhous, global \rangle_3\}$
	$PT_{(2,2)}$	$\{\langle emiss, global \rangle_2\}$
d_3	$PT_{(3,1)}$	$\{\langle greenhous \rangle_2\}$
	$PT_{(3,2)}$	$\{\langle global, emiss \rangle_2\}$
d_4	$PT_{(4,1)}$	$\{\langle carbon \rangle_3\}$
	$PT_{(4,2)}$	$\{\langle air \rangle_3, \langle air, antarct \rangle_2\}$
d_5	$PT_{(5,1)}$	$\{\langle emiss, global, pollut \rangle_2\}$

Algorithm

1.DP=null;

2.for each document d∈D$^+$

3. let PS(d) be the set of paragraphs in d;
4. SP=SPMining(PS(d),min_sup);
5. d=null;
6. for each pattern pi ∈ SP do

7   p={(t,1)|t ∈ pi};
8   d=d+p;
9   end
10   DP=DP Ü {d}
11   end
12   T={t| (t,f) ∈ p,p ∈ DP};

13   foreach term t ∈ T do
14   support(t)=0;
15   end
16   foreach d-pattern p ∈ DP do

17   foreach (t,w) ∈ β(p) do
18   support(t)=support(t)+w;
19   end
20   end

Inner Pattern Evolving

In IPE we identify the noisy documents. It is used to reshuffle the weights using the offenders. Sometimes the positive pattern are going to be incorrectly identified as negative pattern, then we see which pattern could be a reason to give rise this error. Such a pattern is known as offender.

$$\Delta(nd) = \{p\epsilon DP | termset(p) \cap nd \neq \emptyset\}$$

We estimate the threshold and calculate the minimum support. A threshold is usually used to classify documents into relevant or unrelevant categories. If any noisy patterns occur they are going to be removed and therefore the shuffling method is finished.

$$\text{Threshold(DP)}=\min_{P\epsilon DP}(\textstyle\sum_{(t,w)\epsilon\beta(p)} support(t))$$

By shuffling their weight, the uncertainties contained in these offenders can be evaporated.

3   Conclusion

Many data mining techniques have been proposed for fulfilling the various information tasks. These techniques include association rule mining frequent itemset mining, sequential pattern mining, maximum pattern mining, and closed pattern mining.

In this work, an effective pattern discovery technique has been proposed to overcome the low-frequency and misinterpretation issues for text mining.In this paper the d-pattern and inner pattern evolving is used to effectively discover and update the patterns. The d-pattern is used to calculate the support and evaluate the weights.In pattern evolving we tend to estimate the threshold based mostly and update the discovered patterns. The latest version of RCV1 is taken.

References

1. Aas, K., Eikvil, L.: Text Categorisation: A Survey. Technical Report NR 941, Norwegian Computing Center (1999)
2. Joachims, T.: A Probabilistic Analysis of the Rocchio Algorithm with tfidf for Text Categorization. In: Proc. 14th Int'l Conf. Machine Learning (ICML 1997), pp. 143–151 (1997)
3. Lewis, D.D.: An Evaluation of Phrasal and Clustered Representations on a Text Categorization Task. In: Proc. 15th Ann. Int'l ACM SIGIR Conf. Research and Development in Information Retrieval (SIGIR 1992), pp. 37–50 (1992)
4. Zhong, N., Li, Y., Wu, S.-T.: Effective pattern discovery for text mining (2010)
5. Li, Y., Zhong, N.: Mining Ontology for Automatically Acquiring Web User Information Needs. IEEE Trans. Knowledge and Data Eng. 18(4), 554–568 (2006)
6. Wu, S.-T., Li, Y., Xu, Y.: Deploying Approaches for Pattern Refinement in Text Mining. In: Proc. IEEE Sixth Int'l Conf. Data Mining (ICDM 2006), pp. 1157–1161 (2006)
7. Wu, S.-T., Li, Y., Xu, Y., Pham, B., Chen, P.: Automatic Pattern- Taxonomy Extraction for Web Mining. In: Proc. IEEE/WIC/ACM Int'l Conf. Web Intelligence (WI 2004), pp. 242–248 (2004)
8. Baeza-Yates, R., Ribeiro-Neto, B.: Modern Information Retrieval. Addison Wesley (1999)
9. Cortes, C., Vapnik, V.: Support-Vector Networks. Machine Learning 20(3), 273–297 (1995)
10. Yan, X., Han, J., Afshar, R.: Clospan: Mining Closed Sequential Patterns in Large Datasets. In: Proc. SIAM International Conf. on Data mining (SDM 2003), pp. 166–177 (2003)

A Fast Auto Exposure Algorithm for Industrial Applications Based on False-Position Method

B. Ravi Kiran, G.V.N.A. Harsha Vardhan

Anil Neerukonda Institute of Technology and Sciences
Visakhapatnam, Andhra Pradesh, India

Abstract. In this paper we implemented an Auto-exposure algorithm based on the False-position method in order to correctly expose the leather samples. Though we are doing this for the leather industry, we can directly use this false position based, auto exposure algorithm for the natural scenes. The main reason for choosing the False-position method is that, it converge the root values quickly when compared to bisection method. The implementation of our auto exposure algorithm is performed by using the point grey research programmable camera.

Keywords: Auto exposure, False position.

1 Introduction

Nowadays the people want everything to be automated in a quick and reliable manner. The Auto Exposure algorithm is one such method for the cameras to find the proper exposure of the scene .We are considering the shutter speed and gain to adjust the camera for proper exposure of the scène. There are many auto exposure algorithms with different types of methods [1], The usage of the numerical methods is one among them.

The numerical methods are used to find the roots for different types of equation. The numerical analysis methods are used in the auto exposure algorithm to find the values of the shutter speed and gain of the camera as at which the scene can be properly exposed by considering them as variables to find the roots.

In the leather industries to detect the defects of the leather sample, the leather sample images have to be correctly exposed but the varying lighting condition in the industrial environment is the main problem. So an auto exposure algorithm has to be used to capture those leather samples with a proper exposure according to the lighting conditions. The camera should identify the correct exposure values in a faster way, if the camera takes more amount of time to adjust itself to the situation then some of the leather sample images are not captured by the camera so the camera has to adjust to the lighting conditions quickly.

The lighting conditions for a scene are classified into three categories such as underexposure, Overexposure and Proper exposure. These three are based on the amount of light that is captured by the camera for the scene. For under exposure the

S.C. Satapathy, S.K. Udgata, and B.N. Biswal (eds.), *FICTA 2013*,
Advances in Intelligent Systems and Computing 247,
DOI: 10.1007/978-3-319-02931-3_58, © Springer International Publishing Switzerland 2014

information of the low lighting regions are lost as they appear dark in the image, similarly for overexposure the information of the high lighting regions are lost as they appear brighter in the image because more amount of light is captured by the camera. The optimal exposure captures both the low lighting and high lighting details without any loss of information.

To obtain the optimum exposure for the under exposed scene the shutter speed and gain are increased, this is the adjustment to the camera for under exposure , similar we decrease the shutter speed and gain for high lighting scene to obtain the optimal exposure for the over exposed scene. This process of adjust is performed in the auto exposure algorithm until proper exposure is obtained. To obtain the optimum exposure one methodology is the numerical analysis methods .There are many numerical analysis methods, The Bisection method is one basic method which starts by identifying the initial roots and calculating the bisect value of them and by repeating the process of bisecting with new approximation and the initial root basing on the function value of the approximation. Similarly in auto exposure the initial exposure value is identified and verifies the type of exposure basing the on the luminance value of the scene. The new approximation is obtained by bisecting the previous approximation with the initial root; This process is repeated until proper exposure is obtained. The advantage of the bisection method is its reliability, but the disadvantage is it takes number iterations to bisect to the optimum value, as the interval to bisect will decrease so it takes more amount of time to get optimum exposure value.

To overcome this disadvantage the false position method is used, as it finds the optimum value faster than the Bisection method. The secant method has the problem of divergence of the root ,this problem is also solved by the false position method as the initial values are considered so it will reliably leads to the optimum exposure value ,.So thus the false position method is fast and reliable to find the exposure values.

2   False Position Method

The false position method [6][7] is one of the numerical analysis method, which is used to find the root for a non-linear equation(curve). The root of the equation is a point where the curve cuts the x-axis when y equals 0.

The false position method [6][7] is one of the numerical analysis method, which is used to find the root for a non-linear equation(curve). The root of the equation is a point where the curve cuts the x-axis when y equals 0.

The basic formula of the false position method is as in formula (2) where **approximation$_i$** is the next approximation and**a_i, b_i** are the initial roots with function values**$f(a_i), f(b_i)$**

$$approximation_i = a_i - \frac{f(a_i)*(b_i-a_i)}{(f(b_i)-f(a_i))} \hspace{2cm} (2)$$

So this root finding formula can be used for the auto exposure algorithm to find the shutter speed and the gain as in formula (3) and (4) respectively [4] for under exposure scene.

For under exposure scene, the next shutter speed approximation S_{k+1} of $k + 1^{th}$ iteration is calculated by using Maximum shutter speed S_{max} ,Shutter speed approximation of Kth iteration S_k, Desired mean luminance value $mean_{pro}$ which the proper exposure mean luminance value i.e. equals to 120, Mean luminance of the scene using K^{th} iteration approximation $mean_k$, as in formula (3).

$$S_{k+1} = S_{max} - \left(\frac{(S_{max} - S_k) * mean_{pro}}{mean_k} \right) \tag{3}$$

Similarly the next approximation of the gain $Gain_{k+1}$ of $k + 1^{th}$ iteration is calculated by using Maximum gain control $Gain_{max}$, Gain control approximation of Kth iteration $Gain_k$, Desired mean luminance value $mean_{pro}$ which the proper exposure mean luminance value i.e. equals to 120, Mean luminance of the scene using K^{th} iteration approximation $mean_k$, as in formula (4).

$$Gain_{k+1} = Gain_{max} - \frac{(Gain_{max} - Gain_k) * mean_{pro}}{mean_k} \tag{4}$$

Over exposure scene, the next shutter speed approximation S_{k+1} of $k + 1^{th}$ iteration is calculated by using Minimum shutter speed S_{min} ,Shutter speed approximation of Kth iteration S_k, Desired mean luminance value $mean_{pro}$ which the proper exposure mean luminance value i.e. equals to 120,Mean luminance of the scene using K^{th} iteration approximation $mean_k$, as in formula (5).

$$S_{k+1} = S_{min} - \left(\frac{(S_{min} - S_k) * mean_{pro}}{mean_k} \right) \tag{5}$$

Similarly the next approximation of the gain $Gain_{k+1}$ of $k + 1^{th}$ iteration is calculated by using Minimum gain control $Gain_{min}$, Gain control approximation of Kth iteration $Gain_k$, Desired mean luminance value $mean_{pro}$ which the proper exposure mean luminance value i.e. equals to 120,Mean luminance of the scene using K^{th} iteration approximation $mean_k$, as in formula (6).

$$Gain_{k+1} = Gain_{min} - \frac{(Gain_{min} - Gain_k) * mean_{pro}}{mean_k} \tag{6}$$

The mean luminance value for the settings (shutter speed and gain) of the K+1th iteration will be repeated until the proper exposure of the scene is obtained. The formulas (3) and (4) are used to find the exposure approximations for under exposure scene. Similarly for overexposure the formulas (5) and (6) are used to find the exposure approximations.

The following is the Algorithm for auto exposure using false position method:

Input: The scene that is to be captured with proper exposure
Output: The scene with proper exposure values.

Steps

1   Start
2   Identify the initial exposure value as the initial root
3   Capture the scene with initial exposure value
4   Check whether the scene is properly exposed or over exposed or under exposed basing on the mean luminance value.
        4.1   If the scene is under exposed then calculate the next approximation of the exposure value using the formulas (3) and (4) , else go to step 4 (b)
        4.2   If the scene is over exposed then calculate the next approximation of the exposure value using the formulas (5) and (6) , else go to step 4 (c)
        4.3   If the scene is proper exposed then go to step 6
5   Capture the scene with new approximation values and go to step 4
6   The proper exposure values are displayed
7   Stop

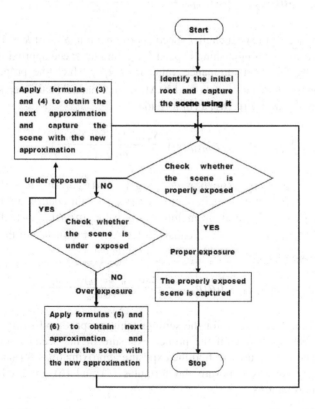

Fig. 1. Flow chart of auto exposure algorithm using false position method

The above figure 1 represents the flow chart of the auto exposure algorithm using the false position method, which is a diagrammatical representation of the algorithm mentioned above

3   Experimental Results

In this section we are comparing the false position based auto exposure algorithm with the bisection based auto exposure algorithm.

The figure 2 represents the proper exposure obtained by using the auto exposure algorithm implemented by using the bisection method with mean luminance value 101.9

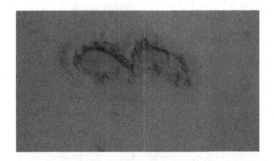

Fig. 2. Proper exposure of the leather sample using Bisection method

The following are the results obtained by using the above algorithm mentioned in the flow chart, which is the auto exposure algorithm using false position method.

Fig. 3. leather sample at iteration 1 with shutter speed 4000.0ms and gain 17.0dB, which is an over exposure with mean luminance 254.8

Fig. 4. leather sample at iteration 2 with shutter speed 1882.4ms and gain 8.0dB, which is an over exposure with mean luminance 254.1

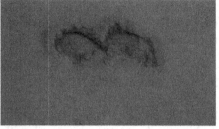

Fig. 5. leather sample at iteration 3 with shutter speed 887.7ms and gain 3.8dB, which is an under exposure with mean luminance 69.9

Fig. 6. leather sample at iteration 4 with shutter speed 1345.1ms and gain 5.7dB, which is an under exposure with mean luminance 82.8

Fig. 7. leather sample at iteration 5 with shutter speed 1355.1ms and gain 6.7dB, which is a proper exposure with mean luminance 132.3

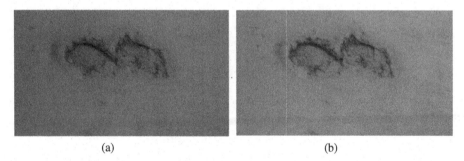

(a)   (b)

Fig. 8. proper exposure images of bisection and false position methods

The Bisection method took 10 iterations to find the proper exposure of a leather sample in a particular lighting condition, whereas false position takes only 5 iterations to find the proper exposure for the same sample in the same lighting conditions. The figure 8 compares the proper exposed images of the bisection method and false position methods as in figure 8.a and figure 8.b respectively.

The figure 8.a represents the proper exposure obtained by using bisection method and the figure 8.b represents the proper exposure obtained by using the false position method.

The usage of the false position for the auto exposure algorithm than the bisection method reduces 4-5 iterations to converge to the proper exposure value. The false position method is quick and reliable to use in the leather sample industry to find the proper exposure than the bisection method and the secant method.

4   Conclusion

An auto exposure algorithm has been implemented using false position method which is quick and reliably estimates the appropriate exposure values to correctly expose the scene. The false position method reduces the number of iteration to converge to the proper exposure value which reduced the computational cost. The false position method is simple and has less number of computations to perform, so this algorithm can be easily used on the devices which contain less amount of memory available such as cell phones, surveillance cameras etc. The extension of this work is also accepted in the 8th International Conference on Robotics, Vision, and Signal Processing & Power Applications (ROVISP 2013) Penang, Malaysia.

Acknowledgment. We are very much thankful to AICTE, New Delhi, India for funding us to develop this algorithm for leather defect detection mechanism. We are also very much thankful to Professor Ramakrishna Kakarala for his valuable support and suggestion, which helped us a lot to develop this algorithm.

References

1. Muehlebach, M.: Camera Auto Exposure Control for VSLAM Applications. Autumn Term (2010)
2. Vuong, Q.K., Yun, S.H., Kim, S.: A New Auto Exposure System to Detect High Dynamic Range Conditions Using CMOS Technology. In: Third International Conference on Convergence and Hybrid Information Technology ICCIT 2008, November 11-13, vol. 1, pp. 577–580 (2008), doi:10.1109/ICCIT.2008.45ccc
3. Vuong, Q.K., Yun, S.-H., Kim, S.: A new auto exposure and auto white-balance algorithm to detect high dynamic range conditions using CMOS technology. In: Proceedings of the World Congress on Engineering and Computer Science. IEEE, San Francisco (2008)
4. Cho, M., Lee, S., Nam, B.D.: Fast auto-exposure algorithm based on numerical analysis. Electronic Imaging 1999. International Society for Optics and Photonics (1999)
5. Weisstein, E.W.: Method of False Position. From MathWorld–A Wolfram Web Resource, http://mathworld.wolfram.com/MethodofFalsePosition.html
6. http://www.ff.bg.ac.rs/Katedre/Nuklearna/bookcpdf/c9-2.pdf (link referred on June 19, 2013)
7. http://users.encs.concordia.ca/~kadem/CHAPTER%204.doc (link referred on June 19, 2013)

Legacy to Web 2.0

Hemant Kumbhar[1], Raj Kulkarni[2], and Suresh Limkar[3]

[1] Department of Computer Engineering, SVPM's COE, Baramati, India
[2] Department of Computer Science Engineering, WIT, Solapur, India
[3] Department of Computer Engineering, AISSMS IOIT, Pune, India
{hemant.kumbhar,sureshlimkar}@gmail.com,
raj_joy@yahoo.com

Abstract. In recent years web applications allows users to interact or collaborate with each other in a social media dialogue, in contrast to websites where very few users view the static content that was created for them. A rapidly growing trend in web applications is the development of user interfaces through Rich Internet Applications. Among other capabilities, RIAs offer high interactivity and native multimedia support, giving them a major advantage over legacy web systems. In this paper we propose a model which converts legacy web application into web 2.0 application. Conversion to the Web2.0 is twofold, i.e. automatic as well as manual. Our proposal focuses on the user interface adaptation of existing web 1.0 applications (Legacy) to Web 2.0 and add new RIA features with some good enhancement in legacy web system by web assessment.

Keywords: Legacy, Web 2.0, RIA.

1   Introduction

The development of Web sites and applications is increasing dramatically to satisfy the market requests. The software industry has to suffer under the pressure of a very short time-to-market and also has to face extremely high competition. The growth of traditional HTML-based Web applications (Web 1.0 or Legacy) has been based on Web methodologies coming from the Web engineering community [6].

In the internet era, web applications (WA), are playing a major part for allowing the enterprises to reach a leading position in the market place [7]. Although several methodologies have been proposed to develop WAs so far, existing methods and techniques, with a few expectations, do not support adequately all the activities require the engineer a WA, or they are tailored for websites, the main consequences of this lack of support is the low quality development documentation that is generally produced.

The Web 2.0 is commonly associated with web applications that facilitate interactive information sharing, interoperability, user- centered design, and collaboration on the www. Web models are generally based on the notion of encapsulated layers that divide objectives into different levels according to goals.

S.C. Satapathy, S.K. Udgata, and B.N. Biswal (eds.), *FICTA 2013*,
Advances in Intelligent Systems and Computing 247,
DOI: 10.1007/978-3-319-02931-3_59, © Springer International Publishing Switzerland 2014

Through this division [6], model developer can specify data structures using the data model, links and relations using the hypertext/navigation model, the web application appearance using the presentation model, and so on. A large body of research and development covers integration at the data and business logic layers, but few investigations examine it at the presentation layer. However, the complexity of activities performed via web user interfaces (UIs) continues to increase, and multimedia support as well as ubiquity is becoming fundamental in a growing number of Web 1.0 applications. Consequently, many developers are converting their applications to Web 2.0 UIs by using RIAs technologies [31]. Web models can't cope with all the new features that RIAs provide, though, so it's not only important to develop Web 2.0 applications from scratch but also to integrate the rich Web application user experience over existing Web 1.0 applications. Flexible web site generation, restructuring and maintenances can be achieved by database centric approach by associating web semantics with the modeling constructs of the ODMG object model [13].

The RUX-Model offers a method for engineering the adaptation of legacy model-based Web 1.0 applications to Web 2.0 UI expectations. It uses existing data and business logic from the Web application being adapted and provides a UI abstraction that the method transforms until the desired RIA UI is reached[6][7]. So we require bringing Web 2.0 to the Old Web [2]. Currently, most web development practices do not follow systematic methodologies; rather, information structuring and management is based only on developer's knowledge and practical skills [21], [5]. An approach based on guidelines to support the analysis phase and aims to establish a systematic engineering step, which lead a website developer to a high quality and accessible developed website. The issue of interest is related to site structure, conformance to web design rules, content and performance. By analyzing the web site we can suggest many good improvements in the legacy or existing websites[25]. They categorized the criteria and devised methods to measure them; also they have examined the Websites of interest, assessed the results and drawn conclusions, in order to improve our Website look and performance. As suggested in [26], guidelines for small selection of web design, usability, and accessibility related results of research, most of them derived from Human Factors International (newsletter). So if we adopt 15 guidelines given in [26], the web accessibility will be very good. Some empirical results provide an empirical foundation for web site design guidelines [27].

This paper is organized as follows: First, Literature survey in Section 2. Then Section 3 our proposed system, Section 4 implementation details. Section 5 results. Finally in Section 6 conclude the paper.

2   Literature Survey

In 2002, attempt was made for reengineering existing large- scale application to the web [11], the legacy application has to be provided with an (additional) interactive web interface that (a) decorates the outputs of the system with HTML mark-up and (b) translates the (e.g. form based) inputs via the web browser into the (legacy) system's APIs.

Generic web view transformation and template storage is also helpful for fast development of website. One proposal is to restructure an existing web site by adapting them to a controller- centric architecture [19].

The reverse engineering tool WARE[18]is also developed, whose main purpose is to provide a support to the recovery, from existing Web Applications, of UML diagrams dealing not only with static content, but also with more challenging dynamic content.

The concept at the base of RIA to take full advantage of their new capacities, and proposes an integrated Web Engineering approach based on the WebML [7], [20] and the RUX-Model conceptual models [6], [7]. Adapting web 1.0 User Interface to web 2.0 User Interfaces through RIAs based on RUX-Model is implemented [8], [3], [29]. Using a schema-based clustering technique, extracts a navigational model of web applications, and identifies candidate user interface components to be migrated to a single-page AJAX interface [5]. It is tool for Migrating Multi-page Web Applications to Single-page AJAX Interfaces.

A Template-based Method for Theme Information Extraction from Web Pages[1], FiVaTech: Page-Level Web Data Extraction from Template Pages[3] and TEXT: Automatic Template Extraction from Heterogeneous Web Pages[4] are the attempts for extracting data from web site. Website analysis and evaluation is done for improvement in current website[9],[10],[15],[17].

Legacy Migration to Service-Oriented Computing with Mashups is proposed in [32]. Web 2.0 Toolbar: Providing Web 2.0 Services for Existence Web Pages [33]. 30 automated evaluation tools to help designers to improve their sites. Are suggested[28],

The recovered models are specified by referring to the Ubiquitous Web Application (UWA) design methodology[16]. Also Control- centric architecture construction is suggested by [18], for getting abstract view of the site. On the same way we can extract MVC from web site.

3   Proposed System

The proposed work defines a process, including reverse engineering methods and a supporting software tool, to understand existing web applications to be maintained or evolved and also of the process for reverse engineering Web applications with different characteristics, and highlighted possible areas for improvement of its effectiveness, with the help of web assessment.

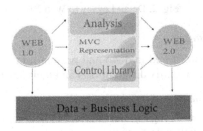

Fig. 1. Overview Legacy to Web 2.0

4   Implementation Details

Proposed work is as shown in block schematic is Fig.2.

1) Web Assessment and Improvement Suggestions

Currently, most web development practices do not follow systematic methodologies; rather, information structuring and management is based only on developer's knowledge and practical skills. EWAT (Empirical Web analysis Tool) is an empirical web parameters driven tool [5]. This research tool attempts to do the empirical analysis of web.. This approach aims to establish a systematic engineering step, which lead a website developer to a high quality and accessible developed website. The issue of interest is related to site structure, conformance to web design rules, content and performance. This tool provides empirical evidence on page composition, page formatting and overall page characteristics. Like this we can do web assessment of existing web site and suggest.

2) db4o

db4o is the open source object database that enables Java and .NET developers to store and retrieve any application object with only one line of code, eliminating the need to predefine or maintain a separate, rigid data model.

In this legacy to web convertor db4o is used to store the legacy website.

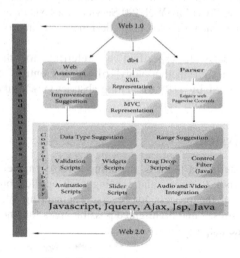

Fig. 2. Detail Legacy to Web 2.0

3) XML Representation

This module reads website stored in db4o and present it in terms of XML file. XML representation is useful to identify different tags and their attributes with values.

4) MVC Representation

This is one representation of legacy website to MVC architecture. Hence with the help of this module we will able to view legacy web site in MVC and also able to separate out data manipulation, business logic, navigations in legacy website.

5) *Parser and Legacy Page wise controls*

This parser is implemented on the top of HTMLParser Version 1.6. It   reads legacy web page(html, jsp, asp, aspx). Parse it to separate out controls in it. We are focusing specially on form tag and all input tags. Then control list gets prepared for that page and forwarded to Control Library.

6) *Control Library*

a) *Data type Suggestion*

This module of control library works like intelligent system. According to the control list prepared by parser, this module suggests possibly data type for input controls. For suggestion this module uses dictionary. Dictionary is probable list of input control names generally used in web page like "txtbookingdate" of date type, "txtfirstname" of alphanumeric, "txtpincode" of Numeric type.

b) *Range Suggestion*

This is the module which adds more intelligence to input controls by attaching range of values to it. For example suppose booking is available for week only, and then entered value can be checked at client site against range fixed. It will help minimize the network traffic.

c) *Validation Scripts*

If legacy web page doesn't has validation, as per the suggested or manually finalized data types this module inserts validation scripts in legacy web page.

d) *Widget Scripts*

This module helps to give rich user experience (web 2.0) to the user by inserting different widgets to legacy web page. For example, we can insert calendar widget, weather widget.

e) *Drag drop Scripts*

Adding jQuery drag drop effects to legacy web page gives web portal feel to the end users. End user also gets opportunity to move and place different blocks of web page as his/her desire.

f) *Control Filters*

These filters are for finding out input controls with known names, reading attributes having desire values.

g) *Animation Scripts & Slider Scripts*

In this module facility is given to run third party software for creating animations in flash or with jQuery. Animations created with software can be placed in the web page at user's desired place.

i) *Audio and video Integration*

Here is given the facility to insert audio and video in legacy page.

7) *Data and Business Logic*

Data and business logic is kept untouched throughout the process of converting legacy web to web 2.0. Because proposed work in an attempt to focus on the user interface adaptation of existing web 1.0 applications (Legacy) to Web 2.0 and add new RIA features.

5   Results

Experiments were carried out to check the feasibility and the effectiveness of the proposed approach. In a first experiment, simple web pages were collected from net , different metrics are applied to check the site structure and performance of those web sites. Web analysis results are as shown in table I. In second experiment we collected a complete, small size, WA. We stored them in db4o database which given the very easy to manipulate storage structure. Then we retrieve the stored web from db4o to XML format. Also revived data from db4o is used to separate out model, view and controller structure of the WA. And finally in third experiment, we collected some website (generally the web1.0) and run our control library for converting it to Web 2.0. Below are some screen shots of our control library execution and changed pages.

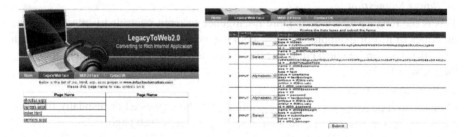

Fig. 3. Legacy website pages list   **Fig. 4.** Automatic Data type suggestion for web page controls

Fig. 5. Legacy (i.e) Original web page   **Fig. 6.** Web page After adding calender automaticaly

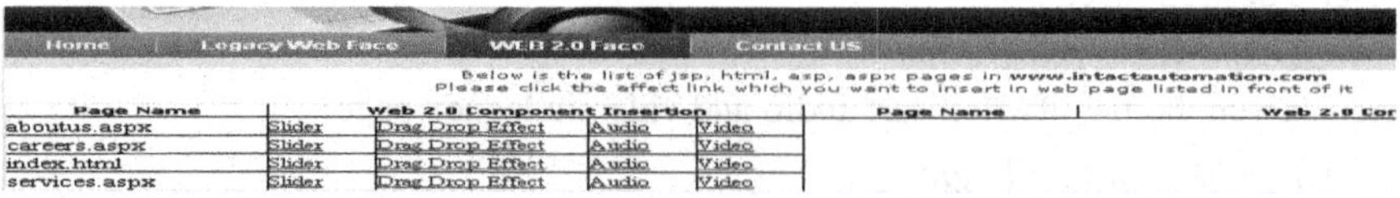

Fig. 7. Automatic effects addition to webpage

Table 1. Sample Results for web Analysis Module

Sr.	Guideline	www.solaskargr	www.phobos-	www.baramatiindust
	Control Flow Result			
1	The home page footer should include the URL for the home page	0 no. of pages have home page footer url	1 no. of pages have home page footer url	2 no. of pages have home page footer url
	Look And Feel			
1	Every page should have a title.	1 pages have not a title	1 pages have not a title	0 pages have not a title
	Accessibility Result			
1	Avoid animated gifs	3 animated gif present	0 animated gif present	16 animated gif present
	Reliability and Authenticity			
1	Error messages must be written in clear and	0 no. of pages with clear error messages	0 no. of pages with clear error	0 no. of pages with clear error
	Error			
1	Error messages must be written in clear and understandable language	0 no. of pages with clear error messages	0 no. of pages with clear error messages	0 no. of pages with clear error messages
	Testing			
1	Do the website have alternate text tags under graphics?	0 no. of graphics without alt tags	0 no. of graphics without alt tags	0 no. of graphics without alt tags

6   Conclusions

The results of the conducted experiments were satisfactory and encouraging: they showed the feasibility of the approach and a good effectiveness. Thus the proposed system is very helpful for development of web2.0 application from legacy web application rapidly and more interactively. Proposed system gives guidelines for good website development or enhancement in terms of performance and structure

References

1. Linaje, M., Preciado, J.C., Sanchez-Figueroa, F.: Adapting Web 1.0 User Interfaces to Web 2.0 User Interfaces through RIAs (2007)
2. Alt, F., Schmidt, A., Atterer, R., Holleis, P.: Bringing Web 2.0 to the Old Web: A Platform for Parasitic Applications
3. Kulkarni, R.B.: IEEE paper "Web Application migration towards Web 2.0". In: International Advance Conference on Computing, IACC 2009 (2009)
4. Kulkarni, R.B.: Webrean A Tool to analyse Re-engineering Web Application. In: International Conference on Web Engineering. ICWA, Bhoomneshwar (2006)
5. Ramdas, K.H., Kulkarni Raj, B.: Empirical Web Analysis Tool. In: International Conference at Sandip Foundation, Nashik (2010)

6. Yin, G.-S.: A Template-based Method for Theme Information Extraction from Web Pages. In: 2010 International Conference on Computer Application and System Modelling, JCCASM 2010 (2010)
7. Brambilla, M.: Business Process -based Conceptual Design of Rich Internet Applications. In: 2008 Eighth International Conference on Web Engineering. IEEE (2008)
8. Kayed, M., Chang, C.-H.: Member, IEEE "FiVaTech: Page-Level Web Data Extraction from Template Pages". Transactions on Knowledge and Data Engineering 22(2) (February 2010); Elissa, K.: Title of paper if known (unpublished)
9. Kim, C., Shim, K.: TEXT: Automatic Template Extraction from Heterogeneous Web Pages. IEEE Transactions on Knowledge and Data Engineering, doi:10.1109/TKDE.2010.140
10. Mesbah, A.: Migrating Multi-page Web Applications to Single-page AJAX Interfaces. In: 11th European Conference on Software Maintenance and Reengineering, CSMR 2007 (2007)
11. Linaje, M., Preciado, J.C., Sánchez- Figueroa, F.: Universidad de Extremadura, Engineering Rich Internet Application User Interfaces over Legacy Web Models. IEEE 1089- 7801/2007
12. Linaje, M., Preciado, J.C., Figueroa, F.S.: Designing Rich Internet Applications with Web Engineering Methodologies. IEEE, 11-4244-11450-4 (2007)
13. Spiros Sirmakessis "A Methodology for Evaluating the Personalization Conceptual Schema of a Web Application"
14. Marchetto, A.: Evaluating Web Applications Testability by Combining Metrics and Analogies. In: ITI 3rd International Conf. on, December 5-6 (2005)
15. Zdun, U.: Reengineering to the Web: A Reference Architecture. In: Sixth European Conference on Software Maintenance and Reengineering (CSMR 2002). IEEE (2002)
16. Di Lucca, G.A.: Recovering Conceptual Models from Web Applications. In: SIGDOC 2006, October 18–20, Myrtle Beach, South Carolina, USA (2006)
17. Claypool, K.T.: Re-WEB: Re-Usable Templates for Web View Generation and restructuring. In: CASCON 1998 Poster (1998)
18. Ping, Y., Kontogiannis, K.: Refactoring Web sites to the Controller-Centric Architecture. In: Eighth European Conference on Software Maintenance and Reengineering (CSMR 2004) 1534-5351/04. IEEE (2004)
19. Tsigereda, W.M.: A framework for evaluating Academic Websites quality, Master's Thesis Report in computer science track information architecture (August 2010)
20. Moreno, N., Fraternali, P., Vallecillo, A.: WebML modelling in UML. The Institution of Engineering and Technology (2007)
21. Jarrar, S.: Web Design Guidelines For WSDM, Dissertation submitted in view of obtaining a degree of Master of Science in Computer Science. Vrije Universiteit Brussel Faculty of Science Department of Computer Systems (2001-2002)
22. Di Lucca, G.A.: WARE: a tool for the Reverse Engineering of Web applications. In: 6th European Conference on Software Maintenance and Reengineering 2002. IEEE (2002)
23. White paper- "Enterprise Web 2.0 Modernization Solutions". Nexaweb Technologies, Inc. (2007)
24. Web Design Guidelines for WSDM. Dissertation submitted by Jarrar, S., Promoted by De Troyer, O. (2001-2002)
25. IEEE paper "A Framework for Website Assessment". Presented in "IEEE Melecon", May16-19, Benalmadena, Spain (2006)
26. Meiert, J.: Web Design: 15 Important Research Findings You Should Know
27. Ivory, M., Sinha, R.: Empirically validated web page design metrics

28. Ivory, M., Chevalier, A.: A study of automated web site evaluation tools
29. Preciado, J.C., Linaje, M., Sánchez-Figueroa, F.: An approach to support the Web User Interfaces evolution
30. Muller, P.-A., Studer, P.: Platform Independent Web Application Modeling and Development with Netsilon, Author manuscript, published in "N/P"
31. Hammond, J., Goulde, M.: Rich Internet Apps Move Beyond The Browser. For Application Development & Program Management Professionals (June 2007)
32. Cetin, S., Altintas, N.I., Oguztuzun, H., Dogru, A.H., Suloglu, S.: Legacy Migration to Service-Oriented Computing with Mashups. In: International Conference on Software Engineering Advances (ICSEA 2007) (2007)
33. Hsieh, M.-C., Kao, Y.-W., Yuan, S.-M.: Web 2.0 Toolbar: Providing Web 2.0 Services for Existence Web Pages, 978- 0-7695-3473-2/08 $25.00 ©. IEEE (2008)

A Bio-Chaotic Block Cryptosystem
for Privacy Protection of Multimedia Data

Musheer Ahmad[1], Sonia Gupta[2], and A.K. Mohapatra[2]

[1] Department of Computer Engineering, Faculty of Engineering and Technology,
Jamia Millia Islamia, New Delhi 110025, India
[2] Department of Computer Science & Engineering, GGS Indraprastha University,
New Delhi 110006, India

Abstract. In this paper, we propose a novel bio-chaotic block cryptosystem by combining the features of chaos-based cryptography and biometric system for providing inviolable data confidentiality, nonrepudiation and authentication. The proposed system involves the extraction of wavelet features out of iris biometric template of user, which on chaotic shuffling and mixing with chaotic logistic sequences generates bio-chaotic keys. The encryption is carried out in cipher-block-chaining mode to fetch strong confusion of plaintext data. The proposed cryptosystem offers two-level of security, as the generated keys are under the control of random secret key and biometric template as well. The NIST SP800-22 randomness tests of generated bio-chaotic keys and other simulation tests like histogram distribution, correlation, entropy, etc., of ciphertext data validate the high security performance and suitability of the proposed cryptosystem for protecting the privacy of user's sensitive data.

Keywords: Block cryptosystem, biometric template, bio-chaotic keys, wavelet transform, chaotic system.

1 Introduction

Biometric is defined as measurable physiological/behavioral characteristics of a person, which includes fingerprint, palmprint, iris, hand geometry, voice, face, and knuckle etc. [1]. Biometric based security has become a reliable and effective method for privacy protection, authentication and verification of user's identity and data. These methods have many inherent advantages over conventional cryptographic schemes like they are ever living and don't require to memorize secret keys, as the case is there with conventional cryptographic methods, where users have to memorize the long password strings. The biometrics data is unique, but it is not a cure-all because it involves some risks of being hacked and reused whenever it is sent over the network. Thus, they suffer from biometric-specific attacks [2]. If an attacker can interpret users biometric, then the biometric trait is lost forever and it may bring some frightful outcomes, since the attacker can masquerade as a genuine user [3,4]. Therefore, the most vital task in biometric systems is to protect securely the users biometric to withstand the intruder's attacks [5]. In the last one and half decades, the

S.C. Satapathy, S.K. Udgata, and B.N. Biswal (eds.), *FICTA 2013*,
Advances in Intelligent Systems and Computing 247,
DOI: 10.1007/978-3-319-02931-3_60, © Springer International Publishing Switzerland 2014

chaotic systems are extensively exploited by the researchers to design robust and strong encryption systems for multimedia data. The sequences generated from chaotic systems have inherent features of high sensitivity to initial conditions and system parameters, long periodicity, randomness and unpredictability [6,7]. These features of chaotic sequences are tapped, by the researchers, to protect the biometric template and make them difficult to decipher under various attacks [8]. The biometric-based security systems provide solutions by: (a) protecting biometric template, (b) binding a cryptographic key to biometric and (c) generating a biometric key from a biometric sample [3]. In order to design secure, reliable and robust cryptosystems, the researchers are making efforts to integrate the intrinsic worth of biometrics, cryptography and chaos to provide higher security than the existing systems [3,5,8,9,10].

In this paper, the inherent advantages of biometric systems, cryptographic systems and chaotic systems are integrated to propose a novel bio-chaotic block cryptosystem. The biometric features are extracted out of a biometric sample using discrete wavelet transform and combined with the chaotic secret key in secure manner. The formulated bio-chaotic keys are engaged to encrypt the multimedia data. The bio-chaotic cryptosystem is proposed with an aim to mitigate the biometric and cryptographic attacks. The proposed system offers two level of security and requires to possess the correct biometric sample and secret key for deciphering the multimedia data.

The rest of this paper is structured as follows: Section 2 gives the basic description of proposed bio-chaotic cryptosystem, whose performance is analyzed and discussed in Section 3. Finally, the conclusions are drawn in Section 4.

2   Proposed Bio-Chaotic Cryptosystem

2.1   Chaotic Logistic Map

The chaotic logistic map is employed as a source of generating sequences having high randomness for scrambling the extracted biometric key and encrypting the scrambled key to yield bio-chaotic keys. It also plays a significant role while encrypting the blocks of multimedia data along with bio-chaotic keys. The chaotic 1D Logistic map proposed by R. M. May [11] is one of the simplest nonlinear chaotic discrete systems that exhibit chaotic behavior, it is governed as:

$$x(n+1) = \lambda.x(n).(1 - x(n)) \qquad (1)$$

Where $x(0)$ is initial condition, λ is the system parameter and n is the number of iterations. The research shows that the map is chaotic for $3.57 < \lambda < 4$ and $x(n) \in (0, 1)$ for all n.

2.2   Proposed Cryptosystem

The steps of the proposed bio-chaotic block cryptosystem are as follows:

A.1. Acquire a biometric sample of user's iris and take its 2D discrete wavelet transform at *level* = 1.

A.2. Extract the *cH* and *cV* wavelet sub-bands (having size $m{\times}n$) coefficients. Let A_{cH} and A_{cV} be their average values.

A.3. Extract the binary features matrices F_1 and F_2 from two sub-bands as.

$$F_1(i,j) = \begin{cases} 1 & if \quad cH(i,j) \geq A_{cH} \\ 0 & if \quad cH(i,j) < A_{cH} \end{cases} \quad and \quad F_2(i,j) = \begin{cases} 1 & if \quad cV(i,j) \geq A_{cV} \\ 0 & if \quad cV(i,j) < A_{cV} \end{cases}$$

where $i = 1 \sim m$, $j = 1 \sim n$.

A.4. Combined the two features as $F = F_1 \oplus F_2$. Reshape F to a 1D sequence of biometric keys.

A.5. Take proper initial conditions for Logistic map defined in (1).

A.6. Iterate logistic map for *mn* times and record values $x(1), x(2), ..., x(mn)$.

A.7. Sort the vector $x(1), x(2), ..., x(mn)$ to get new vector $s(1), s(2), ..., s(mn)$. Find the position of values $s(1), s(2), ..., s(mn)$ in $x(1), x(2), ..., x(mn)$ and mark the positions as $P_v = \{v_1, v_2, ..., v_{mn}\}$.

A.8. Scramble the combined feature vector F through permutation vector P_v by shifting the value $F(k)$ to its new position $F(v_k)$, where $k = 1 \sim mn$. Let F_s be the scrambled vector.

A.9. Again, iterate the logistic map with initial condition $x(mn)$ for another *mn* times, record the chaotic values $x(mn+1), x(mn+2), ..., x(2mn)$ and binarize them with a threshold of 0.5 to produce chaotic binary sequence $B_c(k)$.

A.10. Protect biometric key sequence F_s using chaotic sequence B_c as $K_{bck} = F_s \oplus B_c$.

A.11. Decompose K_{bck} into blocks of bio-chaotic keys each of size ρ (= 64, 128, 256) as:

$$K_{bck} = K_{bck}(1), K_{bck}(2), ..., K_{bck}(nok) \quad where \quad nok = floor(mn/\rho)$$

A.12. Read the multimedia image P of size M×N to be protected.

A.13. Reshape the 2D image matrix P to a 1D array of image pixels as

$$P = p_1, p_2, ..., p_{MN}$$

A.14. Convert each gray pixel $p_i \in [0, 255]$ of image P into its binary equivalent and decompose it into blocks of size ρ as

$$pB = pB(1), pB(2), ..., pB(nob) \quad where \quad nob = (MN{\times}8)/\rho$$

A.15. Initialize new parameters as $t = 1$ and $cB(0) = 00.....0$.

A.16. Further iterate the logistic map and capture the chaotic value $x(2mn+t)$, preprocess it to generate a number $w(t) \in [1, nok]$ to randomly select a bio-chaotic key out of *nok* number of keys as:

$$w(t) = [floor\{x(2mn+t){\times}10^7\}]mod(nok) + 1$$

A.17. Encrypt the block $pB(t)$ of plain-image to obtained block of cipher-image as

$$cB(t) = pB(t) \oplus K_{bck}(w(t)) \oplus cB(t-1)$$

A.18. Set $t = t + 1$ and repeat the above steps from **A.16** to encrypt all the blocks of plain-image.

A.19. Execute the inverse of steps **A.14** and **A.13** for cB to obtain a cipher-image of size M×N.

The block diagram of the proposed cryptosystem is shown in Figure 1. The encryption procedure is followed in reverse order for the successful decoding of cipher-image. The user has to provide the genuine biometric sample of iris and the correct secret key to successfully decrypt the cipher-image.

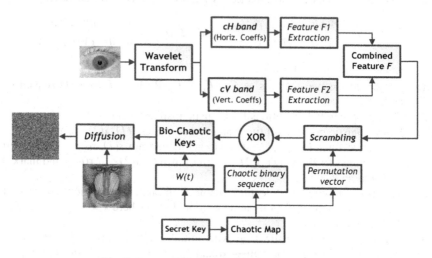

Fig. 1. Proposed bio-chaotic block cryptosystem

3   Results

The biometric iris sample depicted in Figure 2(a) is used to extract the biometric keys. The initial values used for the experimentation are: $x(0)=0.49731$, $\lambda=3.97$, $\rho=128$. The extracted bio-chaotic keys are depicted as an image in Figure 2(b). To quantify the quality of generated bio-chaotic keys, NIST SP800-22 randomness tests such as frequency test, block frequency test, cusum tests, runs test, longest runs test, rank test, FFT test, linear complexity test, serial test, overlapping template test are applied to evaluate the *p-value* corresponding to each tests. If a *p-value* for a test is determined to be equal to 1, then the sequence appears to have perfect randomness. A *p-value* of zero indicates that the sequence appears to be completely non-random [12]. A significance level (α) can be chosen for the tests. If *p-value* $\geq \alpha$, then the sequence appears to be random. If *p-value* $< \alpha$, then the sequence appears to be non-random. Typically, α is chosen in the range [0.001, 0.01]. An $\alpha = 0.01$ indicates that one would expect 1 sequence in 100 sequences to be rejected. A *p-value* ≥ 0.01 would mean that the sequence would be considered to be random with a confidence of 99%. The randomness tests are applied with $\alpha = 0.01$ and results are depicted in Table 1. It is evident that the bio-chaotic keys under examination have passed all the above mentioned randomness tests. The *p-value* obtained in each tests is greater than chosen α, which confirms the high randomness of generated bio-chaotic keys.

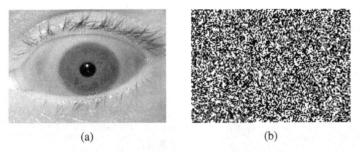

(a)   (b)

Fig. 2. Biometric sample and bio-chaotic binary keys

Table 1. NIST SP800-22 randomness tests results

Test Type	p-value	Result
Frequency Test	0.634418	Pass
Block Frequency Test	0.081754	Pass
Cusum-Forward Test	0.775018	Pass
Cusum-Reverse Test	0.681685	Pass
Runs Test	0.366369	Pass
Longest Runs Test	0.295889	Pass
Rank Test	0.313674	Pass
FFT Test	0.232316	Pass
Linear Complexity Test	0.808827	Pass
Serial Test	1.000000	Pass
Overlapping Template	0.957612	Pass

The standard multimedia *Baboon* image of size 512×512 shown in Figure 3 is adopted for the simulation to showcase the performance of proposed cryptosystem. The encrypted *Baboon* image is shown in Figure 4. It is visually evident from Figure 3(b) and 4(b), that the distribution of pixels in cipher-image is quite uniform.

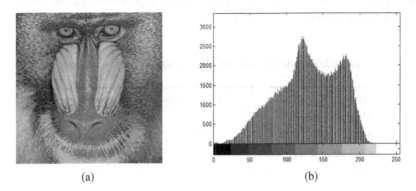

(a)   (b)

Fig. 3. Baboon plain-image and its histogram

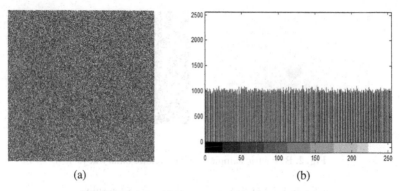

(a)   (b)

Fig. 4. Encrypted *Baboon* image and its histogram

The standard deviation in frequencies of pixel intensities determines the flatness/uniformity of the histogram (distribution of pixels gray-level intensities) of cipher-image. Lower the value of this deviation, more flat/uniform will be the histogram, means it will be more close to the histogram of perfect white noise-image. The scores of this deviation for the two images are found as 868.17 and 32.72. This means that the histogram shown in Figure 4(b) is quite flat and the pixels are uniformly distributed in image of Figure 4(a). The average of gray-level intensities of plain-image and cipher-image come out as 129.69 and 127.49, respectively. The average value obtained for cipher-image is closer to the ideal value = 127.5 (of 8-bits gray-scale perfect white noise image). The pixels percentage difference between the plain-image and encrypted image is found as 99.61%, which shows that the cipher-image is totally distinct from its plain-image. The correlation analysis of adjacent pixels is also quantified. The scores of correlation coefficients for randomly chosen 5000 adjacent pixels of two images are 0.7683 and -0.0019, respectively. It clearly shows that the pixels of cipher-image are highly uncorrelated to each other as the score is very close to the ideal value 0. The plot of correlation shown in Figure 5(b) also confirms the high uncorrelation of adjacent pixels in cipher-image. The information entropy analysis of the images reveal that the image of Figure 4(a) has entropy of 7.99927, which is much close to its ideal value=8 (for an 8-bits gray-scale perfect white noise image); the value 8 corresponds to a perfect random source. The user's biometric sample, $x(0)$, λ and ρ constitutes the complete secret key of the proposed cryptosystem. The statistical measures evaluated are provided in Table 2.

Table 2. Statistical measures of plain-image and its cipher-image

Measure	Plain-image	Cipher-image
Mean gray value	129.68	127.49
Std Deviation	868.17	32.72
Corrélation Coefficient	0.7683	-0.0019
Entropy	7.35786	7.99927

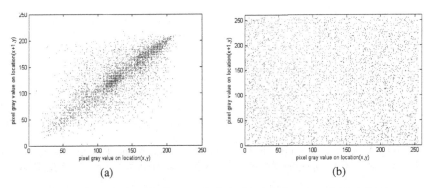

Fig. 5. Plots of correlation of adjacent pixels in (a) plain-image and (b) cipher-image

4   Conclusion

The inherent features of biometric system, cryptographic systems and chaotic systems are integrated in a novel manner to propose a bio-chaotic block cryptosystem. The cryptosystems inherits the features to withstand the biometric-specific and cryptographic-specific attacks. The keys for blocks encryption of plaintext data are dynamically selected out of numbers of bio-chaotic keys for higher security reasons. The proposed system is experimented with standard multimedia image to report and quantify the security performance. The quantified statistical measures demonstrate the high security performance of the proposed system and its practicableness for the privacy protection of user's data.

References

1. Jain, A.K., Pankanti, S., Bolle, R.: Biometrics: personal identification in networked society. Kluwer (1999)
2. Khan, M.K., Zhang, J., Tian, L.: Chaotic secure content-based hidden transmission of biometric templates. Chaos, Solitons & Fractals 32, 1749–1759 (2007)
3. Li, P., Yang, X., Qiao, H., Cao, K., Liu, E., Tian, J.: An effective biometric cryptosystem combining fingerprints with error correction codes. Expert Systems with Applications 39, 6562–6574 (2012)
4. Jin, A.T.B., Ling, D.N.C., Goh, A.: Biohashing: two factor authentication featuring fingerprint data and tokenized random number. Pattern Recognition 37(11), 2245–2255 (2004)
5. Imamverdiyev, Y., Teoh, A.B.J., Kim, J.: Biometric cryptosystem based on discretized fingerprint texture descriptors. Expert Systems with Applications 40, 1888–1901 (2013)
6. Ahmad, M., Ahmad, T.: A framework to protect patient digital medical imagery for secure telediagnosis. Procedia Engineering 48, 1055–1066 (2012)
7. Patidar, V., Pareek, N.K., Sud, K.K.: A new substitution diffusion based image cipher using chaotic standard and logistic maps. Communication in Nonlinear Science and Numerical Simulation 14, 3056–3075 (2009)

8. Khan, M.K., Zhang, J.: Implementing templates security in remote biometric authentication systems. In: IEEE Conference Proceedings on CIS, vol. 2, pp. 1396–1400 (2006)
9. Maiorana, E.: Biometric cryptosystem using function based on-line signature recognition. Expert Systems with Applications 37(4), 3454–3461 (2010)
10. Wu, X., Wang, K., Zhang, D.: A cryptosystem based on palmprint feature. In: International Conference on Pattern Recognition, pp. 1–4 (2008)
11. May, R.M.: Simple mathematical model with very complicated dynamics. Nature 261, 459–467 (1967)
12. Rukhin, A., et al.: A Statistical Test Suite for Random and Pseudo-random Number Generators for Cryptographic Applications. NIST Special Publication, 800–822 (2001)

Some Popular Usability Evaluation Techniques for Websites

Sharmistha Roy and Prasant Kumar Pattnaik

School of Computer Engineering, KIIT University, Bhubaneswar, India
{sroyfcs,patnaikprasantfcs}@kiit.ac.in

Abstract. Usability is widely considered to be an essential attribute for fulfilling the quality requirements of any software product. In context to web based products especially web sites, usability is the process for making the navigational prospects more user friendly and task effectiveness to meet user's satisfaction. If any website fails to navigate a user to his respective goals, then the user loses his interest and considers it as a less usable website. For simplicity, if any software fails to meet user's need and satisfaction then it is considered as less usable and poor productive. This paper aims to survey on some usability evaluation techniques and factors that are helpful for usability measurement of web and intelligence computing.

Keywords: Usability, Websites, Accessibility, Testing.

1   Introduction

In this competitive IT world, web intelligence and computing discipline addresses issues related to quality, delivery and trust, especially Quality of Service (QoS) of any software model or software products, where usability plays a significant role in order to fulfill user's satisfaction and requirements. Usability may be defined as the ease of using any software effectively, capability of learning a new system/ software with user's interest and fulfilling user's satisfaction. While measuring the quality of software products, usability measurement is required in order to identify the usefulness and accuracy of that product. However, nowadays World Wide Web gains its popularity because of fulfilling user's satisfaction by providing different services and application through websites. In this context, usability is of course considered to be a key issue in order to evaluate user's satisfaction, navigational efficiency and general preferences. It has no explicit criteria to measure it because it can only be measured by experts, research scholars and end users. Usability measurement provides several benefits which include, improving the quality of software products, increasing the user's satisfaction with the website and the product, to cultivate the confidence and trust for the websites and the product, improve the efficiency and reliability of software products, to help the users to achieve their goals effectively, reduces errors and increases the user's interest for using the product or the website.

The paper has been organized as follows. In section 2, we presented a brief history of usability. Section 3, provides the attributes which are necessary for web based

S.C. Satapathy, S.K. Udgata, and B.N. Biswal (eds.), *FICTA 2013*,
Advances in Intelligent Systems and Computing 247,
DOI: 10.1007/978-3-319-02931-3_61, © Springer International Publishing Switzerland 2014

applications. Section 4 and 5, cites important notes related to accessibility and description of the usability metrics. In Section 6, we include the different usability evaluation methods. Lastly, section 7, concludes the work.

2   A Brief History

Miller in 2006 [1] carried out a study on usability testing, where he considers usability testing as quality assurance attribute. He mentioned that usability testing should be performed at the initial stage in order to avoid big problems at the end. The author focuses on how good usability testing method can be implemented so as to make the software product faster and to make navigational prospect of a website easier and quicker.

Fang Liu, in 2008 [2] discussed the methods how to use the usability evaluation techniques to evaluate or measure a website. He mentioned that the evaluation techniques have a vital role to give good quality usability evaluation. He focuses on the implementation of the different usability evaluation methods namely, heuristic evaluation, cognitive walkthrough, focus group, questionnaire etc.

Lodhi, A. in 2010 [3] conducted a study on usability testing where the objective of the research is to consider usability heuristics as an assessment parameter for usability testing. To fulfill the aim, the experiment was conducted on some public sectors universities websites where ten numbers of usability heuristics, proposed by usability expert Jakob Nielsen, was chosen for usability assessment and heuristics are considered to be key attribute to access usability of websites. Some academic websites are also rated based on this usability heuristics.

Hasan, L. conducted usability experiment for evaluating nine Jordanian university websites in 2012 [4]. Evaluating criteria was based on some attributes say navigation, architecture / organization, ease of use and communication, design and content. For the evaluation purpose, 237 students have participated to rate the websites by performing the usability test and the observed report states that many students are satisfied with all the above task effectiveness attributes except the design issues.

3   Web Based Usability Attributes

Web usability can be defined as the capability how a naïve user can navigate in a website effectively with the capability of the web application to provide tasks with efficiency and meet user's satisfaction. Famous usability expert Jakob Nielsen, in his book entitled 'Usability Engineering' [5] provides five attributes to measure usability. Thus according to Web usability there is a little difference in the different attributes with respect to the former one [8]. According to Nielsen, the five usability attributes are namely, Efficiency, Learnability, User's Satisfaction, Memorability and Few errors which are discussed below:

Efficiency. Efficiency is popularly defined as how effectively a user can perform in a system with increasing the productivity and learning capacity of a user. Whereas, web application efficiency may be defined as the capability of web application to provide quick link so that a user can navigate easily for fulfilling his requirements.

Learnability. Learnability may be defined as the ease to learn the functionalities of a new system or software. This attribute is most essential for users using the software for first time. Web application learnability is defined as the ease for users to understand the content provided by the web application, so as to navigate easily and retrieve desired information.

User's Satisfaction. User's satisfaction is one of the key attribute that measures the comfort level through the system and is accepted for further use. In context of web based application user's satisfaction is a vital attribute which should make the user feel that it is capable of providing desired information effectively while making the navigation procedure quicker and easier.

Memorability. Memorability is defined as the measure of proficiency with which the user can remember the use of software after a long period of not using it. In terms of web based application, memorability is defined as the capability of user to recollect the specific content with least navigation time when he is not a frequent visitor to that website.

Few Errors. Few errors is a quality attribute which intends to measure the severity of error done by a user and how to minimize those errors. To reduce errors the system should be precise so as to enable a user to do his work without any error and if such errors are present also it should be able to recover quickly. Moreover, every web application should provide least error rate so that the users are provided with correct and specific information without diverting to unsolicited pages.

Neilson probability cited all the important attributes concern to usability. But there are some specific groups of user not getting scope out of it. For example, people who are physically challenged and a group of users who are not having a particular device like laptops, computers. In this work we introduced two attributes which are additional attributes to Neilson model for providing greater flexibility.

Device Independence. Device independence may be defined as the capability of software application to work on a wide variety of devices with little or no modification to achieve flexibility of use. In this IT world, people prefer to use mobile devices at any instant of time to explore and gain more and more knowledge's and to keep up-to-date to the recent technologies. Moreover, every web application should have the ability to function on any devices including mobile phones so that people from any corner of the world can navigate through the websites to gather information effectively and meets user's satisfaction.

Provision for Physically Disabled Person. Physically challenged or disability is an umbrella term which indicates the limitation to mobility, obstruction to perform any activity or any participation restriction. In today's world, communication has become quite faster that everyone irrespective of being physically challenged also, should not be restricted to explore latest information and to keep up-to-date by recent technologies. It will be better if every software product and specially web based applications may provide at best, the facility to be accessed by all people. The web page content developer may include the required information by which the physically challenged people get benefit out of that. The web pages may be furnished with required infrastructure so as to help those groups of users to update themselves. So, usability measurement is very essential in order to meet satisfaction level of physically challenged people.

4   Accessibility

Accessibility evaluation of website may be defined as the degree to which the web application is used by all types of users and technologies. Alternatively, accessibility is the ability of web application to support users to navigate from one page to another and simultaneously retrieve information easily and successfully. W3C WAI has listed a set of tools that can be used for accessibility evaluation of websites. However, this evaluation tool help in broadly determining if a web page conforms to Web Content Accessibility Guidelines (WCAG 2.0), passing the tool based checks does not necessarily guarantee accessibility. They should always be complemented with Manual accessibility testing which is a relatively more accurate method for determining accessibility.

The main goals of WCAG are: To provide text equivalent alternatives of non-text content, providing navigation links and orientation mechanisms for easy access and browsing, proper utilization of markup documents and style sheets to improve accessibility, web pages should support older technologies even when it is implemented with newer technologies, and document that is implemented should be device independent.

5   Usability Metrics

There are no specific techniques or tools so far seem to be implemented to evaluate the user's satisfaction level. However, many researchers have cited that the usability in many cases is dependent on the users who are going to use that website. The following are the metrics for measuring usability of websites or software:

Findability Rate/ Completion Rate. Completion rate is defined as the fundamental criteria of usability metrics. It is typically recorded as binary value (1 for Task success rate and 0 for Task failure rate) whether a user is capable of finding an information or completing a task.

Usability Problem. Usability problem describes the difficulties encountered by a user during his use of any software or website. It describes the probability of problem discovered rates and the undiscovered problems. It forms a key metric for discovering the usability activity impact.

Task Time. Task time is one of the usability metrics that may be defined as the total time taken to complete a particular task by a user. Task time is used to measure the efficiency and productivity rate of a website or software.

Task Level Satisfaction. Task level satisfaction measures the satisfaction level of a user while completing a task. It finds out the difficulty level of a task by asking question to the users after the user has attempt for that task. The difficulty level is measured by comparing the task with other task in the database.

Test Level Satisfaction. Test level satisfaction is calculated after the completion of usability test, to check how many participants have answered the questions asked about the overall ease of use of a software or website. For measuring the Test Level Satisfaction of software we may use SUS (System Usability Scale), which consists of ten items questionnaire with five response actions. And for measuring the Test Level Satisfaction of website we may use SUPR-Q (Standardized Universal Percentile Rank Questionnaire) which consists of thirteen items that measures the four essential items used for making a website successful.

Errors. Errors are defined as any unintended actions, mistakes made by a user while attempting a task. It provides an excellent diagnostic report, if possible should be linked to Usability Inspection (UI) problems.

Expectation. Expectation of users is based on the difficulty level of the task, which is based on the clues of the task scenario, they observe while attempting a task. It helps to diagnose the problem domain.

Page Views/ Clicks. Page views/ clicks are one of the essential usability metric which may be used to find out the task success rate or failure while accessing webpage's or websites. It is considered to be a better measure of efficiency.

Conversion. Conversion is also another measure of usability metrics which is used to measure the effectiveness of a product. It is mainly used in ecommerce. It is also represented as binary measure (0 for not converted, 1 for converted), which is captured in all stages of sales from landing of the product to purchase.

Single Usability Metric. Single Usability Metric is a single summated standardized metric which encapsulates four usability metrics namely: task completion rate, average number of errors, average time on task and post-task satisfaction. This may be evaluated mathematically as: sum of efficiency, effectiveness and satisfaction i.e. Single Usability Metric = Efficiency + Effectiveness + Satisfaction.

6   Usability Evaluation

Usability evaluation is considered to be an essential part of system development process. It includes user's satisfaction while using the product or navigating through the websites and provides the strength and weakness report of the product or website. In broad way, we can classify usability into two ways namely, Usability Inspection and Inquiry of a product, which do not include user's active participation and, Usability Test of a product or website which include the active participation of users [7] and details of each are discussed below.

6.1   Usability Inspection and Inquiry

Usability inspection technique focuses on usability experts/ specialists, and sometimes software developers or professionals to examine the usability of user-interface design. It is performed before usability test to detect or resolve obvious problem. Widely used UI methods are Real-time Walkthrough, Experts Judgments, Heuristic Evaluation, Surveys, Open or Closed Questionnaires, Observation, Plurastic Walkthrough, Outcomes Inspection, Feature Inspection which are described below:

Real-Time Walkthrough. Real-time walkthrough is generally a task specific technique used to identify the usability issues in human-computer interaction system. The evaluator walks from the root and goes through each step and finds out the potential problems or issues that a user may face [15]. The system is redesigned to solve those issues.

Experts Judgments. Expert usability judgement provides an immediate and concrete tactical analysis of the user's experience to make the product or website familiar and also to improve the flaws present in the current system.

Heuristic Evaluation. In heuristic evaluation, usability experts thoroughly go through the user interface design and finds out the problems and compare with the accepted heuristics. It then lists the potential usability issues and provides ranking based on their observations [14]. It provides quick and relatively inexpensive feedback to the designers and helps in corrective measures.

Surveys or case study. Surveys or case study are considered to be structured questionnaire or information that is drawn from the conclusion. It aims to bring out the problems that are faced while using the system or website [13].

Open or closed Questionnaires. Open or closed questionnaires are some set of questionnaires which may be generated from pilot survey or from expert's views to assess or evaluate the usability of any software or website. There are various standard forms of questionnaires which include WAMMI, QUIS, ASQ [12] etc.

Observation. Observation is a process where the evaluators observes users workplace to understand how the users are performing their task, which tool they are using, what kind of mental model the users have about the system, etc. [11].

Pluralistic Walkthrough. In plurastic walkthrough, a group of users or developers or human factors engineers meet together to analyze a set of tasks, and evaluate the usability of a system. Group walkthroughs provide a wide range of skills and perspectives to endure the usability problems [10].

Outcomes Inspection. Outcomes inspection is based on three perspectives: novice use, expert use and error handling. Here inspection is observed on the well defined task scenarios which finally produce predictable inspection results [9].

Feature Inspection. Feature inspection technique focuses on the features of a product or website. Here each feature is analyzed for its availability, understandability and other functionality aspects [14].

6.2   Usability Testing

Usability testing is a technique, in which a product is tested by representative users, and the evaluators use the result to identify any usability problems, collect quantitative data on participant's performance (error rate, time on task etc.) and to determine whether the product or website can fulfill user satisfaction or not. There are few guidelines that are always kept in mind while performing usability test. Some of them are as follows [7]:

Assurity to the User. It has to be ensured to the users that while performing the usability test we are testing the site not the user's ability.

Measures Taken. Two measures should be considered: first one is performance measures, where the success rate, time completion and error rates are considered. And the second one is subjective measures which considers users satisfaction level and comfort rate.

Fact Findings and Analysis. The findings of usability test should be considered if necessary to make any change in the project prototype or website for user's satisfaction.

Alternative Solution to a Problem. The best solution of every project may be found out by using limited resources, within constraint budget and time.

With usability testing the aims is to gather the information's whether the users are satisfied with the product or website or any errors are discovered and there is a need for improvement. There are many testing methods [6] which are as follows:

Hallway Testing. Hallway testing is a general method of usability testing where a group of trained tester is brought to test the product or website in the early stage of design to find out problems which are so serious that a user cannot advance.

Remote Usability Testing. Remote usability testing is a kind of usability testing where the users, evaluators, developers are located in different places and the users can perform the test from their home or workplace. Advantages of remote usability testing are that it reduces the time of travel, reduces the overhead of the organization, provides a natural environment to the users and provides the facility to test the website with different computer configurations thus enables to find out the flexibility of the website.

Expert Review. Expert review is a process where an expert is asked to evaluate the usability of a website or a software product to detect the potential usability bugs. The reports from the experts helps to straighten up the design for usability testing during the early stage of implementation of the website or product and also give suggested solutions for resolving existing usability issues.

Paper Prototype Testing. Paper prototype testing is one of the usability testing which consists of creating hand-sketched and rough drawing of the user-interface model or prototype. Depending on the user performance on that prototype model leads to the actual implementation of that user-interface model. This type of testing is cost effective, reduces the expense of involving many users and also diminishes the overhead of unnecessary coding.

Questionnaires and Interviews. Interviews and questionnaires provide the easiest method to collect structured data regarding user's opinion on performing the task. Basically there are different types of questionnaires and interviews method namely:

Pre-test Questionnaire. Pre-test questionnaire process includes the personal attributes of user for example, age, sex, experience in computer work, etc. which would be relevant and helpful to their performance on the test. Following that, the participant will be provided instruction set how to perform the task successfully.

Post-test Questionnaire. Post-test questionnaire process is performed after successful completion of the task given to a participant. Here, the participants will provide their views and opinions regarding the completion of the tasks. Apart from subjective questions the participants are asked to rate different questions regarding the websites layout, how usability the website is, how easily they have carried out the task, for which task the user finds difficulty, overall impression of the website, websites capabilities etc.

7   Conclusion

Usability is gaining popularity day by day among the industries as well as researchers since it is associated with the user satisfaction. Usability possibly takes the responsibility of measuring the rate of interest of all groups of users during design of a new web intelligence based product. This consideration may be helpful to the designer to improve their design paradigm. However, usability is the attribute that has to be evaluated to measure the intelligence of websites while performing the task in order to achieve the user's needs. In this paper we summarize usability evaluation techniques as different inspection based methods and testing methods which are popularly adopted. This paper focused on our two new usability attributes namely physically disabled person and device independent which adds flavors to product orientation. More evaluation methods and accessibility guidelines are needed for the improvement of usability issues as future scope of work.

References

1. Miller, J.: Usability Testing: A Journey, Not a Destination. IEEE Internet Computing 10(6), 80–83 (2006)
2. Liu, F.: Usability evaluation on websites. In: 9th International Conference on Computer-Aided Industrial Design and Conceptual Design, CAID/CD, China, pp. 141–144 (2008)
3. Lodhi, A.: Usability Heuristics as an assessment parameter: For performing Usability Testing. In: 2nd International Conference on Software Technology and Engineering (ICSTE), USA, pp. 256–259 (2010)
4. Hasan, L.: Evaluating the usability of nine Jordanian university websites. In: 2nd International Conference on Communications and Information Technology (ICCIT), Tunisia, pp. 91–96 (2012)
5. Nielsen, J.: The Usability Engineering Lifecycle. IEEE Computer 25(3), 12–22 (1992)

6. Gardner, J.: Remote Website Usability Testing-Benefits over traditional methods. International Journal of Public Information Systems 3(2), 63–72 (2007)
7. Usability: Improving the User Experience, http://www.usability.gov
8. Nielsen, J.: Multimedia and Hypertext: Internet and Beyond. Academic Press, London (1995)
9. Zhang, Z., Basili, V., Shneider-Man, B.: An empirical study of perspective-based usability inspection. In: Proceedings of the Human Factors and Ergonomics Society 42nd Annual Meeting, Chicago, pp. 1346–1350 (1998)
10. Bias, R.: The Pluralistic Usability Walkthrough: Coordinated Empathies. In: Nielsen, J., Mack, R. (eds.) Usability Inspection Methods, ch. 3, pp. 63–67. John Wiley (1994)
11. Nielsen, J.: Usability Engineering. Academic Press, Boston (1993)
12. Usability Evaluation, http://www.usabilityhome.com
13. Alreck, P.L., Settle, R.B.: The survey research handbook: Guidelines and strategies for conducting a survey, 2nd edn. Irwin, Burr Ridge (1994)
14. Nielsen, J.: Heuristic evaluation. In: Nielsen, J., Mack, R.L. (eds.) Usability Inspection Methods, pp. 25–64. John Wiley & Sons, New York (1994)
15. Wharton, C., Rieman, J., Lewis, C., Polson, P.: The cognitive walkthrough method: A practitioner's guide. In: Nielsen, J., Mack, R.L. (eds.) Usability Inspection Methods, pp. 105–140. John Wiley & Sons, New York (1994)

A Secure Lightweight and Scalable Mobile Payment Framework

Shaik Shakeel Ahamad[1], Siba K. Udgata[2], and Madhusoodhnan Nair[1]

[1] K.G.Reddy College of Engineering and Technology, Chilkur Village,
Moinabad Mandal, Ranga Reddy District-501504, India
[2] School of Computer and Information Sciences, University of Hyderabad, Hyderabad, India
ahamadss786@gmail.com, udgatacs@uohyd.ernet.in,
principal@kgr.ac.in

Abstract. Existing SIP-based mobile payment solutions do not ensure all the security properties. In this paper we propose a Secure Lightweight and Scalable Mobile Payment Framework (SLSMP) using Signcryption scheme with Forward Secrecy (SFS) based on elliptic curve scheme which combines digital signature and encryption functions (Hwang et al., 2005) [5]. It takes lower computation and communication cost to provide security functions. SLSMP is highly scalable which is attributed to SIP for data exchange. This paper uses WPKI, UICC as Secure Element and depicts system architecture and detailed protocol of SIP based mobile payment solution. Our proposed framework is suitable for both micro and macro payments. Our proposed protocol ensures End to End security i.e. ensures Authentication, Integrity, Confidentiality and Non Repudiation properties, achieves Identity protection from merchant and Eavesdropper, achieves Transaction privacy from Eavesdropper and Payment Gateway, achieves Payment Secrecy, Order Secrecy, forward secrecy, prevents Double Spending, Overspending and Money laundering.

Keywords: Mobile Payments, SIP (Session Initiation Protocol), ECDSA, Signcryption scheme with Forward Secrecy (SFS), WPKI.

1 Introduction

Mobile Commerce has become an integral part of today's ecosystem and it is one of the fastest growing sectors in the Internet. Existing mobile commerce solutions do not ensure end to end security. Recently, many mobile commerce protocols using public-key cryptography have been proposed but the signatures generated cannot be considered as qualified signatures as they are generated in the memory of mobile phone which is not a tamper resistant hardware. Transaction level security must ensure end-to-end security with message integrity, confidentiality and non-repudiation. In order to ensure this Wireless PKI is used and is normally implemented on the mobile phones but implementing WPKI in mobile phones has serious limitations such as secret keys stored in the memory of Mobile Phone could be infected by viruses or can be maliciously replaced. So we propose WPKI on the

S.C. Satapathy, S.K. Udgata, and B.N. Biswal (eds.), *FICTA 2013*,
Advances in Intelligent Systems and Computing 247,
DOI: 10.1007/978-3-319-02931-3_62, © Springer International Publishing Switzerland 2014

Secure Element (SE) of the mobile phone, SE is a UICC which is a smart card and is a well trusted and tamper proof device thus UICC card can be used for security critical applications such as Mobile Banking/Commerce. So we need a framework which is fast, scalable, lightweight and should ensure all the security properties. In this paper we propose a Secure Lightweight and Scalable Mobile Payment Framework which is fast, scalable, lightweight and should ensure all the security properties. In this paper we adopt Signcryption Scheme with forward secrecy (SFS) based on Elliptic curve combines digital signature and encryption functions (Hwang et al., 2005) [5] for ensuring security properties and to consume fewer resources. This scheme takes lower computation and communication cost to provide security functions. SFS not only ensures message confidentiality, authentication, integrity, unforgeability, and non-repudiation, but also forward secrecy for message confidentiality and public verification. In this scheme, the judge can verify sender's signature directly without the sender's private key when dispute occurs. This scheme can be applied to mobile communication environment more efficiently because of the low computation and communication cost. Scalability and fastness are achieved by adopting Session Initiation Protocol (SIP) [9] which is an open signaling protocol standard developed by the Internet Engineering Task Force (IETF) in cooperation with many industry leaders for establishing, managing, and terminating real-time communications over large IP-based networks, such as the Internet. Communications via voice, video, or text (instant messaging), may take place using any combination of SIP-enabled devices, such as a mobile phone or a wireless handheld device or PDA. SIP is an application layer peer-to-peer communication protocol for establishing, manipulating, and tearing down communication sessions. SIP protocol is extensible i.e. developers can easily write custom applications for SIP to accommodate video, instant messaging, it is simple to develop and quick to deploy customized applications using SIP. SIP is a refreshing solution that can simplify and enhance your communications capabilities. We adopt WPKI, UICC in our framework. UICC personalization is done as given in [7] and [8]. The rest of the paper is organized as follows. Section 2 presents recent related works. In Section 3, we present the proposed SLSMP framework. Section 4 presents the comparative analysis of our proposed framework with related works. Section 5 presents conclusions and future works.

2   Related Works

Following are the limitations in the existing literature of SIP based mobile payments

In [1] Customer and Merchant has to trust PG, Prior to the start of the protocol both Customer and Merchant should exchange their session keys, It does not ensure non repudiation property. [2] Adopts SET protocol for Mobile Payments which is a not a light weight protocol. [3] Does not ensure non repudiation property and in [4] prior to the start of the protocol both Customer and Merchant should exchange their session keys.

3   Proposed Secure Lightweight and Scalable Mobile Payment Framework (SLSMP)

a) **Customer:** It acts as one of SIP UAs including User Interface, SIP IM module, security module and payment module.

    i) **User Interface:** It provides an input device (e.g., keyboard) for user to type in purchase information and an output device (e.g., screen) to display response.

    ii) **SIP IM Module:** This module supports SIP instant message interactions over internet. It is implemented according to RFC 3428. It allows customer to communicate with merchant by instant messages.

    iii) **Security Module:** UICC is the Secure Element used in this framework which is a generic platform for smart card applications. It has been standardized by ETSI EP SCP (ETSI Project Smart Card Platform). The UICC can host a number of different applications, either from the UICC issuer or from other parties, each defining and controlling its own application(s). It employs General internet security protocols such as (IPSec, TLS and S/MIME). Key management is done as given in [7] & [8]. Messages are sent and received in signcrypted format.

    iv) **Payment Module:** It checks OI and Amount for the selected OI and generates PI encrypted using shared symmetric key between client and issuer.

b) **Merchant:** It is also another of SIP UAs including SIP IM module, verification module and payment module.

    i) **SIP IM Module:** it is similar to the SIP IM module on the merchant side, which allows merchant to communicate with each component by instant messages.

    ii) **Security Module:** It is another tamper resistant Hardware Security Module (HSM) used for keeping merchants private key and other credentials such as PIN and certificates.

    iii) **Order Verification Module:** It verifies the availability of OI and checks the total amount.

    iv) **Payment Module:** It forwards encrypted PI to the Acquirer in order to far ward it to PG.

c) **Issuer:** It works as a SIP UA but located on the server side. It communicates with customer accomplish the transaction. Its main components can be listed as follows.

    i) **Payment Verification Module:** Decrypts the PI using the symmetric key shared between Issuer and Client, Checks the clients account for sufficient funds, Checks if $HOI_M = HOI_C, TID_M = TID_C, Amt_M = Amt_C$, Checks if Timestamps $T_C = T_M$, Checks if nonces $N_C = N_M$

ii) **Archives Module:** Bank Server securely archives digital signatures received and sent by it in its archives, logs all the changes in the certificate status i.e. activation, suspension, termination of suspension and revocation) and all validity confirmations given by OCSP and CRL responder. Log records are cryptographically linked to prevent insider attacks and forging log records (for example backdating the log record). Once a month, cryptographic hash is printed in newspaper. Log record database is backed up on timely basis with three copies stored in different locations.

iii) **SIP IM Module:** This module supports SIP instant message interactions over internet..

iv) **Security Manager/ Module:** The Security Manager is responsible for authentication (using password), signing the message, certificate management, secret key management and distribution. Security Manager encrypts and decrypts the messages, generates and verifies digital signatures, generates shared symmetric keys for all the clients, maintains password and biometric data in its database. Certificate Manager maintains all the certificates of the client. Bank generates and keeps its private key and shared symmetric keys in Tamper Resistant HSM (Hardware Security Module). All the client Mobile numbers are mapped to Certificates and Account numbers. It performs the same function as the security module on customer and merchant.

d) **Acquirer:** It works as a SIP UA but located on the server side. It communicates with customer and merchant to accomplish the transaction. Its main components can be listed as follows.

i) **Order Information Verification Module:** It verifies signcryption message received from Merchant and verifies Hashed OI sent by Client, Merchant and Amount.

ii) **Archives Module**: Same as that of Issuer

iii) **SIP IM Module**: This module supports SIP instant message interactions over internet.

iv) **Security Manager/ Module**: Same as that of Issuer.

e) **PG:** It works as a SIP UA but located on the server side. It communicates with Issuer and Acquirer to accomplish the transaction. Its main components can be listed as follows.

i)   **Order Information Verification Module:** Same as that of Acquirer.

ii)   **Archives Module:** Same as that of Issuer

iii)   **SIP IM Module:** This module supports SIP instant message interactions over internet.

iv)   **Security Module:** It communicates with only Issuer and Acquirer through Private Banking Network which is very secure. Messages exchanged through Private Banking Network are not exchanged in Signcryption Format. So Hardware Security Module is not installed in PG.

Proposed SLSMP Protocol

$$Step1: C \rightarrow M : SIG_{C_M}(MS1), Pubkey_c$$
$$MS1 = (PI)_{K_{ci}}, HOI_C, OI, TID_C, MID, N_c, T_c, Amt_C$$

Client gets all the merchant details (public key parameters, MID (merchant Identity) from his portal and selects items from the portal and makes Order Information (OI) and sends MS1 to Merchant in Signcrypted form.

$$Step2: M \rightarrow A: SIG_{M_A}(MS2), Pubkey_m$$
$$MS2 = HOI_C, TID_C, MID, N_c, T_c, Amt_C, Amt_M, HOI_M, N_m, T_m$$

Merchant (M) decrypts $SIG_{C_M}(MS1), Pubkey_c$ using his private key and gets $MS1$. Merchant sends $SIG_{M_A}(MS2), Pubkey_m$ to the Acquirer (A).

$$Step3: A \rightarrow PG: HOI_C, TID_C, MID, (PI)_{K_{ci}}, N_c, T_c, Amt_C$$

Acquirer receives $SIG_{M_A}(MS2), Pubkey_m$ and decrypts it to get MS2
  a) Checks if $HOI_M = HOI_C$, $Amt_M = Amt_C$,
  b) Checks if Timestamps $T_C = T_M$
  c) Checks if nonces $N_C = N_M$

If all the checks are found to be successful then it keeps a copy of the received message MS2 and authorizes the Order Information. Then Acquirer forwards message to the PG via Secure Private Banking Network.

$$Step4: PG \rightarrow I: HOI_C, HOI_M, TID_C, MID, (PI)_{K_{ci}}, N_c, T_c, N_m, T_m, Amt_C, Amt_M$$

PG receives $HOI_C, HOI_M, TID_C, MID, (PI)_{K_{ci}}, N_c, T_c, N_m, T_m, Amt_C, Amt_M$ from the Acquirer through Private Banking Network which is very secure. Payment Gateway (PG) will perform the following verifications from the message it has received.

  a) Checks if $HOI_M = HOI_C, Amt_M = Amt_C$,
  b) Checks if Timestamps $T_C = T_M$
  c) Checks if nonces $N_C = N_M$

If all the checks are found to be successful then it keeps a copy of the received message and forwards it to Issuer.

$$Step5: I \rightarrow PG := \{AutHorization \ of \ PI\}$$

Issuer receives message

$HOI_C, HOI_M, TID_C, MID, (PI)_{K_{ci}}, N_c, T_c, N_m, T_m, Amt_C, Amt_M$ from the PG through Private Banking Network which is very secure. Issuer (I) will perform the following checks from message it has received.

Where $PI = \{AI, HOI_C, TID_C, Amt_C, ID_C, N_C, T_C, ID_M\}$

If all the checks are successful it authorizes the PI and sends

$$\{AutHorization\ of\ PI\}\ to\ PG$$

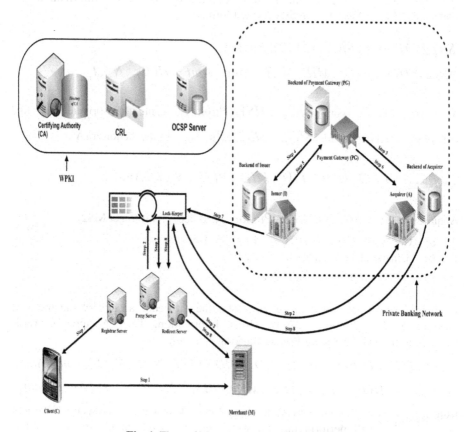

Fig. 1. Flow of Messages in SLSMP Protocol

$Step6: PG \rightarrow A: \{(TID, Amt, ID_M, Success/Failure)\}$

PG informs Acquirer (A) about the success/failure of the transaction.

$Step7: I \rightarrow C: \{SIG_C^I(TID, Amt, ID_M, Success/Failure)\}$

Issuer (I) informs Client (C) about the success/failure of the transaction.

$Step8: A \rightarrow M: \{(TID, Amt, ID_M, Success/Failure)\}$

4   Comparitative Analyses with Related Works

PROTOCOLS \ FEATURES	[1]	[2]	[3]	[4]	SLSMP
Authentication	YES	YES	YES	YES	YES
Confidentiality	YES	YES	YES	YES	YES
Integrity	YES	YES	YES	YES	YES
Non- Repudiation	NO	NO	NO	NO	YES
Key pairs are generated and stored in Tamper resistant device	NO	NO	NO	NO	YES
Are the Signatures generated in "Secure Signature Creation Device (SSCD)"	NO	NO	NO	NO	YES
Ensures Communication Security	NO	NO	NO	NO	YES
Ensures Application Security	NO	NO	NO	NO	YES
Withstands Replay, Impersonation & MITM Attacks	NO	NO	NO	NO	YES
Optimal Consumption of Resources	NO	NO	NO	NO	YES
Uses SIP	YES	YES	YES	YES	YES
Uses WPKI	NO	NO	NO	NO	YES
Uses Signcryption	NO	NO	NO	NO	YES
Ensures Anonymity	NO	NO	NO	NO	YES

5   Security Analysis

Confidentiality, Authentication, Integrity and Non Repudiation: Our proposed SLSMP protocol ensures confidentiality, authentication, integrity and non repudiation properties using Signcryption mechanism.

Order Secrecy & Payment Secrecy: Our proposed SLSMP protocol ensures Payment Secrecy and Order Secrecy. Payment Secrecy is achieved by encrypting the Payment Information (PI) using secret symmetric key which is shared between Client (C) and Issuer (I). Merchant will not be able to decrypt Payment Information (PI) and Order Secrecy is achieved by hashing OI (done by both the Client (C) and Merchant (M)), Issuer will not know about OI thereby achieving Order Secrecy.

Key Pairs Are Generated and Stored in Tamper Resistant Device: We follow the procedure given in [7, 8].

Are the Signatures Generated in "Secure Signature Creation Device" (SSCD): UICC is personalized by the client and client's credentials are stored in the WIM of

UICC. So key pairs are generated and stored in Tamper resistant device (i.e. UICC). Since UICC is a Tamper resistant device, signatures generated in UICC using the credentials stored in the WIM. So signatures are generated in Secure Signature Creation Device (SSCD). So all the signatures generated in the framework are qualified signatures.

Identity Protection from Merchant, PG and Eavesdropper: In order to prevent a merchant from knowing the identity of Client, an anonymous identity is enrolled by the client with CA and Issuer. CA and Issuer know the real identity of the client. Therefore, as merchant, PG and eavesdropper cannot map the anonymous identity with C'S true identity; client's privacy is protected and untraceable.

With stands attacks: Our proposed Mobile Payment protocol (SLSMP) withstands the Replay Attack, Impersonating attack and Man In The Middle Attack because timestamps and nonce (n_c) are included in the messages exchanged thereby avoiding replay attack, the intruder (In) cannot impersonate a client C because Intruder does not have C's private key thereby avoiding Impersonating Man in the Middle attacks.

6   Conclusions and Future Work

This paper proposes a Secure Lightweight and Scalable Mobile Payment Framework (SLSMP) using Signcryption scheme with Forward Secrecy (SFS) based on elliptic curve this scheme which combines digital signature and encryption functions. It takes lower computation and communication cost to provide security functions. SLSMP is highly scalable which is attributed to SIP for data exchange. This paper uses WPKI, UICC as Secure Element and depicts detailed protocol and system architecture of SIP based mobile payment solution. Merely using cryptographic mechanisms, does not guarantee security-wise semantically secure operation of the protocol, even if it is correct. The network is assumed to be hostile as it contains intruders with the capabilities to encrypt, decrypt, copy, forward, delete, and so forth. Several examples show how carefully designed protocols were later   found out to have security breaches [6] (Muhammad et al., 2006). So formal verification of security protocols is essential as it can detect flaws that lead to protocol failure. So we plan to verify our proposed mobile payment protocol using AVISPA and Scyther Tool in the future.

References

1. Zhang, G., Cheng, F., Hasso, C.M.: Towards Secure Mobile Payment Based on SIP. In: 15th Annual IEEE International Conference and Workshop on the Engineering of Computer Based Systems 2008, Belfast, Northern Ireland, pp. 96–104 (2008)
2. Zhang, G., Cheng, F., Meinel, C.: SIMPA: A SIP-based Mobile Payment Architecture. In: Seventh IEEE/ACIS International Conference on Computer and Information Science 2008, pp. 287–292 (2008)

3. Hao, J., Zou, J., Dai, Y.: A Real-Time Payment Scheme for SIP Service Based on Hash Chain. In: IEEE International Conference on e-Business Engineering 2008, pp. 279–286 (2008)
4. Kungpisdan, S., Thai-Udom, T.: Securing Micropayment Transactions Over Session Initiation Protocol. In: 9th International Symposium on Communications and Information Technology (ISCIT 2009), pp. 187–192 (2009)
5. Hwang, R.-J., Lai, C.-H., Su, F.-F.: An efficient signcryption scheme with forward secrecy based on elliptic curve. Applied Mathematics and Computation 167, 870–881 (2005), doi:10.1016/j.amc.2004.06.124
6. Muhammad, S., Furqan, Z., Guha, R.K.: Understanding the intruder through attacks on cryptographic protocols. In: Proceedings of the 44th ACM Southeast Conference (ACMS 2006), pp. 667–672 (March 2006)
7. Ahamad, S.S., Sastry, V.N., Udgata, S.K.: Secure Mobile Payment Framework based on UICC with Formal Verification. Special Issue on 'Future Trends in Security Issues in Internet and Web Applications'. Int. J. Computational Science and Engineering (2012) (in press) (accepted)
8. Ahamad, S.S., Sastry, V.N., Udgata, S.K.: A secure and optimized mobile payment framework with formal verification. In: SECURIT 2012, pp. 27–35 (2012)
9. Rosenberg, et al.: RFC 3261: SIP Session Initiation Protocol (June 2002)

Service Oriented Architecture Based SDI Model for Education Sector in India

Rabindra K. Barik[1] and Arun B. Samaddar[2]

[1] M. N. National Institute of Technology Allahabad, India
rabindra.mnnit@gmail.com
[2] National Institute of Technology Sikkim, India
absamaddar@yahoo.com

Abstract. Technological and overall economic growth of any country warrants a rapid development of the education sector, which is responsible for producing quality human resources for serving the nation and the human society as a whole. Hence, there is a need to make coordinated efforts to disseminate information about the quality of academic details in a simple but detailed manner by integrating modern technologies. Spatial technologies such as GIS, remote sensing and GPS hold potential to remove some of the bottlenecks that hinder the efficiency of this sector. Further, for Right To Education (RTE) easy to use/ perceive spatial information is required for the decision makers. Hence, there is a need to establish a well organised Spatial Data Infrastructure (SDI) portal where each stakeholder can access, use and exchange spatial information for education sector. The present work reports the development of an efficient interoperable Service Oriented Architecture (SOA) based SDI Model for education sector. The developed SDI Model is distributed, modular and allows the publishing of web service descriptions as well as to submit requests to discover the web service of user's interests. The Model supports integration of applications and browsers independent Web Map Service (WMS), Web Features Service (WFS), Web Coverage Service (WCS) and Web Catalogue Service (CS-W) for sharing and exchange of geospatial data.

1 Introduction

The accelerated growth rate of the economy and technology of any country warrants a rapid development of the education sector on which most of the quality human resources produce for servicing the nation. Human resources is an important component for economic and technological growth which contributes significantly to the economy of India. For producing quality human resources, academic sectors play a vital role, which is facing an increased competition on account of globalisation and the level of technology employed thereby necessitating new initiatives for meeting the new challenges. There is a need to make coordinated efforts to encourage people to get better information about the quality of academic details in greater way by integrating modern technologies. GIS, remote sensing and GPS are the technologies which may be used to remove some of the bottlenecks that hinder the productivity and efficiency of this sector.

S.C. Satapathy, S.K. Udgata, and B.N. Biswal (eds.), *FICTA 2013*,
Advances in Intelligent Systems and Computing 247,
DOI: 10.1007/978-3-319-02931-3_63, © Springer International Publishing Switzerland 2014

GIS is a system of computer hardware and software that enables users to store, manage, manipulate, analyse and retrieve large volumes of spatially referenced data and associated attributes collected from a variety of sources [7]. They may be used to create and maintain geographic database and are suited for analysis in planning related activities. The GIS has wider applications in decision making, storage of various kinds of data, bringing data and maps to a common scale for user need, superimposing, querying and analysing the large amount of data and designing and presenting final maps and reports to administrator and planner[12]. GIS can deal with large amount of spatial data at different scales as well as non-spatial data for deriving useful information in maps/ tabular from for better understanding for organised development. The utility and application of GIS for planning of land resources and decision making has become widely popular and are being used for a wide range of applications.

With the integration of web technology with GIS, it gives rich functionality in terms of spatial data sharing on the web. It can provide a real time and dynamic way to represent information through maps on web. So there is a need to establish a well organised Spatial Data Infrastructure (SDI) which is a portal where each stakeholder can access, use and exchange spatial data for social, economic and environmental application. Geospatial Web Service is one of the key technology require for development and implementation of SDI. Design and implementation of SDI is used service oriented architecture (SOA) which it is used for sharing academic detailed information on web. It enables the people to quickly look into the problem and gets the information according to their need.

Fortunately, there are many open source software that can compete the proprietary software in the field of GIS [10]. Developers have created several open source libraries and GIS suites to cope with the flood of GIS data and their formats [8]. The goal of Open Geospatial Consortium (OGC) is to encourage the use of open source GIS standards and development of community-led projects. The available open source GIS software can be used for the works related to database creation, spatial modelling and geospatial web based services [1]. The open source GIS software used for development of SDI include Quantum GIS for creation of geospatial database, MYSQL for storing of security aspects and non-spatial data, ALOV, GeoServer and Apache Tomcat for imparting geospatial web capabilities and PHP: Hypertext Preprocessor, JSP: Java Server Pages and GeoExt: Geo Extension for dynamic server side scripting.

2   SOA Based SDI

SDI provides an environment within which organisations interact with technologies to foster activities for using, managing and producing geographic data. It is the technology, policies, standards, human resources, and related activities necessary to acquire, process, distribute, use, maintain and preserve spatial data [13]. The core components of SDI can be viewed as policy, networking, standards, people and data. These can be grouped into different categories based on the nature of their interactions within the SDI Model [6][11].

Thus, an integrated SDI cannot be thought as having spatial data, value-added services and end-users alone, but instead involves other important issues regarding interoperability, policies and networks also. The development of SDI supports these decision-making functions at different administrative and political levels. The development of SDI at national level in India, i.e., Indian NSDI was initiated in 2000 jointly by Department of Science and Technology (DST) and Indian Space Research Organisation (ISRO) through the establishment of a national task force to prepare an action plan under the aegis of DST. SDI is a web based system developed under GIS environment. Thus, geospatial web service is an integral and important part of any SDI. The basic concept for geospatial data visualisation refers to web mapping of spatial data from various organisation and servers across the World Wide Web (WWW) by following the standards of W3C (World Wide Web Consortium) and OGC (Open Geospatial Consortium) [9].

OGC has defined various standards which are based on a generalised architecture captured, describes a basic data model for geographic features to be represented for web mapping. These standards have been developed to serve specific needs for interoperable location and geospatial technology, including GIS. Web Map Service (WMS), Web Features Service (WFS), Web Coverage Service (WCS) and Web Catalogue Service (CS-W) are the important OGC standards which are frequently used in many applications. Mapping on the Web includes the presentation of general purpose maps to display locations and geographic backdrops, as well as more sophisticated interactive and customisable mapping tools.

The Service Oriented Architecture tries to construct a distributed, dynamic, flexible, and re-configurable service system over Internet that can meet information and service requirements for development of SDI. The key component in the SOA based SDI is geospatial web service i.e. a well defined set of actions. It is self contained, stateless, and does not depend on the state of other services [15].

A geospatial web service is an application that can retrieve data from GIS databases and provide geographic information through a browser interface to end-users [2]. Thus, geospatial web capabilities indicate to a web based GIS which can be modelled using the client-server architecture. A thin client model is where most of the processing work is done on demand in the server and the client does not perform any task other than to display the data on screen.

In development of SDI Model, the major focus has been on SOA based geospatial web service by using spatial data. According to services, it can be grouped data service, processing service and catalog service [4]. Data service is tightly coupled with specific data sets and offers WFS, WMS and WCS. Processing Service provides operations for processing or transforming data in a manner determined by user-specific parameters. CS-W allows users and applications to classify, register, describe, search, maintain, and access information about Web Services [3].

The basic operations in SOA based SDI Model include publish, find and bind. To be able to integrate any services into SOA based architecture, it should provide at least one of the SOA's major operations. Figure 1 shows three major functionalities in SOA based SDI.

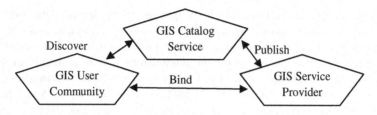

Fig. 1. Three major functionalities in SOA [14]

In SOA based SDI Model, there are three types of key actors i.e. service requester or GIS user community, GIS service provider and GIS catalog service. Catalog service can be called registry service or broker but their main functionality is almost same in most of the applications. Catalog service helps the requestors to discover or find the right services. When a service provider sets up a service over the internet and wants the users to use their service, it needs to publish their services. When a requestor requests a service, the requestors and service brokers need to collaborate to find the right services. After the right service is found, requestor and provider bind or negotiate as to format of the request. Then, the requestor can access and invoke services of the provider.

3   Objective of the Present Work

The main aim of the present work is to development and implementation of SOA based SDI Model. It is proposed to use Quantum GIS for creation of geospatial database, PostGIS and MYSQL for storing of spatial and non-spatial data, ALOV, GeoServer, GeoNetwork, GeoWebCache, WAMP and Apache Tomcat for imparting geospatial web capabilities in terms of WMS, WFS, WCS and CS-W. PHP: Hypertext Preprocessor, JSP: Java Server pages and GeoExt: Geo Extension are used for dynamic server side scripting for development of SDI Model.

4   Methodology Adopted

For development of SDI Model, the main focus has been on the use of a practical approach to explore and extend the concept of SDI in education sector. The developed SDI Model should provide an effective and efficient means of sharing geospatial data and non-spatial data on the web using GIS in a secure way. The proposed SOA based SDI Model in which it follows the basic over view of service provider, service consumer and catalog service.

In SDI Model, it focuses on OGC compliant web services on vector and raster data. Admin is managing the data and gives authority to the different user and Catalog is updated by catalog admin. In catalog, various services are published and ready for service requester i.e. web client and web client discovers the required service. Finally, The required services consumed by the service consumer. This architecture can be fulfilling the most sophisticated workflow for development of SDI Model. Figure 2 shows the layered architectural view that comprises of open source resources.

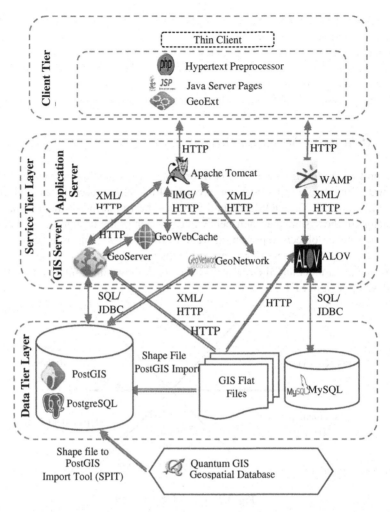

Fig. 2. Layered Architectural view of open source resources

For the creation of integrated geospatial database, the present SDI Model uses Quantum GIS open source GIS software. ALOV, GeoServer, GeoNetwork, GeoWebCache, WAMP and Apache Tomcat have been integrated for imparting the geospatial web capabilities with respect to WMS, WFS and CS-W services. PostGIS and MYSQL are used for storing of spatial and non-spatial data for decision making. PHP: Hypertext Pre-processor, JSP: Java Server Pages and GeoExt: GeoExtension languages have been used for dynamic server side scripting in the framework. In the present work, WAMP and Apache tomcat servers have been taken and then PHP, JSP and GeoExt scripting languages are used for sharing of the geospatial data and non-spatial data and for publication of maps on the web.

The National Institute of Technology (NITs) Network geospatial database and Motilal Nehru NIT geospatial database have been selected to demonstrate the capabilities of developed framework. In the first slave server, the geospatial data and non-spatial data of the old and new National Institute of Technology (NITs) Network geospatial database is stored and in another slave server, the geospatial data and non-spatial data of Motilal Nehru NIT geospatial database is stored. From the master server, by using the scripting language, the geospatial data and non-spatial data of different geospatial database is easily accessed simultaneously.

5   Prototype Development

The prototype development is based on Jacobson's method of Object Oriented Software Engineering (OOSE) for incorporating strong user focus and the time critical nature [5]. The OOSE method involves the formation of models that capture the actors of the system and their behaviour for each of the design stages. The models are made up of objects representing real world entities. This is a natural way for people to describe their environment; and helps in reducing the semantic gap between the developed model and the real world. Figure 3 shows the complete process model for development of SDI.

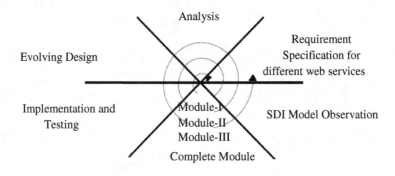

Fig. 3. Process Model for SDI

The process model of SDI is cyclic or incremental in nature and each implementation refines the analysis and design stages through evaluation and testing of a completed module. In Module I, NIT Network geospatial database has been prepared by Quantum GIS Open Source GIS software with the help of political map of India. There are three thematic layers has been created, in which one layer shows the complete India boundary and states boundary and second layer shows the individual state boundary including the location of NIT, and another layer gives the information of particular NIT. Non-spatial database or attribute database contents the information about National Institutes of Technology of India. In the first layer, it has linked with non-spatial data in form of table with ID Code and NIT State name as attributes. In the second layer, it has also linked with non-spatial data in form of table with the attributes as ID Code, NIT State Name and NIT Name. In the third layer, it has non-spatial data in table form with ID Code, NIT Name, NIT Rank, NIT State

Name, NIT Hyperlink, NIT Director Name, NIT Address and NIT Contact No as attributes.

The Module II defines the catalogue services by using Geoserver in which it registers all the desire services. Module III describes the integration of all the services provided by service provider. Initially, for development of SDI Model, it is defined by state diagram for better understanding of the system.

The Quantum GIS, ALOV, GeoServer, Apache Tomcat, MYSQL, PHP and Java Development Tool Kit are used for overall development of the system. Figure 4 shows all the old and new NITs in India.

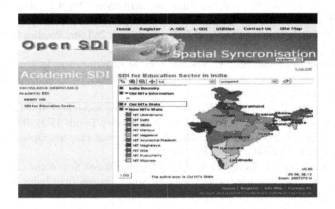

Fig. 4. Old and New NITs in India

6   Concluding Remarks

The present research work is focused at adopting SOA based architecture and OGC standards for creating, accessing, integrating and sharing the geospatial information on the web in education sector. The experience in using Open Source GIS software suggests that various tools and software like Quantum GIS, ALOV, GeoServer, GeoNetwork, GeoWebCache, Apache Tomcat, WAMP, MYSQL, PostGIS &, GeoExt and PHP are available for creation of spatial datasets and implementation of geospatial web services. The widespread use of these open resources in the development of GIS based applications on the web could benefit a vast user community instead of going for costly proprietary solutions and should be encouraged.

The main focus of the present work is to develop and implement open source GIS for SOA based SDI Model. Therefore, the database used for both NIT Network geospatial database and Motilal Nehru NIT geospatial database are indicative and does not contain detailed features. This database may be made more comprehensive in future studies. At present, SDI Model is operational at MNNIT intranet level only. The same may be implemented on the web in future studies.

References

1. Barik, R.K., Samaddar, A.B., Gupta, R.D.: Investigations into the Efficacy of Open Source GIS Software. In: International Conference, Map World Forum, on Geospatial Technology for Sustainable Planet Earth, February 10-13 (2009)
2. Harper, E.: Open Source Technologies in Web-based GIS and Mapping, Master's Thesis, Northwest Missouri State University, Maryville Missouri (2006)
3. Kim, D.-H., Kim, M.-S.: Web GIS Service Component based on Open Environment. In: IEEE Geoscience and Remote Sensing Symposium, IGARSS 2002, vol. 6, pp. 3346–3348 (2002)
4. Li, H., Lu, J., Cai, B., Yao, S.: Study on SOA-Orient WebGIS framework. In: 14th International Conference on Automation and Computing. IEEE (2008)
5. Mall, R.: Fundamentals of Software Engineering, rev. 2nd edn. Prentice-Hall of India Pvt. Ltd., India (2004)
6. Mansourian, A., Rajabifard, A., Valadan Zoej, M.J., Williamson, I.: Using SDI and web-based system to facilitate disaster management. International Journal of Computers & Geosciences 32, 303–315 (2005)
7. Morris Steven, P.: Geospatial Web Services and Geo archiving: New Opportunities and Challenges in Geographic Information Services. Library Trends 55(2), 285–303 (2006)
8. Paul, R.: The State of Open Source GIS. In: The Annual Free and Open Source Software for Geospatial (FOSS4G) Conference (2006)
9. Puri, S.K., Sahay, S., Georgiadou, Y.: A Metaphor-Based Sociotechnical Perspective on Spatial Data Infrastructure Implementations: Some Lessons from India. In: Research and Theory in Advancing Spatial Data Infrastructure Concepts, pp. 161–173. ESRI Press (2007)
10. Raghunathan, S., Prasad, A., Mishra, B.K., Chang, H.: Open Source versus Closed Source: Software Quality in Monopoly and Competitive Markets. IEEE Transactions on Systems, Man, and Cybernetics 35(6), 903–918 (2005)
11. Rajabifard, A., Feeney, M.E.F., Williamson, I.P.: Future Directions for SDI Development. International Journal of Applied Earth Observation and Geoinformation 4(1), 11–22 (2002)
12. Ramachandra, T.V., Kumar, U.: Geographic Resources Decision Support System for Land Use, Land, Cover Dynamics Analysis. In: Proceedings of the FOSS/GRASS User Conference, Bangkok Thailand (2004)
13. Rawat, S.: Interoperable Geo-Spatial Data Model in the Context of the Indian NSDI, Thesis (Master), ITC, The Netherlands (2003)
14. Vaccari, L., Shvaiko, P., Marchese, M.: A geo-service semantic integration in Spatial Data Infrastructure. International Journal of Spatial Data Infrastructures Research 4, 24–51 (2009)
15. Lu, X.: An Investigation on Service Oriented Architecture for constructing Distributed WebGIS Application. In: IEEE International Conference on Services Computing (SCC 2005), vol. 1, pp. 191–197 (2005)

Author Index

Agarwal, Arun   389
Agarwal, Kabita   389
Ahamad, Shaik Shakeel   545
Ahmad, Musheer   527
Albert Singh, N.   59, 111, 119, 129, 165
Amar Pratap Singh, J.   111, 119
Anne, K.R.   363
Anudeep, N.   75
Arora, Shrey   409
Aswani, Amarapini   483
Avhad, Kiran   285

Bandaru, Rajesh   67
Banerjee, Subhashis   345
Barik, Rabindra K.   555
Bhalke, D.G.   155
Bhaskara Murthy, V.   99
Bhateja, Vikrant   451, 459
Bhattacharjee, Anup Kumar   197, 467
Bisoy, Sukant Kishoro   423
Biswal, B.N.   229
Bormane, D.S.   155

Chavali, Rishanth Kanakadri   371

Das, Bishwa Ranjan   147
De, Arunava   197, 467
De (Maity), Ritu Rani   41
Dheeba, J.   119
Dheeba, V.   111
Doreswamy,   435, 443

Fatima, Sameen   371

Gantayat, S.S.   401
Gayathri, G.   483
Gupta, Akansha   409
Gupta, Sonia   527
Gurrala, Jagadish   475

Jagdish, G.   75
Jayaprada, S.   483
Jena, Lambodar   259

Kadam, Megha   293, 379
Kalsi, Aseem   451
Kamila, Narendra Ku.   259
Kavila, Selvani Deepthi   67
Kesavadas, C.   165
Kiran, B. Ravi   509
Kotha, Chaitanya   371
Kranthi Kiran, M.   181
Kuchibhotla, Swarna   363
Kulkarni, Anagha   285
Kulkarni, Anand J.   269
Kulkarni, P.S.   83
Kulkarni, Raj   517
Kumar, Shobhit   137
Kumbhar, Hemant   517

Limkar, Suresh   91, 285, 293, 379, 517

Mahapatra, Kamalakanta   337
Majumder, Koushik   313, 321, 329, 345
Manne, Suneetha   371
Mathuranath, P.S.   129, 165
Megala, M.   49
Meghana, Krishna   371
Metkar, Shubham J.   269

Mishra, Soumya Ranjan   75
Mishra, Sushruta   259
Misra, Ashok   401
Mitra, Ankita   467
Mohankrishna, S.   229
Mohanty, Mihir Narayan   305
Mohanty, Prases Kumar   353
Mohapatra, A.K.   527
Mudi, Rajani K.   25, 41
Murthy, J.V.R.   207
Murty, M. Ramakrishna   207

Naik, Anima   207, 217, 229
Nair, Madhusoodhnan   545
Narasegouda, Srinivas   435, 443

Pal, Sangita   17
Panda, B.S.   401
Panda, Sidhartha   249
Pardha Saradhi Varma, G.   99
Parhi, Dayal R.   353
Parida, Deebyadeep   305
Parwekar, Pritee   409, 417
Patnaik, Prasant Kumar   423
Patnaik, Srikanta   147
Pattnaik, Prasant Kumar   535
Paul, Geenu   59
Pavani, M.   493
Pavani, Palepu   173
Prabha, T. Sashi   173
Pradeep, B.   277
Praneeta, G.   277
Purushothaman, K.V.   59

Raich, Devashri   83
Rajan, C. Cristober Asir   49
Rao, C.B. Rama   155
Rao, K. Raja Sekhara   229
Ravikiran, B.   75
Ravva, Ravi   67
Reddy, P.V.G.D. Prasad   207
Rout, Umesh Kumar   249
Roy, Sharmistha   535

Sahu, Gokulananda   337
Sahu, Rabindra Kumar   249
Sahu, Subrat Kumar   337
Samaddar, Arun B.   555
Samal, Ankit   305
Sardar, Mousumi   321, 345
Sardeshmukh, M.   91
Sarkar, Samrat   313
Satapathy, Mihir Ranjan   305
Satapathy, Suresh.C.   207
Satapathy, Suresh Chandra   217, 229, 501
Sayed, Asim   91
Sensarma, Debajit   329
Sethi, Srinivas   17
Sheela Kumari, R.   129, 165
ShyamVamsi, T.   181
Si, Tapas   197
Simhachalam, Dharmana   25
Singampalli, Jayaprakash   475
Singh, Daya Shankar   33, 137
Singh, Gopal   459
Singh, Prashant   9
Singh, Sapna   33, 137
Singhal, Rishabh   417
Sohail, Aamir   371
Sravani, K.   189
Srinivasu, P.   189
Srivastava, Aastha   451
Srivastava, Atul   459
Sudhan, K. Sai Madhu   75

Udgata, Siba K.   545

Vankayalapati, H.D.   363
Vardhan, G.V.N.A. Harsha   509
Varghese, Tinu   59, 129, 165
Verma, Prabha   1
Vignani, B.   501

Yadava, R.D.S.   1, 9
Yadwad, Sunita A.   493
Yalamanchili, B.S.   363